Enterprise Architecture with .NET

Expert-backed advice for information system design,
down to .NET and C# implementation

Jean-Philippe Gouigoux

Enterprise Architecture with .NET

Group Product Manager: Kunal Sawant
Publishing Product Manager: Teny Thomas
Senior Content Development Editor: Rosal Colaco
Book Project Manager: Deeksha Thakkar
Technical Editor: Vidhisha Patidar
Copy Editor: Safis Editing
Indexer: Hemangini Bari
Production Designer: Joshua Misquitta
DevRel Marketing Coordinator: Shrinidhi Manoharan

First published: May 2024
Production reference: 1170524

Published by Packt Publishing Ltd.
Grosvenor House
11 St Paul's Square
Birmingham
B3 1RB, UK

ISBN 978-1-83508-566-0

www.packtpub.com

To the memory of my father, Jacques, who died just before reaching his magic age of 88. Thanks for teaching me what industry means, both in terms of hard work and in doing things the right way; you inspired a lot of this book.

– Jean-Philippe Gouigoux

Foreword

Any complex organization, whether it's a service company or an industry, requires rigorous management, made even more complex due to its intertwined various facets. Facing digital transformation pitfalls while increasing performance and productivity needs to be conducted as a whole, from strategic to operational and technical levels, involving working with a systemic approach thanks to the information system field.

I have known Jean-Philippe Gouigoux for more than 10 years. Both passionate about the challenges posed by the digital transformation of organizations, we have combined our respective academic and business expertise to offer formalized feedback and recommendations through international scientific publications on the migration of legacy systems to microservices-based architectures and, more recently, on business-IT alignment issues. The main lesson learned is that business needs and best practices must always drive technological changes. In his book, Jean-Philippe, thanks to his extensive technical experience as an architect and manager and as an author and consultant, provides valuable advice on how to design viable information systems. The relevance of his book is that it covers the business aspects of the information system and, at the same time, an enlightened and practical introduction to software architecture and application aspects in .NET and their deployment.

This book is a goldmine for all developers who want to see beyond code and for all architects who want to see beyond software architectures, to work toward the achievement of their organization's success by creating efficient and sustainable information systems leaning on business-aligned services.

Dalila Tamzalit

Senior Associate-Professor, Nantes University.
PhD. Habilitation to Direct Research.

Contributors

About the author

Jean-Philippe Gouigoux started programming in 1985 and has never stopped since then, working as a developer, then an architect, an R&D manager, and today, the chief technical officer of a group of software editors. Coming from a mechanical engineering background, his main career goal is to apply a rigorous approach to information technology and try to make information systems as modular and standardized as mechanical systems. JP has written more than 10 technical books and has spoken at around 100 conferences and on university courses, teaching architecture, testing, service-oriented design, performance in .NET, Docker and Kubernetes, and so on. He has been a Microsoft MVP since 2011.

Big thanks to Guénaëlle, Clémence, and Mathieu for supporting my dedication to writing. My motivation has always been to make you proud.

About the reviewer

Guillaume Collic is a passionate software developer who loves software fundamentals, the craftsmanship movement, functional programming, **Domain-Driven Design** (**DDD**), and all approaches that help make the domain clearer and IT aligned with business. He was a Microsoft MVP from 2013 to 2019. He currently works as a staff engineer at Agicap. When he's not working, he spends time coding and sharing with communities (board game enthusiasts, makers, developers...).

Table of Contents

3

Reaching Business Alignment   33

4

Dealing with Time and Technical Debt   73

5

A Utopic Perfect IT System   101

Part 2: Architecture Frameworks and Methods

6

7

8

9

Exploring Domain-Driven Design and Semantics   173

10

Master Data Management   191

11

Business Processes and Low Code   229

12

Externalization of Business Rules   265

13

Externalization of Authorization   291

Part 3: Building a Blueprint Application with .NET

14

Decomposing the Functional Responsibilities   339

15

Plugging Standard External Modules   377

16

Creating a Write-Only Data Referential Service   425

17

Adding Query to the Data Referential Service   467

18

Deploying Data Referential Services   499

19

Designing a Second Data Referential Service   531

20

Creating a Graphical User Interface   555

21

Extending the Interfaces   595

22

Integrating Business Processes   625

Part 4: Validating the Architecture

23

Applying Modifications to the System   681

Preface

Thanks for acquiring this book! It was born out of the need to industrialize Information Systems. Software is everywhere but, apart from a few high-cost applications, it is full of bugs and costs a lot in maintenance, and most developers spend half their time reinventing the wheel or fighting against the Information System to add some new features.

This is my harsh observation from more than 25 years in the field and a few years working specifically on digital transformation for many different organizations. These diverse experiences convinced me that the root problem is always the same: the software industry does not actually follow the rules of industrialization, which are to decompose systems into smaller ones and join modules together with standards-obeying interfaces.

Working together with academics helped me formalize these problems of business/IT alignment, along with their sources and possible solutions. But most of my customers have told me that I stood out as a consultant because I was able to not only provide boards of directors with a clear transformation plan but also to accompany it with advice on semantics all the way down to the API contracts, and even some hands-on help in the code for the hard bits. They said this made the whole partnership more tangible and provided visible business results, so this is how I wanted this book to be: some theory first because we need to switch the way we think and create Information Systems, but some code as well because too many of them are not efficient today.

Having seen the positive effects on the business efficiency of many companies, from a small lawyers firm to industry/logistics enterprises and to some huge farming cooperatives, I am convinced applying industry concepts to software design and architecture will become a more and more cost-effective approach, particularly since software and data are becoming the competitive backbone of most organizations nowadays.

Using norms and standards for API interfaces, externalizing functions such as authentication, authorization, electronic document management, and, most importantly, thinking in terms of business functions rather than technically, is of utmost importance if you want to transform an Information System from a cost center to a strong asset of an organization.

This method has helped many companies and government-owned agencies, and I hope it will do the same for your Information Systems!

Who this book is for

This book aims to help you create or evolve Information Systems architecture in such a way that it becomes more flexible and provides more business value to your organization. Since it is theoretical, business-oriented, and technically applied with an application in .NET, it can be read at many levels and by many profiles:

- Board directors and chief officers can read the first chapters in order to understand digital transformation and how it can be strategized. Finding out about the most common problems in Information Systems should help them invest in the right places and prioritize work, taking into account not only features but also technical debt reduction when justified.

- Tech leads, architects, and R&D managers will also benefit from reading the next chapters that make the strategy more tangible and explain best practices in structuring Information Systems, choosing tools and methods, and applying alignment in their job.

- - Finally, developers or hands-on architects and technical directors will continue reading the book's last chapters that really put the previous chapters into practice by coding the different concepts, plugging external services, and orchestrating API and webhooks together to create a small but working sample system.

What this book covers

Chapter 1, The Sad State of Information Systems, explains what prevents most Information Systems from providing as much value as they should. It also proposes some methods to measure the efficiency of an Information System, focusing on what causes are behind most of the problems observed.

Chapter 2, Applying Industrial Principles to Software, starts by explaining what characterizes the principle of industrialization, and then proposes a way to apply a transformation to Information System management. It focuses on the importance of standards and norms to reduce complexity, which is quite high in most virtual systems.

Chapter 3, Reaching Business Alignment, brings a layer of theory to the problems analyzed in the previous chapters, explaining Conway's law and some other principles, patterns, and anti-patterns in business/IT alignment. This chapter also introduces the four-layer diagram, which will be used throughout the rest of the book.

Chapter 4, Dealing with Time and Technical Debt, describes the concept of technical debt and explains how Information Systems often start quite well adapted to their context but slowly drift apart from efficiency over time. It also reflects on how to adopt an Agile approach to architecture.

Chapter 5, A Utopic Perfect IT System, is about setting a vision of an absolutely perfect Information System that would be completely aligned with business needs and where changes would be as easy as the original design. Even if this utopic system will never exist in reality, it will help you understand how externalizing functions and strictly separating responsibilities can make existing Information Systems more efficient and ease their evolution over time.

Chapter 6, SOLID Principles from Code to Systems, starts a more technical set of chapters. It discusses the SOLID principles, which usually apply to code, and adapts them to systems architecture. We also present the example Information System that will serve as an applied example for the rest of the book.

Chapter 7, C4 and Other Approaches, lists some of the most well-known architectural methods in software design and architecture. It analyzes the differences and common points between them, giving you some ways to choose which is best suited depending on the context.

Chapter 8, Service Orientation and API, starts with a historical description of all the techniques that have been used to implement the concept of a service, explaining what people expect from the concept. It will explain how standards and pivotal formats are useful in services.

Chapter 9, Exploring Domain-Driven Design and Semantics, is about the importance of having a clear definition of the semantics of a business domain. Though this may sound a bit distant from the notion of software architecture, it actually is at the heart of the method of business/IT alignment.

Chapter 10, Master Data Management, provides a structured view of the management of data in a well-aligned Information System. Master Data Management is much more than persisting data in a database. It is about defining its life cycle and the associated governance and ensuring a progressive evolution of data management, particularly when some legacy data repositories are already in place.

Chapter 11, Business Processes and Low Code, explains the place of business processes in an Information System, and how they can be described formally using standards such as BPMN. Though this is still a theoretical chapter, it describes the tools that can be used in order to implement business process management in an Information System.

Chapter 12, Externalization of Business Rules, closes the loop on the three parts of the ideal IT system, after data and process management. It explains what a business rules management system is and shows some examples of implementations.

Chapter 13, Externalization of Authorization, explains the concept of identification and authorization management and shows how it can be externalized to the benefit of the evolution of the whole Information System. Paradigms such as **Role-Based Access Control** (**RBAC**) and **Attribute-Based Access Control** (**ABAC**) are also explained with examples.

Chapter 14, Decomposing the Functional Responsibilities, is where we start applying the knowledge from the previous chapters to an example; we'll get our hands dirty with the code and services. The chapter starts by using the four-layer diagram to model the system and then prepares the technical specifications for the features that will be better considered in external applications.

Chapter 15, Plugging Standard External Modules, takes these specifications into account and proposes some off-the-shelf software solutions for the different features that have been identified, such as Apache Keycloak for IAM, MongoDB for persistence, Alfresco for electronic document management, and so on.

Chapter 16, Creating a Write-Only Data Referential, focuses on the business domain data management of the example Information System. It shows how to code an API server in .NET that will serve the data needed in a way that as far as possible respects standards and all the principles of business/IT alignment that have been described previously.

Chapter 17, Adding Query to the Data Referential Service, builds on the previous chapter and adds some features to the data management system that has been put in place, but also shows how it can be tested in an industrial manner. Standards such as the Open Data Protocol are used in this chapter to improve the implementation.

Chapter 18, Deploying Data Referential Services, shows how to deploy the services that were created in the previous chapters and how to plug them into the IAM. Since security is such an important and complex subject, this is the subject of this dedicated chapter.

Chapter 19, Designing a Second Data Referential Service, uses all the principles presented in the previous chapter to build a second API implementation, this time complexified by the fact that this service has to use webhooks and create a local cache of data in order to reduce calls to the first data referential service. Though it does not add any new principles, it does show how loose coupling can be realized in practice.

Chapter 20, Creating a Graphical User Interface, adds some graphical interfaces to our example Information System, which only exposed API endpoints for now. After showcasing a web-based single-page application and explaining how the concepts and recommendations of business/IT alignment also apply in the GUI domain, it details connecting the data pagination mechanism with the backend, observing a rigorous separation of responsibilities.

Chapter 21, Extending the Interfaces, builds on the previous chapter and applies some industry standards to the GUI that was previously built, namely by automating tests on it but also by showing how importing data can be automated to migrate a system from its legacy data management. It finishes by showing how a good separation of components can help, with the right technical stack, to quickly build a mobile application.

Chapter 22, Integrating Business Processes, closes the set of chapters corresponding to the example application by explaining how business processes can be integrated into it through the GUI, a BPMN engine, or a low-code/no-code modern orchestration tool such as n8n. It finishes by explaining when a dedicated service is necessary to implement orchestration or choreography.

Chapter 23, Applying Modifications to the System, sums up the whole idea of the book by showing how the Information System that has been built can now support all kinds of changes by being impacted by side effects or losing its ability to evolve further. In this chapter, we will change the data structure, the GUI, and the business rules, particularly authorization, and even adjust the business processes, hopefully demonstrating that the initial planning of the architecture makes all this easy.

To get the most out of this book

Docker has been used in order to reduce the required tooling as much as possible. Since all images are available online (even the custom ones created for the example information system, available at `https://hub.docker.com/repositories/demoeditor`), all you need to work with the application is Docker (see `https://docs.docker.com/engine/install/` for the installation instructions). The image versions follow the branches of the code.

If you want to debug the application and make some changes to it in order to follow the instructions from the book, then you will also need .NET 8.0 SDK (https://dotnet.microsoft.com/download), Visual Studio Code (https://code.visualstudio.com/download), and Git (https://git-scm.com/book/en/v2/Getting-Started-Installing-Git). Postman (https://www.postman.com/) will also be used to quickly inject data.

Software/hardware covered in the book	Operating system requirements
.NET 8.0 SDK	Windows, macOS, or Linux
Visual Studio Code	
Git	
Docker	
Postman	

If you are using the digital version of this book, we advise you to type the code yourself or access the code from the book's GitHub repository (a link is available in the next section). Doing so will help you avoid any potential errors related to the copying and pasting of code.

Download the example code files

You can download the example code files for this book from GitHub at https://github.com/PacktPublishing/Enterprise-Architecture-with-.NET. If there's an update to the code, it will be updated in the GitHub repository. This GitHub repository is made with branches that correspond to steps in the building of the examples in this book: https://github.com/PacktPublishing/Enterprise-Architecture-with-.NET/tree/main/DemoEditor#versioning.

We also have other code bundles from our rich catalog of books and videos available at https://github.com/PacktPublishing/. Check them out!

Conventions used

There are a number of text conventions used throughout this book.

Code in text: Indicates code words in text, database table names, folder names, filenames, file extensions, pathnames, dummy URLs, user input, and Twitter handles. Here is an example: "Imagine we use a link between a book and an author entity."

A block of code is set as follows:

```
{
    "rel": "author",
    "href": "https://demoeditor.com/authors/202312-007",
    "title": "JP Gouigoux",
    "authorMainContactPhone": "+33 787 787 787"
}
```

Any command-line input or output is written as follows:

```
catch (Exception ex) {
  transac.Rollback();
  throw new ApplicationException("Transaction was cancelled",ex);
}
```

Bold: Indicates a new term, an important word, or words that you see onscreen. For instance, words in menus or dialog boxes appear in **bold**. Here is an example: "Once connected, you will be presented with the welcome page interface, which you can return to at any time by clicking on **Business Central**, or the home icon in the top-left part of the screen."

> **Tips or important notes**
> Appear like this.

Get in touch

Feedback from our readers is always welcome.

General feedback: If you have questions about any aspect of this book, email us at customercare@ packtpub.com and mention the book title in the subject of your message.

Errata: Although we have taken every care to ensure the accuracy of our content, mistakes do happen. If you have found a mistake in this book, we would be grateful if you would report this to us. Please visit www.packtpub.com/support/errata and fill in the form.

Piracy: If you come across any illegal copies of our works in any form on the internet, we would be grateful if you would provide us with the location address or website name. Please contact us at copyright@packt.com with a link to the material.

If you are interested in becoming an author: If there is a topic that you have expertise in and you are interested in either writing or contributing to a book, please visit authors.packtpub.com

Share your thoughts

Once you've read *Enterprise Architecture with .NET*, we'd love to hear your thoughts! Scan the QR code below to go straight to the Amazon review page for this book and share your feedback.

https://packt.link/r/1-835-08566-0

Your review is important to us and the tech community and will help us make sure we're delivering excellent quality content..

Download a free PDF copy of this book

Thanks for purchasing this book!

Do you like to read on the go but are unable to carry your print books everywhere?

Is your eBook purchase not compatible with the device of your choice?

Don't worry, now with every Packt book you get a DRM-free PDF version of that book at no cost.

Read anywhere, any place, on any device. Search, copy, and paste code from your favorite technical books directly into your application.

The perks don't stop there, you can get exclusive access to discounts, newsletters, and great free content in your inbox daily

Follow these simple steps to get the benefits:

1. Scan the QR code or visit the link below

https://packt.link/free-ebook/9781835085660

2. Submit your proof of purchase
3. That's it! We'll send your free PDF and other benefits to your email directly

Part 1: Business-Aligned Architecture and the Problems It Solves

This first part of the book is not technical at all and aims to provide a good understanding of what business-/IT-alignment is all about. It explains the difficulties and shortcomings in existing information systems and how some industrialization approaches may help overcome these problems. This is useful for anyone confronted with digital transformation. People with an IT system that does not deliver on its functional promises is hard to evolve, and is not cost-effective should also get a better understanding of why this is the case. After explaining causes and symptoms, this part of the book proposes a theoretical approach in the form of a utopic system that would be free of all these shortcomings and will guide our work in the next parts.

This part includes the following chapters:

- *Chapter 1, The Sad State of Information Systems*
- *Chapter 2, Applying Industrial Principles to Software*
- *Chapter 3, Reaching Business Alignment*
- *Chapter 4, Dealing with Time and Technical Debt*
- *Chapter 5, A Utopic Perfect IT System*

1

The Sad State of Information Systems

Before jumping to solutions, it is essential to share a thorough diagnostic of a situation. In the case of information systems and, more generally, computer use, any user knows of the term "bug" and has experienced the frustration associated with malfunction, sometimes with a high personal impact (loss of personal data, consequences regarding revenue, and so on). For companies, a malfunction in IT can have harsh consequences since they depend more and more on computers to realize their business operations, hence their financial objectives.

After defining what an information system is and explaining how its efficiency (or lack thereof) can be calculated, an attempt at classifying the causes of such problems will be exposed. As for solutions, this will be the subject of the rest of this book. But for now, we must understand what is going wrong with information systems, how this happens, and – more importantly – why.

In this chapter, we will cover the following main topics:

- What is an information system?
- Why is software building still a craftsmanship, with good and bad consequences?
- How the efficiency of an information system can be evaluated
- How to classify the different impacts that can happen on information systems, as well as what their causes and consequences are

What is an information system?

Before talking about the state of information systems, it might be useful to give a clear definition of what an information system is and even what a system is.

A **system** is a group of items that operate together to reach a common goal. This is the basic difference between a system and a union of individualities: the parts of a system work together toward a vision.

An **information system** (we will sometimes abbreviate it as **IS**) furthers this definition as a group of items that share information to reach a common objective. Strictly speaking, an IS is not necessarily made of software, even though most of what we will talk about in this book is about computerized information systems. And, even in the most sophisticated IS, there remains a non-neglectable part of information that is not software-contained. This is a situation that we will talk about, but for the main part of this book, we will put ourselves in the hypothesis of software-based information systems as they are now pervasive in almost every company and organization.

So, an IS is understood as a set of software tools that operate toward a goal. This goal is typically designed as a business process for most of the companies that own the system. It should be noted that software always depends on hardware, but this tends to be more and more hidden in the background and the IS is more and more considered as the software means, organized together to reach a business goal, by implementing functional processes efficiently.

A quick history of information systems

When dealing with general subjects, such as the quality of information systems, it is always interesting to have a look at the past and analyze the evolution toward the current situation.

If we follow the definition stated previously, information systems only appear when at least two entities collaborate. In our hypothesis of software-based information systems, this means at least two computers have been connected. This means that initial room-sized computers that are operated on-premises should not be considered systems, though they were often called "large systems" (the items that were assembled, in this case, were computation mechanisms, short-life memory, and long-life memory).

Let's go forward in time a little bit and talk about IBM's client-server mainframes: those are the first ones that we can consider as information systems since there were client stations connected to a central computer, with information flowing between them. The protocol was proprietary, which made it easier for IBM to reach the high quality of the service provided by these systems (lots of them are still in operation nowadays). Since a unique implementation of a protocol was defined by the same team, compatibility, and interoperability were a very light issue. This is legacy, but when the system is working and when modernizing it is highly risky, the businesses logically make the choice of not moving anything (this is the first rule in information systems management: if it works, do not touch it).

Fast-forward again and we are in the nineties. This is the era of **personal computers** (**PCs**). Though there is a worldwide attempt at keeping compatibility between the machines, anybody using computers at this time who was unable to use a given piece of software, because it did not support the embedded video card, knows this was partly a failure. Of course, things have greatly improved since then, with **Video Electronics Standards Association** (**VESA**) international standards being created for video cards and screens, such as VGA or VESA, and lots of other norms that make it possible, in the present time, to change components of a PC without breaking down the whole system. Still, the toll on information systems has been quite high: how can we expect networks of machines to work well together when even a single machine's components are hard to assemble? Proprietary formats, different encodings, an almost complete absence of strong data exchange protocols: everything led to a difficult situation, and only high-budget companies could operate complex computerized systems with the help of experts, who would twitch about anything from jumpers on an electronic board to compilers parameters to make the system work.

Fortunately, Y2K came with the internet expansion and its radical approach to normalizing the exchanges between computers. Now, every single computer that abided by TCP/IP, HTTP, Unicode, and other internet-born standards could exchange data with another one anywhere in the world, regardless of its hardware and operating system implementation. This is one of the biggest steps forward in the history of IT and the root definition of what our "modern" information systems are today.

A few software layers were added to this to make it easy to reuse functions in a system. It started low, with local reuse of code through routines, then libraries and components. With the advent of network capacities, this evolved to distributed components, web services, and finally **service-oriented architecture** (**SOA**). This is also the time when n-tier architectures were put in place, establishing a first layer of responsibility separation inside the software application, between **graphical user interface** (**GUI**) management, the exposition of functional services, the implementation of business rules, and persistence management.

The last bit, SOA, has led many companies to costly failures, and the vast majority of attempts at putting such architectures in place resulted in important financial loss and abandonment of projects. The technological step was too high to operate smoothly, and lighter alternatives have appeared to remove local difficulties of the SOA approach:

- **Standardized message-oriented middleware**: To counter the proprietary exchange protocols put forward by large software companies operating on SOA and using it to lock their customers in their system

- **The REST approach**: To lighten the weight of SOAP/WSDL web services and the weight of all associated norms

- **Enterprise Service Bus**: This technique is used to reduce the importance of the middleware and reach a "dumb pipes" paradigm where the software applications participating in the system would be able to communicate with each other without needing a central piece of software that would necessarily become a single point of failure

As the presence of a few reference books (Sassoon, Longépé, and Caseau) show, the best practices to design a strong and evolution-capable IS were already available in the late nineties, though not very well known. But it was during the 2000s that these practices came to a larger share in the community and SOA and other service-based approaches flourished, leading to the microservices architecture in the beginning of the 2010s. This set of practices is still considered as the reference architecture at the time of writing, though we will see that not all its recommendations should be applied without a strong analysis of its usefulness in the studied context. As we will see, the granularity of services is key to obtaining an efficient IS. But for now, we will talk about software building in general and try to understand the current limits of this so-called "industry."

Software building – still craftsmanship

The precise definition of an information system, as interpreted in this book, has been given together with a brief history of its evolution. This history is not only very short but shows many recent evolutions, most of them radically different from the previous state of the art. This very rapid evolution is a sign that information systems design is not something that can be considered stabilized and completely understood.

There remains a large part of craftsmanship in the design and deployment of a software information system. Craftsmanship has its advantages: human attention to detail, custom-tailored functions, uniqueness, and more. It also has numerous drawbacks, such as high costs, difficulty in evolving in a controlled way, dependence on a few creators, and many others. These drawbacks outgrow the advantages in modern companies for which the information system has become the operational backbone.

Well-crafted information systems are an evolution from arbitrarily evolving ones, and there is nothing to be ashamed of with a craftsman's job, but the way forward today is toward an industrialized approach to information systems. This is what this book is all about.

Craftsmanship as opposed to previous lack of quality

Craftsmanship is used in several domains of IT where opposition is established with older, more arbitrary, and self-organized methods where used. For example, numerous IT conferences include the word "craftsmanship" in their names as a statement of their will to address quality and heterogeneity issues.

Information systems have long appeared simply because pieces of the system were put together and linked without any reflection on the whole system itself. This is generally what happens with so-called "point-to-point integrations," where connections are made between different software modules while considering only the source and the destination of the link, without any map of all the links that exist, sometimes reproducing an already existing link or reversing the initially intended direction of the functional dependency between the two modules.

Such systems that are born without anybody thinking of the whole functioning have very few chances to remain stable for long. In the rare cases where only a few connections are created, the system can operate fine, but we all know that IT evolves very quickly, and business needs are added all the time ("the only stable thing in a system is its need to evolve"). If nobody has a global view of the whole system, there is no way that its overall evolution is going to develop ideally. It would be pure chance and Murphy's Law states that, if something can go wrong in a software system, it most definitely will.

Software craftsmanship involves a will to not let the system create itself and develop in the wrong direction, but to take extra care of the software quality and methods to build time-enduring and evolution-capable systems. When applied to code (where it mostly happens), software craftsmanship includes test automation, a refactoring approach, quality indicators monitoring, and many other methods and techniques that go beyond minimal practices.

We could argue that these practices are contrary to craftsmanship, but this is only true in established industries, where craftsmanship lives in opposition to standardized industrial production. In the world of IT, industrialization has not happened yet. The word "craftsmanship" has been used by artisans of IT who pride themselves on analyzing problems and solving them with code more elegantly and sustainably than usual slack in their analysis and in the realization (many of us have seen code without any analysis documentation and a single unit test, and yet that has been sent into production just because "it worked"). Craftsmanship – as the word is used in the IT movement – could even be said to be the very first step into industrializing the software domain since this is the first resolute action into making code clean and not letting it derive into what is commonly named in the IT jargon as a "big ball of mud."

It might look like I oppose craftsmanship and industrialization, but one isn't better than the other: they are simply two phases of the development of a domain – in our case, software. Personally, after 38 years of programming with the desire to do things cleanly, I consider myself a craftsman; my goal with this book – and the past 15 years of my architect career – is to humbly help make another step so that software gets out of its infancy problems and becomes an adult industry.

As a side note, a subject of discussion among software architects is about our working domain to be able to ever become an industry. I tend to think that the amount of creativity needed to be a good software engineer will make it impossible to fully industrialize software production, but that a lot of things should be industrialized to reach this mature, adult phase, where IT will finally deliver its complete value.

A word about "emerging architecture"

The concept of emerging architecture is that, in some human constructions (mainly software applications), architecture is not established upfront, but appears gradually with the construction of the object. This concept is often used together with Agile methods, where the software construction is iterative and not done in a single succession of phases, such as in the "V-cycle" method. Instead of a single design/development/test series of steps, Agile software development loops many times over these steps, each time building on the previous cycle to gradually work toward a final vision that may even evolve during the steps.

In this case, each step involves the minimal design activity necessary to realize the cycle, in a **Keep It Simple** mindset. Thus, there is no initial complete vision of the final architecture, which sometimes can be seen as a serious limitation of Agile methods, but is at the same time their force, as they continually adapt. There are ways for this to go awry, though, and it mostly happens where participants in the project expect the architecture to emerge naturally. This confusion is often seen in software developments where no architect is involved and where developers believe individual best practices, goodwill, and craftsmanship will have a quality impact on the outcome. The reality is that these practices will positively affect the individual results of each Agile step but will not guide the overall architecture in any sound place since there is no long-term direction.

Therefore, it is very important to understand that emerging architecture does exist but necessitates an active engagement in making it happen gradually. It naturally happens at the module level, where a single developer carefully refines and refactors the code. But to work at the system level, there needs to be the same level of engagement.

Craftsmanship as opposed to an industrialized approach

In the previous section, we have opposed craftsmanship to lack of quality intention in the code and thus shown how positive craftsmanship can be. Here, we will point out its limits when opposed to industrialization. Craftsmanship bears the idea of a highly skilled individual operating carefully, where time spent is of little importance compared to quality.

Though craftsmanship – in its noble sense of dedication to hand-polished, high-quality work – is worth praise, it also states that the level of maturity of the domain is still low. When reaching maturity, disciplines tend to separate different steps of the work, automate some of them, standardize practices and tools, and overall become more efficient, even bringing quality to a higher level that cannot be obtained with an individual human approach alone. Moreover, industrialization normalizes the whole process and makes it possible for anyone following the norms – and not only highly skilled people – to reach this high level of quality.

This is what has been done in all industries (and this is even why we call them this), and it is the natural evolution human workers try and reach. In the software field, predictability, quality, and time-to-market are sought-after qualities and an industrial approach is necessary to reach them.

It is essential to point out that there is nothing wrong with a non-industrial approach. Software is not an industry yet. After all, bridges have been built for more than 4,000 years, so it is quite understandable that this has become something of a controlled and well-established way of working. Software building, on the other hand, has only been there for a few decades and still is in its infancy.

But the important information here is that craftsmanship, despite all its advantages, is the step before industrialization, and lots of information systems owners nowadays really crave an IT team that reaches this next step. Stakes are huge for some of them, where competitive advantages come mostly from the information system. It has been said that "all companies today are software companies" and this stresses once more the importance of the information system and the absolute necessity of reaching higher levels of quality.

Concept of technical debt

Technical debt is a concept that has been created to explain, through a metaphor, how the low quality of software development can negatively affect its future development. In financial debt, you must pay regular interest, depending on the amount you have borrowed. In the software side of the metaphor, buying some time by cutting corners and lowering overall quality will have to be paid regularly, so long as the low-quality module remains active. Bug correction and maintaining the module will take some time, and the team will not be able to spend on new features that have value for the users. The lower the quality of the module, the higher the "rate of interest" – in our case, the time spent on maintenance. In the worst case, the quality of the software is so low that all available money/development time is spent only on paying for the interest of the debt (keeping the application running), which means that no money remains available to reimburse the debt/correct the software, let alone paying for features with more value.

This concept will be discussed at length in *Chapter 4*, but since we are talking about craftsmanship, now's a good time to explain the link between both concepts right away.

Craftsmanship is often seen as a software development approach where technical debt is maintained at its lowest level possible, sometimes being virtually nonexistent. A good craftsman developer will take pride in having zero-defect software, 100% coverage via automated tests, with a fully automated integration and deployment system.

The industrialized approach that was opposed to craftsmanship results mostly in an improvement of the overall quality, but also goes beyond in how it deals with technical debt. Where craftsmanship tends to have an individual and binary approach toward technical debt, as opposed to careless development that lets it potentially go unleashed and out of control, the industrialized approach manages technical debt. The financial metaphor still stands: instead of refusing any debt altogether, a well-thought operator will carefully manage their capital, borrowing if the advantages are indeed higher than the costs of the loan. Compared to a craftsman, an industrially-oriented developer will be more aware of the importance of time-to-market and assume a controlled level of technical debt if it helps them reach users before their competitors, thus bringing benefits that will partly be affected in reducing the technical debt afterward.

The long-standing comparison of software to mechanical systems

Comparison of software and mechanical industries (in most known cases, the automotive industry in particular) is such a habitual position of contenders of both sides that it has been worn out. Also, comparisons do not necessarily bear logic, depending on the way they are realized. For example, in the famous *Microsoft versus General Motors* meme, comparing a single car with a single piece of software brings strange conclusions from each side (*If the car industry evolved as quickly as software, we would have cars driving 1,000 miles per gallon"/"If car industry worked as the software industry, a car would crash inadvertently every 1,000 miles*"). The underlying error in this ill comparison is that the level of complexity is not the same on both sides. If you were to compare a single car created in an industrial factory, the cost should be compared to the single operation of a functional process in an information system, since only one function is individually operated (transporting a few people in the case of a car, calculating a payroll – for example – in the case of the software application). If we turn the problem the other way around and want to compare the internal operation of the software with its tens, maybe hundreds of thousands of lines of code, the right comparison would be with the car factory itself as it has the necessary modularity to change its car model production, and it also contains thousands of functions and many more moving parts to perform such a moving task.

In short, wrongly-calibrated comparisons can be very misleading and of little practical interest. But comparing information systems building to the design of a car factory, for example, is closer to the reality of the complexity of each system and can provide interesting insights, so long as you contextualize them carefully. As stated previously, comparing information systems to bridges can be pertinent if we stay on the criteria of the age of the industry. How could IT, an activity that is only a few decades old, reach the same level of maturity as an activity that has been developed over several millennia?

We will stop this comparison for now, but you will be introduced to several comparisons between the software "industry" and more traditional industries that better deserve this qualification throughout this book. We will try to keep the comparison as helpful and legitimate as possible. Again, there is no judgment in saying that IT is not completely industrialized yet: some systems undoubtedly are, and some are not. And this should not be taken against anyone as industries took many human generations to reach their current level. Software-based information systems simply have not had the time to do so. This book is about the means to help go in this direction, taking into account experiments from other industries, while keeping in mind that comparisons may sometimes be misleading and should be used while paying attention to their applicability.

Now that we have established what an information system is and the many infancy problems that it may have because the field is not industrialized yet, a sound engineering approach is needed to evaluate this lack of maturity.

The efficiency of an IS

Engineering is about making things in a controlled way; information systems will not escape this much-awaited transition as their efficiency could be improved. To do so, we need indicators, measures, and a method to get the results on a given information system. Fortunately, there is a consensus on this in the field of expertise, as we will see now.

The measure of efficiency of a system

Since a system is a group of items working together toward a goal, as defined previously, measuring the efficiency of the whole means more than just knowing the efficiency measures of each item and summing them one way or another. It is often said that the value of a good information system is much more than the value of each of its moving parts and that the most positive effects come from their interactions.

A good way to measure the efficiency of an information system is to evaluate how much time and money it helps save on the functional processes it supports. However, this is quite complicated to measure since a single IS generally operates several processes, and their respective importance and costs should be evaluated. If you consider that processes are also partly human-operated and that the gain of efficiency does not come only from the software part but from the way the functional teams use the application, how they have been trained, what investment has been put in the hardware for performance, and so on, it starts getting difficult to evaluate the efficiency of a complete system with these metrics. Furthermore, the output may be closer to a return-on-investment calculation than that of an efficiency metric.

This is why the efficiency of IS is often evaluated through simpler, more attainable metrics, namely the percentage of cost used for maintenance. This simple ratio allows us to know how much money is put to keep the system working (this is an operational cost) versus how much has been put into designing it and making it better (which is an investment cost). As functional features are what is asked of the system, the more maintenance there is to keep it operating, the less money is related to getting value out of the system. Of course, there are many points of view on software efficiency, and we are going to browse a lot of them together in this book. But maintenance is simply the easiest way to start evaluating the state of an IS.

The cost of maintenance

Maintenance may be the field where software is still lacking the most. Methodologies have arisen on the subject of software design and development that are starting to help turn craftsmanship into a more industrial approach. But design and development are just the beginning of the journey for a piece of software. Once in production, the application has to be deployed, patched (sometimes while functioning), improved, and maintained in the long run. This is where bad design, technical debt, and high costs of maintenance show the shortcomings of the design phase.

Again, the picture is not all that bad. After all, there are proven methods to upgrade software applications while running, which is quite an achievement if we compare this to its mechanical counterpart of changing a car's wheel while running. But an analysis of the total costs of maintenance of ISs shows that this is only the tip of the iceberg and the general situation is quite awry.

A study from Gartner dated around 10 years ago and based on 3,700 companies indicates that 72% of the budget of IT is used for maintenance. You read that correctly: almost three-quarters of the cost of an IS is affected not by designing and improving it, but by adjusting and keeping it working while in production. Again – and with all the care that should be taken with such a comparison – imagine a house that, once you have paid for its building, would cost you three times as much every year just to prevent it from collapsing! This shows the sad state of information systems today.

You will find many other studies on this, but the Gartner one may be the most well-known. To give just another confirming point of view, another study from the same period showed that "roughly 20% of technology-supported initiatives are failures, and that number increases with the size and complexity of the project. However, unsuccessful projects often show unmistakable symptoms of failure before execution even starts."

The good news is that we are not condemned to go on this way. In its infancy, car manufacturing had the same problem, with a huge maintenance cost compared to the initial production toll. The first cars were not industrialized but hand-built and their yearly maintenance was almost as costly as the car itself, with tires that had to be changed every few thousand kilometers, oil that had to be added so often that dedicated tanks were present in addition to gas ones, and engines had to be revised by mechanics every few months.

Today, when you look at maintenance metrics in industrialized goods, the situation is magnitudes better. A brand-new car from a trustworthy company will happily drive tens of thousands of kilometers before its first garage review and, with the improvement of engine manufacturing, adding oil is something of an old story nowadays. Another example is oil filters, which are so standardized that two sizes cover almost all of the European market of individual cars.

In factories, where statistics of maintenance are keenly observed because the financial yield highly depends on them, we have numbers available to quantify this in terms of maintenance cost. Generally speaking, the costs of a factory are divided into three groups, the ratio of which is important to determine the efficiency of the whole:

- **Fixed expenses**: These are the expenses that are not correlated with the production volume. Typically, the cost of the building (through debt reimbursement or lease, depending on whether the company owns or rents the premises) does not depend on how much you produce. To improve efficiency, you are looking at reducing these costs, but augmenting your production level will also help you because it will reduce their relative impact on your revenue.

- **Value-adding related expenses**: These are the expenses that are directly related to your production. The more you produce, the more you will pay your suppliers for the goods they provide you with. You will get more benefits, except if you are selling below the revenue price, which is not only forbidden in most countries but is generally a very bad business decision. All in all, these expenses are considered good expenses because the higher they are, the more money comes in as well!

- **Maintenance expenses**: These are expenses that you need to plan for your production tool to continue operating smoothly. Maintenance expenses are a hard beast to tame because, if you neglect them for small economies, they might come back at you and bite strongly (a little economy on cheap lubricating oil and your million-dollar machine may fall apart in a few years). Maintenance costs have two drawbacks. First, contrarily to fixed expenses, maintenance expenses grow with production (the more your machines operate, the more maintenance they need, and this continues exponentially as old machines need more maintenance). Second, the opposite of the second kind of expenses, they do not add visible value to your product: if you include a better engine in the cars you sell, the associated value is immediately perceived; if you buy better oil to maintain the machines that are used to build those engines parts, no customer will ever realize it, let alone pay for it.

With that, we understand why factory managers are very attentive to maintenance costs: they are the "bad" expenses and the difficult ones to manage. Fixed prices need less attention, simply because... well, they are fixed! And value-adding costs are not such a problem because, when they increase, that means your business is growing. So, maintenance is the key and most factory managers will be judged by the owner or their boss on their maintenance metrics. If you do not control maintenance and the costs are soaring, they will be replaced by another manager. If they are too restrictive on the maintenance budget, very costly failures may start to appear, and, again, they will be considered responsible for the mess.

Statistics help managers find the right way in this difficult equation by comparing it to what has been done in the past in similar situations. For example, in heavy industry factories, a fair repartition of cost is considered the following:

- **Fixed expenses**: 10% of the budget

- **Value-added related expenses**: 85% of the budget

- **Maintenance expenses**: 5% of the budget

There are, of course, tolerated differences in these numbers, but a maintenance ratio higher than 7% has to be justified and, beyond 10%, this is a strong alert regarding the maintenance not being in control anymore.

Now, let's go back to the Gartner studies, where the admitted average of budget used to "keep the lights on" by the 3,700 IT leaders interrogated was 72%. Again, the comparison seems negative to the so-called IT industry: 10 times worse! But there are a few circumstances that must be considered. First, the natural and optimal sharing of costs in IT is necessarily different since the design is only made in the mind and does not need expensive material prototypes like in the heavy industry. Also, the fixed costs share is becoming lower and lower due to the availability of "as a service" artifacts. When, a few decades ago, buying a huge computer had a large toll on the budget, particularly in the first years when it was only partially used, cloud operations allowed us to buy computer power as needed, which made these costs fall into the investment costs group.

So, we could consider that the numbers are not directly comparable. Nonetheless, the IT industry has a problem with maintenance costs. And as engineers, the question that immediately pops to mind is: Where does this come from? The next section will hopefully show you that the causes are – most of the time – easy to establish since some generic errors are made when designing information systems. These are at the root of the observed shortcomings.

Examples of the prohibited evolution of IT systems

So far, things may been a bit theoretical. Most information systems are created without a defined plan and global, architected vision and this is reflected in the cost of maintenance – and thus the total cost of ownership – of the whole system. But what does this mean in practice? Is it that bad?

You may have heard expressions such as "spaghetti dish" or "data silos." In the first case, the modules of the IS are so intertwined that it becomes impossible to touch a given part of the system without causing side effects on another one. In this case, evolution becomes complicated. The second expression is associated with modules of an IS that are so tightly separated from each other that they cannot share common data. This generally leads to duplicated data, loss of quality, and sometimes contradictory processes in the whole system. These are just a few examples of the designation of generic problems that can happen.

The following sections dive much deeper into such mishaps and detail what chains of reaction make information systems get slower, harder to operate, and, in the worst cases, stop them from working completely. As an architect for almost 10 years and then a consultant on information system evolution for small to very large companies, I have observed enough hindered information systems to be able to create a classification of what went wrong and how this can be analyzed. This experience, shared with a research lab in France, led to several scientific papers where this analysis was formalized and business/IT alignment anti-patterns were documented. Some of them will be detailed in the upcoming sections.

Classification of the causes

The following diagram has been extracted from a scientific article that was published in 2021 – *Business-IT Alignment Anti-Patterns: A Thought from an Empirical Point of View* – and that Dalila Tamzalit, LS2N / CNRS, and myself presented at the INFORSID conference in June 2022:

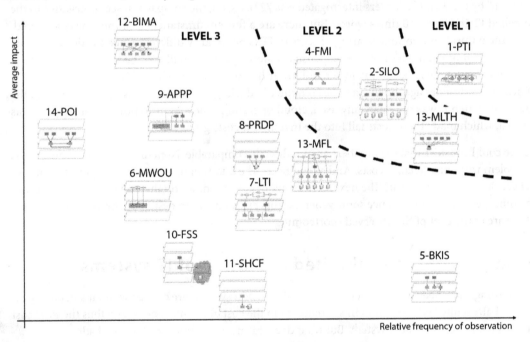

Figure 1.1 – Classification of business/IT alignment anti-patterns

The small diagrams are a visual code to identify the different patterns of misalignment, the meaning of which will become clearer in the next chapter. For now, only the position of the different blocks is to be defined.

The horizontal axis expresses the frequency of finding the pattern in information systems that have been studied. Admittedly, those were ISs with problems, since consultants like me were hired to work on them. However, a decade of experience on almost 100 organization/industry systems from many different contexts showed that this is – sadly – something extremely current in ISs and that high-quality, rapidly evolving, and cost-efficient ones are extremely scarce. These are reserved for very small companies with limited needs or very large and rich ones that have known from the beginning that their IT system was their spine and brain and invested accordingly.

The vertical axis of the schema evaluates the impact the anti-pattern has on the functioning of the system. The higher the position of the anti-pattern, the more it hinders the correct working and/or evolution of the IS.

The consequence of this classification mode is that the anti-pattern at the top right position (**LEVEL 1**) is the most impactful and most often observed. This case corresponds to when the business processes are directly implemented in the software layer of the system. In *Chapter 3*, we will consider the decomposition of IS on four different layers, but for now, suffice it to say that the good alignment between these layers is the most important source of quality and that the four-layer diagram is of such importance that the symbols of the 14 anti-patterns documented are based on it.

The following three anti-patterns, as shown in the top-right corner of the preceding diagram (**LEVEL 2**), are a bit less spread and harmful than the first one but account for a lot of observed difficulties. They correspond to the following three cases:

- **Feature with multiple implementations**: This leads to different business rules, depending on which software is used, and obvious mistakes as consequences

- **Silos**: These cause duplication of data and additional work, together with errors due to synchronization or lack thereof

- **Monoliths**: Heer, a single software application concentrates so many business functions that its evolution is complex and it becomes a bottleneck for the whole system and sometimes even a blocking point

The remaining 19 anti-patterns (**LEVEL 3**) are less observed and/or less dangerous for the evolution or correct functioning of the information systems, but their knowledge can help us spot them in maps of systems and improve the situation.

The following diagram (*voluntarily blurred to protect customer information*) shows how a quick map of an information system can help us visually find the "hot spots." In this case, two applications that received lots of "hard-coupled" (we will come back to this notion of coupling in *Chapter 4*) streams of data caused evolutionary problems, particularly since one of them was obsolete and the other one difficult to evolve for commercial and regulatory reasons:

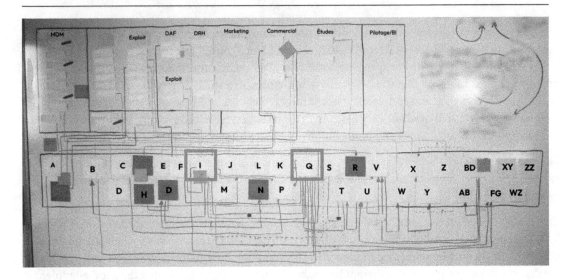

Figure 1.2 – A hand-made map of an information system, revealing
the high coupling for two software applications

The story behind this diagram and the reason it is displayed here, while not readable, is that when designing this map of an information system, a non-technical board member came into the room and immediately pointed at the two notes with a lot of red wires pointing to them, saying "I think I know where the problem is." It made us realize that we did not need to continue mapping more precisely as the existing analysis was clear enough that we could start acting on the main coupling problems. It also was discovered that the two applications were also the ones with the highest obsolescence.

It is outside the scope of this book to study all the groups of anti-patterns that can cause problems in an information system as the subject is the architecture of such systems and we will mostly concentrate on how to avoid these problems right from the design phase. Nonetheless, if you're curious, you are invited to read the *Business-IT Alignment Anti-Patterns: A Thought from an Empirical Point of View* paper (refer to the *Further reading* section) for an academic and more formal presentation of the classification.

The classical symptoms of a blocked IS

Behind the causes, there are symptoms of ISs that include some (and sometimes many) of these anti-patterns. If you're experienced, then you will certainly be familiar with some of them:

- The testing time of a software application or a process increases exponentially. Release time increases, reaching sometimes up to several years

- A version impacts customers or users who are not concerned with an included evolution or feature as it is on a function that they do not use

- Some side effects of an IS module cannot be explained (well-established bugs, but also effects on performance, non-predictable behavior, and more)

- Low levels of satisfaction for internal information systems, loss of market share for applications that participate in external/customer information systems

Rate of renewal of large systems

When such impacts happen and become larger and more difficult with time, the perceived solution is often to throw the existing IS away and build a new one, which is often called the **Big Bang approach**. Not only is this the most expensive way of tackling the problem, but it happens to also be the riskiest one as the bugs that have been found in the existing legacy applications are bound to reappear for some of them or be replaced with new ones, making it extremely slow to reach a satisfactory situation. There is even a high probability that, by the end of the rewriting process, the new system will also be far from the expected behavior since business needs have changed again in the meantime.

This is why it is always better to gradually improve an existing system, leaning on its good behaviors and progressively improving the modules that cause the most trouble for users.

Depending on the level of quality of an information system, the following can occur:

- It can be in such a fine condition that all evolutions can be realized without any side effects on functions that are not concerned

- It can require a few days' work to implement a simple new working feature and deploy it

- It can present some couplings that make it longer, but not more difficult, to do so

- It can be so full of problems that implementing new features is complicated and necessitates dedicated project management and impact analysis

- It can be in such a state that evolution is almost impossible or at the price of a higher instability of the whole system

The evolution of ISs is often based on phases with a duration of a few years. After 2 to 3 years, and in the best cases 5 years, the future of the business is so difficult to predict that planning the evolution of the underlying IS does not make any sense. In highly volatile activities, even a plan of 2 years may be considered too long.

Of course, many ISs cannot be fully corrected in only a few years and several plans will be necessary to re-align them. In this case, the approach is the same as in an Agile software development project: a first step – though much longer, generally around a semester – is realized to realign the most urgent problem, at which point objectives are re-analyzed and a following re-alignment step is undertaken, and so on, in a continuous improvement approach.

Although much more tempting in some cases (and in particular very appealing to non-technical profiles that do not grasp the difficulties linked to stabilization), the Big Bang approach is rarely the solution, and if you have to deal with an inefficient information system, you will most likely have to plan for a step-by-step evolution. While you will be doing such a change in an IS, this will quickly come in handy with legacy modules. And though the IT industry is quick to consider older technologies as garbage (just type any technology name, followed by "is dead," into a browser and you will realize the extent of this), a responsible approach to IS evolution is to observe respect toward the legacy. The reason it became legacy is that it has been providing value for a long time.

Summary

In this chapter, we discussed what an IS is and why, despite all the expertise and care that can be given by true craftsmanship in software design, the system that links all these applications together can present many issues, mainly in terms of maintenance costs and the ability to evolve in time and rise to the challenges of new business processes and requests for features. Many symptoms can alert us of the state of a given IS but they all boil down to one main reason: IT has not reached the state of a truly industrialized domain since it is still a very recent human activity compared to other actual industries.

In the next chapter, we will talk about how industrialization principles can be applied to software. This can be summed up in two actions: cutting down complexity and standardizing the interfaces.

Further reading

- Longépé, C. (2019). *Le projet d'urbanisation du S.I. – 4th edition.* Dunod/InfoPro. EAN 9782100802432. https://www.dunod.com/sciences-techniques/projet-d-urbanisation-du-si-cas-concret-d-architecture-d-entreprise-0.

- Sassoon, J. (1998). *Urbanisation des Systèmes d'Information.* Hermès. EAN 9782866016937. https://www.fnac.com/a270920/Jacques-Sassoon-L-urbanisation-des-systemes-d-information.

- Caseau, Y. (2011). *Urbanisation, SOA et BPM – 4th edition.* Dunod / InfoPro. EAN 9782100566365. https://www.dunod.com/sciences-techniques/urbanisation-soa-et-bpm-point-vue-du-dsi.

- Gouigoux, J. P. & Tamzalit, D. (2021). *Business-IT Alignment Anti-Patterns: A Thought from an Empirical Point of View.* In E. Insfran, F. González, S. Abrahão, M. Fernández, C. Barry, H. Linger, M. Lang, & C. Schneider (Eds.), Information Systems Development: Crossing Boundaries between Development and Operations (DevOps) in Information Systems (ISD2021 Proceedings). Valencia, Spain: Universitat Politècnica de València. https://aisel.aisnet.org/isd2014/proceedings2021/managingdevops/3/.

2

Applying Industrial Principles to Software

This chapter explains what can be done to make IT a real industry, and this begins with applying the main principles of industrialization, namely cutting complexity into small pieces and then standardizing the modules, and in particular their interfaces. We will make a comparison with the development of cities, where the normalization of water pipes, electricity, and other interfaces has allowed for continuous evolution.

In this chapter, we will explain the very concept of industry, as this is a very often used name, but not necessarily every time with a precise understanding of its meaning. We will also learn about how industrialization works by cutting complex problems into small ones and then making the small ones simple and repeatable, principally by means of standardization. We'll also understand what benefits can be drawn from such an approach, in particular in information systems.

We'll cover the following topics in this chapter:

- What is an industry?
- Management of complexity
- The benefits of standards and norms
- The urbanism metaphor of information systems

What is an industry?

In the previous chapter, we compared craftsmanship to industrialization, hopefully showing that, while the former has nothing to be ashamed of, the latter is its natural evolution in time. All industries start with artisans and, with the work becoming more and more controlled and repeatable, potentially end up as real industries where the artisans have gradually converted to engineer competencies and jobs. Most people understand this without any need for explanation because it can be seen in many day-to-day experiences. When one, for example, first attempts at realizing a new task (say, cutting hair), the first trials are not comparable to the ensuing ones. After a period of time, the process starts to become more regular (the hair is cut fine, and the initial customer who accepted to be your *guinea pig* does not complain anymore). Given enough training, one gets expertise in the field and develops a routine (the hair is cut at a defined length and with the expected shape, in a way that can be precisely reproduced in a future haircut).

How about trying to formalize what is behind industrialization, though? In other terms, how do we characterize what constitutes industrialization? As we saw, there is the concept of being reproducible, which means there is a measured norm that should be met. Also, this norm is shared between all knowledgeable people in the field, which means it becomes a standard. In our example with haircuts, there are names for haircuts, and everybody in the field knows what "trimmed" or "shortened" means, which makes it safe for customers not to leave the hairdresser with a hairstyle they did not expect. Also, from one hairdresser to another, one can expect to obtain a globally similar result once stated using the right vocabulary. We obtain a homogeneous quality in this way.

In industrialization, just like in hairdressing, there is also the concept of addressing small parts of the whole. Except if you would like to look like a well-known person, you will not describe your haircut as a whole but describe small parts that form a complete hairstyle. For example (although not a good one, from an aesthetic point of view): long at the back, short on the top, tapered on the side. Addressing the parts instead of the whole and cutting a complex problem into small, simple ones that can be solved simply is the basis for industrialization, engineering, and even problem-solving as a whole.

A good hairdresser may be a fine artisan, but once you can get a similar haircut from many professionals in the field, it has simply become an **industry**. In the following sections, we will apply this definition to IT and show how industrialization happens there. In order to do so, we will first dive a bit deeper into what really stands behind the concept, in terms of actions to realize.

The two roots of industrialization – modularization and standardization

After a simple example as an introduction to what we are going to call industrialization when applying it to Information Technology, we are going to dive a bit deeper into two associated movements in the concept, namely cutting big problems into small ones, which can be called **modularization** (as we expect small modules of a whole system) and solving these small problems with a normalized approach to reach homogeneous quality, which can be called **standardization**.

Modularity to reduce complexity

Before explaining the concept of complexity, let's see with another example how it relates to modularity. This time, we will use a mechanical comparison by analyzing the different modules of a car. A modern car is a feat of engineering, gathering so many parts in a sophisticated manner that is it virtually impossible for a person alone to build such a system. When one decomposes a car into modules, there are definitely clear-cut, well-separated modules that each have a purpose:

- The engine will provide power to displace the car
- The body will protect the driver and passengers
- The wheels and driving train will transform power into motion
- The chassis will hold the other modules together in a rigid way, and so on

The engine alone still is quite a beast, but we can decompose it further, into sub-modules:

- The injection system will bring gas into the chamber
- The pistons will transform the explosion into linear motion
- The crankshaft will convert linear motion into rotary motion
- The lubrication system will ensure the system does not degrade due to wearing, heating, and so on

Again, the complexity has lowered, and, if we go one step further into decomposing modules, the lubrication system can be described as the following:

- A pump ensures oil circulation
- The oil performs cooling and allows for frictionless movements
- The oil filter removes small debris that could otherwise increase friction and wear, and so on

This time, we have reached such a small level of complexity that almost anyone could act on these modules: adding oil can be done by anyone owning a car, provided they know where to pour it; replacing an oil filter is as simple as unscrewing the old one and screwing a new one into the same place.

Modularity, really, is the art of cutting complex things into small parts that are easier to manage. If modularization is done well, complexity decreases at each step. Imagine we had separated the engine as left and right portions of it: we surely would not have made it easier to observe and maintain. Indeed, modularity is not simply the cutting of the system; it is the art of cutting it in an intelligent way so that complexity decreases. But how can we do that? This is where the experience of artisans and the help of a long history of making comes into place, providing enough expertise to know where the system should be and what will make it simpler. The first engines surely did not have an oil filter, but after some time being obliged to remove all the oil from the engine after a few hundred kilometers, filter it, and pour it back into the engine, it became obvious inserting a filter into the engine oil flow was the clever thing to do. If we try to summarize this in just one sentence, modules should be carved out after functions. The oil filter is there because, in the process of lubrication, there has to be a filtering function. It does make sense to assign this function to one and only one module.

Standardization to ensure modularity is helpful

The relationship between the different modules, the way they fit together, and how they interact are other criteria that must be taken into account. Cutting down is not enough: if one wants the whole system to function, defining smaller modules is the first step, but once created, they must be put back together to reach the global goal. This is where the way the modules have been cut is important, and we have explained previously that it should follow functions.

But how about putting them back together? If modules are aligned with functions, how can we make sure they fit together well? In fact, the problem is quite simple to explain and sometimes extremely hard to solve, needing large engineering teams to do so: we have to ensure that the common function they share is exactly the same. If two functions have to be reassembled, that means they have a small connecting sub-function in common, which is generally called the interface. This interface has to be defined in a similar pattern on both sides.

Let's take the example of our oil filter again: it has been separated from the rest of the lubrication system and engine for its definition, but it also has to be put back in the engine system to operate and participate in the higher level of function, namely providing power to the car. To do so, it has been explained the oil filter has to be screwed back to the position in the engine, and this is where an interface will be needed. This interface is simply a screw thread: the oil filter will present a threaded oil and the engine a threaded growth at the place where the filter has to be placed, with of course a hole in it, allowing oil to flow in and from the filter. The interface itself is defined with functions:

- It should provide a stable attachment
- It should be tight enough to be oil-proof
- It should allow enough fluid circulation, and so on

We are one step forward, but there remains another step to do in order to reach industrialization: the interface must be standardized, which means all the preceding functions should be specified in such a way that replacement is easy and each provider can participate in the higher-level module simply by knowing the interface. In our example, the oil filter, in order to participate in the engine system, has to adhere to the following:

- Use the precise thread diameter (for European oil filters, it is a 20-millimeter diameter, with a thread step of 1.5 millimeters)
- Oil-proofness is ensured by a circular joint that is 62 millimeters wide in the same standard
- The capacity to retain debris based on fluid circulation, and thus the duration of the filter use, is determined by the volume of the filter, which comes in two standard sizes, and so on

Here is a very low-grade schema of how an oil filter is attached to a car engine: a threaded hole in the oil filter adapts to a threaded piece of metal with a hole in it that is on the external part of the engine, for easy access:

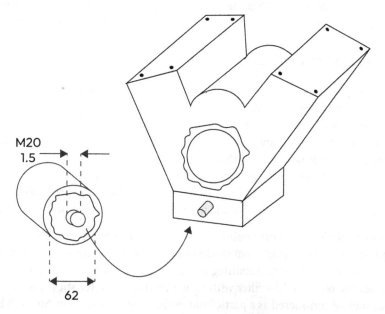

Figure 2.1 – Schematic positioning of an oil filter on a car engine

There we have it! If we now go back to the explanation, we have modules that are so standardized that one can buy them anywhere and they will have the same interfaces, although their inner functions may be different; modules put together will each have their function but provide a higher-level, more sophisticated function to the global system they form together. A few steps more and the whole system operated by the many modules and submodules will have a complexity that could not have been addressed without this industrial approach.

To take another example that is more associated with our day-to-day experience, small batteries, and chargers are currently the target of a push to standardization by governments. This is particularly visible in the European Community, where USB-C has been pushed down the throat even to massive opponents such as Apple. Large companies have been using many different non-compatible connectors and chargers for decades, leading to a huge waste of electronic systems and complexity in the everyday lives of the users, forcing them to juggle many different pieces of equipment. This law is already having some results in making charging a phone less complex for the public.

Talking about complexity, we will have to define more precisely what is behind this term, and this is what we are going to do in the next section.

Management of complexity

The word *complexity* refers to the quality of something that is composed of many different parts. It is often confused with *complication*, which brings the meaning of something hard to understand. Most information systems are complex, and how this complexity is handled can make them complicated.

Different types of complexity

The concept of complexity was introduced previously when talking about how to reduce it by cutting large, difficult-to-operate systems into small ones that are easier to deal with. In this section, we will come back to this concept of complexity and start by stating that there are two kinds of complexity, namely the **intrinsic**, functional one and the avoidable, **accidental** one. The first one comes from the function itself, and if a module is to provide this function, it cannot do less than this. The second one is everything that is added when implementing the function to make it operate, and that cannot be considered as purely necessary for the function itself. Of course, the whole deal will be to reduce as much as possible the second one, since the two add up and the first one cannot be reduced by definition.

In our example of an oil filter for a car engine, the folding of the absorbing paper inside the filter is intrinsic complexity, because the different stacks of paper and how they form a complex path for the oil are the way the filter functions, retaining the heavy metal particles in the foldings of paper, while the oil reaches the output of the filter with cleaner, purer characteristics. The metal casing of the filter, really, cannot be considered as a participant in the filtering operation. Sure – it is helpful to hold the paper sheets together and facilitate manipulation, but it does not participate in filtering: this is accidental complexity in the oil filter.

Information systems are filled with accidental complexity, and considering that the smallest text notes application nowadays uses thousands of lines of code and megabytes of memory, this only starts to show the problem.

Computer science as a way to deal with complexity

It may sound weird that complexity has reached such a level considering that computers have been designed with added productivity in mind. After all, early computers were built to strongly accelerate calculations that otherwise would take days, weeks, or even months, and required careful double-checking in order to avoid errors as much as possible. The cost of investment in creating a computer was huge due to increasing complexity at first (designing modern computers and electronic chips is one of the most complex endeavors of our civilization), but the using of the computer to quickly produce accurate results for many problems would largely pay for the investment.

A lot of work currently done by computers today is of a high level of technical complexity: displaying high-resolution real-time pictures from 3D modeling in games, performing long calculations such as discrete Fourier transforms or Monte Carlo simulations, and so on. Lots of these operations could not be realized with the same level of accuracy and low error level by humans or even large groups of them. Thus, we can consider IT has helped reduce complexity.

Information systems and complexity

But at the same time, and this is particularly true for people like me who have been in the software field for more than 30 years, it just looks like computers actually did not bring the incredible functional advances that the huge increase in computing power would lead us to believe. GPUs are millions of times quicker, but games are only a few times better. Personal computers are hundreds of times more powerful, but vocal typing is still far from perfect, and word processing has basically not changed, with new features being – most of the time – useless at best, and bloatware at worst.

It just happens that, along with the capacity of computers, we have asked them to do more and more. And while some of these additional operations are bringing new value (optimization of mechanical models, capacity to simulate complex physical models, and so on), a lot are non-value-adding features (larger screens, infinity of nuances of colors) that really make for additional comfort but, in **line-of-business (LOB)** software applications, do not bring anything to the functional value.

To summarize, it looks very much like accidental complexity has grown almost at the same rhythm as computational power, and thus, the remaining power has brought very little performance in handling intrinsic, business-oriented complexity.

The concept of "as a service"

Luckily, there is also good news on the evolution of information systems, and the "as a service" approach is one such example. The "as a service" approach means that something of value is provided to the user without the material part. **Infrastructure as a service (IaaS)**, for example, brings you memory and CPU without the hardware part of the computer; that is dealt with by someone else, generally the cloud provider. **Software as a service (SaaS)** provides you with working software that you can call with a simple web browser without having to worry about prerequisites, installation, purchase of licenses, and so on.

If we consider this approach with respect to the concepts of complexity exposed previously, we can say that the goal is to reduce accidental complexity to almost zero by providing not even the function alone but only the results of the function, which is the service requested. If almost nothing of the surrounding artifacts remains; only the outcome of the software-assisted procedure is obtained. For example, in IaaS, infrastructure as a whole is not what is needed per se by the buyer: one does not crave physical computers consuming space, needing local temperature control, racks, and so on, but has to go through this accidental complexity to obtain CPU power, RAM usage, storage space, or network bandwidth and connectivity.

The "as a service" concept has considerably diminished the perceived complexity of information systems. Of course, there is no free lunch, and the overall complexity is still there (it has even increased). But the separation, not modules by modules this time, but functions by functions, has established a clear cut between the high technical complexity that is handled by the provider of the service and the low complexity that remains for the user. The financial transfer from the latter to the former is explained by the fact that the user gets a great advantage of focusing only on value-adding, business-oriented complexity. How the provider of the service gains a financial interest in handling higher technical complexity (which would be accidental for a mere user, but is standard business complexity for the provider) comes from the fact that they are an expert in it, handle large volumes for many users, and apply scale-related cost savings. In the end, everyone benefits from a clear cut of complexity, which can also be described as a separation of responsibility and task specialization, which is consubstantial with industrialization, as explained previously.

Link to a minimum viable product

Lots of people working with Agile methods know a famous picture illustrating the concept of a **minimum viable product** (**MVP**), from Henrik Kniberg, who created it in the mid-2010s (`https://blog.crisp.se/wp-content/uploads/2016/01/mvp.png`): it shows a first line of product evolution from a wheel to two wheels linked, then to two wheels and a body, and finally a car. During this evolution, a smiley frowns all the time and is only satisfied at the last step. In the second line of the image, the steps are replaced by a skateboard (sad smiley), then a bike (neutral smiley), followed by a motorbike (reasonably happy), and finally a convertible car (extremely happy smiley).

It has been studied a lot and is a great description of the concept of the evolution of a software application from MVP (the skateboard) to a full-fledged project (the car on the right). Lots of imitations do not carry as much meaning because they miss a few details. For example, some of them end up with the same car on the two lines, which is completely wrong as Kniberg purposefully showed different cars at the end of the two processes. The whole story is perfectly explained at `https://blog.crisp.se/2016/01/25/henrikkniberg/making-sense-of-mvp`, and I am certainly not going to paraphrase it, but rather try to make a link with what was written previously about the "as a service" approach.

What is the service that is talked about in that famous picture? Does it have a car? No – owning a car or driving it is merely a side-effect on the service itself, which is "going from one point to another." Using an MVP will help us collect feedback as quickly as possible on the actual needs of the users. Now, if we go to the extreme of the "as a service" approach and consider displacement of the person (and possible luggage) as the one and only request, science-fiction-like teleportation would be absolutely perfect! And we are a bit more reasonable with possibility and price; as Kniberg says, maybe the most basic approach should be to provide the user with a bus ticket.

This would also be a valuable MVP, but that would be forgetting the fact that an MVP does not mean the designer does not have the final destination in mind: we provide the skateboard to collect feedback (for example, "stability is important") while still having in mind that we want a car in the end, maybe because the initially expressed need is that of autonomous travel.

What is of uttermost importance – and we will come back to why the cars are not the same in the end – is that, while taking into account feedback, stability was important in this example and the design quickly evolved to a bike, which is more stable and easier to stay on. But this is not the only feedback that was received. For example, the fact that the vehicle was not covered was not really an issue, and the design evolved into a bike and then a motorbike, which has no wind or rain protection. In the end, the proposed car has no roof: not only because it is simply not requested, but because the interest in driving with hair in the wind may have arisen from the feedback loop. If one had created the car directly, maybe the customer would not have thought of an open roof. But asking for continuous feedback has shown an additional desirable feature (not a need, though, but just a "bonus") that would have not been detected otherwise.

This is what companies talk about when they express their desire to "delight their customers." Most of us engineers do not immediately get it because we tend to see problems with one optimized solution that derives from initial specifications, but the best solutions bring value to customers that they would not even have thought about initially. And guess what? Since all companies are generally good at creating the expected features, these unexpected and delightful features will be the ones your customers will use to differentiate your service from your competitors!

Now that the concept of complexity should be clear, we will propose a first approach to how to reduce it.

The benefits of standards and norms

The first section of this chapter, *What is an industry?*, started talking about standardization and how it is essential for modularization to make sense. Let's imagine the contrary and a system that has been arbitrarily cut into several smaller ones, without reflection on the way to define these parts, how they interact, and how they each can be replaced by improved versions. The result would be that modules could not be designed without knowing the whole and could not be replaced by existing modules since the way they are glued to the rest would not already exist. At best, this would only make the whole problem a bit easier to address; at worst, the added difficulty of putting everything back together would largely overcome the reduction of complexity with respect to addressing the whole system at once.

This is the reason why the cutting interfaces of the modules and their standardization are so important, and why we will dedicate the next section to stressing this with additional examples.

Docker, containers, and OCI

The **Docker** technology is a great way to talk about norms and standards because its very name starts with a metaphor for an industrial concept that prospered through standardization, namely freight and shipping containers.

Until the 1950s, freight transportation was not standardized at all, and filling a ship with freight was quite a craftsmanship: packages came in all sizes and weights, some of them being soft, some of them being hard. The way to bind them together so that they did not move during transportation was customized at every different shipping. It was extremely hard to correctly fill a vehicle since there was little chance all packages would fit nicely to occupy all space while keeping fragile and lightweight packages at the top and heavy, solid ones at the bottom. If you then add the problems of load balance, humidity, or temperature effects that could transmit from one package to another, and the occasional last-minute package that was too heavy to put on top of the other ones and forced the dockers to unload part of the shipment and rearrange everything, you start to get an understanding of what a complicated job freight shipping was at that time.

Meet Malcolm McLean, who devised in 1956 a shipping system based on wood boxes that could be easily transferred from trucks to trains and boats. After only 10 years, in 1967, this great idea was used so much that the **International Organization for Standardization (ISO)** defined three standard sizes of "containers." Despite road/train/sea transportation activities being a huge, worldwide business, after only a few decades, virtually every operator on earth uses standard-sized metal containers that allow optimization of the whole logistics chain and have the following benefits:

- **Ease of loading**: Any expediter can get their hands on a container and load it at their own rhythm, then contact a transporter and have them carry the container, without any risk of refuse because they cannot take care of a particular shape.

- **Improvement in the handling of goods**: Since the metal boxes have standardized sizes and corners with handling holes, there is no use anymore in changing manipulation tools that are needed to press the package (and potentially break its content). Now, the prehensile tools simply lock the four corners of the container and lift them. The speed of handling is also improved as there is no need to handle different packages one by one: the machines lift a container – that carries many different packages inside – as a single unit.

- **Optimized storage**: Industrial containers are plain, parallelepiped-shaped boxes. Their stacking wastes almost no place but the width of the walls. Today, large ships are sized around containers for size optimization.

- **Interchangeable material**: Containers have become such commodities that there is almost no question of property. A container can be easily repaired or replaced by another. A container basically never travels empty. Some of them have traveled several times around the world without their initial buyer seeing them again.

The Docker name and logo clearly state the philosophy behind the technology: the term *docker* refers to the job of loading freight on ships, and Docker's logo shows a whale carrying containers on its back. The link is quite obvious to the transportation business, and the company wants to become the equivalent of industrial containers for application shipping.

Just as with industrial shipping containers, Docker containers provide standard-based advantages:

- Whatever is inside the container (Java process, .NET web, Python script, NodeJS API, and so on), the external interface is exactly the same, and one can simply type `docker run` and have the container up and running.

- Once put inside a Docker image, an application can be shipped to any place on the planet through registries and it will be executed in the same manner, whichever country it is used in.

- **Independence from the underlying architecture**: Docker containers do not know or care whether they are operated on a Windows machine, a Linux server, or even a Kubernetes cluster. Since they have a standard size, they can fit anywhere.

With all these nice features, Docker quickly became the de facto standard for application deployment. Docker itself could even have become the definitive standard, but there were a few shortcomings, and a higher-level, more widespread standard appeared a few years later: **Open Container Initiative (OCI)** created a low-denominator, but undeniable standard that every container technology (Docker, but also other technologies, though less known) adheres to.

Containers have undoubtedly industrialized and strongly improved the way applications are deployed. The rise of microservices is strongly related to container technologies since the deployment of numerous small applications would have been extremely complicated with the old approach of manually setting up dependencies and resources for each application. Some even say microservices architecture appeared only because Docker allowed them to exist.

Docker is an example of how a technology that normalizes a given software-related function (in this case, application deployment) can have a huge impact and replace, through a single standardized approach, loads of proprietary, manual approaches. But this is not the only time this happened in the industry...

Another example with IAM

Identity and access management (**IAM**) is another field of IT where normalization has brought a great deal of help and positively changed a difficult situation in the last decades. Remember when each and every software application had its own user management and passwords? Let alone different, not compatible, ways of handling groups and authorization management, and so on. Such a mess... Everyone in the field was glad when the first approaches of **single sign-on** (**SSO**) appeared, and the **Central Authentication Service** (**CAS**) implemented it in readily available software. Identity and authentication providers made the field more complex to grasp for beginners, but avoided hundreds of thousands of badly-designed IAM systems, replacing them with online, always-accessible identities.

Security Assertion Markup Language (**SAML**) quickly became a standard, and tools such as Shibboleth help diffuse the capacity of handling in a correct, open source, manner. More recently, **OpenID Connect** (**OIDC**), OAuth 2.0, **JSON Web Token** (**JWT**), and other standardized approaches basically killed any discussion on the best way to identify, authenticate, and authorize accounts, accounting for new features that needed to be taken into account and now covering virtually any needs in the field. Keycloak is a production-ready, standard-based, open source application that can act as the glue between standards, which means we now have all the tools to really deal with IAM in a standard way. The benefits are such that companies not using these approaches yet will be obliged in the next years to take steps to do so, as security issues are going to make it mandatory to stop trying to deal with IAM on proprietary, fragile implementations.

There again, the function of IAM has become a commodity owing to standards and an industrial approach of separating modules, each of them with its own responsibility:

- Identification deals with who the accounts and individual owners of accounts are, with all associated metadata. **Lightweight Directory Access Protocol** (**LDAP**) and **LDAP Data**

Interchange Format (LDIF) come to mind as standards for this responsibility, but **System for Cross-domain Identity Management (SCIM)** also can be used, as well as extensions such as SCIM Enterprise Profile to incorporate organizational charts, for example. JWTs can be used to carry this data in a normalized way.

- Authentication is about proving the identity of accounts. OIDC, of course, comes to mind, but **Fast IDentity Online (FIDO)** and **Universal 2nd Factor (U2F)** are standards related to authentication as well, introducing physical devices to improve authentication management.

- Authorization is – once identity is established with proof of authentication – the way to deal with what the person is allowed to do in the software (or, otherwise, remembering that information systems are mostly, but not only, about software). **eXtensible Access Control Markup Language (XACML)** is an XML-based norm for this, but there exist also more recent approaches such as **Open Policy Agent (OPA)**.

In conclusion, IAM is another example of how information systems have positively evolved once the recipe of industrialization has been applied: dividing this complex subject into clear-cut, separate responsibilities, and then applying norms and standards to each of them.

The last part of this chapter will make an analogy with other systems that may be highly complex and use lots of standards, namely the cities in which lots of us live. And I am not talking about smart cities, where software serves urban management, but just cities in their organization emerging over time.

The urbanism metaphor of information systems

Why did I spend so much time and use so much text talking about technologies that became standards and had a great impact on the ease of use and capacity of the evolution of information systems inside of which they are used? Well, because what has been done for application deployment and IAM can be done for any function in a software system. There may not be an undeniable, internationally approved standard behind every functionality you need to operate in your system, but deploying a locally approved standard will provide the exact same benefits inside the perimeter of your own information system.

This approach of industrializing an information system by cutting it into zones and standardizing the interfaces between them is the best approach to keep it in a functional state of health over time. Depending on the context, you may hear about "business/IT alignment," "enterprise architecture," or "urbanization of information systems." The third expression refers to a metaphor where the information system is compared to a modern city:

- **The organization follows hierarchical zoning**: Large zones are dedicated to housing, commerce, or industry. Inside these zones, one will find neighborhoods that define a smaller portion of the zone. Finally, blocks articulate buildings together inside a neighborhood. One will find the same hierarchy inside a well-groomed information system with large business domain zones (for example, administration), inside of which specialized direction will appear (let's say human resources), and finally blocks of functions (in our example, hiring management).

- **Fluids are standardized for a city to operate correctly**: If firefighters had to adapt to different pipe diameters in different parts of the city, there would, of course, be a problem. The same goes for electricity, water, waste pipes, and so on. That may sound crazy today since all this has been perfectly normalized for decades, but at the beginning of the 20th century, a city such as Paris had several different electricity companies, some of them operating 110 volts, some 220 volts, some in **alternating current** (**AC**), some in **constant current** (**CC**), some at 50 Hz, some at 60 Hz, and most of them with different plug formats.

- A large city always evolves, and work in progress in the east of the city is meant to have as little impact on the life of inhabitants of the west side. The same goes for information systems where change is the only constant, and the impact on one piece should be as much as possible without impact on other applications. Town architects provide a global vision and direction of evolution but the day-to-day changes of the city are organic and can happen because of normalization. Well-established information systems can do the same.

Sadly, it seems that enterprise architecture is not very widespread. This partly comes from the fact that this is a complex activity; but it also comes from a lack of knowledge and information spread, against which this book proposes to humbly provide a remedy. I will try in the next chapters to show that industrial and standardization approaches in information systems can bring a lot of value and radically reduce stiffness and difficulty to evolve for most information systems and that the knowledge and practices to reach this are far from being as complicated, as most IT architects think.

Summary

This chapter explained at length the concepts of industrialization and standardization and then explained how they can be applied to the field of software and computer science. Lots of information systems nowadays have difficulty evolving, as stated in the previous chapter, and industrialization, though a recent field in computer science, is a way to strongly improve their efficiency.

In the next chapter, we are going to get a bit more practical, starting from the – admittedly theoretical – material in this chapter, and explain methods to put the industrial approach in place in information systems. The most known approach that will be presented is called "business/IT alignment." In a few words, it states that the structure of IT must reflect the structure of the business processes the information system is there to help.

3

Reaching Business Alignment

Following the first chapter explaining the problems information systems globally face and the second one on the general theory of industrialization, it's time for some actual battle-proved methods! Though we will not get our hands on the code or deploy software before *Part 2* of the book, this third chapter is much more applied and will show the principle of what is called business/IT alignment. The idea behind this principle is that the software system should reflect the structure of the business domain it aims to automate as much as possible. In a way, this is applying **Conway's law** (which will be explained) backward, using it to obtain the desired result. In practice, it is important to know the map to rule on the ground, so we will be using an information system mapping technique based on the four-layer diagram pushed forward by **CIGREF** (short for **Club Informatique des Grandes Entreprises Françaises**), among other organizations.

In this chapter, we'll cover these topics:

- Software for business and the principle of alignment
- Conway's law applied to application and systems
- Introducing the CIGREF diagram
- Using the four-layer diagram
- Patterns and antipatterns of alignment

After describing the method and drawing similarities with the **TOGAF** (short for **The Open Group Architecture Framework**) framework or other methods, we will apply it to a sample IT system so that you really can benefit from it as soon as you have finished this chapter. Finally, we will see best practices, but also anti-patterns in business alignment. Just like any other method, business/IT alignment using the four-layer approach has its advantages and its limits. It is especially important to know them to apply the method as efficiently as possible and also to know how to use it to determine when and where there are alignment problems in the information system under study.

Technical requirements

As said in the introduction, this chapter will be more practical than the two previous, theoretical, ones. This means a reading prerequisite—since we will talk about methods of analyzing information systems, you should have previous analytical contact with them at least. Sure, everyone uses them nowadays, but you will need a bit more than just experience using them, in particular some knowledge about the different parts they are made of. Nothing fancy here, but you need to understand the difference between software and hardware and that information systems are generally there to automate business processes, which are sets of human and computer-based tasks organized toward reaching a goal. You will also need to be able to recognize the different parts of such a system. If we call them systems instead of simply software applications, this is because they are more complex and made of several modules. You need to understand this and be able to tell which parts the system is made of.

You will also need to be able to classify these parts of the system. Are they classified by function or by some more concrete, IT-related criteria, such as their position on the on-premises servers versus in the cloud? Are they autonomous or do they communicate a lot with other functions, and if so, with which interfaces and protocols? This is certainly just general knowledge or common sense for most of us, but this is something you will need to be able to read this chapter nonetheless. This will help you to point out critical questions. For example, when talking about interactions between parts of the systems, are we talking about business dependencies or concrete, IT-related streams of data? To give you a better understanding of the difference, let's go over some illustrations. An example of business dependency is when the ordering system depends on the customer's list. Indeed, we can record a company with our orders, but it does not make much sense to record an order without knowing the customer who buys the products listed in it. On the other hand, an example of IT-related data streams is when the ordering system accesses the customer's database to propose an existing record.

Software for business and the principle of alignment

At this point of the book, it should already be quite clear, but it does not harm us to recall that we are only talking about professional information systems. In short, we put ourselves in a case where software really means business: applications that are mission-critical for companies, information systems that help production for a commercial company or a non-profit organization, etc. All recommendations that follow would not make any sense on small systems and would be way too complicated if applied to a simple piece of software application.

This being said, the hypothesis is that, since there is a business, there is good knowledge of it; we know who the participants in the business are, what the stakes and objectives of the business are, which strategy is being conducted (even if it is not 100% clearly defined, as happens in a lot of companies), etc. This last one is of utmost importance; there is no need to design an information system if the strategic direction has not been defined.

> **Important note**
>
> It should be emphasized that the definition (even if not completely precise) of the business strategy is an absolute prerequisite to an analysis of enterprise architecture and the mapping of the information system. As an analyst with many years of experience with information systems, I always refuse a project if I realize in the first meetings that there is no company-level strategy, and I recommend anyone to do the same. If you are reading this chapter thinking about how you are going to define an information system and realize there is no business-defined strategy, you are better off stopping reading right now and coming back when this essential information becomes clearer. Trust me on this one—if you realize the upcoming steps before the vision of the objectives of the information systems is (at least globally) clear, you are going to waste a lot of your time and do more harm than good.

Why is all this so important? Because you are going to use the definition of your target business to design the information system to be aligned with it. Technology should always be at the service of the users, so knowing the business beforehand is what is going to drive the design of the information system (again, there is no need for a perfectly detailed strategy, but at least a vision or a direction). This is what is called business/IT alignment or, in the context of this book, simply alignment. We will see that this is the only stable method you need to obtain a sound and future-proof information system, free from the problems that have been described in the first chapter. But before we dive more into the method of achieving business/IT alignment, let's just take a look at other methods that are not business-related but technically driven and learn where they apply and what their limits are when designing complex information systems.

The jungle of technical recommendations

If you are a true software professional and worry about the quality of your deliverables, you most certainly have read a lot about technical methods that help in addressing the issue and improving your software skills. You have heard about **V-cycle**; **Agile methods** to organize software teams; **extreme programming**; **test-driven development**; **behavior-driven development** for practices; programming patterns to improve code structure; development-specific **key performance indicators** to follow the quality of code; and so many more that a complete chapter would not be enough to describe them all.

Though there is interest and things to learn from most of them, their sheer number shows the limits of such methods: they are only true in given contexts (otherwise, due to the number, some would oppose others) and, sadly, most of the people writing about them often forget to define this perimeter because they are interested in explaining the technique itself. The more the technique has helped them to lift a given obstacle, the more they will tend to present it as an essential, go-to recommendation. In extreme cases, the person recommending the practice does not realize the small size of the context operated and will consider this practice as universal, encouraging others to use it without limit.

This is of course where the critical thinking of the reader is expected, but at the same time, the reader should logically be someone who knows less about the domain of expertise of the writer, and it may be difficult for him/her to spot the limits of the content. "A little knowledge is a dangerous thing", and the internet is polluted with people who just learned a new trick and will gladly expose it to the community as the solution to everything. This enthusiasm is understandable, and I have certainly done this in my blogs or general training activities, but that does not mean there is no solution and, again, the way forward is to improve the critical thinking of the reader.

An example with KISS, DRY, and WET

Let's take an example of some practices that you have certainly heard of: **KISS** (**keep it simple, stupid**) and **DRY** (**don't repeat yourself**). The first one states that simplicity should always prevail when creating a software implementation. This is particularly true in Agile methods since additional features or feedback from the users will certainly force the rewriting of the code. The second one implies that code should never be repeated and that similar blocks of code should be put into a unique function called from the different places in the code where the same function is needed.

Before any further analysis, we should note that there seems to be a similarity, or at least quite a strong link, between the two recommendations. After all, if we reduce the code repetition, we make things simpler (or at least they may appear so, but this is the subject of the following). We may thus question the use of the two approaches, but again, software engineering is not an industry yet, so every craftsmanship has its own tools and uses. Fair enough…

But the real point of analysis of these two methods is their context of application. As with most technical best practices, they do not blindly apply to everything and their use should be carefully pondered. Sure, it makes sense to unify a simple function to display a warning when you find it many times in the same class, but how about the same situation with two classes in distinct software applications where the warning label slightly differs? If the text is not the same, maybe this is because the warning is on a situation that is not the same, so we should analyze the condition on which they are called. However, as the software modules are not the same, the variables will not necessarily be the same, so it is going to be difficult to analyze the similarity of the situations prompting the warning dialog display. And how about coupling? If we decide to keep only one code, which application module should have it? Or should we make another module that stores the code of the dialog box? And in this case, what about the fact that the life cycle of two applications now influences the versions of this common library, which could certainly become an issue? Sometimes, unifying code can do more harm than good.

This kind of discussion has brought lots of reflections, and a new good practice has been proposed by the acronym **WET** (opposite to DRY), standing for **write everything three times**. Indeed, the hesitations exposed mean that, to find the right decision, it is beneficial to wait and gather some more clues on the actual similarity of the contexts of use, and the creator of the method proposes to write the code three times before thinking about unifying it. This is a sound approach, as it avoids the "black or white" approach of the DRY principle and opens a whole domain of a gray area corresponding to the actual truth: it depends on the context. Writing the code, writing it a second time and observing the similarities, then writing it a third time and analyzing the return on the investment of unification certainly is a sound approach... but does this make it a law that every programmer should obey? Of course, no—again, critical thinking applies, and the three times may not apply to you. Maybe you will need five, maybe you will decide the limit will be time-based and not based on the number of occurrences (wait for one year of maintenance, for example). It is up to you, and I suspect three times was chosen partly because it made for quite a humorous opposition to the DRY principle.

All this is to say that lots of technical, code-related methods that are used in software development are sometimes presented as hard truth or generally applicable best practices, but, most of the time, they are just principles that apply to a given context, and can (and must) be bent to fit others.

The particular case of internal toolboxes/frameworks

Before we switch to solutions on the issue just explained (and do not worry, there are some), a most-observed example of how best practices applied without a thorough context analysis can lead to bad situations is the development of bespoke company frameworks. During the writing of this chapter, I stumbled upon a great article from Aaron Stannard (`https://aaronstannard.com/dry-gone-bad-bespoke-company-framework/`) that happens to perfectly reflect my own analysis on frameworks after 25 years of programming with or without them, creating some and cursing myself to have done so, adoring some others and realizing they brought a huge value to the software I was responsible for, etc.

Aaron Stannard explains in this article how the DRY principle applied too strongly has harsh consequences on teams that develop specialized frameworks to unify coding and, in the end, they obtain results that were the opposite of those that were expected in terms of boilerplate code reduction, scale capacity, and quality of code.

Some frameworks bring huge value to software applications and information systems, and you will easily locate them:

- Everyone knows them and is happy to use them

- Newcomers consider the framework helpful in making them more productive and quicker

- All users can describe in very few sentences what the framework does

- The framework has very little dependencies and can be used easily

- It is so important that developers will go as far as doing the thing they hate to keep it going: documenting it

On the opposite side, there are frameworks in code that hinder the programming process:

- They have been created by a few experts and the rest of the team does not use them

- Newcomers tend to write the equivalent code directly and question the fact that the framework would have helped them be more productive

- They sometimes do so many things that none of their features are stable

- They bring some other constraints (operating on just one type of database, requiring script modules, elevation of privileges, etc.)

- Their functioning is in the mind of the experts (often, a single person, which is dangerous for the company), and training their colleagues on the use of it is somewhat of a challenge.

The difficulty in knowing where you are between these two ends of the spectrum is linked to the fact that the creators of the framework will, of course, be biased about their "baby". They would always overestimate the time saved by using it or forget or overestimate the time to maintain it, the time to teach it to others in the team, the risk that the company is taking by putting an important dependency in the hands of one person, and so on. If you have to evaluate the use of a framework, you should put all these factors in a spreadsheet and coldly evaluate the return on investment in the short, middle, and long term.

Finally, the modern programming platform has now brought so many tools that the very existence of frameworks can be questioned. For example, **.NET Core framework** version 8.0 has such a large ecosystem and **Base Class Library** that all purposes for which frameworks were created ten years ago have simply disappeared:

- Object/relational mapping is taken charge of by **Entity Framework**

- API request/response with mapping to objects is handled by **ASP.NET Web API** and integrated **JSON/XML** serialization

- Monitoring is taken care of by the logging stacks that can be plugged into any third-party listeners

- The consistent page description is done with **Blazor**, with style handling included

- Deployment on mobile applications is realized through **Multi-platform App UI** (**MAUI**), which also unifies with the Windows frontend, and so on

So my advice on this question of the frameworks is to wait as much as possible before creating a framework and trying to avoid it as much as possible (most of the time, admit it, you want to create the framework not because it is good for your business, but because it is fun to code). If it has to come, it will do so, and this is where emerging code architecture (we will come back to this notion soon if you have not heard about it) makes sense.

The only stable guide: aligning to your users' business

After this long digression, we come back to our initial consideration of the difficulty of knowing when to apply best practices and so-called "principles" in software. Yet, we desperately need guidance for the architecture of information systems because the stakes are too high and we have seen how bad the impact can be on business if we fail at this. So how do we know? Is there a sound and stable way that always applies? Some actual laws that we will be indeed able to truly rely on when designing something as essential for a company's future as the structure of an information system? Yes, there is, and it is not a technical rule, but a method for technical decisions: always relate them to the business of the information system. This is the root of business/IT alignment.

In this chapter, I am going to repeat this rule and explain it in many different ways, as it is so important for a correct information system architecture—what drives the design of the information systems is the structure of the business. This is particularly true for the concepts, and creating a successful software backbone for a company always starts with a perfect understanding of the business. What is a customer? What products do we sell? What are the main processes of the company that ought to be computerized? Most of these questions sound trivial, but they are only so because human brains can adapt to the background context.

Let's take the very first question: what is a customer? This is so obvious for anyone in the company that the question is rarely asked, and it would sound quite ridiculous if any employee questioned it. "Customers are companies we do business with". All right, but how about individuals? Yes, sometimes we also deal with individuals; this is the difference between B2B and B2C. And what if we question the term "doing business with"? What frequency and volumes are we talking about? Sure, any volume may be taken into account, and a person buying a simple bolt from you may be considered a customer in the exact same way as a company acquiring thousands of them a month. But how about time? Would you consider someone who has bought a product from you twenty years ago still a customer? No, certainly not. How about one year ago? Yes, of course... In this case, where is the limit? Ten years? Five? Decisions, decisions, decisions!

About this previous example, you may find companies where managers do not agree on the exact definition of a customer between marketing and commerce... So, how would you expect a dumb computer to decide this? The hard truth is that, if you want to replace some tasks with software and data streams, you will have to make everything absolutely crystal clear for it to work. This is where most of the information systems fail—they have not been designed with a perfectly accurate view of the business, leaving some details to the implementation or use of the system by humans. It may work for some time when humans compensate for the missing knowledge in the computer, but sooner or later, there will be trouble. In the best case, the system will never be as efficient as expected. In the worst case, people leaving the company with their compensating knowledge will cause the system to come to a halt.

What does this example show us? First, one should be extremely clear with the business definitions when it comes to making a software information system deal with them. The second is that, very often, these concepts are business rules, which means they are not perfectly stable but depend on how the business is run. The definition of the customer may change in time if your boss decides one day that the list of customers should not present the ones that have not bought anything from you in the last three years, where this duration previously was five years. If this situation is bound to happen from time to time, it is of course of utmost importance that its impact on the information system is limited. Putting this logic in a unified line of code or, even better, in a parameter is a great move in this case.

On the contrary, storing the list of customers and their definitions in a single table will cause great coupling because the definition (name, address, contacts, etc.) will not change when the "customer status" does. If removing a company from the customers' list means you have to delete the entry from your databases, there may be an impact on other functions that still need this data (for example, guarantee management, accountants, etc.). We will come back to this example in *Chapter 9* with some more details on how to model it correctly, but for now, please remember that business thinking must always guide how your software concepts will work. Alignment is not something that arises from nothing or goes both ways—it is driven by functions, and the software should blindly follow the business domain ontology.

In fact—and this will be the last bit in this section—sticking to the business domain reality should even go one step further and in particular stick with it *in time*. This means that your model should always accommodate time since the business always varies. Change is the only constant in life; businesses cannot escape this and rather are highly dependent on it. With ever-changing business rules, more and more complex organizations, and higher functional sophistication, the information system must be designed from the beginning to accommodate change. Management of time is so important that the next chapter will be completely dedicated to it.

Digression on the digital transformation of companies

While we are talking about software representation of the business domains, we might as well expand on this and observe how software has "eaten the world" (*Why Software Is Eating The World*, Marc Andreessen, 2011) and why this active digital transformation relates to the business alignment concepts talked about previously. A schema is worth a thousand words, and *Figures 3.1* and *3.2* should explain the main difference digital transformation makes to how we operate.

Before the digital transformation, the human operator was at the center of the business operations and operated both in the real world and on the computerized vision of the business (or parts of it, as we saw earlier):

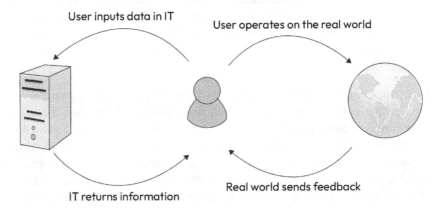

Figure 3.1 – Schema for when the human operator was at the center of business operations

The digital transformation brings a whole new approach where software becomes the main tool of operation for the human and operates in the real world on behalf of the human user. The hardware and software systems receive orders and signals from the human interfaces and the sensors, translate them, and send them back to the human operator (through graphical user interfaces) as well as on the concrete reality (through mechanical operators or other ways):

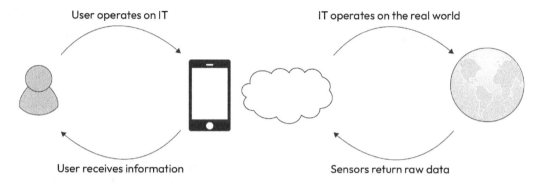

Figure 3.2 – Schema for digital transformation

This has a particular consequence that IT is now at the center of the picture, interacting both with the real world and with humans. Depending on your point of view, computers helped humans to not get their hands dirty with direct interactions, or they took us away from direct interactions with the world and all the risks associated with potential bias and the wrong representation of reality. Social networks are the paramount manifestation of these negative impacts and, as software engineers, you should always be aware of this kind of risk and design systems accordingly, as their impact on the real world is now well established and people working in the field should keep a moral, responsible approach.

I hope this made the digital transformation and the interaction between humans and computers clearer. Let's now concentrate on how humans can impact the way software is organized and talk about something called Conway's law.

Conway's law applied to application and systems

We talked a lot about the limits of so-called "laws of software" in the first section of this chapter, so you may wonder why I will now spend several paragraphs talking about something that, at first sight, may seem similar. Nothing could be more different... Conway's law is a true, stable guide for information systems design, as it does not state a recommendation but draws a theory from multiple observations and lets one decide its own conclusion on the subject.

Melvin Conway stated in 1967 that "any organization that designs a system will produce a design whose structure is a copy of the organization's communication structure". In our case study, which is information systems, this means that the architecture of the resulting system will reflect the structural organization of the team defining it, and this would imply the following:

- A team with a strong separation between frontend and backend will produce a system where these two software functions are indeed independent

- A team where people are grouped depending on their business domain knowledge will produce an information system with clear-cut business-aligned services

- A team of only one person would give birth to a very cohesive, monolith-like system

- A team with no or low communication between its parts would create a system where modules do not correctly interoperate with each other

This last example may look like an extreme, but, sadly, this is the case with most information systems because the teams are generally composed with a single piece of software in mind, resulting in many applications composing the system without the interoperation having been thought out in advance. Thus, links are established in an upon-needed, point-to-point manner, resulting in brittle links and inefficient systems.

Since this initial empirical observation from Conway, the eponym law has been verified many times and, though it cannot be demonstrated like a mathematical law can, it is nowadays considered something largely reliable. It is considered such a strong law that system designers have started using the law to structure teams in ways that would bring a desired shape to the resulting system. In this approach, the law is not only seen as a consequence but as a helping tool to shape the system as needed. What I am talking about comes from my own experience but also has been formalized under the name of the **inverse Conway maneuver** since many other software engineers have had the same approach. Martin Fowler, for example, puts this law forward (`https://martinfowler.com/bliki/ConwaysLaw.html`) and even draws a relationship with **domain-driven design** (**DDD**). This will be further explained in an upcoming chapter specifically aimed at explaining the importance of semantics—related to the concept of ubiquitous language in DDD—in designing the parts of an information system.

Using the inverse Conway maneuver, one can influence the resulting design of a system by working on the communication and structure of the teams working on designing it. This is a great way to achieve the much-desired business alignment that we have talked about since the beginning of the chapter. By defining the teams alongside the business domains and giving them business-related concepts to talk about with each other, the resulting system will be made of modules with clear-cut functional responsibilities and well-structured interoperations between modules, favoring the long-time evolution of the resulting system.

Influencing the alignment of a system is great, but, most of the time, one will be exposed to an existing one, with the only possibility of understanding its state. This is where we need a way to analyze the existing alignment, and this is where a dedicated method of diagramming is useful.

Introducing the CIGREF diagram

As we have seen, business alignment has much to see with vocabulary and the right expression of concepts that are related to the functional domains. For those of us who are more comfortable with schemas, there exists a more graphical way to visualize this alignment, and it is called the **four-layer diagram**.

In France (where I am from), it has been popularized by the Club Informatique des Grandes Entreprises Françaises (a large French company club for software architecture), but this is a very widespread way of thinking and no ownership has been claimed on this idea, at least that I know of. The concept is quite simple and is about separating the different levels of an information system, each using the levels I will present to work. At the top of the diagram, one will find the business processes that the system serves, and one level down are the business functions that are needed for this. These two layers are purely functional and are not even related to software; some tasks and functions could be realized by humans without any impact. The two technical layers at the bottom are respectively software and hardware, the former using the latter. This way of specifying a system (or rather its outermost structure) can be schematized as follows:

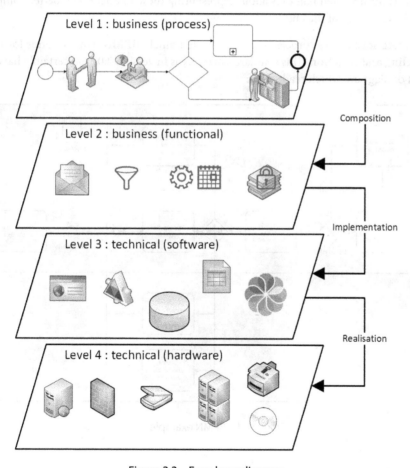

Figure 3.3 – Four-layer diagram

Let's dive a bit deeper into each of the levels in the following sections, and in particular show how each level's content is represented in detail, as this schema is only symbolic and does not contain the actual contents of each layer.

> **Important note**
>
> In the rest of the chapter (and in subsequent parts of the book), we will refer often to these four layers or levels by their number, starting from the top (**Level 1**, **Level 2**, **Level 3**, and **Level 4**). The next section details **Level 1**, and so on…

Process level

A business process is a series of tasks organized toward reaching a defined objective. Since we talk about IT, at least some of them will of course be automated, but there can be man-activated tasks as well. Processes are used to structure a business or any organization's activities, and **business process modeling** (this is the accepted name) is about representing these to document, better comprehend, and improve the efficiency of the entity.

There is a standard for business process representation, namely **BPMN**, which stands for **Business Process Modeling and Notation**. This standard currently is in version 2.0. You certainly have already seen this kind of diagram, which reads itself:

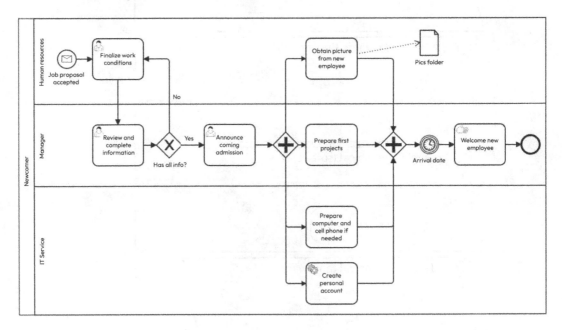

Figure 3.4 – BPMN example

In this very short and undetailed example of a new process in a company, you will find the most used components of the BPMN standard:

- Tasks are boxes, generally filled with text starting with a verb describing the activity represented. Icons inside the task may specify whether it is automated, has user inputs, is entirely manual, etc.

- Arrows between tasks indicate the flow of information in the process and, as a consequence, the order in which the tasks are processed.

- Diamond-shaped boxes intervene in this stream of information to incorporate complexities such as choices or parallel activities.

- Events are represented by circles. A process will always have a start and one or more ends. It can also have intermediate events and all those can reflect types of events, such as time-based, message-based, etc.

> **Tip**
>
> For more information about BPMN, I strongly recommend studying the BPMN poster available at `http://www.bpmb.de/index.php/BPMNPoster`.

Though Layer 1 of the CIGREF diagram is mostly based on BPMN diagrams, as they are the standard for business processes representation and the main subject of this layer, one can also find in this layer some additional information that goes together. For example, business domain-related rules can be specified as **DMN** (**Decision Modeling Notation**) and are added to this first level, as they have impacts on the business processes themselves. DMN is, by the way, a "sub-norm" included in the BPMN standard.

Depending on what one tries to obtain, the layer one map may be very coarse and with few details. This is an example from an animal genetics company I advised, which already had a complete ISO-9001 diagramming of its processes and simply needed to know which parts of the information system related to which business process (intentionally blurred for confidentiality reasons):

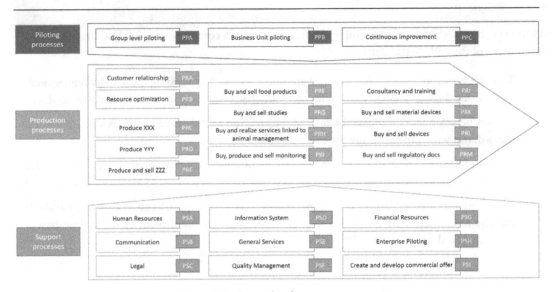

Figure 3.5 – Example of a processes map

Since this customer of mine did not have any problem with the processes themselves, their representation remained very light during the whole alignment project. The only subtlety in this was that operational processes had been separated from support processes and piloting ones.

Contrarily, the following diagram is only one of many processes in layer 1 of a map I created for another customer, this time an organization where IT problems mainly came from a lack of definition in the processes (and, of course, when objectives are not clear, it is hard to have an efficient IT system). In the following diagram, the readability of the text is not intended, as I only want you to look at the overall process diagram:

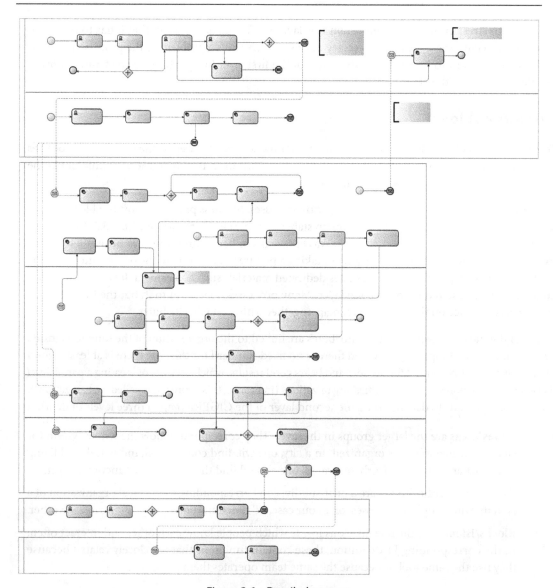

Figure 3.6 – Detailed process

Layer 1 is of course the most important level of the map of your information system because nothing depends on it. All other levels depend on the one above, but the processes layer has to be designed from scratch, purely based on the business domain knowledge. And, as explained, if processes are not clear or badly designed, this will of course affect the software and, in time, the efficiency of the whole information.

It is hard to stress enough how important this layer is. This does not mean that it has to be drawn with many details; these are only necessary once someone spots a problem in the process and needs to know it in depth to improve or correct it. But the first layer has got to be correct and cover the whole perimeter of the organization.

Functional level

The second level of the CIGREF map indicates what functions can be used by the tasks of the processes above to realize them. This is still a functional layer, but this time, the organization is different as the atoms of this layer are about who can do what.

There are indeed functions that do not vary with processes. Business processes must be able to change according to strategy, but some things remain stable. In our example shown in *Figure 3.3*, to welcome a newcomer to a company, one of the tasks was to take his or her picture. This may be useful for other things in other processes, but the very act of taking a picture is a function of the organization, may be realized by a given person, and necessitates dedicated materials such as a camera. It is thus a function and has to be registered in this second layer. Many processes may point to it, but the function will remain in one place on the second-level map, together with its stable attributes.

Some of the most important of these attributes are linked to the organization of the functions inside the organization. Things may vary, and there is no fixed standard for this, but there is at least a good metaphor that has been used for decades and bears good results, which consists of creating a parallelism between information systems and cities' organizations (in France, this approach is called "urbanization"). This metaphor leads to decomposing the second layer of the CIGERF map in three levels of depth:

1. **Zones/areas** are the larger groups in the layer. They correspond to how the whole system (or city, in our metaphor) is organized. In a city, one will find commercial, industrial, and living areas that are generally clearly separated; one should find the same in the functional layer.

2. **Quarters/neighborhoods** are a finer subdivision corresponding to the local organization of a system, with people, businesses, or, in our case, IT functions potentially talking to each other.

3. **Blocks/islands** are the fine divisions inside which people know each other and interact often. In the corresponding IT definition, these contain functions that are closely related because they use the same tool or because the same team operates them.

I am often asked what kind of decomposition should be used, at least for the top-most level. This is a difficult one, as the **business capability map (BCM)** (as this second layer of the CIGREF map is often called) is most of the time simply non-existent in companies. Lots of them have a clear view of their processes due to ISO 9001 certification and the relative standardization of the approach. Companies can also track down their software layer, at least for the biggest blocks for which they are billed. But in the middle, this BCM is quite often forgotten, and we will see a bit further that this is the source of loss of misalignments in the systems.

The fact that this layer is forgotten is in itself a big part of the problem in many information systems, as leaders may favor the process layer (the objectives) and put less effort into the BCM (how to realize them). Yet, a vision without a plan is just a wish, as the proverb says, and a well-crafted functional organization of the system is a great step forward to success.

In this metaphor with urban development, the business capability map is cut in the same way that would be associated with the composition of a large city; just like there are industrial, commercial, and living areas in a city, there will be a global organization in an information system, often in the following five areas:

1. **Master data management** is where the most important data of the information system is managed (customers, products, etc.). This data will be used by many parts and actors of the system. As such, it deserves dedicated governance (a clear definition of who is in charge, with what practices and tools, etc.) and to have its own dedicated zone in the second layer of the map. When governance is not clear on such important data, it often happens that different groups will duplicate them, which is not only costly but can bring complicated problems when exchanges need to be realized.

2. **Shared tools** are not data but still commonly used functions that one will refer to in many other parts of the information system and most processes' tasks. They are often classified in a dedicated zone, where one will find office automation tools, content management software, identity and authorization management, and so on.

3. **Externally oriented functions** (sometimes called "collaborative") are for all functions that are about interoperation or exchange with functions that are outside the scope of the information system itself. This is typically where one will find an extranet or commercial websites, connections with partners, etc.

4. **Governance/piloting** is the zone where functions used to supervise the system itself will be found. Reporting functions will be placed there, as well as key performance indicators, high-management functions, and so on.

5. **Business-oriented functions** are the last, but not least, area. This is where a company will organize all the functions corresponding to its core value and operational domain. If you are a company that builds mechanical parts, you will find there all the functions associated with engineering, production, stocks, selling, maintenance, and installation. If you work in e-business, there will be buying and selling functions, logistics, web operation, security, etc. There will also be some groups of functions that support the core business, such as human resources, legal, and administrative functions, which are general to most companies. When the entity is well organized, it is evident (and proof of good alignment) that the quarters in this zone correspond to the different directions in the organizational chart. Conversely, if it is hard for you to tell whether a given function is under such a direction or service, this may be a sign of a lack of alignment.

The next figure is an example of a BCM that follows the five zones principle, with the main one decomposed following the organizational chart directions. Again, this is just a commonly observed pattern and not a recommendation whatsoever. One should adjust the BCM decomposition as one sees fit to obtain alignment. Again, in terms of the text readability of the following diagram, the text content is not important but only the structure:

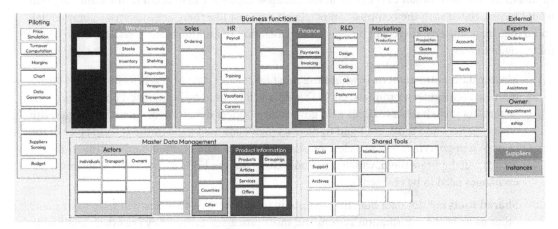

Figure 3.7 – BCM example

For the anecdote, these five areas are also found in the French government's information system BCM, with the four supporting areas surrounding and the business-oriented functions in the middle, which are, in this context, separated by the different ministries the French government is composed of. I've highlighted the separations with black lines, as that is what I intended to show you (the readability of text is not intended):

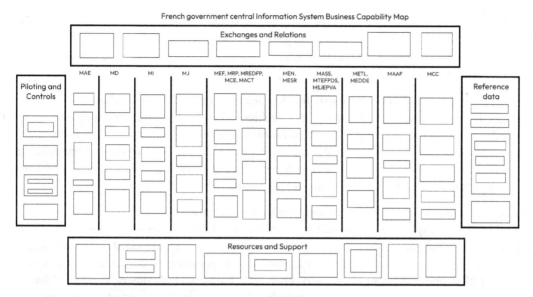

Figure 3.8 – French government BCM

We will come back to the importance of this second layer of the CIGREF. As stated, the first one indicates the vision and what value brings the company to the market, but the second is about how to operationally implement this vision, and this is where things are lacking in disorganized information systems. In the dozens of information systems I have been able to watch or help refine, the BCM is *always* the least-controlled layer. This absence of correct handling of it is the root cause of the information system not giving satisfaction.

Software layer

With the next two layers of the CIGREF map, we enter the realm of technology. For now, processes and functions were purely business-related and one could very well use them for a computer-free information system. But this is not what we have today, and the subject of analysis in this book. We will work on computer-based, at least partially automated systems here, and the two technical layers that the top ones rely on are about implementing the function with computers. Their separation is extremely easy: the third layer is everything that is "soft", which means immaterial, virtual, and non-concrete, whereas the fourth layer groups everything that is concrete. To say it as simply as possible: if you can touch it (computer, network cables, appliances, data centers —even if the walls do not belong to you), it will go to the next layer that we will cover a bit later. If it is technical but not "touchable", such as a software application, a database, a stream of information, an API implementation, etc., it belongs in the third level in the CIGREF map, which we will detail right now.

This third level is in general quite easy for the companies to create in its first implementation: looking at bills and asking people what software they use is generally enough to find out the 80% most important uses of software in your system. But, even if exhaustive referencing is not the goal, this may not be enough for two reasons.

First, there may be some "hidden" software that has been bought by a service without the company knowing (this is called **shadow IT** and can be a problem where maintenance or strong ownership is needed). There is a risk that they will not appear on the map if they are strategic, and this might become a problem also if people using them suddenly leave and the associated functions collapse without anyone understanding why. If you have heard stories of a company having software problems after a key person retired, this is what we are talking about.

The second problem is that software is not only applications but also data, and data is usually harder to locate and follow in a system. Sure, you can locate databases with their commercial licenses or the IP and port they use. But you will find data in so many other places, such as dreaded Excel worksheets. Again, who has not heard of an Excel workbook that was so important for the company that everyone knew about it? In one of the companies I accompanied in business/IT alignment, there was "Serge's Excel file", which everyone kept telling me about when I was trying to figure out where the source of truth was for the articles and prices for the company. It turned out that, in this company of almost a thousand employees, there was no governance at all on product information management, and this person called Serge, at some point when he desperately needed the information, took the job of collecting the data from commerce, administration, and engineering and putting it together in an Excel workbook, trying his best to follow the changes, new products, end-of-life dates, changes in price, etc. As this was not his primary job, he had little time to do so and the content of the file was neither complete nor free from errors. But since this was the only source of data available, everybody quickly copied Serge's file or referred to it with server links. The managers never considered the fragility of this approach to a hugely important source of data (maybe even the primary referential of a commercial company together with the list of customers), and guess what happened when Serge eventually left the company? The system slowly decayed because it was nobody's job to maintain the one-person, non-documented work. Information became dirtier, orders started to be false, prices could not be adjusted because most people using the information in the file or in the connectors that had been created on it had no clue where the data was coming from, etc.

One may object that what I am talking about is not part of the third layer but relates to the second one, and indeed, master data management and data governance and ownership have to be detailed on the second level. But in this case, I wanted to show that a poor technical implementation (which definitely belongs to the third layer) and a lack of understanding about where the data in its software form stands were the root of the problem that went right up to the first level of the map and derailed two of the main processes of the company, namely production, and sales.

Categorizing the content of this third layer really depends on a lot of factors. Some companies with strong internal IT and programming capacities will tend to have the applications and data grouped by the technical team that operates on it. I have seen others grouping the software layer by technology, as their main concern was to operate the technologies internally even though they were—for the most part—bought on the shelf. In some cases, software can be sorted by editors. And there are many other ways to sort it. In the next example (once again blurred for confidentiality reasons), the cutting has been done using functional domains, as the company was quite big and the software application and data's responsibility had been affected by the business directions and services (which is quite a good practice, as software should always be at the service of the functions):

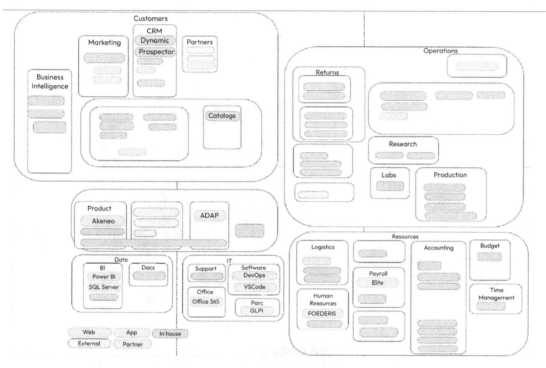

Figure 3.9 – Software layer

Hardware layer

As explained before, the hardware layer lists and organizes everything from the information system that is concrete. After all, the important bits of computerized systems are data and virtual functions that dramatically accelerate the processes, but we should never forget that, even though in a remote cloud location nobody really cares, all this is realized by electrons flowing in electronic chips and board, with power supplies, cabling, hard disks, and screens somewhere.

This layer is nowadays very standardized and under control and, in dozens of information systems where I have analyzed shortcomings in alignment and performance, virtually none of them exposed this limitation due to a hardware problem. In fact, it is so rare that in most of the CIGREF maps I have done, the fourth layer is very thin, with almost no details, or sometimes it's not even represented at all. For example, if we take the big picture of the different layers I have shown you as samples, it so happens that nothing was represented for the hardware layer:

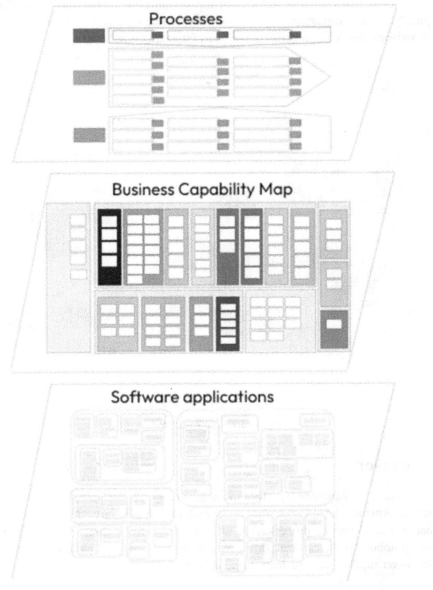

Figure 3.10 – Three layers only (the readability of the text is not intended)

Luckily, something that company owners know very well will always remind us of the existence of this layer, namely their costs. And even if the machines are more and more virtualized and made invisible, the financial costs are still there. With time, the ecological impact of information systems and data centers eventually also becomes a part of the equation, making this layer more visible as well.

If you have to draw a hardware layer, you will find lots of excellent diagramming systems that distinguish between server types, can provide dedicated icons and metadata so you can list hardware atoms separately, etc. All in all, again, this is a very controlled and standardized layer, which certainly explains why we rarely have to work on it when talking about IT alignment, other than referring to it to complete the software costs and balancing the sum to the benefits expected from the first two layers.

To provide an example of a hardware layer, here is a chronological series of such a layer diagram in a company that progressively externalized its IT (the orange bits were servers operated directly by the IT service):

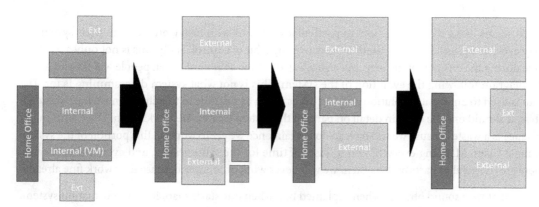

Figure 3.11 – Hardware layer

Due to the relatively low importance of the representation of this layer (and not of the layer itself, of course), we will not go deeper into descriptions of how to use the diagram to map a hardware layer. The groupings, also, are quite obvious; most of the time, they are based on data centers with the physical location on the top, separate physical servers, then virtual machines, etc. Networks are also represented with standard symbols and, all in all, diagrams are generic for this layer.

Using the four-layer diagram

The principles of the CIGREF mapping should now be clear, so we can see how to use this technique to improve alignment and, thus, the efficiency of the information systems. As said, the first action to take when taking control of a system is to create a map of it. Otherwise, there is simply no way to be comfortable with handling such complex sets. This means of course that creating the map for the existing state of the information system is the very first action to take, and a CIGREF map is great for that. But lots of questions still remain on how to do so. This section contains a few battle-proven pieces of advice on how to use the mapping technique.

As an important note, do not worry if you do not know precisely how to use the CIGREF method by the end of this chapter. For now, I will just show you how to draw it and how to spot problems of business/IT alignment in it, but the rest of the book will present lots of other examples of the CIGREF map in action for many different purposes. This means it will hopefully become clearer how powerful it is in analyzing and structuring an information system and you should not feel concerned if you do not know yet precisely how it is going to be used in practice.

What should we map?

First of all, one should never map the entire system in detail. It is, of course, essential to have a coarse map covering the whole perimeter of it to understand what we are dealing with, but only the parts of the system that need attention should be mapped in detail. That may sound obvious, but there have been so many times with teams I've advised where I realized, a few weeks after explaining the method, that they had diagrammed absolutely everything in the information system. Thus, this advice needs to be put forward.

It just seems like a natural reaction, in particular for teams with heavy problems in their system, to map everything, as it gives a sensation of recovering a bit of control. Sadly, this is not only a waste of time during the creation of the map, but also a waste of time afterward, when people will try and adjust the diagram following the evolution of the system. This is not what system diagramming is for. The map is used to anticipate evolution and force it as much as possible in the direction we have decided. Thus, it should only be done in detail on parts of the system that we are working on. Diagramming in advance is a waste because we will have to do it again once we get to work on this portion of the system in most cases. Mapping everything is a waste of time for the same reason, and even more, because there are (hopefully) parts of the system we will never work on, simply because they work fine already!

Again, that may sound obvious when explained but, when one starts mapping an information system, there seems to be some kind of frenzy mapping happening and I have seen many times well-educated and experienced people realizing that they had worked for nothing (the best example I have is an organization with thousands of employees where the IT team had diagrammed a process for establishing badges to access the internal restaurant; this diagram was of course never used and certainly not even read by anyone else than its creators).

The following schema visually explains how a map should evolve in time, with parts of a system being detailed only when needed:

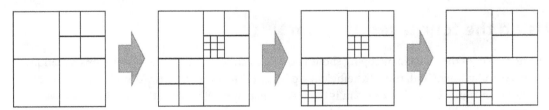

Figure 3.12 – Mapping evolution

One of the other consequences of this—and more subtle—is that the parts of the information system that have been diagrammed in detail should, after the diagram has been used to improve them, not be updated anymore. In Figure 3.12, this is what happens to the quarter that had been refined in the second step. In the fourth step, it is not detailed anymore. That may sound counterintuitive; indeed, once the effort has been made to detail this quarter, why just abandon it? The same reason applies: if this quarter is now well organized and alignment has been reached between business and IT, there is a chance that either we will not come back to it or, if we do, the map will have changed. So why bother wasting time on it?

How to start drawing a four-layer diagram

Starting with an empty sheet is always difficult; it is even more difficult when the IT problems are such that the activity of the company is impacted. To ease starting the mapping activity, it is sometimes easier to use paper or a whiteboard to give a first view of the system. I like showing the following "first draft" of an information system I have been called to analyze because of the anecdote that comes with it:

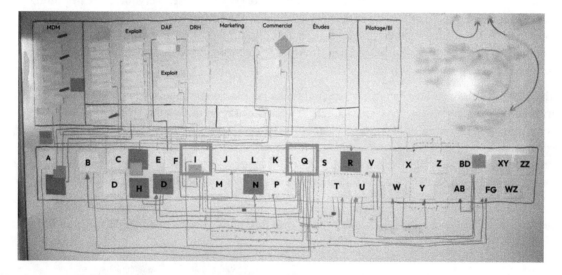

Figure 3.13 – Using CIGREF

This industrial company had difficulties in stabilizing the streams of data for scientific analyses and communication of the results to their customers. We (the IT team and myself) started spending a few hours drawing the four main processes, the incomplete but close enough BCM, and the applications involved in them together with the stream of data. At that point, we were about to evaluate the frequency of the exchanges that were drawn to determine the weakness. Then, the director of the company came by and, looking for a few seconds at what we had done, pointed his finger at two of the Post-it notes representing applications and said "Well, the problem is apparently here". This person did not have any technical or IT background whatsoever but, seeing that lots of streams would go in and out of these two entities, he immediately understood that they were holding back the efficiency of the system and, in particular, its ability to evolve. It turned out after additional analysis that one of the applications was obsolete and the second one had a complexity problem. The system was thus redesigned to improve and the main steps were to integrate a new application for the first one and to create a superset of APIs exposing the legacy functionalities in a cleaner way for the second one.

This shows the power of good information system mapping, as it helps all actors—and not only technical-savvy ones—understand what is going on in their IT, which now is, almost everywhere, their main working tool.

A generic difficulty when aligning information systems

One of the problems that almost always happens in the system and has an impact on its map is that confusion is frequent between the processes in layer 1 and the dependencies in layer 3. Most functional people tend to think that their IT system implements processes exactly like they have been designed; how naive of them… Software is not necessarily made in-house and, when bought on the shelf, there is almost no chance that it perfectly fits the company's process. Of course, everyone at the beginning of a new software project will swear that they will abide by the editor's logic and adjust their process to keep the solution generic and avoid costly specific customizations. But the reality in most of these situations is that IT will end up integrating the software hammering circles in square holes and the result will not be clean.

This lack of distinction also happens when functional people express themselves in technical terms without fully understanding the consequences of what they say on the system. Come on, we've all at some point had managers or big bosses talk about ESB or ETL as if they were able to write the Camel route themselves in XML! This overconfidence in IT's simplicity also brings interesting statements of project duration once a BPMN is drawn and functional authors are convinced that it can simply be executed, just like it was a WS-BPEL schema with endpoints already existing.

Let's take an example and imagine you have been provided with a BPMN schema such as this (the French labels do not matter; simply observe the structure of the diagram and the fact that all tasks are manual):

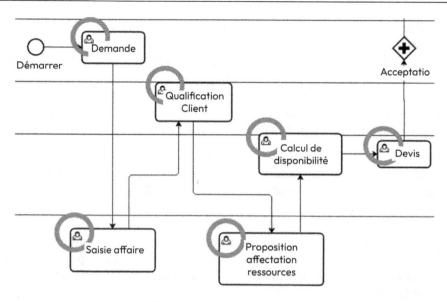

Figure 3.14 – Manual process

Now imagine that the diagram of data streams corresponding to the realization of this process, as given by the IT, is something like this:

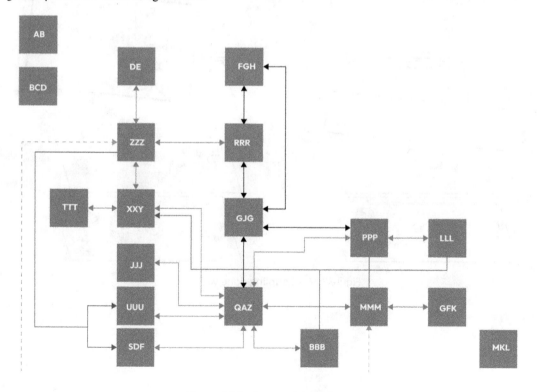

Figure 3.15 – Process in software

It does not need any technical expertise to realize that there is no relationship whatsoever between the tasks in the process and the streams of data that have been created between the software applications and databases. So, a fair part of the job of mapping the information system will be about drawing relationships between the tasks and the corresponding streams of data. In the end, you may realize that some of these streams are incomplete, needing others to compensate for the lack of data. You may find streams that bring too much data into software that should not even have the authorization to see it. You may even find streams of data that simply serve no known purpose anymore.

To relate the two layers, you will have to create a strong business capability map, as layer 2 will be used as an indirection between the processes and their software implementation. This is why BCM is so important and, if you operate on information systems, you will realize that this is often missing, simply because it is lesser known than processes and software/hardware. Yet, BCM is essential to reduce coupling and favor the evolution of the system by providing a way to create dependencies without being stuck by technical implementations that are hard to change once set in place or having to change the IT because a process has been modified in the way the company operates. The BCM, in this case, acts as an indirection layer that makes possible the evolution of layer 1 with limited impact on layer 3 and vice-versa.

The following schema sums this up, and, by the way, I often tell my students that if they only remember one slide from my course on IT alignment, it should be this one:

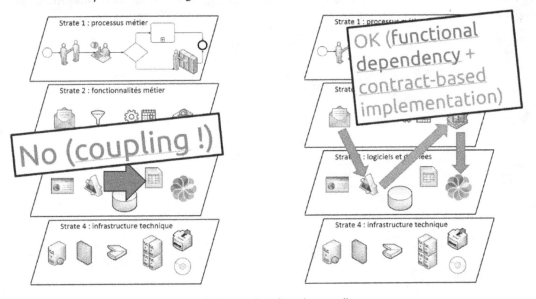

Figure 3.16 – Coupling done well

Evolving the information system in time

Of course, after mapping the existing state of the information system, the next step is to improve it, and this calls for a strategy with multiple steps to make it more realistic. Again, the CIGREF map is here to help by explaining clearly what needs to be adjusted at each step and showing how the dependencies need to be taken care of. If one changes a function by plugging it into a new and improved software application, all tasks depending on this function will need to evolve to take advantage of this, except of course if there is 100% compatibility, and in this case, we can consider that the function itself does not change.

The goal at the end of these steps is to reach a state as close to (and cost-effective) a realistic and nicely aligned system as possible, which could be schematized as the following:

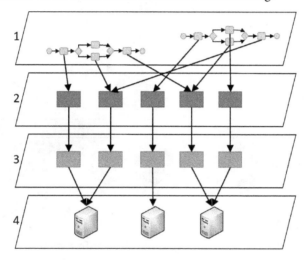

Figure 3.17 – Ideal alignment

Of course, this may sound ideal, and, most of the time, we will never reach alignment for the complete system. But local alignment can be reached, as shown in the next schema, which I exported from one of my customers who managed to align one of their most important support processes (namely, handling newcomers, as this was a company with high structural turnover) in such a way that the efforts needed would drop by an estimated factor of ten and the duration of the process would reduce to a third of the initial measured time. This was achieved via the automation of some of the tasks, but the alignment made this automation possible, as the process was entirely manual beforehand due to a clear lack of structure in the IT (the company was in a non-technical business and put a limited budget in its IT, only realizing after it reached an almost desperate state how important it had become for their activities). The following diagram is blurred for confidentiality reasons; we intended to showcase only the flow of the process:

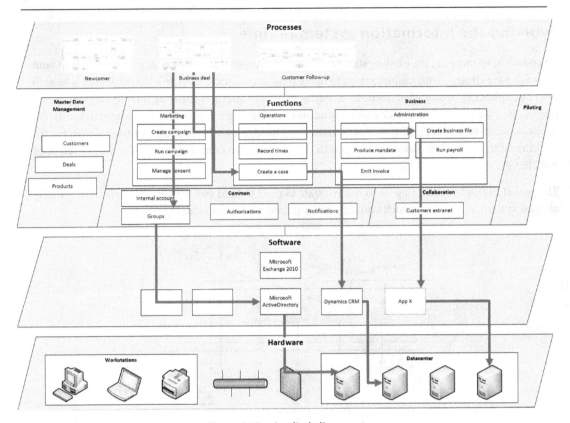

Figure 3.18 – Applied alignment

We will come back in the next chapter to the evolution of the information system in time. For now, just keep in mind that the CIGREF map has to be used to establish the actual situation but can also be used to model a desired future situation and every step in between.

The four-layer diagram method for service providers

You may wonder about the use of the CIGREF map when buying software as a service and whether it means that we do not draw anything in the hardware layer. The answer comes back to the use you want to have of your map, as you do not map for the pleasure of having something complete but because your use case needs you to be precisely aware of how your information system is working in its details. This means that, if your interest is in software alignment functions, you have absolutely no use in knowing where the servers are physically, and, in this case, there is no use in drawing anything in the fourth layer; you will simply leave it empty. On the other hand, suppose one of your concerns is about the locality of your data because your board has a constraint on data sovereignty. In this case, you will indicate the precise region in which the data centers supporting the software are located in the fourth layer. This way, it will be easy to spot something that is outside the allowed regions.

This comes back to a very important piece of advice: only map what you really need. It is very easy to get carried away and map lots of things that you actually do not really care about. In particular, the hardware layer is generally quite easy to map because automatic network exploration applications can help and good system administrators often have agents on every machine deployed for security and software inventory. So, people tend to have a very precise drawing of the fourth layer, even when their problem is about function to software alignment. In this case, you'd be better off replacing the whole hardware layer with a few general blocks—typically the data centers used, including your own server room—and labeling them with the associated cost, as this is (in this case) the only information that will help you.

Talking about services, an equivalent question can be asked the other way around, namely about the CIGREF representation for a software editor or integrator that would provide a software service to its customers. How should we represent this? Is the cloud used for the customers a part of the information system of the editor? Should it be drawn in a particular way?

Again, the notion of the usefulness of the map should drive our response to this. Imagine the problem at hand—and thus the reason why you establish a map—is that you have a problem of coupling between your internal functions and your offer for your customers. This can be a security problem because ransomware on one side could easily propagate to the other. This may also come from the accounting team, which does not know which machines and services should be billed to customers and which costs should remain internal. It could also come from system management difficulties, such as the fact that shutting down a supposedly internal server eventually had an impact on your production. In this case, the right approach would be to draw the current, unique information system and then to draw the target map, which is composed of two distinct information systems and a precise representation of the interactions remaining between them (for example, sending the usage data from the production information system to the internal information system so that accounting can establish billing for the different tenants).

This is typically where you would use the zone for "externally oriented functions" of your business capability map (layer 2 of the CIGREF representation). In the production information system, you would find in this zone the function for "reporting data usage per tenant" or "sending total API calls per tenant". And in the internal information system, you would find, of course, in the "business" zone/"accounting" quarter, the function to handle this data and calculate the bill for the tenant. Another example would be functions such as "request tenant access blocking" or "archive tenant" that you would find in the "externally oriented functions" zone. They would typically be called by the internal information system to instruct the production information system that a customer has not paid the bill and should be at least blocked, and maybe later, completely removed.

Another example of a link between the two systems is, of course, when the internal software-production workflow has produced a validated, complete new release of the software sold to the customers (this is the main role of the software editing company). There has to be a link somewhere since the information system exposing this software to customer tenants will use this deliverable to update its services at some point. One of the best ways to establish this link while keeping very low coupling is to create a container registry that will be filled with images coming from the first system (with the right tags, of course) and consumed by the second information system by pulling the images it needs to expose in the tenants.

The only remaining question is where the registry should be placed, and the answer—if you need a very stable one—is to have one on each side: a registry that centralizes all the production from your continuous integration as a software editing company on one side and another registry that serves as an image cache on the other side. This makes it easier for your continuous deployment as an integrating company to keep on creating tenants even if the first registry is not accessible anymore. This clean separation can even be used to implement some high-level rules, such as "only contained images with the STABLE tag should be put into production", by caching only those in the second Docker registry.

One might argue that, since there are calls between the two systems in this case, that might mean they are a single system and should be represented as so. Again, the map is not here to reflect the full reality of the world but to help you carry out your duty in information management. If the orientation you wish to have is a good separation of concerns (and for security reasons, it should be), then your map should represent your goal, as it will help you in doing everything that is needed to reach this objective.

Finally, another argument could be opposed to this vision, by stating that, today, every information system on the planet has some kind of connection together, may it only be by the internet network, that covers almost every local system. Also, when companies buy others, they connect their information systems, sometimes in such a tightly knitted way that they become a single system in the end. Again, this only depends on your strategy, so the CIGREF map should simply be aligned with the vision.

Patterns and antipatterns of alignment

Several years of consulting on information systems have led me to observe that most problems relate to a few misalignments in the system, themselves belonging to only a few patterns. Having worked for quite some time in a limited business domain, it was a surprise to realize after a few years when I started working with agricultural cooperatives, chemical risk analysis companies, lawyers societies, and other companies within very different fields, that these patterns (or rather, antipatterns since they cause problems) were the same everywhere.

Dalila Tamzalit, who is a researcher at French CNRS, took up with me to classify these antipatterns and document a method to find them and exploit information to better align information systems that suffer from them. This led to an article published in the International Conference on Information Systems Development in 2021 (available at `https://aisel.aisnet.org/isd2014/proceedings2021/managingdevops/3/`). You will find in the next section a summary of some of the information that could help in managing business alignment.

The sad reality of alignment

First, it should be known that most information systems, as was explained in *Chapter 1*, suffer from basic problems that limit their efficiency. In terms of a four-layer diagram, these problems can be summarized as this:

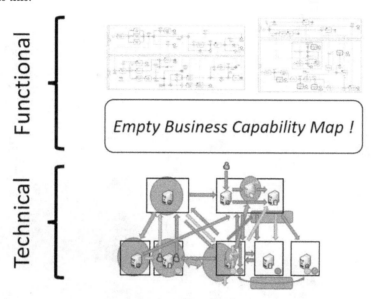

Figure 3.19 – Common problems

Processes may be well understood, since there is rarely any BCM, but the corresponding realization of the processes is often done with point-to-point ad hoc interop that quickly brings the system to a "spaghetti dish lookalike" where streams of data happen in an uncontrolled way.

What we can aim for

It was already stated, and it may sound logical, but we do not aim for a completely aligned system. A well-designed system for two business processes may be as simple as the following, where a bit more than ten well-adjusted streams of data implemented the full business needs with the same amount of applications (the majority of them already exist, in this example, and being simply correctly plugged) and almost no additional hardware:

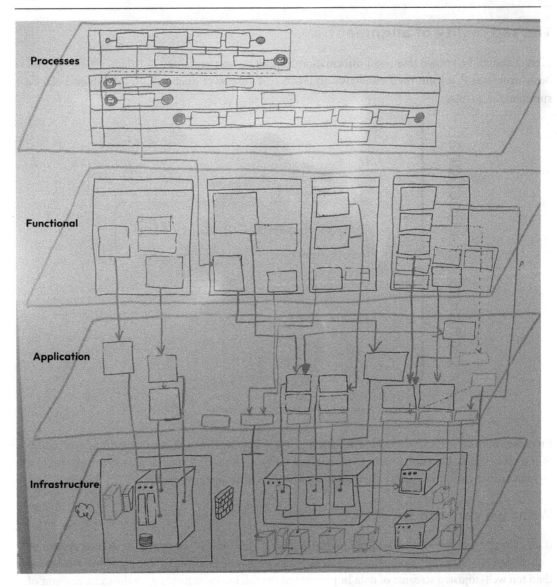

Figure 3.20 – Good alignment (the text is blurred for confidentiality reasons)

The main antipatterns in alignment

Now the goal is clear, let's come back to our alignment antipatterns and present the main four of them (we only explain what they are, and how they can be fought against and reduced will be covered in the rest of the book):

1. **Purely technical integration** happens when a process is not designed in the first layer but directly implemented into the software. The consequence is that any change in company strategy, domain-based rules, or even simple optimization of the process will lead to a change in software. This is the root cause of hearing "We do not evolve quickly enough, as we are dragged down by IT" or "It is not possible to have this business function due to software limitations" (and its variant "adding this new data attribute in the whole software chain, from interface to reporting, will take six months and need a new release of four applications").

   The symbolic representation of this antipattern is the following:

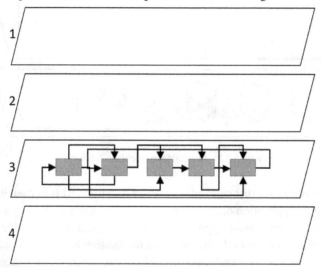

Figure 3.21 – Antipattern number 1: Pure Technical Integration

2. **Applicative silos** appear when two parts of the organization have worked on their IT needs without talking to each other. The resulting system shows the result of this as two independent systems in the diagram. There may be some cases where complete isolation is thought to be important (human resources, finance, other highly confidential zones), but, from experience, there always comes a time when links have to be established between the different zones. This may come as a harsh reality in these cases, as data has been completely duplicated, uses different formats, or uses technologies that have not been chosen to ease interoperation, etc. The main risk in this case is that data sources are simply opened to the other area, which will cause major authorization problems. In one of the worst cases I have seen, full HR data had been made available to the ERP by a trainee to implement the reimbursement of travel fees,

which was of course a major breach of confidentiality and exposed the company to potential GDPR issues until it was corrected.

The symbolic representation of this antipattern is the following:

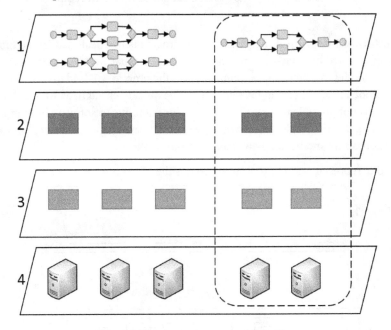

Figure 3.22 – Antipattern number 2: Silos

3. **Monoliths** are applications that concentrate a lot of functions. This in itself is not necessarily a problem, as a given application may implement everything that is used by a business domain. The problem arises because these applications also implement functions that should be shared or already exist in other parts of the system. Data duplication is a huge problem in information systems, as they never correspond from one side to the other, which makes it hard to know which source is the closest to the truth, resulting in bad decisions or false computations.

The symbolic representation of this antipattern is the following:

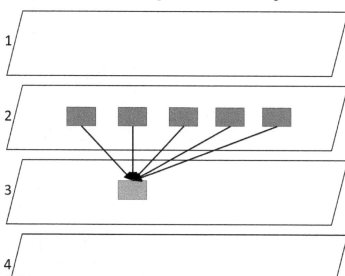

Figure 3.23 – Antipattern number 3: Monolith

4. **Functional multiple implementations** are a problem because the different implementations have almost no chance of working in a compatible way. One can easily understand that, if a financial budget summary is computed in a given way in an application and in a different way in a second application that is supposed to do the same thing, it is difficult to take intelligent actions to manage the company. One such case I witnessed at a newspaper company showed a different number of readers depending on which application was queried, with deltas that were so large that, in some situations, the newspaper could not know without additional calculation whether they were gaining or losing readers.

The symbolic representation of this antipattern is the following:

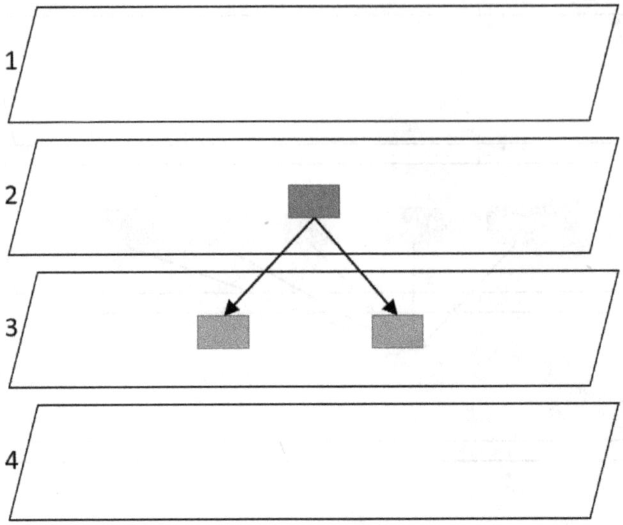

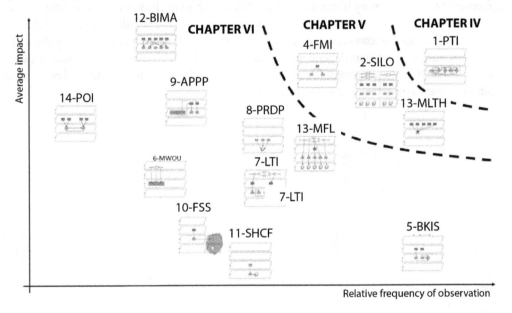

Figure 3.24 – Antipattern number 4: Functional Multiple Implementations

Some other antipatterns and a proposed classification

Only the four most important antipatterns have been shown, and the whole set of **business/IT alignment antipatterns** (**BITA**, in its proposed short form) is the following:

Figure 3.25 – Antipatterns classification

To use this classification to improve alignment in the existing information system, each of them comes with a structured identity card with the following information:

Visualization:	Short name: Pure technical integration		
	Description: many items in 3 depends on several items in 3		**Effect on evolution**: strong
			Effect on usability: none
	Typical cause: IT system organization led by technical needs instead of functional		**Ease of recovery**: complex
			Time to recover: medium
Characterization: software applications call each other directly or through the use of other software, without any level of indirection or contracting.			
Consequences: this causes high degree of coupling, since modifying a given application may have impact on many others, and thus many other functions or workflows depending on them. The system gets rigid and fragile to any evolution.			
Additional hints: coupling is typically caused by direct software calls to APIs, PL/SQL methods, COM functions, use of triggers. Middleware, when used inappropriately, also causes coupling, only displacing it in another software process.			

Figure 3.26 – ID card for an antipattern

A full explanation of how the antipatterns appear and—more importantly—the standard actions to correct them can also be found in the complete article cited before.

You can view the full article on GitHub, which will show you the amount of such information available on the first BITA: `https://github.com/PacktPublishing/Enterprise-Architecture-with-.NET/blob/main/Business-IT%20Alignment%20Anti-Patterns%20A%20Thought%20from%20an%20Empirical.pdf`

This taxonomy, coming from experience, has led us to propose a structured way to improve information alignment that can be summarized in the following BPMN diagram:

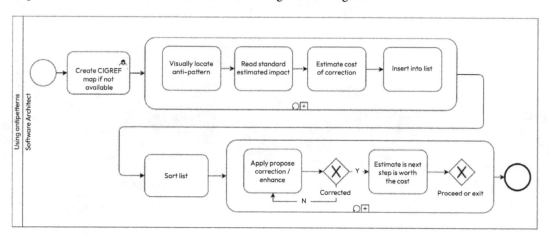

Figure 3.27 – Proposed method to detect antipatterns in an Information System

I encourage you to try and apply this method to your information systems of study and send feedback. This will be useful to the academic community and hopefully for the other readers as well.

Summary

This quite long chapter showed how a formalized diagramming technique can be used to obtain a better understanding of an information system and also document the evolutions it will have to follow. A map is a prerequisite to control a given ground, and a schema is worth a thousand words, which makes this step a must in any information system architecture activity.

Since IT is quite a specific context, we needed a dedicated way of drawing a map of an information system, and this is what the CIGREF map is about. Its four-layer arrangement helps separate the business-oriented aspects (processes and atomic functions) from the technical aspects (software and hardware).

This way of representing an information system also helps in visualizing its alignment, by checking that the third layer (the software bits) is well adjusted to the second layer (the business capability map) it implements. The business/IT alignment is the most important measure of quality and capacity to evolve for complex information systems and a feature that must be sought.

The next chapter will extend the present one by taking into account the dimension of time. We have quickly talked about this in a previous section, explaining that the CIGREF map could be used to document the actual state of an information system, but also the desired future states it should progressively reach (the Big Bang approach is never a practical alternative). But, as you will see, time appears in many other aspects of information systems and can be quite a difficult parameter to handle.

4

Dealing with Time and Technical Debt

An information system is like a living organism: it always moves and changes. Yet, most of them are designed "one shot," without thinking of its adequation to the business in time, but only of its capacity to handle the business needs at the moment it is designed. Lots of IT problems can be related to time.

In the previous chapter, we talked about business alignment and the importance of basing the structure on business concerns. This must also be applied to taking time as a parameter of the equation for a good information system: if the business feature is a one-shot/disposable one, its technical implementation will be no more complex than a prototype, quickly coded and soon discarded after use. On the other hand, for a feature that will be used for decades in production, you must carefully hone the design and polish the implementation, with as few moving parts as possible, since a good architect knows that maintaining such a module will eventually cost much more than its initial development (see, for example, `https://natemcmaster.com/blog/2023/06/18/less-code/` on this issue). The code's quality and its ease of maintenance (hence the developer's dreaded activity of documentation) will be much more important than the capacity to quickly deliver the feature for time-to-market reasons.

This chapter will analyze this problem of time adequation in information systems since most of them are designed with a time-fixed goal and rarely with evolution in time. This is why the majority of them see their performance quickly degrade in time and also why when their construction is too long, the result is not even in conformity to the expressed needs, hence the emergence of Agile software. The concept of technical debt will be explained, as well as the notion of coupling. Hopefully, by the end of this chapter, you will have improved your critical thinking of what needs a **Proof of Concept** (**PoC**) approach and what needs strong, evolutive, design.

We'll cover the following topics in this chapter:

- The impact of time on systems due to functional change
- How the Agile approach aims to solve the problem
- The concept of technical debt
- An experience-proven blueprint method for information systems

The impact of time on systems due to functional change

One of the most difficult things to do when designing systems is to consider time. After all, it is already hard to picture how something complex should look at a given moment. Taking time into account requires an additional depth of thought, which can make this difficult. Moreover, time variations happen in every aspect as the system evolves, but it should also take into account time in its functioning, just like another variable.

A bit of fun with ill-placed comparisons

Here are a few sentences you may have heard in your everyday life concerning industries other than computers:

- The repairman changed my car engine this weekend; it is now ready for another 10 years
- I took generic pills instead of the commercial brand: they are less expensive and I did not notice any difference
- Since we started regularly maintaining the furnace, we haven't had any failures during the winter
- The dimensions of the parts evolved a bit, but we just had to change the parameters on the CNC machine; there was no need for the machining expert for such a small change

Now, let's try and transpose this to the IT industry and see if we can hear the same expressions without at least a wry smile or a smirking face:

- We changed the ERP this weekend; everything seemed to work quite well on Monday morning
- I used some free software as a replacement for my commercial suite: less expensive and since it was 100% compatible, everything works exactly as before
- Since we regularly maintain our information system, we never have any major failures or bugs
- The business people need to adjust the system due to regulatory systems but, since it is just business rules, they do not need the IT team for such a small change

Do any of these sentences sound realistic? If you happen to have the slightest experience in information systems, you will know that they aren't, and even sound humorous. Nothing could be further from reality than these utopic sentences. The equivalent sentences should be more like the following:

- Management has decided to change the core ERP; we expect the information system to have a stabilization period of at least 6 months, and the initial project with analysis, deployment, and training will certainly take at least a year.

- I switched to open source to eliminate license costs, but since I had to adjust most of my processes, I lost a few functionalities and experts are hard to find on this technology, I am not sure the TCO will be lower in the end.

- Due to new cybersecurity compliance rules, we made the entire IT team push updates to software applications that are disseminated everywhere in the information system; we hope most of the servers are covered but know that we still have a high level of risk on the employees' workstations.

- "This new GDPR will oblige us to release a brand-new version of the software and adjust most of the data streams in the information system; IT will certainly spend most of its non-maintenance time on this for the next 6 months.

These versions are way more realistic but sound like a desperate evaluation of the situation. Where does such a catastrophic capacity of the information systems come from? As explained in *Chapter 1*, information systems are not industrialized yet. But if you pay closer attention to these sentences, you will realize that they all bear the notion of evolution in time. And this is what makes them sound silly. If time was taken out of the equation, they may look fine:

- The ERP we currently use works correctly

- I am using free software and it works as expected

- I am using the latest version; everything seems to be fine (for now)

- We have set up the software with initial rules (and we hope we don't have to change them in the future)

In short, IT can work and provide great services but most of the time, this is when time passes and IT has got to evolve when problems arise.

Consequences in the software world

The preceding comparisons may seem anecdotal, but they bear some reality as change is the only constant in life, and thus in information systems. A fun story I once heard stated that a perfectly stable information system is possible but needs three components: a human, a computer, and a dog. The computer does the work, the human feeds the dog, and the dog protects the computer from the human touching it.

Again, despite the humorous approach, there is some truth in this joke: the fact that the perfect system can be considered so because it is stable (the dog preventing the human from causing change, and consequently chaos). The computer can do a perfect job because it does not need to change what it has been programmed to do. Humor put aside, the consequences of time and evolution on information systems can be described formally and various concepts are associated with their different categories, all of which we will describe here:

- The first concept associated with time in IT – and this one is well-known by anyone using computers – is the notion of software upgrades. As time goes by, a piece of software, whatever quality efforts are put into its design and development, will have to undergo regular changes of versions and at least security patches to remain fully operational. A software application is a complex system in itself, with sometimes millions of lines of code. If we go on with comparisons with the mechanical industry (sorry about coming back on this one, which certainly comes from my academic background in mechanical systems), that means a standard industry-grade application is closer in complexity to a commercial airplane than a standard automobile. No wonder it has to be upgraded and adjusted during its lifetime, just like airliners necessitate heavy maintenance. The difficulty comes from the fact that most applications are designed in a way that is not as modular as we would expect, and unexpected dependencies happen very often, making the application work as a cohesive entity. If you take a Rafale war aircraft, the engines can be changed by two mechanics in a few hours because the whole plane has been designed with this constraint. How about your ERP software? Is there any way you could switch the authorization engine in a few hours? Most certainly not... This is why the majority of software applications have a limited life expectancy: after many version upgrades, the overall quality always degrades and, in time, the application becomes less adapted to the business. Sure, some applications stay in business for more than 10 years, sometimes 20 or even more. But ask the users if the reason is that the software is perfect and if they love it and you will always get the same answer: the piece of software is still here simply because it is way too dangerous to try and remove it!

- The second kind of impact of time on information systems is not on the software part but due to the business itself. As stated at the very beginning of this chapter, information systems are live entities, and they evolve continuously because business itself evolves. New strategies, regulation changes, large company reorganizations, fusion with an acquired company, selling of a business unit, and so on – there are so many factors that can move around the uses of the information systems that there is almost no way that they can remain stable for a long time, even in very stable business domains. In addition, on top of law-related regulations, many business rules are specific to companies and this makes it hard to produce applications that can truly pretend to reach the "one size fits all" state. Even with the best intention to keep things simple, companies often end up tweaking the software applications they bought, or integrating them with dedicated connectors or custom code, for them to comply with their way of doing business, simply because it costs less (at least at first) to do so, rather than reorganizing the corresponding function. But this is a trap and is where time comes into the game again: as time goes by, this specificity will cost more and more. First, every new major version of the application may make it fail, and money will be spent on keeping the specific code compatible with the new version. Most of the time, this was not fully budgeted, which means the overall cost grows and grows in time, sometimes ending up costing much more than initially adjusting the processes to the software. There is also a part of psychology in it: functional experts would feel bad adjusting their way of doing simply because some piece of code from an external editor thinks it is better to do it another way. What do they know about their job?

- The third link to time in information systems is neither on software nor on functions, but on how the software is adapted to the business functions. This can be done through integration, customization, adjustment of application parameters, adapting the way the application interoperates with other parts of the software, and more. The link to time is a bit more subtle here, but all these ways remain mostly specialists' jobs. And since experts are rare, it is very common that this step in a software project takes more time than accounted for. Changing a parameter is quick but analyzing all possible impacts in a complex system needs a good understanding of it (we talked about the need for a map of the information system in *Chapter 3*) and can take a lot of time. This is one of the reasons why ERP projects – a well-known example – take so much time in a company (despite everything a commercial can tell you about it, there is no way in practice to reduce this time below 6 months at the very least). Another consequence of this is vendor lock-in: as more and more parameters get changed from their default values, as more and more connectors or integrations are added to the system, it becomes more and more difficult to change the software for other vendors. After a given amount of time, the application is so engrained in the system data streams that customizing a new application would take a huge effort (particularly since documentation is not the best asset of these projects), hence the stopped evolution of some IT capabilities.

All these have the consequence that the company using the information system is, in a way, dispossessed from its business processes as there are so many ways IT stands in the way and can prevent rapid evolution. Sure, IT helps automate processes and, once set in place, can provide interesting gains. But the effort to make it work – and in particular to keep it working in time – might not make it very interesting (remember the Gartner statistics showing that 70% of the IT budget goes into maintenance alone!)

Finally, **technical debt** is also a concept that is strongly associated with time passing by. As it turns out, it is very much like entropy and tends to always grow with time. But this one is so important that we will analyze it in its very own section later in this chapter. For now, we are going to look at how Agile practices can help us deal with time.

How the Agile approach aims at solving the time problem

Agile has lots to do with time management, so it may help us deal with the time issues around information systems. To explain how, we will go back to what Agile is and then observe different ways it can solve the time-based complexity we need to tame.

A metaphor to explain Agile

Agile is about taking the time factor into account. In a V-cycle development process, everything is planned and time passing by is only supposed to make things go forward in the process. The Agile approach recognizes that time is itself a factor in the project, and it appears everywhere:

- There is time negotiation because quality should not be compromised, and adding resources does not make a software project get quicker ("five cooks do not bake a cake in 10 minutes instead of 50"). So, the only way to adjust to hazards is to increase the time or reduce the functional scope (which comes back again to increasing time if the client still wants the complete initially-requested perimeter to be realized by the end of the project).

- Time is one of the major decisions in organizing an Agile project: how long should the sprints be if you work with Agile? What cadence should be used if you use a Kanban approach? At what frequency should we organize stabilization sprints? How quick should continuous integration be to be efficient? How sustainable in time is the rhythm used by the team?

- Filling a sprint is a negotiation around time available, and how estimates of time to be taken by backlog tasks should be done and added up to fill the sprint.

One of the best metaphors I have found to explain Agile software development to my customers or students also talks very much about time. The idea is to compare two ways of shooting an arrow at a target: the usual way is to aim at the target, carefully accounting for the wind and the distance to the target and, when everything is fine, shooting the arrow and hoping there will not be a sudden gust of wind, that we have estimated correctly the angle, and more. Guess what? If the target is far enough, hitting the bull's eye is pretty much a question of luck, in these conditions. This is what the V-cycle is about: carefully planning for the project development in time, considering as many initial conditions as possible, and eventually launching the project, hoping that nothing will get it out of target... Sadly, there will always be external changes in conditions, customers changing their minds, a team being sick, an important dependency not being released on time, and so on.

Meet the Agile way to hit the bull's eye at every shot, or at least with a considerably higher probability: you must take the arrow in your hand, walk to the target against the changing winds, correct your path if the target moves, and eventually plant the arrow in the target when you're close enough. Sure, walking to the target with an arrow in your hand takes way longer than the flight of the arrow once it's been shot. But are you sure it is going to take longer than hitting many arrows in the wind to eventually have one reaching the target, not even in the middle? The difference lies in the conditions of the project. If everything is stable, no external dependencies are there, and you are in a fully controlled context, maybe planning everything in advance will be a bit quicker than adjusting step by step. However, the vast majority of software projects do not belong to this utopic situation. Most of them are developed in extremely changing contexts, with hazards everywhere.

Back to the concept of emerging architecture

In *Chapter 3*, emerging code architecture was quickly cited and I said we would come back to this. Now is the right moment to do so. Since we talked about Agile development and we are in the middle of a discussion about time, let's look at two things that are closely related to emerging architecture. This concept is about achieving a good architecture **without aiming at it in advance, with schemas and plans**, by refining the architecture along the development of a software project and refactoring the code structure at every step of iterative development. Without aiming at it in advance... does this remind you of something? This is the metaphor we used previously to explain the Agile approach to hit the target of a software project. Again, time is the concept that allows us to reach an agreement between architecture (in its meaning of structuring in advance) and the impossibility of knowing a complex business domain before working with it (thus needing to advance with the arrow in your hand and readjusting the path). This opposition and its resolution are so important that they need a dedicated section.

Apparent opposition between architecture and Agile methods

A decade ago, when I started understanding the principles of Agile software and applying them to the technical team I was leading, it was hard for me to understand why a typical Scrum team would be developers, testers, a product owner, and a Scrum master. How come there was no architect in there? As it was my business card title at this time, I took it personally. This was a bit disturbing since, at the same time, I was realizing the huge value Agile had compared to the old way we were using by then.

After discussing this with many Agile leaders who brought the concept to France, I eventually gave a conference in 2013 specifically on the subject of how to bring together architecture and Agile methods (French version: `https://www.infoq.com/fr/presentations/concilier-architecture-et-agilite/`). After exposing the many contradictions and how an "ivory tower" architect would have a hard time in short iterations, I ended up explaining a possible way to conciliate the utility of "seeing in advance" and "acting in short iterations and adjusting the vision". Like most patterns, which are not invented but discovered independently by many people, this concept of emerging architecture is simply the result of any work trying to erase the contradiction stated previously.

Again, time is the great equation solver here: **architecture and short iterations are not opposed if you set the time horizon of architecture to only a few iterations.** This way, the probability that the target moves a lot is strongly reduced, and the architecture remains useful because it helps structure the development of these few iterations.

This solves the difficulty for the architect as their job remains necessary, even if it changes quite a lot if you consider the job is to think long shots ahead. But then again, even before Agile, architects in ivory towers (imagining a long time ahead without a grasp of reality and giving plans to the teams... that will not follow them) were largely seen as pointless.

It also helps us understand the concept of emerging architecture, which states that, if refactoring is done correctly at the end of each sprint, the final structure of the code will be perfectly adequate for the functional needs... just like a perfect architectural vision (a long-shot of an arrow in the center of the target, in our metaphor) would have done in pure theory (but is practically impossible except for very small projects).

In addition to time, semantics also helps remove the contradiction exposed previously. The word *architecture* is used in two different ways:

- Architecture as *the emerging global shape of a project* is about the structure of the code produced by the team

- Architecture as *the act of envisioning a structure* in the application (or even higher, in the information system as a whole) is about trying to reach this structured state by initially thinking and acting on the system

This means that this theoretical contradiction can be overcome. But that does not mean that there is no practical impact, and I will show you one because it will help us go back to the notion of aligning the technical aspects with the business ones. But before that, I will add an external analysis.

The position of famous architects

Just like for any scientific discipline, we software architects save a lot of time "sitting on the shoulders of giants," which in our case involves reflecting on the state of the art established by real experts of the subject. Martin Fowler is certainly one of the best references in software architecture. On this question of opposition between "hack, code, and fix" and "big upfront design", I strongly recommend reading the excellent article at `https://www.martinfowler.com/articles/designDead.html`. The willingly-provocative title *Is Design Dead?* hides the real background subject, which is precisely what we're talking about here.

Martin Fowler's response to the opposition to architecture is simply to **only apply design to increase the capacity of the system to evolve**. As usual, there is no "true or false" answer between the two extremes, namely **eXtreme Programming** (which explicitly admits its extreme character) and **Big Up Front Design** (which often does not admit or even recognize its extreme character and quickly produces "ivory tower architects").

This is where the job of the architect becomes an art as they need to proceed with good skills to obtain a subtle balance using some up front design without imposing unmovable limits, but still providing healthy guidelines that will truly help developers produce faster features *in the long term* (and not get blocked by software entropy, as described in Martin Fowler's article).

Since, in addition, change is the only constant in software development, there is no use in writing what will be out of target due to functional changes in advance, but that does not mean the architect's job is removed: on the contrary, it is about **easing the evolutions in the future**. Architecture is not about UML or code frameworks but about guidelines on the way the system should be structured: what are its fixed points, what are the articulations around which it moves; where should we focus on quality, where can we afford disposable code since the business evolves so quickly that investing on solidity is pointless? Sometimes, heavy architecture efforts can be justified precisely to accommodate an important module to change very frequently. This is, for example, the case of using a **business rules management system** (we will come back to this in more detail in *Chapter 5*).

The same applies to coding patterns: the mere fact of working correctly, and refactorizing your code continuously, will naturally bring patterns in your code, even if you did not know them in advance (I told you that craftsmanship was not dead!). An excellent bit of evidence of this, if you have not experienced it already by yourself in a coding activity, is that, when you read a lot of code (which very few people like to do, even though most great writers have been hungry readers beforehand) or when you follow groups of students, you will realize that these patterns are often discovered anew. This is the very definition of patterns since they are universal and, however, they're found, if you structure your code correctly, the context will make you end up using the right pattern for this precise problem.

Thinking ahead with function contracts

Previously, I talked about a practical example to illustrate the opposition on long-term and emerging architecture that we talked about before letting the expert (Martin Fowler) speak. The example I am talking about comes from my own professional experience and is a diagram I have created for a complex scenario of interoperation between several applications to implement a functional process. When analyzing the business needs at the beginning of the project (or rather their initial expression, because they change alongside the project), I created the following stream diagram:

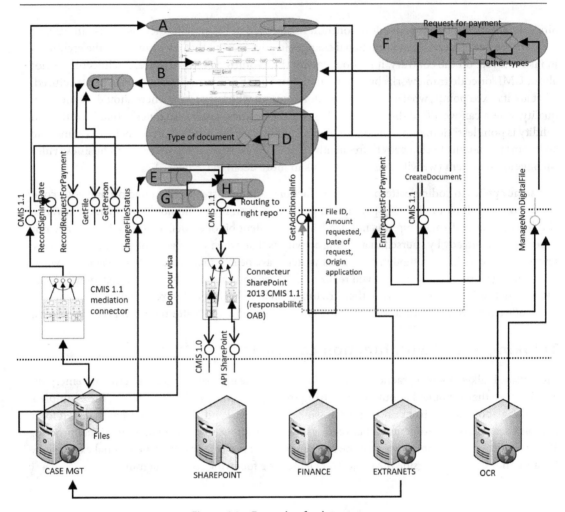

Figure 4.1 – Example of a data stream

Following *Chapter 3*, you should normally recognize the duality between the function and the software implementation, and in particular, the essential recommendations that dependencies should point to level 2 instead of directly to level 3, which would result in point-to-point interoperation, creating a "hard coupling" of software solutions (we will come back to this expression shortly if you do not know what it means).

Even if we use machine symbols, the bottom layer is really about software servers, hence layer 3. The intermediate zone is also part of the software layer as it contains connectors destined to transform API calls into proprietary calls if needed. The orchestration layer at the top shows how functional tasks expressed by API contracts are put together to create fine-grained processes. It can mostly be considered as layer 2 in the CIGREF map, with some touches of layer 1.

The reason I'm showing this schema is twofold. First, it illustrates when architecture goes too far and how emerging architecture can help save time: I drew this when I was a young architect, not fully aware of the concept of emerging architecture and it went too far. Sure, it helped to have a project vision, but in the end, almost none of the connectors and data streams worked the way I envisioned they would and time was wasted there.

Secondly, this schema provides more detail on the interaction between layers 2 and 3 of the CIGREF map, showing they are about API contracts (and we are not talking about API implementations with code, just contracts, which are lists of functional capabilities expressed in precise, technical terms). This is where it becomes particularly interesting because, despite the technical implementation not being (at all) the one envisioned and despite the orchestration of the streams having changed many times since the original blueprint, **it turned out the API contracts have remained unchanged for many years.**

I cannot tell you how much satisfaction it has brought me to realize this a few years after the project. When thinking about the code, my initial vision was a failure and a very small portion of it had been implemented. When thinking about orchestration, the way the APIs are glued together has changed due to modifications in the processes and the business rules. But the API contracts I designed with the business-knowledgeable people were still there years after, basically unchanged, and their extreme alignment with the business had allowed all these changes to happen with very limited impact in code or customization.

In short, code architecture should be limited to a few iterations. But business alignment architecture is an investment worth doing, even at the very beginning of the project, because its value will not decay.

This is why contract-first API design (again, I am not talking about API implementation, but solely the definition of the contract) is so important: the contracts can be established with pure functional knowledge, leaving every technical aspect outside, thus ensuring a very stable base for the definition of business modules and their dependencies that will serve as a strong foundation for the software implementations to come afterward.

We will come back to this notion of contract-first thinking in *Chapters 8* and *9* but for now, please just keep in mind that the time horizon has an impact on architecture in its two meanings. For technical architecture, the time horizon should be limited for "the arrow to reach the target." But for the functional architecture, the time horizon does not stand any limit because in this case, what we are doing is getting to know where the target is. This is a prerequisite at even just thinking of hitting it!

The concept of technical debt

Technical debt may be one of the most discussed concepts related to IT management in the past decade. As the quality of software is a fundamental goal of this book, it is something we have to clearly describe.

The general definition of technical debt

If you are reading this book, you are certainly interested in software quality and how to architect things correctly, so you most certainly have a good grasp of the concept of technical debt, or at least have been exposed to it. Nonetheless, I will give a quick definition of it so that you can try and formalize it.

> **What is technical debt?**
>
> Technical debt is the amount of accidental complexity you have allowed to enter your project, together with its increase in intrinsic complexity.

Let's decompose this a little. When you develop a software project, it always aims at producing a function on a given business domain perimeter. There is an inner, definite, stable complexity that comes from the domain you are trying to address: it is much simpler to print an address on an envelope than to optimize the flights for an international airport.

Or is it? Just as a side note, take great care to always know exactly what we are talking about when business needs are expressed, and if you cannot, do not hesitate to formulate caveats about what can and will be done. Explaining the functional needs with a simple sentence does not necessarily mean that the objective is itself simple, and much can be hidden behind this. In this example, you should immediately have the reflex of asking how the addresses should be printed, if there are different envelope formats to support it, what the international address is, if there are some norms and standards to respect, how the data will be provided, and so on...

This first type of complexity is often called **intrinsic complexity** because it comes with the functional business domain you want to address and there is no way to escape it. Short of doing less than your client expects, there is no way you can reduce this. This does not mean you should take all this complexity right away: remember the Agile approach to cutting projects into small, manageable chunks that will be dealt with one at a time ("How do you eat an elephant? One bite at a time"). And if your customer wants you to go all the way to treat the complete business domain function and complexity, you will simply add as many sprints as needed to reach the desired level of intrinsic complexity. It will only take longer and, thus, be more expensive.

Now, let's cover the second type of complexity: the accidental one. For this, let's take the first example of expressed functional need, namely printing addresses on envelopes. To keep this short, suppose we only have to print standard addresses with four lines on a standard A5 format envelope, that the data for the addresses is provided in any format we want, and that the hardware part (a special printer for envelopes) is taken care of. How could we, as developers, implement this requested function?

One of the simplest ways that comes to mind is to use the fusion function of an Office Word application, consuming the XML data from the integrated assistant, and saving the file for future uses by the client.

But there is (way!) more than one way to implement a function with software and you may very well find yourself using a Java application that has been created from scratch to read addresses from any format, create a PDF document, and send it to the printer. It is not that complicated, but there are already more moving parts than in the first technical solution... and their maintenance is yours, instead of the Office editor's! You will have to take care of the Java runtime with a version compatibility issue. The PDF generation can be a little tricky too. And maybe the developers will have left a few TODOs in the code, indicating that some corner cases need to be solved in the future. In the end, though not extremely complicated, this solution is more technically complex than the first one we proposed, though it reaches the same objectives in functional terms.

The delta is what we call accidental complexity in that this is the complexity that – contrary to the functional one – could have been avoided. It is sometimes confused with technical complexity, but this is not the right wording. There will always be some kind of technical complexity to implement a function: concrete execution cannot come from nowhere and there must be some kind of software to execute a function. Accidental complexity is, as its name suggests, the level of technical complexity that sits on top of the minimal necessary effort to implement the functional need. Therefore, it is considered an accident because things could have been done without it, and it is there because of external, unwanted reasons.

Causes of technical debt and its relation to time

What are these unwanted reasons? Well, there are so many of them that it would be difficult to list them all: laziness, lack of time to reach the appropriate quality, lack of training, the tendency that we all have to use a well-known technology rather than one that would be a better fit but would have to been learned first ("when all you have is a hammer, every problem looks like a nail"), lack of technical watch, with the result that we simply do not know about the mere existence of a better way to do things, and more.

On top of these reasons, there is another one that has a deeper reach, namely that most technical experts, deep inside of them, actually *love* complexity. I often compare developers to gases (without malice: I am one myself, have been for decades, and still fall out in this trap): they will always occupy all the space you give them.

This is another similarity with thermodynamics since I was talking about entropy as a metaphor for technical debt.

Let me give you a (not so unrealistic) example. Form a million-dollar team with software experts, asking them to create a function that calculates the sum of two integers, and you have a high probability that none of them proposes to simply use Int32.Add. They will work under the hypothesis that you know what you are doing: since you set up such a great team and budget, you must have elevated goals of creating a high-performance function to add integers with virtually no limits in size, working in all conditions with predictable outcomes.

This is because developers are engineers and rarely business people. If they were, the first one you contacted in the team would tell you that you need no other hire and that they will take the whole job by themselves for only half the million dollars. Following this, they would put together a complex piece of machinery that simply calls Int32.Add, makes you wait for a few months to hide the extreme simplicity of it, and delivers the end product to you afterward.

One of the key findings is to always give some boundary constraints to your developer teams; otherwise, they will add accidental complexity, sometimes even a large amount of it... and it is always a pain to know your best customer's business process is blocked by a bug in a function that was added "in case we need it in the future" by an overzealous developer.

This first set of reasons for technical debt is quite critical of the developers but wait – there's something to say about functional people too!

How about laziness in explaining the precise need? Lack of communication with the developers? Extreme reliance on them to figure out the technical complexity (previously, I talked about the danger of a one-sentence business needing expression: it often hides a lack of understanding by the requester itself)? Lack of availability to test the results and adjust the functional request? The list could go on, as well as picking on our beloved product owners. It is not because they are not technical themselves that functional actors cannot cause accidental complexity!

All these changes in the functional definition of what should be done have a huge toll on technical implementations: we do not change code like we change a wheel on a car! There are links between the functions, and the overall complexity of code escapes the human brain's capacity after a few lines only. So, if indications change all the time, the result will undoubtedly be some low-quality code cutting corners on quality, full of "temporary" workarounds (who are we kidding? We all know they will stay there until the application end of life), TODO indications for a hypothetical colleague to magically appear later and refactor the dumb incomplete code into a marvelous elegant new version, and so on. I am not even talking about dead code that will bloat applications forever, simply because the complexity – and lack of documentation – just makes it so risky to remove it and create side effects...

Technical debt's relation to time

Why, why, why is there so much junk code in today's software industry? Well... time, again. Technical debt is another concept that's strongly related to time. Try to operate a root cause analysis on the aforementioned symptoms and, after a few consecutive "whys," you will almost always reach the same answer: "not enough time." Lack of junior training? We do not have the time. Lack of product owner availability? They do not have the time. Absence of documentation? We do not have the time...

Time also appears in another way in technical debt: as mentioned previously, technical debt, like entropy, always grows. And like entropy, there may be some special places where disorder is locally reducing, but this is always by consuming energy and growing disorder in other places, which makes entropy grow in the whole system.

Technical debt is the reason why there is such a small portion of software applications that can still run fine after more than a decade. Casual observers may think this is because software is a world of fast changes, but when you think of it, change is not that quick. Java was the thing of the 90s, .NET came a decade later, the 2010s saw the use of JavaScript rise to things it was not made for, and the 2020s marked a few attempts at new languages with none marking its time for now – not such a tremendous rate of change... So, why do we change software so quickly? Simply because they are so full of technical debt we cannot maintain them in a costly manner anymore!

This is another link to time that technical debt exposes: as time goes by and technical debt grows, the time toll it takes on the project by slowing down development goes up. This is why technical debt is called this: just like financial debt, you must pay the interest, so long as you keep some borrowed capital. The higher the capital borrowed (the depth of your technical debt, which is related to the number of times you have been cutting corners in your development process), the higher the interest (the additional time it takes to add a feature to your application).

And since we're talking about a linear relationship between the level of debt and the time it eats away at software development, that means there is a rate, just like in financial loans. Now is a good time to analyze this rate in more detail.

Debt or usury?

Many of us make loans, at least once in our lives, to buy a house. And it makes sense to pay a few percent of the amount obtained compared to the advantages we can draw from them: owning the house at the end of the loan, not paying rent anymore, and more. Depending on the economic context (and individual preferences also take a great part in the choice), there may be some cases where it is better to rent and not buy but, in the longer term, building some capital always wins.

However, this only holds true because the rate is small enough! How comfortable would you be if you had to borrow money at a rate of 10%, 20%, or even 50% interest per year? In this case, of course, nobody would ever borrow some money as, 2 years later only, the loan would have cost as much as the capital: in this case, you are way better off holding off your purchase for 2 years and buying cash.

Except there are cases where you cannot do this. Sure, you can rent a place to live instead of buying it. But how about when you need money to eat or have a temporary shelter because of a hard incident in your life? Without regulation, the banks could increase their rates as much as they want, and, in some cases, you would be obliged to take the loan anyway because your life depends on it. In this situation, you would have to reimburse this when your situation improves, but the rate is so high that it would eat up everything you have saved and you would end up contracting another loan, in a never-ending poverty loop. This is to avoid the situation that, for centuries, banks and even individual actors performing financial operations have been limited by governments through what is called the **usury rate**.

If you do not know this financial term, usury refers to lending money at such a high rate that it makes it practically impossible to reimburse the capital. Society improvement is why it is now illegal in most countries, where maximum rates are fixed. For example, in France, at the time of writing, the usury rate was 5.33%, which means that banks are not allowed to lend money at a rate higher than this value.

Now, let's go back to technical debt and evaluate the rate at which we borrow. This isn't very hard to find since the Gartner study on the cost of maintenance in information systems has already been cited twice previously: it is a flabbergasting 70% of the IT budget! Okay, this does not account for a 70% interest rate in technical debt, because you should also count the benefits the IT systems offer the company and the cost of doing otherwise. I will let you do the calculation as this can vary depending on your organization's context. But all in all, there's a chance that you will reach a figure that you would, by no means, tolerate from your bank on a financial loan, and that will be *way* higher than the usury rate.

So, why should we tolerate this? Reasons for the situation have already been cited previously; now is the time to lay a few solution paths to get rid of excessive technical debt. This is what we will do in the coming section.

Ways to balance technical debt

People often talk about fighting technical debt or suppressing it. This is not the right wording as it bears the meaning that technical debt (and thus accidental complexity) should be downed to zero. This sounds like a hard thing to obtain as perfection costs a lot of money: in fact, your goal should not be removing all technical debt but rather keeping it in control, just like you should not try and find a 0% interest rate loan (you never will) but rather find the right rate that allows your project to be more cost-effective, balancing the interest rate, the amount and duration of the loan, and more.

So, a bit of technical debt is acceptable. If you have to roll out this feature in time for the yearly seminar with all the customers, who will care that you did not put logging in place for the occasion? The only really important part is that you have placed a ticket in the development tool and the product owners agreed that it will be done in a coming sprint, before putting the feature into production. If they come back on their promise and try to delay this "technical" ticket and make a fuss about it, remind them of the consequences, send an email explaining how this will impact the future, ask for their written agreement of responsibility, escalate it to the big boss... whatever it takes to get this feature back on track! Otherwise, it will be your responsibility that technical debt starts to grow.

Of course, the best way will always be simply to not let the technical debt slip away. Sure, it may sound easier to say than to do, but knowing how the problem can arise is already a great step forward. Remember that the mere concept of "technical debt" was not known or formalized in the 2000s; now, even non-technical managers working in IT or software development may have heard about it. This is already a big improvement and lets you make an educated point to them, explaining how reduced delays and lack of training or time for documentation and quality will end up in slow development in a few months. Again, if you are a technical lead or a CTO, controlling the technical debt is one of your first and most important duties.

But you may be in a situation where the technical debt of a software application is already high – either because you have let it slip in the past (silly you) or because you are responsible for a piece of software that was already in a bad state. First of all, make it abundantly clear – if it is not perfectly known by stakeholders – that the situation is bad: you wouldn't imagine how long bad software teams can invent pretexts for their inability to deliver, and you do need to have the capacity to improve the situation. If you accept the job but do not quickly issue warnings about the unstable situation, everyone will assume the software is fine. And you will not be able to alert about its state later, because, as you are a technical expert, everyone will logically consider that you should have seen it before, particularly if it is such a mess as you describe. You might even find some inconsiderate former owners ready to swear that the software was perfectly fine before they handed it over to your team!

In this case, establish a map of the technical debt in the application (using the CIGREF four-layer map, not forgetting that technical debt can even come from badly designed processes or ill-defined functions, with incorrect governance). There might be some places where a bit of technical debt is acceptable. There will be some others where it eats most of the maintenance time and budget and where it has to be urgently taken care of. When evaluating the risks of correcting this, lots of people who have participated in the initial mess will tell you that the impacts will be so high that trying to correct the software will not work, and maybe even that they already tried and it failed. In this case, make your best estimations and ask for managers to decide and take ownership. When announcing that rewriting the concerned feature will cost 100,000 dollars and will come with a 20% risk of impact worth 200,000 dollars more, everybody around the table will certainly frown... but if you also explain that this software has cost the company 40,000 dollars every year for the last decade, thus already sending 400,000 dollars down the drain, the decision makers will be quick to perform the calculation and give you a go.

This means you may reimburse some of the capital (in the metaphor of financial loans) by removing technical debt, even if it is generally difficult to explain the benefits of this to the managers. After all, the business impact is not immediately perceivable and no customers are complaining that the software doesn't work. So, again, you really must put a strong case together by evaluating how much time technical debt costs you, what features could have been done for your customers' delight at the same time, and how much time it will take to reimburse the debt and bring the application to a sound level of quality, without forgetting the impact analysis – there is always a risk in "changing the engine while the car is running."

The Big Bang temptation

How about the Big Bang approach? You know – throwing it all away and restarting a new, fresh, clean product. The dream of all engineers... **If you think about it, this is not the right situation. If it happens without you being able to prevent it, then this is for the best.** Let me explain this: the Big Bang approach, however seductive it may be, is *never* the right one. If you have a technical debt problem in your software, this is because your development process is wrong. So, if you start another application hoping it will be better than the previous one without fixing the process, you will simply lose a few years and reach the same state. If you know what was wrong in the process and have corrected it, the application will improve, so there is no use in throwing it anymore.

Does this mean that the Big Bang never happens? No, of course. Even if it is not a good idea, people still try to do it... and fail. But a clean slate is such an appealing idea that application owners, even though poor in marketing in all other aspects, will go to great lengths to obtain an agreement and budget from the stakeholders. They will do so by promising increased performance, providing better time-to-market for future features and ease of improvement, and so many other qualities that participants will eventually wonder why this had not been proposed earlier. And again, they will fail. This is not me saying this, but a return on experience that you will find from about everyone with experience with these kinds of projects.

There are a few exceptions, like in any rule, and some Big Bang projects have indeed succeeded. I have noticed this in projects where the Big Bang was not intended by the teams but experienced through the fact that the old project simply collapsed down on its weight. In France, where I live, we have had many cases of such huge government projects where the failure was so big that there was nothing to be saved from the project and new software companies had to restart from scratch. The "Louvois" project (managing soldiers' salaries) comes to mind, with millions of euros being thrown away with the project. To come back to what I said before about shared responsibilities in these events, technical problems were abundant in this project, but there was also a huge lack of functional feature descriptions and almost no cutting down of the project, which led to this industrial catastrophe.

The different types of coupling

Coupling between applications is the last concept we well talk about concerning time... Just like technical debt is a little too much of something that must exist (technical complexity, to implement functional complexity), coupling is the name given to a dependency that is a little stronger than what's needed from a functional point of view.

Let's consider an example: you want to send a contract with an electronic signature to one of your customers. Of course, the contract module will have a dependency on the electronic signing application, as well as on the module providing the information you need about this customer (namely, their financial or legal contact email address). But there are dependencies and dependencies...

Imagine that the first one has been very well designed, and you can simply send a contract for a signature by calling an API in your information system that will then take care of everything. You do not even need to know what company will do the actual work, nor how this will be legally binding: you simply call the API and if it returns an OK (`HTTP 200`, in technical terms), you are fine. This kind of dependency is a low-coupled one: things may change in the implementation, with your company preferring another electronic signature vendor, or routing the documents to sign in different ways, depending on who is calling the API: you do not care as you simply call something such as `https://mycompany.com/document-sign` with a `POST` command. This is all you know; whatever happens behind the scenes is not your concern. You still depend on the function's completion to get your contract signed but this dependency is very flexible and you may never have to change anything in the way you call the function; the coupling you are subject to is low.

Now, let's take the second dependency and imagine something at the other end of the spectrum: you need to get the email of the financial or legal contact for the customer, and to do this, you must know the customer ID. Sadly, this is not the same identification as the one you use internally in your service. So, first, you will have to call the service in charge of customer references to know the exact identification to use. When you have this information, you will have to dive deep into a folder shared through its IP number and go down the folder structure, starting with the year the consumer was recorded (it looks like it's the first two numbers of the identifier you got, but you're not sure), then a folder named `Contact`, and final a folder with the kind of contact, namely `FIN` for finance and `JUR` for legal. There, you will eventually find a Word document in which you will have to skip through some useless information before you finally get to page 2, where you find the email address you have been looking for all this time.

This sounds far-fetched but this is a real-world example I have seen during my consulting years (admittedly, though, it was one of the worst information systems I have ever seen in the 15 years I was working on customer systems). And we're not even done yet! Some customers had multiple identifiers; and when they were deleted and re-entered in the database, their identifiers were recovered... but their data was in the folder corresponding to the renewal year instead of the initial creation year. At some point, the `Contact` folder was renamed `CONTACTS`, ruining the few attempts at automating information recovery, and the codes for the types of contacts were changed. Finally, the Word documents in the folders evolved in their format and the email address was not to be found at the same place, leaving people wondering if the new location contained the right data or if it was about new email information. All this useless complexity made this dependency on an email address one with a huge strong coupling (again, this is the worst example I have seen).

Of course, low-coupling is generally better but, just like for technical debt, it is OK to have some coupling and the important thing is to control it. There might be some places where extremely strong coupling is not a concerning issue. For example, tight coupling to a budget structure is usually not a big deal because these structures are regulatory imposed and they change with a frequency calculated in decades. So, it's no big deal if you have to radically review your processes and software applications in this case. On the other hand, you will want low coupling if you use a commercial dependency from a provider that tends to increase their license price when they realize you do a fair deal of business with their tools (you most certainly know of such editors). In this case, showing your CEO/CFO that you have made it possible to switch providers with a few changes of parameters and a small migration procedure will make you their beloved CTO partner as they will return to negotiations with the supplier having extremely strong arguments and an easy escape door if the latter will not change its pricing policy.

Finally, coupling is also related to time. There are lots of kinds of coupling, but there is a category of coupling that is chronologically measured. If, to proceed with a task in module A, you need a piece of information that comes from module B, the dependency is synchronous (and you will certainly find it is implemented through a synchronous call such as an HTTP `GET` call). If your process in module A can continue freely after it has called a function in module B, the dependency is asynchronous (and you will certainly see it as, for example, a `POST` API that returns a callback URL that you may call afterward to see if the job is done – or even better, you can register a webhook to get informed once the job is terminated; this will send you another URL you can contact to get the result of the external task). In *Chapter 17*, we will come back to this approach and, in particular, explain the difference between orchestration and choreography and when to use each of them – as always, there is no true or false here and the right technology depends on the exact functional need and its context. Ideally, both approaches are used in a good information system, each in the context where they are the preferred solution.

At this point, you should now have a clear understanding of technical debt and how it affects the evolution of information systems. As time goes by and technical debt cumulates, IT assemblies slow down in their evolution and sometimes their sheer functioning because of technical debt. What can we do about this? Well, this is the subject of the next section.

An experience-proven blueprint method for information systems

In this last section, I would like to explain a method I have been using to create blueprints of information systems evolution that I have been perfecting in the past few years with several industrial customers. There's nothing special in there, nor particularly innovative, as it is just applying common sense to reach function targets... but it is formalized enough so that I can imagine you finding value in the description of the steps used.

The full method is quite elaborate and would need a dedicated book, so I will concentrate on a case that is precisely the subject of the chapter, namely dealing with time and technical debt. The example I'll use will involve extracting a monolith software application ridden with technical debt but sadly used as the core of the customer company's business (yes, a worst-case scenario). A multi-year planned blueprint had to be created to extract this dependency while limiting the impacts on the daily business. The following sections explain how this has been done, concentrating on the method and not showing the exact actions, to maintain the confidentiality of the customer.

It all starts with... mapping!

As explained in *Chapter 3*, we have to start with a formalized mapping of the problem, and the CIGREF map was established around the perimeter to be studied. As the problem was on the functional and software layers, the processes were not represented at all and the fourth layer regarding hardware infrastructure just skirted over because all we needed there was the global cost of associated machines. The result was the following structure, where you can spot the top-left large white square in the software layer (this was the monolith subject we studied):

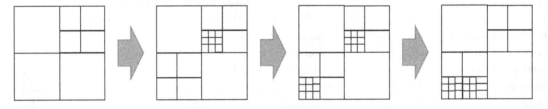

Figure 4.2 – The evolution of the preciseness in maps

This is something so important that I want to stress it again: **only the parts related to the study have been mapped**. Remember this graph from the previous chapter?

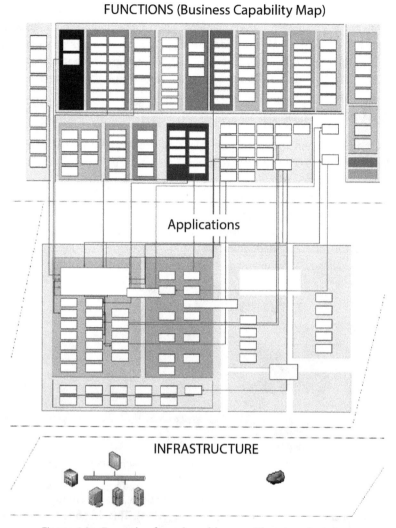

Figure 4.3 – Example of a real-world map with reduced content

It was applied there because, with 20 software applications or so, it was only a small part of the whole information system of this company. As for the business capability map, it was more exhaustive, but this is only because we needed the whole perimeter for another project. As you can see, only a few of these functions were related to the software applications under study (following the lines between layer 2 and layer 3).

Back on the giants' shoulders, I will simply recall that what Martin Fowler states about classes applies perfectly to functions and software applications mapping: Martin recommends not to draw all classes in UML diagrams, only important ones. He then goes on to explain that the main problem with diagrams is that people drawing them try to make them comprehensive. Diagrams should help us understand concise and clear information, while only the code should be the source of comprehensive information.

Finding the atomic actions

Since extracting the monolith at once and changing it for a new application was not possible (remember, "No Big Bang... ever"), we had to devise a way to progressively extract functions from the application and migrate them, step by step, and with limited impact on new, modern implementations. But in which order? This is where the map will help, showing the modules and their dependencies. From here on, I will use arbitrary schemas to better explain the approach, even if they deviate from what happened in this project. Let's imagine that the three important features we need to "rescue" are based on five software modules with the following dependencies (beware, there are functional implementations – links between layer 2 and layer 3, and technical couplings – links inside layer 3):

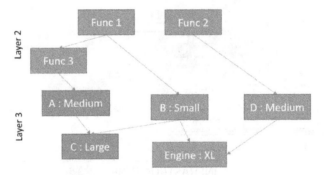

Figure 4.4 – Simple example of functions and their implementations

Once we've done this, we can use the information provided to draw two chronological approaches (this is where the time relationship is the most present in this section). This first one could be, for example, to try and get the first functions out as quickly as possible (time-to-market strategy):

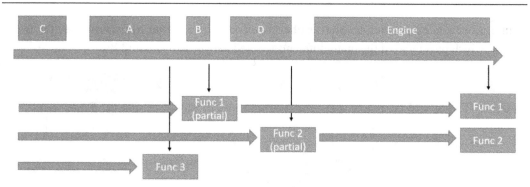

Figure 4.5 – Scenario with time-to-market priority

In this case, the duty of rewriting the engine that features (or functions) 1 and 2 eventually depends on is left for later, but at least function 3 is quickly available in its new implementation. The problem with this approach is that, since it leaves the technical debt for later, the development process will remain slow and features will be harder to release until the end of the project.

This calls for another approach where technical debt will be addressed first. The advantage is that features will flow quickly in the future. However, the shortcoming of this alternative approach is that it will take some time before we see the result (and as explained previously, this is where you'd better have a strong business case to convince the stakeholders to finance this investment). This second approach can be seen in the following diagram:

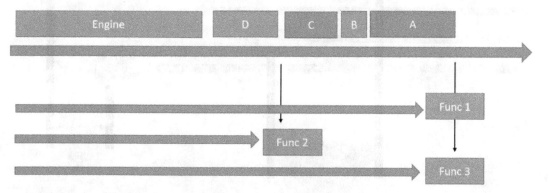

Figure 4.6 – Scenario with TCO priority

To sum up the advantages and drawbacks of each approach, it is better to superimpose the two diagrams, which puts forward the main differences, noted here by letters A and B:

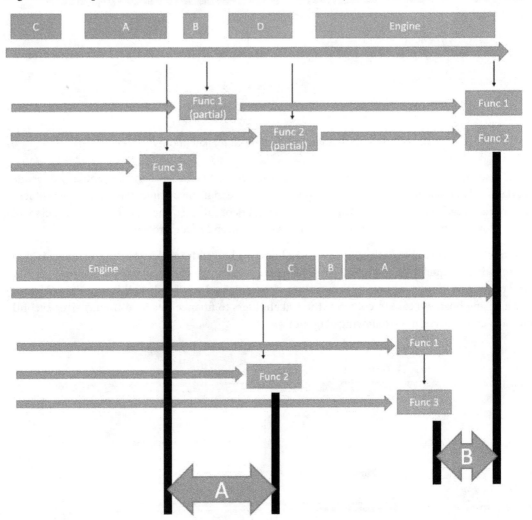

Figure 4.7 – Graphical difference between the two scenarios

The delta marked A shows the difference between **initial** time-to-market. This happens to be an important criterion in lots of businesses because customers need to be truly convinced that you are going forward. Whatever the trusty relationship you have with your customers – or internal users, by the way – it is hard to let them without any demonstration of what you have been doing for months, and business owners know it. In the first scenario, feature 3 arrives sooner on the market, and may then participate in the financial investment for the rest of the project. In the second one, the first feature to be released not only arrives later but is not the same, which can make a great difference depending on what is most important for the users.

The delta marked B shows the difference in total time spent on the project: while scenario number 2 shows the first results less quickly than scenario number 1, it solves more technical debt at the beginning of the project, which will make for easier and quicker developments for the remaining time. This is something that should be taken into account because a development team costs a lot of money. Depending on the complexity of the project, this delta may become very important (note that in the preceding diagram, the scales are completely arbitrary and do not represent anything representative as it depends on your project).

Now that we have established the two basic scenarios, we will dive into something a bit more realistic.

Prioritizing the actions based on business criteria

It is a used trick, but presenting two extreme alternatives often helps in getting stakeholders to choose an intermediate approach. There are good chances that the decision will not be between the two extreme approaches that have been explained previously but for a compromise somewhere in the middle. But then again, how do you adjust the cursor? What criteria can you use for this?

Some criteria immediately come to mind for any business-savvy software architect, namely gross income/ turnover and profitability rate. Often, I let the third criterion be something vague that stakeholders evaluate in terms of importance to the strategy of the company (they know more about this than any IT person). What is important at this point of the project is that these criteria should be evaluated fast, in a "planning poker" mode, and that they remain limited in number. I use levels of one to three stars for each of the criteria, and no more than three criteria as a whole. This is generally enough to find out what the best scenario is.

To come back to something a bit more realistic, here is an example of such a decision table that I have created with another company I advised (some entries have been deleted or renamed as they would have revealed too much):

Projet	# processes improved	Impact on productivity	Impact on growth	Project size	Delay	Risks
HA on IP Phone	***	**	*	M + costs	6 months	Low
Changing MDM governance tool	*	***	*	XL	3 years	If technology changes
Suppliers interop streams (4 connectors)	*	***	*	M + costs	fin 2021	Low
Increase robustess on production streams	***	**	*	L	6 months	Depends on competencies
				XL	12 months	Average
Logistics BI	*	*	*	S	6 months	Needs an expert in [software]
				M	18 months	Low
Transport optimization reporting	*	**	*	S	6 months	Needs an expert in [software]
				M	18 months	Low
Connector with supplier [X]	**	*	**	S + costs	April 2021	SLA + needs security audit
Webapp to follow transport	***	*	***	M	Mid 2021	Delays
Cash management tool	*	**	*	S + costs	6 months	Financial risk
Automatic lettering for customer billing	***	**	*	S + costs	6 months	Low
Manufacturing Execution System	*	***	*	L + hardware	6 to 12 months	Low (monitoring?)
KYC project	***	***	**	L	Mid 2021	Perimeter is still blurry

Figure 4.8 – Prioritization of projects through the chosen criteria

Interestingly, in this case, you will note that the criteria were not the "standard" ones proposed previously. Here, we have the following:

- The number of processes improved (in layer 1 of the map)
- The estimated impact on overall productivity (this was an industrial production company)
- The estimated impact on growth (they were operated on a rapidly consolidating market, where small companies were bought by bigger ones, so quickly growing was of utmost importance to flourish)

Also, a column was added to the traditional estimated weight of the project and delay estimates, namely for the impact and risks associated with the project (which is a sound approach and should be done for any such projects). To stress the fact that ease of elaboration of such projects is much more important than completeness and exact respect of the methodology, here is what the table initially resembled, after the 2 hours of analysis directly on a whiteboard.

You can find this whiteboard image on GitHub: `https://github.com/PacktPublishing/Enterprise-Architecture-with-.NET/blob/main/Whiteboard%20image.jpeg`

A few last words on time, semantics, and alignment

Why a complete chapter on the importance of time in software development? You might have wondered this and I hope that I have convinced you, not of the importance of time, which is a given in any project, but rather of the fact that lots of problems in software development and information systems design should be watched across the prism of time management.

It often happens that, taking it for granted as it passes by without anything we can do about it, we forget about time in our activities of design, not taking into account the perspective of the future. How many information systems were born ill because they were designed to solve today's problems, without thinking of how business will evolve in the future? Technical evolution, sure, we can handle them, and we do more than we should sometimes by preparing the system for the next framework, the future technology, and so on. But this is not the issue! The important thing is the business domain functions: they are changing as well and the information system has got to remain aligned with the business, not to some technical evolution that may be a fad in a few years.

Time is also often forgotten when we tend to analyze concepts as stable, simply because we do not use a broad enough time spectrum to explain them. Lots of concepts evolve while seemingly being extremely stable. In *Chapter 3*, I explained how the concept of "customer" is a business rule and not a stable entity: a customer can be anyone who bought something in the past 12 months for commerce while being anyone who has done so in the past 24 months for marketing. And maybe maintenance will have a list of customers based on who has got a running guarantee. These rules, like any business rules, can change in time. At some point, maybe the big boss will get tired of commerce and marketing not communicating the same numbers and evolution rate of customers and will force them to adopt a common definition; maybe the guarantee duration will change, which will affect the maintenance list of customers. Who knows? One thing is sure: if you have not considered time, you will have trouble.

A small anecdote – that I hope you will find revealing of the importance of time – to conclude this chapter: I happened to advise an organization that, through fusion with another one, had to change their logo. Since these organizations were subject to political changes after elections, the people there wanted to make it easier to adjust their logo in the future because this operation was extremely long and boring: the logo had to be changed in every message template from every application of the information system that would produce documents, and we were talking about thousands of templates and months of work to adjust the logo everywhere. The initial approach was to propagate a server share path everywhere: this way, changing the logo would automatically happen everywhere when the files were modified. Luckily, we had time to think about it a bit more and it was decided to switch the approach to a URL exposing the logo resource in different formats through content negotiation, and using an additional URL query parameter to indicate the reference time for which the logo was to be sent. This way, most applications who simply created documents along the way did not have to care about passing this parameter and would get the latest logo bitmap resource; but for the few legally-constrained applications that had to be able to fusion a letter at a given time in the past with the exact pixel-perfect form, it remained possible to do so without the new logo appearing on a old letter, which would have cause trouble. Today, this organization has equipped itself with an archive-capable content management solution, which now solves the problem in a much better way. Technical evolution took care of the functional need in a new way. Again, it was only a question of time!

Here's one last note about the importance of time and technical debt in information systems: there can be things even worse than technical debt, in what we could call "functional debt" or, more precisely, "semantics debt." This will be the subject of *Chapter 9*. But for now, we are going to end the theoretical part of this book with something that is extremely theoretical: a perfect information system!

Summary

In this chapter, we have shown how time influences everything in the analysis of an information system since it is a living entity that evolves throughout its use. Time is so current in our daily lives that it is often forgotten when structuring an information system, but it is the key to designing it in such a way that it will stand the test of time and have a reduced TCO.

The Agile approach is one of the first methods that dealt with this time-related reality, and it has radically changed how software is created and handled. The same approach can be applied to systems as a whole set of applications working together, which means the technical debt can be handled globally and kept under control. This notion of technical debt is not well named and, as a future chapter will show, semantics is very important, so I would recommend keeping this in mind at all times and, if possible, adjusting the naming, as was proposed in this chapter.

The rest of this book contains lots of recipes to help you reduce this technical debt or at least keep it to an acceptable level. But in the next chapter, we will try and imagine a perfect information system, with little or no debt or coupling.

5

A Utopic Perfect IT System

Remember the joke in *Chapter 4* about the perfect information system being formed by a computer, a human, and a dog? There was some truth in this because humans change all the time (their minds, their way of doing, the rules they follow in business, what they want to buy, and so on), while computers are happy with repeating, stable, well-defined tasks. Of course, the two do not mix well, hence the presence of the dog to prevent the human from making a mess of the computer work.

But who created the computer first? Human, of course. So, the joke was not about the fact that humans should not touch computers at all, but rather create them and then let them do the work without changing anything afterward. Of course, this would only be possible if the first attempt was, through some kind of a miracle, perfect.

In this chapter, we will cover the following topics:

- The concept of an ideal system
- Data management in the ideal system
- Rules management in the ideal system
- Processes management in the ideal system
- How close can we get to this utopic system?

This chapter will describe what would resemble this 100% ideal information system, with an ability to adapt to the ever-changing business changes and still perform in terms of speed and robustness, as well as energy efficiency. Such a system was imagined by Dominique Vauquier in his fundamental book *Le système d'information durable - la refonte du SI avec SOA*. The three main entities constituting the ideal information system are **master data management (MDM)**, **business rules management system (BRMS)**, and **business process modeling** (we are going to explain these in more detail shortly). The next few sections will explain them one by one, and this chapter will end with an analysis of how this utopic system could be created in reality and what would prevent this.

The concept of an ideal IT system

Before we detail the different parts of such an ideal system, I will explain a bit more about the usefulness of such a model. When things are a bit clearer about the goal we want to reach, we will detail the different parts of the concept and try to make them more concrete.

The use for an ideal structure

"Ideal" or "utopic" are terms that engineers generally have a strange relationship with. Though much of their thinking is done inside hypothetical contexts to make an advance in theory, they know that practice will always be quite a long shot from the theory and, depending on how they are versed between theory and practice, engineers can easily fall into many traps:

- Staying purely in theory risks that their thinking is never applied in the real world, which ends up in a loss of time and also an impossibility of truly knowing the value of their work.

- Not having enough theoretical background to make advances in their field; remaining purely in the practice of the job might help refine some areas but seldom bring strong, domain-changing, revolutionary ideas.

- The worst of all: switching between the two but never making real links between them that enrich the practice from the theory and confirm the theory from the practice. This capacity of bringing the two together is, in my opinion, what makes the best engineering structures, may they be companies or even countries that teach and organize this capacity at the highest level.

This is why there is a place in such a book, otherwise extremely turned toward practices with lots of returns on experience from many real-world information systems, for a bit of theory and ideal thinking.

The origin of the design

The structure of an ideal system, as we are going to describe here, was first coined in a *Le système d'information durable - la refonte du SI avec SOA*, by Dominique Vauquier, one of the best-known enterprise architects in France, and proponent of the PRAXEME method. I was lucky enough to receive training more than 10 years ago and its explanations were so clear that I immediately bought his books, which quickly became some of my references for my consulting work in information system industrialization.

The principles behind this idea were that, knowing lots of information systems, it was possible to describe a kind of meta-system that would encompass all the necessary features with only a few generic modules that could be customized to implement the business-related aspects. Each of these three modules would be extremely general and not related at all to any particular business, which explains why there is so little in the whole system. These three areas are as follows:

- Storing data and making it available (roughly speaking, this is what gets done by databases)

- Executing business rules to make decisions (this would be where software applications consuming the data come into the game)

- Implementing complex processes by orchestrating many small tasks (though very approximative, this is what making applications work together is about)

Dominique Vauquier was indeed able to point to any business feature as a composition of these three different technical functions. After working with many industrial information systems in small and large companies, he considered that the three modules together would be able to handle every imaginable feature of any given business domain. Though this is not a structure that has been largely used for now, it may be just a very anticipated glimpse at what would become industrialized and well-controlled information systems a few decades from now. At least, this is – from my humble point of view – the best candidate for such a structure with so much ahead-thinking.

Using a vision to define a target

Why should we talk about an architecture that is so futuristic and that we do not have any proof of its capability to be applied in practice? Can this lead to overthinking or over-engineering? If you ask yourself these questions, you are very right to do so, and this chapter is not about pushing you to use this tool for your next information system architecture.

Just like any pattern or architecture tool, it is principally there to give you some ideas and, if some parts of it are fit for your system, it will be great that it helps you go forward and accelerate your thinking.

But having an "ideal" or "utopic" vision can also be helpful to give a global orientation to your realization. Remember the metaphor of Agile development, where we do not shoot an arrow but walk with it toward the target and plant the arrow with our hands when the target is reached? Well, to do so, we agreed that we need to know what the target is like to find and reach it. And sometimes, the target shape can be hard to imagine. Sure, a business-oriented mind will give you the best idea and you should remain functionally-focused. But what happens if the business ideas are quite blurry? Maybe a "soft," adaptable information system would allow you to start working while things get a little clearer. Starting building the system will even help the business owners think better about what they want. And if you have an "ideal" system that can very quickly adapt to their minds constructing the business idea, it may be the kind of system you need for such a project.

Of course, there will be an extra cost of being completely generic (in performance but also time for initial setup), but if this is the kind of situation you are in, the capacity to adapt to a moving business target by simply customizing parameters of the information system may largely overcome these drawbacks. And nothing prevents you, once the business vision is settled, from re-implementing some now-stable parts with specific, optimized bits of code, where needed.

Now that the vision is clear, we are going to dive deeper into the three parts that constitute this ideal IT system, namely the data management module, the rules management module, and, finally, the process management module.

Data management in the ideal system

MDM is the very first part of the ideal system. Before talking about how the data is processed and used, it is logical to explain what form the data will take and how it will be managed. This is why we will explain the concept of MDM in a utopic system before discussing BRMS and BPM.

Data is the blood of information systems

Data is the most important component of any information system. Some people may even think they are purely about data since the very term "data" has been so hyped in the past decade.

Admittedly, it is hard to imagine an information system without at least one database to store the data, since it would mean there is no knowledge of any business event that stays there. It is possible to imagine some pieces of information that are completely transient and not stored, but a whole system made of such particular cases sounds impossible to fathom.

This means that data should be taken care of first and foremost, and this is what the first module of the ideal generic information system is about. The feature is called MDM. The term *master*, in this case, refers to the most important data in a system, those used by most of the participants in the system. But in this particular case of a utopic system, every single piece of data would be considered as such and placed in a single module that deals with this responsibility of "MDM."

Data as the 21st-century oil – really?

Before we dive a bit more into what would be such a module, a small additional note on the importance of data: you may heard the expression "oil of the 21st century" regarding data. This is to stress the point that data has become such an important part of commercial organizations nowadays that it can be compared to oil, which has brought a huge part of the industry improvements in the 20th century.

Data is used by many industrial companies to follow their machines, optimize production, and relate it to selling and stocks, in short virtualizing the industrial process in a "digital twin" to better control it. But for some industries, data may even be the raw material itself, and lots of digital-native companies nowadays make their money purely out of data collection, refining, and reselling, particularly for advertisement activities.

Oil exploitation, which has happened since the 20th century, has allowed industrial production to be accelerated dramatically in almost every domain (heavy industries, trains, fertilizers, agriculture machinery, and so on) but also brought to life many of the biggest companies to exploit and produce oil itself. This comparison is fair, although care should be taken in keeping a few important differences in mind.

First, the equivalent of the oil-cracking refinery column has not been invented yet for data, or at least not in its standard form. Sure, some business intelligence tools and big data approaches can help in some cases, but the rate of failure of data projects (estimated between 70 and 80%, depending on the studies) shows that we are still far from industrialization in oil-refinery plants. To this day, data remains partly, for most companies, such as oil, was for land owners at the end of the 19th century: pollution. Sure, there is value in it, but it is completely out of exploitation, and there is no standard way of extracting this value.

Let's also not forget that data remains based on actual oil: data centers consume a huge amount of energy (soon, 8% of the energy on the planet will be taken for digital uses, which is more than air travel and will soon reach car transportation). And though some data center owners pretend to use renewable energies (but just buy compensating activities, which does not reduce global oil consumption), most of them are still heavily dependent on oil for energy. In addition, the production of servers and network hardware, the transportation used for them, and the exploitation functions are large consumers of energy and, eventually, oil.

Therefore, it is very important to keep in mind that data may indeed become the oil of the next century, but we are in this situation at the moment. The data management approach described in this chapter may help you reach it, particularly by using governance to improve data quality, which remains the number-one cause hindering data exploitation.

A really "know it all" system

I will not go into too much detail in this chapter since MDM is the subject of a dedicated chapter later in this book, and it will be explained together with a real example, which will hopefully make everything easier. For now, I will just propose a comparison of the MDM with a "know-it-all all" system. Indeed, a well-made MDM will not only know the state of entities at present but will know the whole story of each entity and remember their different states and modifications, from cradle (creation) to grave (archiving or deletion). Due to this, **persistence** is the main responsibility of an MDM module.

However, there are some other responsibilities:

- **Ease of research**: This goes with persistence because keeping data without making it available afterward would not make much sense. This particular responsibility might be delegated in its implementation, typically when the MDM application uses an indexation engine, for example (full-text search is a complicated matter that justifies cutting a responsibility into several more atomic ones and trusting dedicated modules with each).

- **Origin of reporting /business intelligence**: Just like data search, reporting is an important responsibility that MDM modules can trust other implementations with, for example, a data lake.

- **Ability to handle multiple versions of entities**: As explained previously, a true MDM should know **every** version of a data entity, not only its "latest" state (the quotes are there to stress the point that this "latest" concept is something of a pain in lots of systems, particularly when they are distributed, because it depends on the consistency of data, and thus transactions, optimistic/ pessimistic locks, and so on).

- **Capacity to manage validation for attribute values**: Though this might be a shared responsibility with the BRMS, which we will discuss later in this chapter, a small amount of validation can be done by the MDM itself to ensure its basic consistency. This should not be confused with data correctness, which depends on its purpose (a single piece of data can be fit for business use but not for another one) and can vary well in time, depending on the data versions but also the changes of business rules.

- **Business rules**: Some entry-level business rules, which only use data that is present inside the MDM system, can also be added to the MDM, though most of them are concerned with validation, as stated previously.

- **Handling of history**: Alongside storing successive versions of data, the MDM has to be able to make it possible to easily retrieve these versions depending on the time targeted, browse through the history, find who has made which type of change on the data, and more.

If this doesn't seem clear, don't worry – the example provided later in this book of an MDM storing a person's data will help you understand how all these responsibilities are used and implemented.

Relationship with CQRS

If you are into data architecture, the separation of concerns between storing versioned modifications of the data and searching the different states of it might have been identified by you as the principle of **command and query responsibility segregation (CQRS)**. We are indeed talking about the same principles, even if CQRS goes way further than just separating the two responsibilities and proposes a technical approach on how to make them work together.

CQRS is too big of a subject for it to be treated in this book together with all the other subjects we have to explain, but it perfectly fits the separation of responsibilities of an MDM module, and I strongly encourage you to use this approach when creating MDM implementations. The example we'll cover in the next few parts of this book will use a CQRS approach, though in a very limited and simplified implementation. Again, this is an edition choice as adding a full Kafka engine would have made it difficult to concentrate on the application itself and would get us too far from the subject of this book.

The need for data quality

Coming back to the oil metaphor, having sand in it is a real problem because oil has to be purified enough to enter the plant. Otherwise, not only the quality of the output products would not be as good, but this could also damage this extremely expensive piece of industrial equipment that the refinery column is in. This is why sweet crude oil is one of the best types of oil, and also the reason why gas coming from bituminous sands is so costly to refine (in addition to a catastrophe for our environment).

In IT, there is a strong equivalence to data as clean data is a great product that will allow precise reporting on the business activity, quick insights on problems that may appear, and overall better control of the company. Discuss this with anybody in the industry and everyone will agree that the quality of data is of paramount importance... Yet, it is estimated that more than half of the job of a data scientist is to clean data. I have seen several companies where this proportion dangerously increases to around 80%. I say "dangerously" for many reasons:

- First, paying a high-salary profile to realize such low-intelligence tasks is a waste of money

- It is also generally a waste of time as good data scientists will generally leave the job in the next 6 months if they have to work more on cleaning and assembling the data than "making it talk"

- Finally, it shows a problem up the pipe because the IT system that sends this dirty data has not been correctly designed, which means that there are good chances that these reporting issues are not the only ones

How come everybody agrees that data cleanliness is important but the situation remains so awful? Well, it just so happens that data management is generally not about the technical part (which is quite easy, with the number of tools we have nowadays) but about the organizational part: who is responsible for which data? Who has the right to collect it and update it? Which group decides on how data is cut and which data goes into this or this service responsibility? All these subjects are a part of what is called **data governance**, which, despite its huge importance, is mostly not dealt with by many companies, though they have all the technical tools to store data. Again, this is the same root cause as usual: technology has never solved an organizational problem, but editors are so good at making you believe this… and, as a company owner, you want so much for it to be true! But no, you will have to do the uninteresting job of data classification, finding data owners and data stewards, implementing regular meetings to follow up on the decided data processes, and more. Creating your MDM module now is precisely the right time to do so as its success heavily depends on these actions.

Designing a generic "referential"

There will almost always be an internal designation of the software application that stores the main entities of an organization, for example, products. Most of the time, people will call this by the name of the technology – for example, the "products database" or the "products Excel file" – but this is a bad habit as it couples the functional concept with a technical, software, concept. Not only can these two evolve separately (and the "products file" becomes the "products database") but they should be able to do so as much as possible, and thus be kept separate.

The "products list" is already a bit better from this point of view, but my favorite (because it is very close to an equivalent word in French, which is my native language) is to call this a product "referential." This name bears the important notion of it assuming a status of reference for the data contained within while being also possible to see it as a referential in space, which comes back to the idea of a map of the information system, just like we would have a geographical map, with coordinates inside a referential.

> **Important note**
>
> The naming could be much worse, as I already cited anecdotally in *Chapter 3*, with one of my customers calling the article's referential simply "Serge's file." This led to a lot of confusion as it was hiding the uttermost importance of keeping this file up-to-date, which was not even Serge's job officially!

Most of the time, so-called "referentials" are dedicated to a single type of entity (products, sales, customers, and so on) The idea of the utopic system described in this chapter is to have a single piece of software for any type of entity. We won't dive too deep into the utopia in the rest of this book and will maintain a more standard approach with one implementation of referential for a given business domain. This also allows us to adopt the best technology for each case (the "best of bread" approach), rather than a "one size fits all" approach, which is interesting in theory but extremely complex to make work in the real world.

Choices of implementation for an MDM

Love it or hate it, Excel remains the simplest MDM you can have, and in many small companies, a well-organized process using Excel can already go a long way, provided you have removed the most important shortcomings of a good organization (a centralized file shared with rights for different users rather than copies everywhere, strict organization of data quality, and so on).

Of course, there will be a level of complexity that will necessitate a dedicated implementation. That sounds crazy, but there are very few software applications dedicated to MDM. There is an offer from Microsoft based on SQL Server called Master Data Services, but its use sounds very limited, at least in my area of knowledge. Following a discussion with Microsoft, the product is indeed abandoned, and the successor is a partner product called **Profisee** working together with **Microsoft Purview** to provide data governance functions (`https://learn.microsoft.com/en-us/azure/architecture/reference-architectures/data/profisee-master-data-management-purview`). Semarchy (`https://www.semarchy.com`) sounds like a new and interesting approach to integrated MDM but I have not given it a sufficient try to recommend it yet.

I have seen some companies using headless CMS systems such as Strapi, Sanity, Cockpit, Prisma, and others to create backends that can serve as entry-level MDM. But this generally lacks a good data versioning system and implementation of governance remains the job of the integrator. All in all, this is a very technology-based approach and, as stated previously, an MDM is much more than simply storing data.

This is roughly it for an off-the-shelf approach, and admittedly this does not go a long way. Every correct implementation of an MDM I have seen to this day has been a dedicated development. Some business applications, such as ERP, sometimes have a good referential for products or customers, but most of them miss the alignment on business.

> **Tip**
>
> It may sound logical that an ERP may not be business-aligned as it tries to be generic to many business domains. Nonetheless, when you see a top five ERP system proposing two separate domains for customers and suppliers, ignoring the possible duplication of data for most of the existing companies, this shows that the problem is not only the variety of businesses, but the very approach of these ERP companies, which consider that modeling the business is 100% the customer's responsibility, whereas it should be a shared responsibility based on written, forward-compatible, standards. But this would go against vendor lock-in, which is not in their interest, and this is why the only way forward is for the customers to choose an implementation that accepts standards for business entity representations, including developing these standards in the (rare) cases where they do not exist.

What remains – and represents the vast majority of running MDM systems – are dedicated applications coupled with a database (and sometimes an indexation engine, and even a data lake, as explained previously). They are custom-developed for a given business use by internal IT or external software companies. This is a huge waste of time and money if you think globally as many companies have needs that are extremely close to each other. However, this is just how it is for now due to the lack of business standards used in the software industry.

Now for the second part of the ideal system: after talking about data, we need to discuss business rules.

Rules management in the ideal system

Business rules are predicates that apply to data to help implement real-world decisions made that are necessary for the business. For example, you may state that a customer cannot be sent a product so long as they have not checked their bank coordinates. This is a business rule as it can be expressed without any signs of software implementation: this could be done manually by someone in the plant checking with accounting via phone that the rule has been respected via p before sending the package to the customer.

Rules as the nervous network of the information system

If data is the blood of the information system, business rules are its nervous and muscular networks: they use the blood to realize some activities in a given way. A business rule in our metaphor could be that "if you feel a burning sensation on your finger, the arm must retract." This rule is implemented in our spinal reflex system, which uses the sensors in its fingers to send the information/data through the nerves, resulting in the muscles from the concerned arm being unable to be controlled.

In IT, the implementation of business rules is mostly what are called applications. This is generally where business rules are contained. In our example, there must be some code somewhere in the ERP that alerts the package preparation system that the customer has not passed the condition of having a verified way of payment. The reaction of the packaging system may be that this order will not be processed, or maybe that it will be processed in advance but the resulting physical package will be retained before delivery. The actual implementation is the responsibility of each module, but the business rule remains the same: "No customer delivered with their orders so long as we have not verified we can get paid,"

Business rules also contain all the calculations, from the most complex to the most mundane, that you will perform in the information system. Let me go back on this because there is something to say about the right separation of responsibility: **no client (GUI or application calling an API) should deal with business rules, which always have to be implemented server-side**. This is certainly something that you know about and sounds logical: after all, it is obvious that such important bits of functional reasoning must be centralized to ensure that they are applied everywhere the same way. But look at any client for a few minutes and you will find loads of functional decisions that are made locally, sometimes with the best reasoning. For example, think about calculating a net amount from the gross amount and the VAT rate. Sure, the VAT rate does not change very often in an application's lifetime, and the way the calculation is made is itself very stable, so there should not be any big issue with trusting the client to do this computing, and we are getting better performance since we avoid a roundtrip to a server.

Alright, but you can always imagine ways that the business rule is going to change. As already imagined, the VAT rate can change. Also, if we are talking about net amount calculation for a multi-line order, you must handle multiple VAT rates in a single entity. The question boils down to risk management: if you know that you will never hit these particular cases, that's alright. If you doubt the possibility they will appear in the future, you should first ask your product owners about it. If they are not sure about it, you will have to compare the costs of different approaches:

- You can hard-code the business rule in all clients and hope it doesn't evolve. If there are not that many of them and you can easily upgrade them, no big deal.

- You can prepare for a possible change by making the rate a parameter of the applications you use (when they support it); this will already be better easier in case of adjustment, without a huge upfront cost.

- If you foresee changes in the business rules in the coming years and you know adjusting the software will be complicated, maybe you should put the rule in a server-side, centralized, application. Sure, it will be a bit slower, but client-side cache can help and you will be future-proof.

- If you are in a situation where lots of business changes are going to happen, and many business rules are going to be dealt with, then you'll reach the point where investing in a centralized piece of software dedicated to managing your business rules makes sense.

BRMS as a dedicated tool for business rule handling

Now, let's concentrate on this last hypothesis, where you need what is called a **BRMS**. Again, it is very important to state that this is not a module you will need in every information system and the simplest ones do not justify such an installation effort. But beware that most information systems start simple and become more complex. The real difficulty is that, once you are in production, it is going to be harder to replace existing rules with a BRMS customization since all calls will have to be changed if they have not been unified behind an API contract. This is why this utopic information system makes sense: it shows what you should do (and does not hide the size of the investment) if you are serious with your information system and do not want to take any bets on it. If you know from the start it is going to grow and your activity will become industrial, then you should start with something that might seem overly complex and expensive at first but that will pay back hugely in the future. If you are not convinced about this, simply read all the returns of experience of startups that have failed because they have not invested in their information system, though their idea was customer-approved, their pricing was fine, and they had a market: this is one of the most common reasons for startups to fail, but also bigger companies (the latter, though, generally do not provide any feedback).

Common implementations of business rules

How would you go about implementing a BRMS? It may make you smile but, again, Excel is the simplest implementation for a BRMS, most of the time without people knowing it. One accountant will certainly tell you (if you ask) that they do not use any BRMS for their job. But at the same time, they will boast about a hand-crafted Excel workbook with all the calculations (and thus business rules, whether regulatory or company-wise) they need in their day-to-day job as well as for the monthly closing period. What is this if not a BRMS? In fact, due to the relatively rare use of a "true" dedicated BRMS, it simply happens that business rules are contained in Excel files for most small companies.

Within larger companies, there are more business-line applications, such as ERP, or even bespoke software when the companies have internal software development capacity. In this case, the proportion of business rules integrated into the application rises, but this is not necessarily for the best. Why do lots of business-oriented people love Excel spreadsheets despite their shortcomings? Simply because they have full control over it and do not depend on the IT department to make changes, add functions, and so on. Start implementing a business rule inside some application and the coupling problem immediately pops up:

- Is it possible to change the rule without recompiling, testing, and deploying the application again?

- If the rules can be customized, is it possible to do so via a functional operator or do you need the IT to implement the change (or in the best case, train them)?

- How do we deal with rules that must change at a particular date and time (for example January 1 at midnight for lots of accounting-related or business-wide reporting rules)?

- Is it possible to have two concurrent versions of the business rule, or are we obliged to stop the software at the end of the year to make sure nothing is done before somebody changes the rule at the beginning of the new year (do not laugh: it happens very often…)?

Despite these potential problems, implementing business rules in dedicated business applications is already a nice step forward, particularly if this is done server-side, preferably as the implementation of a well-documented, contract-first API. This solution is less prone to errors than Excel spreadsheets, which might be disseminated across the organization with different implementations of rules, resulting in a mess. It also tends to provide a first level of governance of the business rules since different services are responsible for "their" software applications, and thus for the business rules implemented (or customized) in them.

Still, these implementations are not representative of a BRMS in the sense of our utopic system, namely a unique one that would contain all the business rules of the company. And if you want to go all the way to a future-proof information system, you will rather use something such as Drools (an open source package, declined in JBoss Rules by Red Hat) or IBM Operational Decision Manager. These pieces of software will generally provide the following features:

- The execution of business rules, which provides what is called a decision engine (**DMN** stands for **Decision Model and Notation** and is an accompanying standard of BPMN). If you want to know more about this important bit of a BRMS, I recommend the excellent documentation page from Drools, available at `https://www.drools.org/learn/dmn.html`.

- Storage for the definition of these business rules, which sounds obvious but isn't when you add the requirement of storing the whole version history of the business rules.

- Use of decision tables, links between events and actions, and, in the most sophisticated cases, inference engines that are going to mix many different rules to find additional conclusions (an overly simplified example would be that "all purchases need an invoice" and "an invoice must include the buyer's address in the header," to "all purchases need the address of the buyer").

- Some of them provide a graphical editor for non-technical people to be able to adjust rules by themselves. Though editors claim anyone can use them, they remain quite technical and require some training.

- Together with this, a sandbox is very useful for people to test their business rules adjustments before putting them into production.

- Depending on the performance needed and the variation of the calls, a cache mechanism can be a good addition to a BRMS.

> **Tip**
> Open Policy Agent is a BRMS that is dedicated to authorization rules. We will show its use in *Chapter 12*.

Finally, after talking about data and business rules, we are going to add the third and last module to complete our ideal IT system.

Processes management in the ideal system

Our ideal system handles data and can make decisions on it with business rules, but what makes it tick? In *Chapter 4*, we stressed that everything is a question of time, but nothing has the role of putting the data and rules in controlled, time-based, motion. This will be the role of the process engine.

Processes as the brain of the information system

Coming back to our metaphor of the human body, you may have guessed that business process management, as the module that orchestrates the actions of applications containing the business rules over the data, can be associated with the control tower that is the brain. As our brain coordinates the actions of each muscle together with receiving the signals coming from our five senses to achieve complex results such as catching a ball, typing on a keyboard, or dressing, the processes describe the complex tasks that must be accomplished by each part of the information system to reach a global objective, and the BPM engine coordinates these tasks, effectively calling the modules in charge (or rather, the API, which themselves are implemented by technical module; remember the important point we explained in *Chapter 3*?).

If the utopic information system is made of one MDM and one BRMS that contain all the data and all the business rules for each of the business domains, the BPM is the module that makes everything work together, by interoperating all of the functions.

Implementations of a BPM

Before talking about technical implementations (and this should be a reflex by now), we have to try and find an accepted standard for our feature. Luckily, an indisputable one exists that we have talked about previously: BPMN. So, if you are looking for an evolutive way of executing processes, where the engine can be completely decoupled from the design, you must use BPMN. I have tested editing BPMN 2.0 on a given editor's GUI and executing on another editor's execution engine, and it works well, which says something about the maturity of the norm.

Of course, not all information systems need the kind of sophistication we are talking about, but all of them use a central orchestration system at some point, even if large parts of it can be manual and processes might not even be designed in BPMN, but only known in their heads by the participants of the information system (again, this does not mean that everything is based on computers). And guess what? When we do not need any complexity at all, we will find our old friend Excel!

Tip

At this point, I know you must be convinced I am an absolute Excel fan. However, I'm not and I have seen my fair share of huge functional problems because of using Excel for shared data, not controlling its spreading in the organization, and more. I am fully aware of the way it prevents people from working together by making it so easy for almost everyone to have a little part of the information system. But at the same time, I am not in favor of forgetting everything spreadsheets have brought to IT and the empowerment of business specialists. Also, it is not my fault that most information systems nowadays are so unsophisticated! So if you have to use a crude tool to do the job… so be it! And Excel will long remain the "Swiss Army knife" of information systems. You will be better off with a complete toolbox but if you can only have one tool, take this knife rather than the circlip-removing tongs or some other exotic and highly-specialized utensil. Going back to the software, better alternatives for MDM, BRMS, and BPM exist today than an Excel spreadsheet. However, if you are aware of Excel's drawbacks and control them, its versatility and ease of use will speed you up in many parts of the information system. In particular, though MDM and BPM are much better outside Excel, I have often set up some "low-quality BRMS" by putting complex calculations in Excel and plugging the spreadsheet into APIs with great business success. Yes, this is somewhat of a hack, but so long as it is hidden behind an API, there is no problem with a temporary, quick, and dirty implementation… and the business users LOVED it!

How close can we get to this utopic system?

So far, we have described the contents of an ideal information system, as described by Dominique Vauquier in his seminal book. This assembly of MDM + BRMS + BPM will also be found in some other sources, such as the "sustainable information system" described by the SHIFT project in its approach to the digital transformation toward a more ecologically friendly IT industry, or the **Agility Chain Management System (ACMS)** approach.

Of course, this is an ideal vision, so the question is not to use it as a blueprint that actual information systems should all be based on, but how close we can get to it when designing new information systems or improving existing ones.

Favoring contexts

Various context elements may favor the use of the approach discussed in this chapter:

- **A brand new information system**: If you are lucky enough to start from scratch with a newly-created organization, the legacy will not hold you back and this is a huge factor that can make an ideal information system come to life.

- **Degree of sophistication**: This will not be enough, of course, because as stated previously, there is a place for such an ideal system only when it is justified by business. If your information system has to be highly evolutive because business rules and markets are changing very often, data structures are unclear, and processes must be adapted quickly, then this is another factor that can favor the adoption of the architecture explained previously.

- **Investment approval**: Once again, this factor is not enough because the business may need such a highly sophisticated information system but the investors might not be aware of this. For lots of CEOs, IT is just a cost center. But if you understand that information systems have become the backbones of most industrial activities, you have one more positive factor.

- **Capacity of implementation**: Everything seems fine, but do not underestimate the difficulty of implementing such an information system. This approach is so different from what most professionals know that you will have difficulty finding people able to deploy it. You will have to convince them, train them, and adjust them to your way with very little existing feedback and return on experience.

Admittedly, these are a lot of factors and chances are you will not have them now. For people my age, there is even a chance that we'll never see such a perfect alignment that would make this utopic structure the most used one. But it exists – its advantages are obvious and the accelerating rate of changes and digital transformation of the industry will, without any doubt, make it a more and more observed option. A technical watch also shows that, in the past 3 to 5 years, the subject of MDM has advanced and some advanced companies have put them in place, together with an internal developer platform, bringing them some very interesting results. In short, we are at the very beginning of the trend, but the chances that the wave will grow are high.

Approaching the target, the Agile way

The fact that such ideal information systems are not mainstream yet and will not be the standard in the near future does not mean that they have no interest. It would be a bit like somebody working in an Agile team who would not want to know what the vision of the final software is, arguing that they are only interested in what will be done in the current sprint!

An ideal – a vision – is also there to provide direction, somewhere to guide our trip. Remember the metaphor of the arrow? We must know what the target looks like to locate it and reach it; otherwise, we are just randomly walking around, hoping for the best. In conclusion, please do not discard this concept because it is not immediately practical: knowing it can help you make choices in the future between different directions of your information system because you will feel closer to this long-term, ideal, direction that will bring you better value. Chances are you will never reach the ideal state, but hopefully, the direction taken will bear more value and flexibility to your information system. The state of information systems nowadays is so poor that having something above average is already a huge advantage in the era of digitalization of all industrial processes.

Summary

In this chapter, we presented what could be an ideal information system. We talked about all the problems that can arise in software industrial deployments, particularly concerning their evolution, so it was quite logical to show not only punctual solutions but also how the problem could be addressed globally. Of course, a fair part of this approach remains utopic, but this exercise is interesting in the fact that it shows that only three modules, if very well built, can serve any business needs. The separation between data, business rules, and business processes is something that should be kept in mind even if the information system you work with is far from being ideal.

This chapter was quite conceptual, but this concludes the first part of this book, which was meant to cover a lot of theoretical background. The next part of this book will cover more practical methods for architecture and, to be fully articulated between theory and practice (which is the main premise of this book), the third part will provide source code to show you how to implement the concepts and approaches of sound architecture that we have talked about. In particular, the three main modules that have been explained here will be shown in a sample application with a limited functional perimeter, but with all the sophistication needed to grow to a full-fledged information system if needed:

- The MDM will be based on a NoSQL database, with a back office server in ASP.NET that exposes API-designed contract-first

- A separate BRMS will be used for authorization management

- A BPM approach will be held to show how to make processes adjustable, with a low-code implementation

To summarize, in the next seven chapters, we will use battle-hardened methods to design the sample application by using examples that apply to parts of the use cases we will need. This, in the following five chapters, will be used to implement the sample application, module by module. Finally, the last three chapters will be about deploying and adjusting this sample application, just like if it were in production. As time passes by, adjustments need to be made to keep it aligned with the functional business needs.

Further reading

A sustainable information system - progressive refurbishing of the IS with SOA, by Pierre Bonnet/Jean-Michel Detavernier - Dominique Vauquier - Hermès / Lavoisier - November 2007 - ISBN-13: 978-2746218291

Part 2: Architecture Frameworks and Methods

After *Part 1*, which was quite theoretical so it could be understood by readers without a technical background, *Part 2* of the book gets a bit more practical and explains how some principles can be applied at the architectural level of an information system to make it better. This is where we will talk about methods to design low-coupled services with a good alignment to business. General software principles will be covered, but also wider approaches such as the externalization of authorization management and business rules in general. The correct approach to managing data in a business-/IT-aligned information system will also be detailed.

This part includes the following chapters:

- *Chapter 6, SOLID Principles, from Code to Systems*
- *Chapter 7, C4 and Other Approaches*
- *Chapter 8, Service Orientation and APIs*
- *Chapter 9, Exploring Domain-Driven Design and Semantics*
- *Chapter 10, Master Data Management*
- *Chapter 11, Business Processes and Low Code*
- *Chapter 12, Externalization of Business Rules*
- *Chapter 13, Externalization of Authorization*

6

SOLID Principles, from Code to Systems

Starting from this chapter, we are going to step away from the theoretical part and, while we are not yet starting to code (this will start in *Chapter 13*), we will start applying the theory to designing a small information system made of several applications. We will decompose the different functions, show how they help produce business process outcomes and create the software behind these functions. To do so, we will design the different components and the API contracts for the services involved, and think of how the data should be designed and governed. And in *Chapter 13*, we will use all of this design phase to actually implement the sample information system.

Of course, this information system will be reduced in perimeter and complexity, but the exercise has been designed to include most of the important decisions that should be made. You will find strict responsibility separation, nice separation between processes and functions, decoupling of the service through APIs, standardization of the contracts, best-of-breed approach, adapting the software stack to the functions desired, independence between the software and the hardware, and lots of other principles.

In this chapter, we will start designing our demonstration system by thinking of the functions it should expose. To do so, we will use the **SOLID principles**, extending them to information systems. SOLID is the acronym composed of the first letters of the five essential principles of software development, which are the following:

- **Single responsibility** states that one module should only do one thing
- **Open/closed** differentiates between open to evolution and closed to modification
- **Liskov's** principle explains how substitution should work
- **Interface segregation** follows with how contracts should be strongly aligned to business functions
- Finally, **dependency inversion** deals with coupling and how it should be done, at the inverse of what seems natural in most cases

It happens that these principles, often applied to software applications, actually apply to every software system and constitute a great way to design their different modules. We will thus use them to design our sample information system. But first, we need to describe the business requirements for this system.

Describing the sample information system requirements

Before any kind of analysis, we will imagine what the system owner would want from it. Of course, as we explained, time is a very important constraint in information systems design, which has a long life, and we will simulate the fact that we do not know everything about the requirements at first. In particular, the last chapters of the book will simulate the fact that new requirements arise for the imaginary company owning the information system, and explain how the system will adapt to it. This point is particularly important because the main goal of this book is to show how a system should be created or adapted so that its evolution is simpler in time.

To make the exercise as realistic as possible, while keeping it simple for it to be contained in a single book, we will imagine the company, the users of the information system, their business, the data they manipulate, and so on. This is what we are going to do in this first section.

The company and its business

The company, which we will call **DemoEditor** for this purpose, would be an editing company that contracts with individual authors for the writing of books and then sells these books. We will imagine that this is quite a small company (less than 50 persons) and that its current information system is extremely reduced, mostly composed of a standard Office 365 organization providing them with email capability, basic SharePoint document management, an externalized website, and lots of internal functions being implemented through Excel workbooks.

While this remains a comfortable situation, because the information system has not turned into a spaghetti dish due to a long accumulation of point-to-point interoperations, degradation of legacy software applications, and so on, it still shows signs of inefficacy. The multiple copies of the Excel workbooks make it difficult for the employees to have a clear view of the pool of authors and the state of the writing of the books. Also, as the process of writing is not uniform, the director of the company complains about not being able to find clear statistics about global advances or delays in book delivery.

The business is mostly about finding the right subjects for the market, choosing the right author, following up on the writing of the books, and organizing the right sales process.

The users and actors of the information system

The 50 persons are, mostly, book editors. Then come the sales team, a bit of administration, and the director. For this simple example, we will consider that all the printing and distribution of the books is outsourced to another company and that DemoEditor concentrates only on the editing process.

The book editors' job is to find authors, find book subjects, and match the right author to the right book. Then, they follow the writing process and make sure that the quality is there.

Then, it is up to the sales team and their job is to find indirect customers, which means libraries or book-selling organizations, as DemoEditor does not sell directly to readers. This means that commercial people actually sell the books by numbers, and not by units. Though we will not deal much with this part in our demo software system, it would be important in a real situation.

Finally, the director needs to keep numbers in check through reporting and statistics coming from the sales team and the editors. The smooth running of the company heavily depends on the right deadlines for the book as much as the quality of the writing, the match between the subject and the author, and the expectations of readers and book-selling companies. This means that the director has to measure all of these indicators, and the information system is of course expected to provide them. Asking editors or salespersons to fill in Excel sheets every week does not make sense, as this is time lost for their real jobs.

Data manipulation

As you may have imagined, DemoEditor's information system will have to manipulate data about authors, books, and sales, plus some additional statistics drawn from these primary data. The author will be known by their identity, a little contact information, maybe their bank coordinates to pay them royalties, and certainly information on their skills. Books will be registered with business-wide reference numbers, their titles, their summaries, and other information about the content. Sales will basically be the number of books sold to booksellers, with the associated date and maybe sales conditions.

Reporting data will be everything that can be used to apply business intelligence by the director on sales, authors, and books: how many books are sold for each category, what the time trend of selling them is, how many editions authors can work on for a given book before sales slow down and the novelty effect does not work anymore, who is their best salesperson, which bookseller returns the least number of books or reorders some the quickest, and so on. The reporting data is definitely bound to time, and not only by the fact that reports evolve in time, but also by the fact that reports should show the movements of business in time as well as geographically.

The stakes of the information system for the company

DemoEditor is a small company, which means that employees can "fill the gaps" and do a little bit of everything. While this is an advantage in some cases, meaning that they are agile and adaptive, it also means that they do not tend to do things in an industrial, repeatable way. Excel spreadsheets may be copied over and spread out in different versions all over the company instead of a unique, reference version being kept on the network. Also, data sales are spread around the different salespersons because they tend to be competitive toward each other and it is thus difficult to homogenize quantity discounts (the price is fixed) as well as customer listings.

As the commercial pipe is not very formal, some leads become prospects and then customers without the salesperson being able to really give statistics on how long and how much effort this takes them. The director really has a hard time knowing whether the company would sell more if they hired more salespersons. The choice of authors for the right books is also an issue. Generally, the editors have a good grasp on the market and know quite well which subjects should be written about. But the pool of competent authors is quite restricted and the authors are mostly known for the books they have already written. Most of the time, the editors do not know which other technologies these experts know about and there have been times when a book has been contracted to a new author, after taking a lot of time to find them, only for the editor to realize a few weeks later that one of the good authors who's already written several books for DemoEditor actually had the right skills for the new project. An updated, shared, and efficient knowledge of author competencies is important for DemoEditor.

The situation as seen by the person put in charge of improving the system

You are asked by the director to come and help with the information system. Everyone in the company knows IT could be more efficient and help them better but, as they say, they are not specialists in it. Since there was no internal IT guy, they did as best as they could, but they realized the "do it yourself" spirit of a small company can only go so far, and they had to get someone to give some structure. The director is also anxious about increasing the size of the company before this effort is done; otherwise, it may add more problems than growth.

As the business processes go on and the budget is not extensible, you are tasked with *changing the wheel while the car is running.* There may be some small stops in the IT system, but not for long periods. The data has got to be cleaned but the databases of authors and books have got to remain available along the way since they are everyday tools for most of the company's employees. The director does not care much about its reporting being unavailable, or even destroyed, since it is not very useful at the moment and most of the numbers were false anyway.

In the next chapters, we will put ourselves in the skin of an engineer asked to carry out this foundational task and design the different components of the renewed IT system, and decide how its services should operate and what business domains should be designed. After that, we will implement all this and progressively transform the information system. But for now, we have to transform the theory learned in the previous chapters into principles that will guide us along the way. SOLID principles are a great set of principles that apply perfectly.

SOLID principles and application to systems of whatever scale

SOLID principles are important principles that apply to software applications, but they happen to also apply very well to software systems in general, so we will be able to use them to structure our project. We are going to explain the five principles one by one, together with how they apply to the transformation requested by DemoEditor and the design of its new information system. As this is a book about information systems and not software development, even though we will eventually build some implementations, I will not describe the principles from the coding point of view, but only briefly their main idea and, with more details, their translation into systems design.

Single responsibility principle

This principle states that a class, or in our case, a module of an information system, should do one and only one thing. This is quite wide as a definition but it can be narrowed down a bit by stating that an entity should have only one business reason to change. If the same class should be upgraded when there are changes in the author's management and the book's management, there is a problem regarding this single responsibility principle and the class should be decomposed into at least two smaller ones.

This principle is obviously easy to translate into an information system where it applies directly to modules, whether they are services, components, or other grains of entire systems (we will come back to the management of granularity in the next chapter). Each entity in the system should do one thing and only one. If this is observed from the point of view of the software applications composing the system, that means that each application should be in charge of only one business domain of the system. Since we manage authors, books, sales, and so on, we should indeed find one application for each of these. This notion of a business domain is not very precise for now but again, we will come to this in a future chapter, namely *Chapter 9*, detailing the **domain-driven design** approach and the concepts of domain and bounded context.

For now, let's just agree that a business domain needs its own application. If you are thinking microservices, yes, this is the track we are going to follow but bear with me, as this "micro" qualification is not always necessary and we would rather talk about "services" (with a clearer definition in *Chapter 8*).

This first principle may sound very simple (and it is in its expression), but its implication can be very profound. To give just one example of the complexities we will have to deal with in our sample application, let's unroll the case where a service depends on another, like in the case of the relationship of the "author of a book." As said before, author and book management are two separate responsibilities. But how should we deal with the relationships between both? Is it another service? In any case, when someone reads a book entity from the service, how should the author of the book be retrieved and displayed? We can ask the same question the other way around: if someone calls the API on a given author, how should we display the list of books this author has participated in writing?

Diving into this last scenario is useful to get a better grasp of the concept of responsibility. Imagine that there are two separate services, each with its own database since they are supposed to be independent. Now, a user calls the API with a GET word on a particular book, let's say `https://demoeditor.org/api/books/123456`. The module is indeed responsible for sending the book title, ISBN/EAN number, and some other attributes of the book. How about the information about the author? This is where the principle of responsibility helps to draw the line. The editors would tell you that, most of the time, when they get the information about the book, they need to know the author, but only by their identifier and some main data, such as their first and last name. This is the responsibility of the book service. And if you ask your product owners again (the editors, as they are the ones who will be using the information system), and they need some more data, they will turn to the `/api/authors` service to get it, using, of course, the identifier provided by the initial answer from the `/api/books` service. Thus, that data is the only responsibility of the second service.

Each reader who knows about the principles of good database design is certainly suffocating already, considering that this approach necessitates data duplication. Indeed, since `/api/authors` is responsible for the whole data for the authors, including of course their first name and last name, that means that, if `/api/books` is responsible for providing, upon request, the identifier, first name, and last name of the author of the given book, the rule of non-duplication is broken! And this is where the concept of responsibility is interesting and should be dug into deeper. How about considering the following share of responsibilities?

The `/api/authors` service is responsible for providing always up-to-date data about an author, including their first name and last name. This means that it is the reference source of truth for the authors: anyone who needs the latest, best-so-far information about an author should turn to this service, which will be responsible for providing it on time. Since it is the reference for this data, the service would certainly provide it with a value date, as the data may change over time. For example, an author could change their last name after getting married; this should be tracked by the author service, as it is responsible for the data about authors and their integrity.

The `/api/books` service is responsible for providing the same service for books, which means the same level of engagement on the book title, identifiers, and so on. But when talking about the author of a book, this is a relationship to another reference service, so the only data it is responsible for in terms of service is to point at the right entity in the other service. And this raises two interesting questions.

The first one is functional: is the link supposed to simply point to a given author, or is it supposed to point to the value of an author *at a given point in time*? This necessitates answering some business rules: if an author gets married between the first and second editions of a book, should the author's name as it appears on the book change? And if so, how should the registered copyright be adjusted? And is the same true for a simple reprint of the original edition of the book?

The second one is more technical: if the book service stores the link to the author service and the latter is not available when needed, what happens? If a copy of the "usual" data (first name and last name) has been stored in the book service, no problem since it is now *independent* of the second one. But this comes back to the functional question again: if the name should not be changed on the book after an author gets married, no problem, and it is even better, as the local copy will prevent the difficulty of reaching the author's service with the date of value to retrieve the "old" data. And if the name should evolve, it may be better to temporarily fall back on the old one rather than providing just the machine-readable identifier; only editors would know…

I hope this example, though complex, has shown you what we mean by the word *responsibility*. It is admittedly complex but, remember, everything we talked about was linked to business complexity and is not accidental. Indeed, talking about value date and the importance of history in reference to data management may sound overly complex because it is not very often dealt with in current information systems. But this is a real problem and this lack of reflecting on the actual functional reality is a problem as it prevents reflecting all business rules! That does not mean that the software that will be constructed based on this reflection will take all this complexity into account. In a true Agile way, you will certainly start with something very simple. But this deep understanding of the functional complexity ensures that the software will be easy to evolve in the future and that you will not be stuck at some point of implementation because the software is not aligned with the business.

Open/closed principle

The **open/closed principle** starts with a paradox in its very expression, which makes it strange at first sight: how can a module be open and closed at the same time? The comprehension of this principle is very important to create systems that will evolve because it states what should remain closed and what should be open to change for this evolution to work as smoothly as possible.

When applied to object-oriented programming, the open/closed principle states that a class should be open to extension, but closed to modification. Encapsulation and private members are used to prevent any instance in the program from modifying directly the state of another instance; otherwise, it would be very difficult to follow what happens when executing the program. Even debugging would be complicated if there is no way to track what class modifies the state of another one. This is why a class keeps its members private and only opens some public functions to allow only some changes in its state, in a way that is controlled by its own code, following its own rules. This is the closed part of the principle.

But a class, in general, is not marked as `final` to let another class inherit from it and specialize the functions that are marked as `virtual`. The inheriting class can also add some data members, in addition to accessing the one from the inherited class that is marked as `protected` (or, of course, `public`). Again, the class controls what can be overridden and what cannot, but at least it is open to an extension of its behavior by another class. This is the open part of the principle.

Applying this principle to information systems does not imply a big change in reflection, as the services replace the classes, and the techniques to extend or protect only vary from the practical point of view. If we continue with examples where services are REST APIs, we can draw a parallel between the members of a class and the data that is persisted by the implementation of an API contract: nothing but the service can modify this data, as only the implementation has access to the persistence used (a database or anything else). Of course, some API methods may allow some specific modifications to be carried from the API clients, but the API implementation controls this and applies business rules to ensure the modification is carried as it wants (or maybe rejected, by the way). This is the closed part of the principle applied to a module of information systems, and it is quite obvious in its resemblance to the application in a class.

The implementation of the open part of the principle on an API is trickier, as the behavior of an API can be extended in many ways:

- One of the ways to do so is to create an API that will extend the initial API contract. Mechanisms exist in the OpenAPI grammar to implement polymorphism and it is also possible to aggregate types in such a way that the new one contains the initial type and all its content.

- Another way is to create an API that replaces the exposition of the old implementation, but still relies on it for all the standard data, and then provides its own data in addition. If done carefully, the extended API may even be fully compatible with the initial API contract, since it only adds new data (and if it just passes the initial data without changing any behavior, it even complies with the Liskov substitution principle, which we will see shortly).

- A third option is to use an API gateway to expose the updated contract and implement the mixing of data coming from the original API and the additional data coming from another service dedicated to its storing and manipulation. This approach is a bit closer to the principle of inheritance.

These three approaches are schematized as follows:

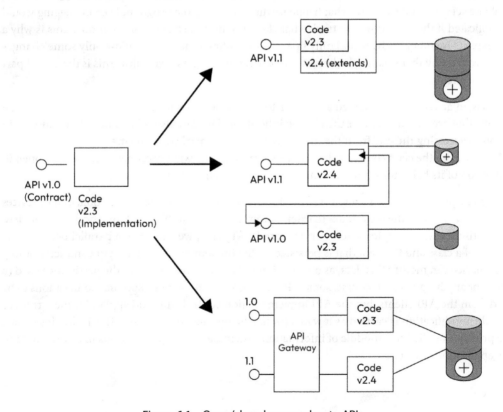

Figure 6.1 – Open/closed approaches to APIs

Liskov substitution principle

The **Liskov substitution principle** requires that, when an entity is replaced by another one that enriches or specializes its behavior, it should do so in a way that is compatible with the way the base entity functions. I have found that the easiest way to explain this is through the following example. Imagine an object class, that we will call `Display`, that would implement a `print` function that sends black text on a white paper (through a printer, a screen, or anything else; no importance here). Let's now suppose that a class called `ColorDisplay` specializes in the `Display` class and proposes a new function signature called `print` again, but accepts a parameter named `color` that allows the user to specify any color they want. How should the parameterless `print` function behave? The inheriting class will certainly point to its new and improved `print` function. And in this case, what should be the default color passed to this function? If you answer "black, of course," you know what the Liskov substitution principle is about not surprising the user of a class, and ensuring them that the behavior is as expected.

The same goes for services inside a modular information system. Again, we will use APIs, as this is the standard way to decompose modules in such a system. When an API contract is provided to you, it states what method can be called, with which verbs, and through which URL; it also states the exact name of the attributes that you can send or that you will receive. But you may very well respect the letter of the contract API without respecting its spirit and this is what the Liskov substitution principle is about.

Let's translate our previous example from object-oriented programming into API services, and imagine one can use version 1.0 of `/api/print` to process some text that will be sent to the device in black color. If using version 1.1 of an `/api/print` API, with the support of `/api/print?color=[HEX-VALUE]`, we will definitely expect that pointing our old client to the new API results in the production of black text. In terms of the stream between the API contract and its implementation, this can be pictured as the following:

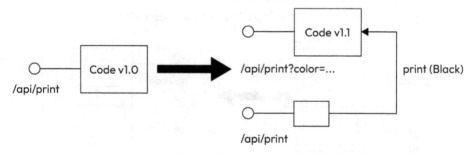

Figure 6.2 – Liskov substitution

Interface segregation principle

The **interface segregation principle** recommends splitting the interfaces in such a way that an implementing class does not have to implement a behavior that it will not use. Again, as the goal is not to dive into object-oriented programming but just to draw a parallel with information systems, we will only give an example to explain this approach. Let's imagine that a Rectangle class implements the IShape interface. This means that Rectangle would have to implement the getSurface and getPerimeter methods but also some methods such as drawShape and so on. If the only interesting thing in the program is to compute the mathematical characteristics of shapes, then the right approach would be to split IShape into IGeometricalShape and IDrawableShape (i.e., segregate the interfaces) for the classes to implement only the interface they need.

The same goes for API contracts. To come back to our book service, it is better to separate two contracts, one for the book characteristics and one for the book sales characteristics, even if the same implementation will expose the two interfaces, one behind /api/books and the other behind /api/books/sales, rather than force the implementation of all functions by the use of only one contract.

Though it is always possible (and more acceptable in an API than in a class) to respond with a NotImplemented message, the separation into two interfaces also makes it easier to introduce versions. If the sales interface was not correctly defined in version 1.0 and a major, backward-compatible rewrite is necessary, it will be possible to continue exposing the book's characteristics (potentially used by most of the clients) without any change.

> **Note**
>
> **Backward compatibility** is the quality of a new version of an API where all calls used in the previous version work exactly with the same result on the new one.

The interface segregation even makes it possible (and easy) to expose three API contracts, namely /api/books, /api/books/sales, and /api/books/sales/v2:

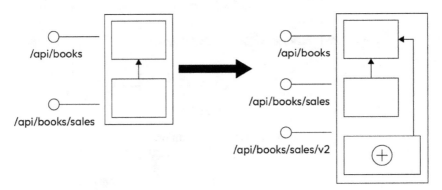

Figure 6.3 – Interface segregation

Though quite obvious in the explanation, this principle is sadly often forgotten in existing APIs. This is particularly so because editors tend to provide functions in the frameworks that generate the OpenAPI contract from the implementation, instead of following the contract-first approach (which is admittedly more complicated, but the only one that leads to proper, business-aligned APIs). Since the contract is automatically generated from the whole source code, the generators do not make any difference between the different methods and resources and produce a single, monolithic contract that does not respect the interface segregation principle.

Dependency inversion approach

The **dependency inversion approach** is named as such because it goes against the usual way of thinking about dependency. Imagine that we have a reporting module and another module that provides the data to be used for reporting purposes. Naturally, we tend to think that the reporting module should depend on the data module in technical terms because this is how it works from the functional point of view. This is a rare case where aligning the technical design directly with the functional concepts is not good enough. If we want to ensure low coupling, we have to make one more step and add some indirection to a common interface that both of the services will depend on:

- Just like the `Data` class would implement `IData`, the `/api/data` service should implement the API contract defined in the data OpenAPI file.

- And like the `Reporting` class would use the `IData` source (and not directly the `Data` implementation) to prevent hard coupling, the `/api/reporting` service would not directly call the `/api/data` service but the URL provided to it by configuration, providing that it implements the data OpenAPI contract. In object-oriented programming, this will typically be done by injection. In a service, depending on the orchestration mechanism, this can be realized through an API gateway, the ingress exposition, or even a (more complicated) service mesh.

The following diagram has already been used in *Chapter 3*, but it is particularly relevant here, as it shows visually the dependency inversion principle in information systems: instead of a dependency of one software module on another, each of the two modules working together points at the same business contract definition (which is purely functional), one to implement it, the other to consume it:

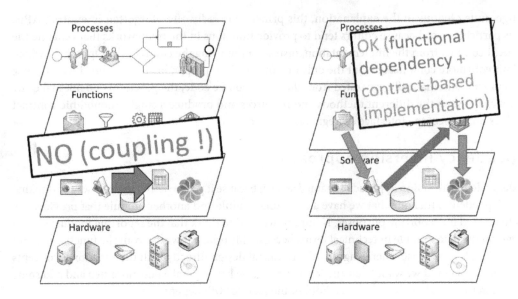

Figure 6.4 – Coupling on the four layers diagram

Let's move on to the next section where we'll analyze the SOLID principle.

Critical analysis of SOLID

Though I have realized in time (certainly like many others) that the SOLID principles apply almost as well to information systems as to object-oriented classes and interfaces, some parts deserve discussion. Indeed, like for any other principle, a strict, not thoughtful, application may lead to problems. And even when the principles apply well to a given context, they may have strong side effects that make them ultimately more harmful than helpful. The necessary measured approach has led to lots of polemics and you must be careful in handling these principles.

Limits of the separation of responsibilities

The first (and maybe most important) principle, namely the one about separation of responsibilities, is sometimes hard – perhaps impossible in some cases – to apply to services or high-level modules of an application. With classes, there is always the possibility to decompose a class without much impact on the whole application, once compiled. Services and modules do not present this ease of composition because they come with additional constraints such as exposition, endpoints, coding interfaces, documentation, integration subsystems, and so on. All this takes a toll on the decomposition, and this is the reason why people who are not careful with microservices and decompose them too much in their systems end up spending more time evolving them than if they had stayed on a monolith architecture. I am not going to talk again about granularity because the subject was already developed in the previous chapters, but this is clearly something that must be pondered and can limit the reach of the principle of separation of responsibilities, or at least its depth.

Another difficulty is not linked to how deeply you can separate the responsibilities, but how intricate some of them are sometimes, even at a relatively high level. This is, for example, a huge difficulty in "micro-frontend" architectures, and I am not talking about the problem associated with how "micro" the frontend component is. The simple fact that visual components present functions, but also have a visual impact, is a huge difficulty in making them independent. This is what is exposed, for example, in `https://jonhilton.net/good-blazor-components/`: by delegating the content of a Blazor component to another sub-component (in the example of the article, using the `ChildContext` property to include a `Card` instance), you indeed externalize the responsibility of displaying the inner part of a visual component. But that does not mean it will not have an impact, as the rendering of the subcomponent will necessarily have an impact on the one above it. Indeed, either the container fixes a size for the child and then takes responsibility for its display, or the container – in a purely responsive design manner – lets the child component adapt its display. In this case, its own size will be impacted by its child, which also ruins the independence that was supposed to be achieved by the principle of separation of responsibilities.

The intellectual solution for this is to consider that each component has the responsibility for its content, but not for its display and that this is the responsibility of the display engine in the browser. It may not be very satisfying because it comes back to a single point of execution for the whole frontend, which we were trying to make modular and easy to evolve, but at least the low coupling between the components has been achieved, though on a limited perimeter. This is one of the many compromises one will have to make when designing information systems and modular applications.

About the polemics on monoliths

This chapter may be a good place to talk about the long-held discussion about "going back to monoliths," a reaction to the drawbacks of microservices that have been very much observed in the discussions of the communities of software architects. In summary, the microservice approach is considered harmful and some people recommend going back to monolith applications. This is sadly another example of how polarized debates are nowadays, because both approaches (and the huge spectrum between them) have value, depending on your needs.

No one serious has ever pretended microservices were the best solution for every architecture. In fact, right from the beginning, most articles explaining the approach insisted that they were only adapted to some specific circumstances (high volumes, frequent modification of the application, clear cut between teams in charge of different modules, and so on). However, some people did not take this into account and now complain that microservices were not the right fit for their case. Even worse, in a binary and not very thought-out reaction, they discard the whole principle of microservices and advertise a so-called "return to the monolith." Any sensible engineer in this case would simply spot the well-known pendulum movement where one extreme chases another one. And the solution in this case is not in either of them or continuously balancing between one and the other, but simply reaching the nice equilibrium between them.

In the case of software applications, what do we call a monolith? An application made of a single process? No, otherwise that would mean that even the smallest application such as `cd` or `exit` should be considered as such. The term has been coined precisely to describe single applications that were in charge of too many business features, which made them too heavy to have a nice evolution or fit the use they were designed for.

The situation talked about in the articles on the so-called death of microservice architecture and going back to monolith simply do not talk about the right subject, which is the granularity of services. Sure, the very small grain of microservices did not satisfy their information system needs. But that simply means the grain was too fine. Going back to the coarsest grain one can have (the monolith) simply displays a lack of depth in the comprehension of the problem. An engineer approach to this would be to search for the right granularity of the services. As said in `https://codeopinion.com/biggest-scam-in-software-dev-best-practices/`, the choice is not only between "Amazon is doing it, so let's do it" and "we are not Amazon so we should not do it" – knowledge of the business and analysis of the context only will tell you where you should place yourself between the two.

It just happens that there is an existing design method called domain-driven design that is the perfect method to find out the right granularity of services for an information system. It also relates strongly to the principles of business/IT alignment, as it states how business domains can be cut into internally cohesive, externally low-coupled modules. Everything we will be doing in the design of our demo application will be an illustration of this research of the right granularity, based on the business functional needs.

Beware of unintended coupling

One last piece of advice on the separation of responsibilities (this first principle is definitely causing lots of discussions!): once you have established a clear cut of responsibilities based on the **business capability map**, you are not done with the separation of responsibilities, as it is very easy to recreate coupling afterward in the technological transformation of the functions.

I will give one simple example of such a problem because it is very representative of the trap I described. Imagine we create our demo application with an API for the books, an API for the book sales (possibly exposed under the same URL, but still independent in terms of contract), and an API for the authors. The director of the company may want, at some point, a report on how the geographical origin of the authors impacts the localization of their sales (maybe an author from Brittany will sell more in their region because their network of contacts is denser around them? Or perhaps it has no impact? Anyway, this might be interesting to analyze). This is a common issue with microservices, which are supposed to have their own persistence. In this case, how can we make links between data, just like we were creating `join` operations when we added a single database? One of the standard approaches is to add a collecting data structure that gathers, indexes, and aggregates data from all the microservices and proposes a dedicated `/api/reporting` service with its own API contract.

This service of course presents a level of coupling toward its source, but this can be made a lower level of coupling, for example, by keeping a local cache or mixing a subscribe approach to the data changes with a direct collection under a lower frequency to ensure that no signal has been lost and the data is reset at a controlled frequency. Also, it presents interesting functions that none of the atomic data services provide, such as indexing, the capacity of dynamic aggregation, and so on. **GraphQL** is a nice protocol to expose such services, and they integrate very well in a **Command Query Responsibility Segregation** architecture.

However, there might be an unintended coupling if the atomic service not only provides data for this reporting service but also starts consuming atomic data from it. And this can happen very quickly because the performance boost of an indexing engine is very appealing to developers. The problem is that this causes a circular functional dependency, which is already quite a problem, but also a higher level of coupling since, once this dependency is created, the atomic data service becomes suddenly coupled to each of the other ones. Of course, the coupling may remain low, but nonetheless, a link has been created and this is – in most cases I have seen – a real pain to come into the system.

If you really need fast aggregated reads in an atomic service, you have to deal with this internally, by adding indexes on the dedicated persistency, for example. This may sound complicated if you see this nice indexing engine right on the other side of the service wall, but this is the price to pay to keep your system free from coupling and easy to evolve in time. Again, this is a compromise and if the architect decides the functional coupling is not a problem, you may save time by doing so. But this has to be a conscious (and documented) compromise.

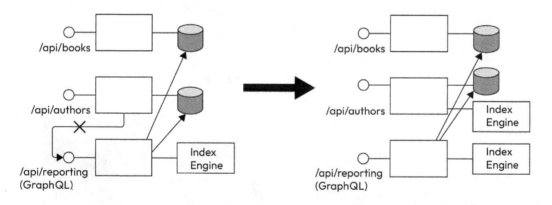

Figure 6.5 – Reducing coupling strength

Wrapping up before going back to the demo

In this chapter, we have learned about the SOLID principles that can be applied to information systems (and not only object-oriented programming, as they were initially aimed at); the single responsibility principle and the interface segregation principle have been used to start a definition of the different

API contracts we will need for our demo application. The open/closed principle will help to keep this API grammar free to evolve, and its evolution will have to follow the Liskov substitution principle for the system to evolve satisfyingly. Finally, dependency inversion has been demonstrated as the core principle behind contract-first API and the capacity to align the software implementations on the business-oriented functions, which is the main goal we seek in this book.

In the next chapter, we will go one step further in the design of our demo application by defining its different components. This will help us to draw a perimeter of the first version we want to create in this book, but also to define more clearly the different parts of the application using a method that adapts to the depth of decomposition.

7
C4 and Other Approaches

In the previous chapter, we provided an initial idea of the business goal of the sample information system (and application software modules) that we are going to build as a practical exercise to follow the principles outlined in this book. Talking about principles, while introducing the demo system, we also explained how the most important principles in object-oriented programming would apply to the design of our system.

Now that the high-level principles have been set, it is time to go a level deeper in understanding our sample information system, and we will take advantage of this to show a few different methods professionals use. There exist many architectural methods and software representations, as we are going to see, and many overlap in lots of ways. In fact, some of them are so close that they really are a way for their author to mark their individual approach, rather than bringing some additional value to the architectural tools. This is the same situation with the JavaScript ecosystem, where there are way too many frameworks. Furthermore, it becomes even worse every day, with a new team thinking they can solve a problem by proposing yet another one. The same thing happens with architectural methods, and we have recently been overwhelmed with **clean architecture**, **hexagonal architecture**, **onion architecture**, and so on.

As the choice of one method over the other is really not important, we will only quickly show these methods so that you can choose which one adapts better to your way of thinking, or simply mix some interesting bits from many methods, as these are merely sets of long-known best practices assembled in a new package name. In the case of our sample information system, the **C4 approach** sounded interesting to provide a first approach on how it is designed, so this is the method that I will focus on to provide more details about. The idea behind the C4 architectural framework is to focus on IT systems from a context level to a container level, then a component level, and finally, a code level. The four levels will be explained as we apply them to our sample system.

Just like with **SOLID principles** in the previous chapter, my goal here is not to explain these methods in depth, as there would be no value in writing what is already available in more detail on freely available and quality resources. My intention is to extract how these methods can be useful to reach a business/IT alignment or at least help represent the questions that are related to this same subject.

In this chapter, we will learn about the following topics:

- The C4 approach
- Clean architecture
- Hexagonal architecture
- Onion architecture

The C4 approach

The C4 approach is about designing a system (or an application) with four different levels of granularity, moving from the context level in which the application is placed, to the container level that will show the different processes composing the application, to the component layer showing the different parts composing the process, and finally, to the code level, where we will find the classes and interfaces used to create the application. Here is the diagram for this approach:

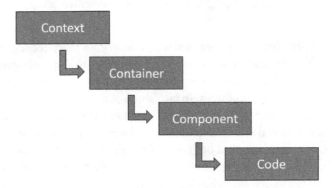

Figure 7.1 – The levels in the C4 approach

Each level details the content of the previous level and gives a more detailed view of the application. The fact that units of the lower level always have to be related to one given unit of the upper layer also helps to correctly share responsibilities and maintain a low level of coupling.

In the following sections, we will draw the four levels for our sample system and software modules to show the methods; at the same time, we will start designing our demo a bit further than the business requirements in the previous chapter allowed us to do. Note that the following diagrams will not necessarily be the exact ones that correspond to the system created at the end of the book; I really wanted to show the complete and realistic process of design, so I am drawing these at the same time as I design the sample application for the book. Drawing the schemas after I got a clear and complete view of the final result would – in my opinion – not be as pedagogical and, in some way, I would have felt like cheating.

The context level

On the **context level**, our system will be used by the three user profiles that were detailed before, namely the editors, the salespersons, and the director of DemoEditor. We will consider that the list of users is already present in the information system as an Active Directory or something equivalent. The same goes for a content management system, which will be considered to already exist for binary document management. Finally, we will consider that a module responsible for the emission of electronic mail is also available in the overall system. It is used to send emails to actors of the system, either to validate some data or to inform them of something. Of course, the authors will be represented in the context, even if they do not use the internal information system directly. Here is a diagram of the context level:

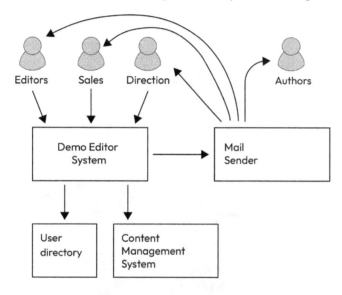

Figure 7.2 – A context-level diagram

We could be more detailed on the content of the arrows, but for now, the diagram is self-explanatory, and the functions used by the different roles were explained in the previous chapter. Our most important center of interest in this diagram is the **Demo Editor System** box, and this is the one that we will detail in the next level of the C4 architecture.

The container level

At the **container level**, we show what the system uses in terms of building blocks. Most of the time, these blocks are separated processes. In our case, since we have decided to have completely independent services, we will have as many containers as API exposition servers. I anticipate a bit on the technical chapters of the book, but we will deploy the system as Docker containers, so there is a complete match between "containers" in the C4 architecture and "containers" in terms of the units of deployment used in Docker or other such technologies.

In the following diagram, the boxes inside the dotted rectangle represent the details of the higher-level box corresponding to the Demo Editor system. We can still represent the other parts of the Context level, but it's best practice not to detail them, as they are outside the scope of our study. We can detail the stream that goes to them (as we did for the **Users directory** arrow, pointing from **Single Page Application**), but this is not mandatory, and we can also keep arrows that simply point from the system as a whole if we do not want to be precise about these streams of data yet (this is what has been done for the mail sender and CMS subsystems).

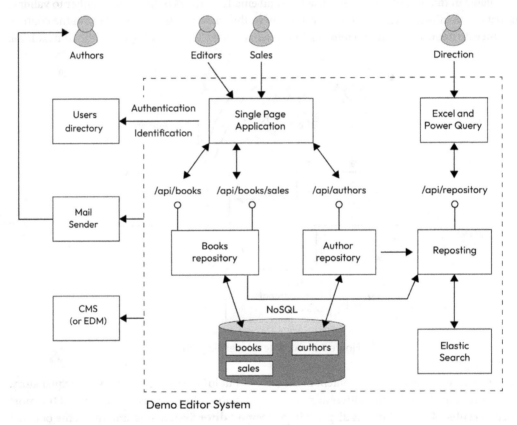

Figure 7.3 – The container level

By the way, at this level of reflection, we start considering what the content of the streams of data could be, and in this particular case, I realized that instead of a CMS, electronic document management could possibly be more appropriate. If we consider that this is an external system and not part of the one we are designing, we would, in reality, not have the ability to change it. However, since there is an interoperation with it, we might be able to influence the protocol used to communicate with it, and this is what our analysis is about. From a practical point of view, I do not know yet, at the time of writing (remember, I want to be as realistic as possible and I have not finished the exercise before writing the chapters), whether there will be an Alfresco container, an Azure Storage, or a SharePoint 365 site to implement this subsystem. All I know is that I will need to pass binary documents and retrieve them, which means something such as **Content Management Interoperability Services** (**CMIS**) streams will certainly be indicated (CMIS is an interoperability standard for electronic documents).

For now, we have not expressed what is inside each of the boxes. We only know that each of them will have its own life cycle as a software application and represent a process and a Docker container (a good practice is, indeed, to have one process per container only so that all things align well). If we go a bit into the development details, it would then be in the right alignment to have a continuous integration pipeline for each of these containers and, logically, a git repository for each of them as well. But I am anticipating the "hands-on" part of the book. For now, let's dive deeper into one of the boxes, namely **Books repository**, and show what it will be composed with, using the third level of the C4 approach, namely the component diagram.

The component level

At the **component level**, we show the different modules and libraries that the executable container will use to accomplish its business requirements. As our books repository will expose two contract APIs, we will need a web server to support the controllers, a client to call the persistence system, maybe together with a repository pattern, a cache component to keep some data locally, and so on. This is what is drawn in the following diagram that corresponds to the third level in the C4 architecture:

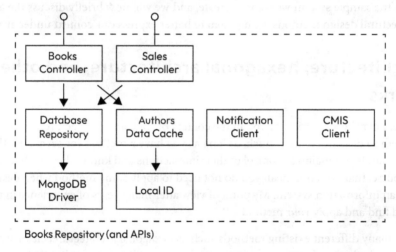

Books Repository (and APIs)

Figure 7.4 – The component level

A component may correspond to different names of artifacts, depending on the platform that is going to be used. In .NET, components may be assemblies, while in Java, they will be JAR archives. Anyhow, this decomposition in components exists in basically all programming platforms, even if it appears in different forms. For example, in some platforms, components will correspond to separate files, while in others, they will be more abstract, through the use of namespaces, for example. In any case, there will be ways to group code-level content into coherent groups, while still with a lower granularity than the higher level of containers, which correspond to entire processes made with several components.

The code level

As the reference documentation of the C4 method (available at `https://c4model.com/`) states, it is important to use diagrams only if they add value. This is something I also stressed when presenting the **CIGREF method** to map the information systems. It is a common reflex for beginners in diagramming to try and map everything in the system as if it were the goal in itself to have a map. The goal is to go somewhere, and the map is just a tool to make it easier. It is of utmost importance to understand this and to only map what is useful. This is why I will jump on this opportunity to demonstrate this and not draw anything at the code level; at this point of the application design, I don't have a clear idea yet of how the classes will be organized. I am sure that the `IDbRepository` pattern will be used and maybe a `UnitOfWork` pattern, but the class organization is not clear yet, so there is no use in trying to draw the diagram yet.

Also, even if it was for pedagogical reasons, drawing such a diagram would not bring much value, since the fourth level of the C4 approach is simply the same as a class diagram from **UML** or **Unified Modeling Language**. We will use these detailed diagrams if we need to in the following chapters, once we progress into the actual making of the system. In the meantime, we are done with the C4 method used to explain the sample system we want to create, and we will now briefly discuss the alternative software architectural design methods, using them to better express our context under analysis.

Clean architecture, hexagonal architecture, and other frameworks

What comes next may be a surprise for you in the context of software architecture – the choice of an architecture method does not matter much, as long as you have a method (or several methods used together, by the way, if you maintain control of their interactions and know when to choose one over another). Of course, that does not mean you do not need to apply some method when designing the architecture of an information system. My point of view after many years of experience in the field is that you should find and apply *your* method.

Also, if you try many different existing methods such as hexagonal architecture, clean architecture, onion architecture, and other domain-centric approaches, you will quickly realize that they offer little more than a different visual approach to the same basic principles, which are highly related to the business/IT alignment we discussed at the beginning of this book and the principles that have been extracted from the SOLID approach, namely the following:

- The business model (*data structures representing functional concepts* and *business rules translated into code*) should be at the core of the architecture. This way, it does not depend on anything, which allows easy changes, fast automated testing, a lack of external versioning constraints, and lots of other advantages. Most importantly, it enables the concentration of the team on the most important thing – business alignment.

- Everything surrounding this business-aligned core should use some kind of indirection to introduce low-coupling and make the evolution of any module of the system easier. Everything depends on the business core module, which itself depends on nothing else. In addition, these dependencies are easy to modify if necessary.

Technical architecture patterns

One last thing before we dive into the details of the methods – here, I will only discuss the architectural methods that can apply to the design of an entire information system. There are lots of methods that deal with the technical architecture of a single application, how it should be structured, how its source code and information flow should be organized, what technical layers should be present, and even, in some cases, recommendations on how to name things.

For example, the **N-tier architectures** (also called **layer-based**) will describe how each set of source code should call another set, from the GUI to the service to the application to the database. MVC, MVP, and MVVM architectures are slightly less linear and describe how each group of source code would talk to only some of them, to ease the evolution and maintenance of the application (with generally an emphasis on their visual part).

These architectures can, theoretically, be used at the level of the entire information system, but that does not mean much more than applying them to all applications that constitute the information system, since they do not say much about how the applications should work together. N-tier architecture recommends that software blocks should talk to each other using the application layer, but since it does not enforce it, lots of old applications interop at the database level, which is a catastrophic coupling in most cases and certainly one of the major reasons why a vast majority of legacy systems are nowadays unable to evolve. MVC is supposed to be applicable at a higher level than just the visual interface, but despite this, it does not reach the level of the entire information system. It may be just a question of semantics, but I personally tend to classify these methods as patterns rather than architectures, since they come from experience rather than pure reflection (and there is absolutely no value judgment in this – we need both experimental and intellectual approaches to be complete).

Since this book is about the architecture of a whole information system, I have focused on the purpose of architectural methods that naturally are applicable at such levels. It happens that domain-centric methods typically respond to this because they convey the very notion of a business domain, which has to be the very first (and unique) way to decompose an information system if you want to get a result that is aligned with a business. All technical approaches can work on the software and hardware layers, but what we need is methods that work at the business capability map level.

Onion architecture

Onion architecture pictures a software unit as several concentric circles containing, right at the center, the business **domain model**! And around it, you will find services (persistence, etc.) and then presentation (a GUI, external interfaces, tests, etc.):

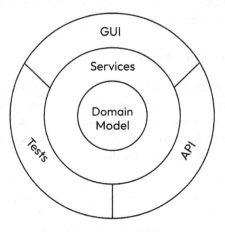

Figure 7.5 – Onion architecture

One of the main rules in onion architecture is that all dependencies should go from the outer side to the inner layers. This way, again, the core business model does not depend on anything, which makes it easy to evolve, while outer circles can be changed provided that they do not impact the business (which is generally much easier than the reverse – imagine basing your entire business model on a proprietary relational database procedures language and trying to change to another database engine).

While this method does not insist on levels of indirection, the principle of dependency inversion (see the previous chapter for an explanation), together with the rule just cited, makes it obvious that interfaces are necessary to implement functions such as persistence. And the same goes, of course, for one layer above.

Hexagonal architecture

Hexagonal architecture represents a software unit in a hexagonal shape, hence its name, with the left part of the hexagon containing interfaces consuming the inside (such as a GUI or automated tests), the right part containing adapters to functional dependencies (such as persistence, notification service, etc.), and right in the middle of it, as you may have guessed, the core, containing the business entities and business rules!

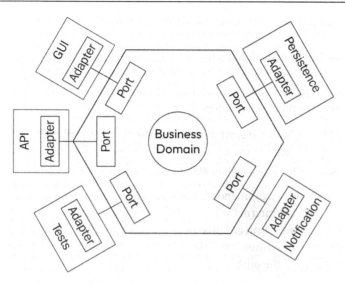

Figure 7.6 – Hexagonal architecture

Hexagonal architecture insists on ports and adapters to organize the communication between the different parts of the application, but this is also a rule in the onion architecture even if it is not displayed in the same way. This helps us remember that interfaces and contracts should always be applied for the dependencies not to evolve into hard coupling, which makes it difficult to evolve the system.

As far as the "*hexagonal*" shape goes, it is pure clutter! I intended to draw three "clients" of the business domain to stress the point that there is absolutely no relationship between the name of the method and a given technical constraint or a particular advantage of the technique. In fact, it is explained in the method that the hexagonal shape is simply used to preserve enough space to display shapes around the central one. How would this be different from circles (used by the other approaches cited previously), ovals, or squares? If the hexagonal shape does not bring anything conceptually enriching, my personal opinion is that it should not be declared in the name of the method.

This may sound like a useless rant against methods, but this is only to make it clear that their value really lies in what they have in common and not their differences. What really matters is the centrality of the core domain, aligned to the business, and the control of the dependencies, but all of this is already there in the SOLID principles.

Clean architecture

Clean architecture involves basically joining the two approaches presented previously while proposing some additional rules. Graphically, it appears very similar to the onion architecture, in concentric circles with the business entities in the middle, and several layers organized around in the core, evolving to the outside in the same direction as dependencies apply.

In this method, you will find almost everything that has already been explained previously in the two other methods, including the fact that the domain model must be clean of any technical dependencies, to address its utmost importance in the system and to bring all kinds of other benefits such as testability.

Inversion of dependency is also recommended, as the separation of concerns. The vocabulary slightly differs, but a quick search on the internet will show lots of blog articles about the extreme similarities between the three architectures. Clean architecture may be a bit more "guiding" than the others, but it is impossible – at the time of writing – to find a documented analysis proving one better, or even widely different in results, than the other.

My personal opinion is that these three approaches (onion, hexagonal, and clean architecture) are so close to each other in the way they work and in their recommendations that you can use one or another without making any real difference in the quality of the architecture produced.

What will bring quality is the profound understanding of why these methods exist and, in particular, the capacity to create modules that are cleanly separated and where you keep control of the dependencies. Cutting concerns and responsibilities without strict rules on their relationship will end up in a spaghetti-like system.

Conversely, following strict principles where the business domain (the most important part) does not have any dependencies and where everything technical is interfaced around it, with an indirection layer in between, is what really matters. And all three approaches have this in common. The rest is pure details, and if you consider, for example, that the hexagonal shape does not bring anything to the method, as explained previously, the existence of several methods really gives the impression that some of them are there just because their author wanted to have their own method.

Summary

The C4 approach was used in this chapter to detail our demo application from four points of view, namely the context in which it is used, the containers it uses, and the components and code that it is made of. Diagrams following this approach have helped us explain what the sample information system we will create is about. We also learned that it is not necessary to create a diagram that completely covers the domain of study, but only for the bits these diagrams have value.

Also, lots of architectural methods have emerged in the past decades. Although they clearly have value in the design of software applications, and thus in information systems, they resemble very much each other in the sense that most recent methods are domain-centric; as such, their value can basically be summarized by the two principles that we already know, SOLID and business alignment principles, namely putting the business functions model first and applying a method to reduce the coupling to dependencies. Both of these make the evolution of a system easier. I do not have any doubt that these methods may have additional value for software organization, and they certainly help you to take a step further than purely technically oriented architecture patterns. However, as far as entire systems are concerned, the C4 approach explored at the beginning of the chapter should – in my opinion – be applied first in a top-down approach, leaving the domain-centric approaches as a way of organizing the third level (the components) of the C4 method.

In *Chapter 8*, we will start defining more precisely the API on which our demonstration system will be based, as these entities are definitely the most important technical aspects if you want a freely evolving application. In the upcoming chapter, we will start by explaining the concept of service orientation and its relation to APIs. In *Chapter 9*, we will use a semantics-based approach to define the business domains of our sample application and translate them into actual API contracts. We have talked a lot about business domains – in the next two chapters, they will be defined, as well as their interactions, for our demonstration system.

Service Orientation and APIs

Now that we have explained many principles and several methods, we are going to move on to chapters that will be a bit more technical and thus will show more examples being applied to our demonstration application. In this chapter, we will explain the notion of *service* from the IT perspective and will try to place services in the history of IT to give you a good understanding of what they are for and what they have brought to the industry. There are still shortcomings, of course, but web-oriented architecture and web services in general bring huge value to the software industry when they are correctly designed (which, sadly, is far from being common).

After this examination of the history of services, we will detail the characteristics of a good service-based architecture (I am not using the expression *Service-Oriented Architecture* for a precise reason, as you will discover shortly) and explain how their current evolution, namely REST APIs, can be of benefit to many software systems.

Finally, we will show how the architecture patterns seen in the previous chapter apply to the definition of services for the demo application. To do so, we will of course use REST APIs since they are at the core of every modern approach to IT system architecture. We will come back to the notion of standards, which was heavily discussed in *Chapter 2*, and explain which ones can be used in our demonstration system. Finally, we will explain what we can do when no standards exist or apply, and this will form the transition point to the next chapter.

In this chapter, we'll cover the following points:

- Looking at the history of service orientation
- Characteristics of a service
- Application to our demonstration system

Looking at the history of service orientation

First of all, let's start with a bit of history. it may be because I am an old-timer and have been programming for the past 37 years, with 25 of them in industrial contexts, but I think it is always interesting to know where we are coming from as this explains a lot of what technologies today have been created for, and what they still miss. This way, not only can we anticipate the shortcomings of certain techniques and software artifacts, but we also avoid the risk of not using them to their full potential, as intended by their creators. Unrolling the history of technologies has yet another advantage: while strolling along this path, you may stumble upon an old but still interesting technology that may better fit your context than the new kid on the block. This does not happen so often, but when it does and you can solve your IT problem with a battle-hardened, yet simpler technology than the tools generally used at present, it can provide you a huge boost in maintenance time and performance. For example, files-based interop may seem laughable to someone using web APIs every day, but in particular, contexts where asynchronous is better, security is not a problem, independence of the process is an advantage, and avoiding the deployment of a web server is a time-saver, they can be the perfect solution.

So, let's start this journey with interop techniques, and in particular begin with why we need them.

The long-awaited reusability

Making two parts of a software entity interop with each other is a concept almost as old as programming itself since it is related to reusability. For a common function to not need to be typed twice, there needs to be a way to separate it from the rest of the code that is different, and make it callable, one way or another, by these pieces of code. This can be easily schematized as shown in *Figure 8.1*:

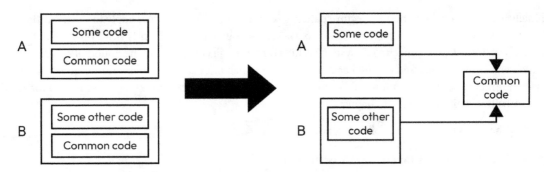

Figure 8.1 – Reusing common code

Code duplication is a problem (although the Don't Repeat Yourself principle may have its own shortcomings and every programming choice is always a compromise), so putting some code in common is generally a valuable orientation. There are lots of ways to organize this, and reusability has been sought after like the Graal in IT for many years, even decades.

Routines and avoiding punching additional paper cards

The first attempt at reusability came from an era that most of you will not even know about, where programs existed in the form of paper cards punched with holes to provide instructions to the computer. This actually came from the Jacquard weaving mechanisms where these paper cards were used to control cloth threads in semi-automated machines to make certain figures and weaving patterns appear in the resulting cloth. Repeatability was already possible by using the same card again and again, but the reuse of a pattern on a given card was done by punching the card many times in the exact same way, resulting in a long manual process and involving the risk of errors. Also, each application, composed of boxes with punched cards in the exact order, had to contain every single instruction. Due to the fixed number of instructions that could fit on a card, it was practically impossible that a card could be reused in another stack. Even if it was possible, extracting the right card, using it, and then returning it to the right position in the stack would have been too dangerous for both applications involved, particularly with regard to duplicating the paper card, even if it was a manual operation.

Then came the idea of routines and sending the code back to a given address of the program to make it repeat some part of its instructions. The notion associated with the infamous (but still worthy) GOTO instruction was born, and it saved many instructions in the programs that followed. But there was a problem: routines could only be used inside a single program. Admittedly, it helped reduce their size. Nonetheless, when creating a new program, it was still necessary to type in the same code – and we're talking about a time when copy-paste simply did not exist. So, there was a need for something better. This is what we are going to explore in the coming sections.

Libraries and the capacity to share instructions between programs

The next evolution of reuse was the concept of **libraries**, where the common code is placed in a different file than the original calling bits of code. This is simple but efficient, as virtually every piece of code—to this day—exists as text content stored in files. Though many other approaches have been attempted to make interop universal, libraries in the form of .NET assemblies with `.dll` files, or Java modules in `.jar` archives, are still the foundations of reuse, and universality is not far when Base Class Libraries like the one from Microsoft are evidence for all .NET programmers. For this particular framework that started under the strict control of Microsoft and progressively flourished into open-source availability, history has been a blessing, since lots of libraries are simply implemented once and for all. Java, on the other side, started as a more open platform but became progressively more closed after the buying of Sun by Oracle. Though things start to unify a bit (at the price of a slower evolution), there still are many libraries to do the same thing within the Java ecosystem. I remember being flabbergasted with my first professional developments in Java, after five years of .NET, by the fact that there was not a single XML parser library, but many of them, each being the best at one thing: Xerces being good at streaming analysis; Xalan recognized as the quickest at fully loading an XML DOM; some other library handling the DTD schemas and validation better; and so on. For someone used to having only `System.Xml` to think about, that was a huge surprise and a big disappointment, since it made the learning curve all of a sudden dramatically steeper than the closeness of the platform and language led me to expect.

Anyhow, libraries are certainly the most widespread approach for reuse in programming platforms and exist in almost all modern languages, be it JavaScript, Python, C or C++, and so on. Though libraries also have their difficulties, not only in the versioning and forward-compatibility areas but also in the ease of copying a file, which sometimes ends up in multiple copies in a code base (which goes against the initial goal of reuse), they remain the go-to approach when reuse is necessary. Of course, as far as interop is needed, they have pronounced shortcomings since they are only usable in their own execution platforms: though bridges may exist, a Java library can only be used by a Java program, a .NET assembly can only be called by another .NET assembly, and so on. This is a problem that has occupied IT engineers for a long time and is where lots of solutions have appeared.

Attempts at general interoperability

The main solution to the aforementioned limitation was to create libraries that contained code in a compiled form, in such a way that the machine code was usable from any caller, whatever language was used to create the calling program, as long as it was also compiled in machine-readable code. The difficulty there was mostly technical: calling one such library was not as simple as using a function name and attributes. Also, there was still a limitation due to the platform of compilation. There was indeed no way to execute a Windows library inside a Linux operating system and vice versa.

Microsoft actually tried to go a bit further in this direction by introducing the notion of OS-controlled components. These reusable units were not directly available as files but as entities known by the operating system itself: their concrete form was still filed, with the .dll extension, but when *registered*, Windows would make the functions available to any program even if it did not have access to the original file. In addition to being a repository for application customization, the registry also stores the necessary information for the **Component Object Model** (**COM**); in fact, it even started its career in Windows 3.11 mostly for this use. Together with COM, a Microsoft technology called Dynamic Data Exchange allowed application components to be inserted into one another. This is what you use today when you open an Excel worksheet inside a Word document and see the menus adapt.

After COM came extensions such as COM+, then for **Distributed COM** (**DCOM**), which was an attempt at breaking out of the perimeter of the local computer and introducing remote execution of components. These did not have the same success as the latest innovation in this vein of components, called ActiveX. ActiveX was a technology built on COM to make it easier to integrate graphical components into applications, instead of just functions. It was even possible to embed these in web applications by delivering them within the browser. At a time when browser security was not as extended as today, it provided lots of interesting features, but the technology is now outdated.

Other technologies for distributed components existed, such as Enterprise Java Beans and all platforms using CORBA, but they had the same limitations as DCOM and did not exhibit the level of low coupling that was initially promised. Version control was left to the maintainer of the platform, no capacity existed for a relationship with the presentation layer, and other shortcomings made for a limited future for these technologies that are nowadays pure legacy.

In fact, interoperation in the form of components might have been too humble to really reach the state of an ever-lasting technology such as ASCII, Unicode, HTTP, and some other norms that are so widely used, including as the basis for new approaches, that they will be around for the foreseeable future of software. Components started inside the perimeter of a single machine and never found a way to step out. A completely universal approach was necessary to make the next step, and it concerned bringing interoperability and reuse to a whole network of computers – and to be worth it, this had to be in the biggest network of all, the Internet.

Using web standards to try and get universal

The next milestone in our history of interop concerns web services, in the general acceptance of the term, meaning providing a service through web standards. The web was the obvious way to make this next step since its foundations of HTTP, TCP/IP, Unicode, and XML were already available and offered a good part of the foundation such a universal interop technology would need.

The first attempt at web services (or what we could call *reusable functions over the internet*) was implemented with technologies such as **Simple Object Access Protocol (SOAP)** and **Web Service Description Language (WSDL)**, and since these standards were alone in the field, they simply preempted the term *web service*, which became the accepted jargon for SOAP- and WSDL-compatible expositions. SOAP was about standardizing the XML content of requests and responses over HTTP to make them look like function calls, with an envelope, attributes with types, possible metadata, and so on. WSDL was the norm used to express the associated contract, in short, the grammar that was supposed to be used in SOAP messages. There were additional norms such as **Universal Description, Discovery, and Integration (UDDI)**, for example, but these fell short of their objectives and quickly declined. This was also the case for lots of the so-called *WS-** standards – WS-Authentication, WS-Routing, and other syntax additions to the web-service grammar that were intended to allow for complementary features.

These solutions gained a lot of momentum in the industry in the 2000s and were at their strongest in the early 2010s. In fact, they generated a whole architecture known as **Service Oriented Architecture (SOA)**. SOA should have remained a generic term, but it has become associated with particular architectures and software. Also, software manufacturers heavily invested in *SOA tools*, making companies believe that central middleware was all they needed to reach interop, while people knowledgeable in interoperability knew that this was only one part of the deal, with semantic and functional interoperability being, in fact, more critical and more complex to obtain than technical interoperability, which is generally the last mile in the process.

This of course led to lots of critics of SOA, and many articles in the mid-2010s announced the death of SOA and its failure to reach its goals. In the meantime, its spread was still huge in the industry and lots of technologies had a rebirth due to SOA.

The steps in middleware

Middleware applications in particular were pumped up by the SOA architecture and, even when they were not based on SOAP and WSDL, they were able to adapt to it. **Enterprise Application Integration**

(**EAI**) was an old dream and supposed that centralized adapters made it possible for many applications in a system to talk to each other, with the EAI platform translating every message from one format to the target format. Of course, its centralized aspect was a **Single Point of Failure (SPoF)** and quite a drawback. If you add to this an update of the EAI bricks, which is needed every time any of the applications change its version, it is no wonder that these customized systems never reached maturity.

Extract, Transform, Load (ETL) is a set of data manipulation tools, but they can be classified as middleware applications, particularly since lots of interoperability streams between applications in common information systems are really pure data transfer, and not business function calls. Of course, this is a crude middleware, but the quality of the data is more important than the sophistication of the tool, and a well-controlled ETL can go a long way in structuring streams of information. Still, ETLs are not completely adapted to digital transformation, and it can be easy to lose control of them. One of the companies I have consulted for had such a messy system of ETL jobs, with more than a thousand of them kicking up every night, that the whole system needed a dedicated tool to orchestrate the jobs in a precise sequence for them to end up in clean data in the morning. With the never-ending addition of new jobs, time started to become scarce during the low-activity periods, and after using simple solutions such as adding more server power and parallelizing what could be parallelized, it reached a point where the whole system would finish its work only after the offices had opened. This, of course, became an important problem that had to be dealt with using radical decisions. To make it worse, the jobs were so interdependent and brittle that there was never a night where all jobs passed and it was necessary to run a few correcting jobs during operations, or—for the riskiest ones—to simply wait for the next night to, hopefully, get the clean data. For informational purposes (and may it serve as a warning as well, since the sheer number of jobs makes the diagram almost unreadable), the following diagram shows a graph of the chronological execution of the jobs. This diagram is meant to be an overview of complex jobs; text readability is not intended.

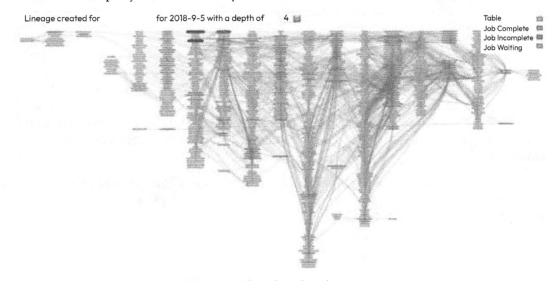

Figure 8.2 – Complex job orchestration

Message-Oriented Middlewares (MOMs) already existed for some time but got a kick from SOA and AMQP, ActiveMQ, MSMQ, and RabbitMQ gaining some market visibility by introducing robustness in message delivery (the WS-Reliability and WS-ReliableMessaging standards never really made it to the top, particularly because reliability needs to be ensured at the application level and not only in the messaging layer). Even in today's architecture, which does not use SOAP web services anymore, a MOM is useful to ensure full-featured transportation of particularly important messages in the system. Some MOM proponents argue that all messages should pass through the middleware, preventing applications from talking directly from one to the other, but this has a toll on performance and we will show that functional standards for messages allow the removal of the mediation layer.

As far as the mediation layer is concerned, MOMs benefited a lot from a standard way of manipulating messages that were defined by Hohpe and Wolf (`https://www.enterpriseintegrationpatterns.com/`), called **Enterprise Integration Patterns** (**EIP**). EIP defines some standard bricks for handling software messages, such as Multiplex, Content-Based Router, Enrich, and so on. By combining these basic bricks of message transformation or routing, a MOM was able to handle almost all possible functional situations. Apache Camel is the reference open-source EIP implementation, and it is used in many middlewares. The term *bricks* is particularly adapted to these patterns, as they can be explained with actual, concrete, Lego™ bricks: I have often used these to visually explain the concepts of software system architecture and in particular, how to make it evolve with minimal impact by introducing a mediation layer with composable actions, each of them handled by a simple assembly of Technical Lego™ bricks, as shown in *Figure 8.3*:

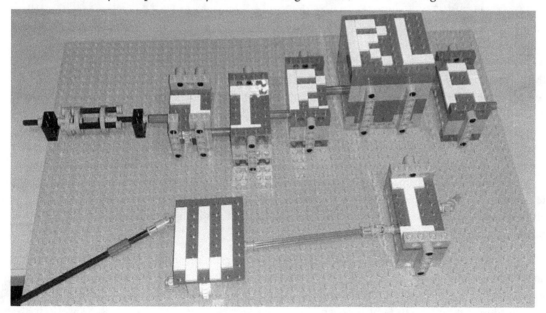

Figure 8.3 – Enterprise Integration Patterns simulated with Lego(TM) bricks

Enterprise Service Buses (**ESBs**) are the natural evolution of MOM and SOA, colliding with the principles of the internet. An ESB integrates all the technologies we have talked about in a system where there is no centralization anymore: the network (in TCP/UDP) is the only thing that remains central and its ability to adapt delivery is used to improve robustness to a node failure. At the same time, the *Store & Forward* pattern is used to make sure that the messages can almost never be lost since they are persisted and only deleted when the next destination has confirmed that they persisted under their control. ESBs had just about everything that was needed to reach the complete functional goal of interoperability in systems that were internet-scaled. But still, they failed, or at least did not succeed as much as would be expected for the ideal solution to such an important problem in the IT industry. In fact, ESBs added all the necessary features, but this was their doom. Since they could make everything, their heavy, complex machinery required extensive expertise to run and maintain.

The most recent evolution – REST APIs

Then came REST, which is a much lighter way of creating web-based APIs, and this changed the ecosystem quite radically again. **REpresentational State Transfer** (**REST**) had been defined before 2000 but became really famous in the early 2010s. In the 2020s, though the part of the legacy software is huge and SOAP web services continue to be exploited, no new project would start based on these old technologies and virtually every new API project is using REST, or at least some degraded, not really "RESTful" approach.

In a few words, REST is about going back to the basic mechanisms of HTTP to allow function calls on the web. For example, instead of sending the code of the operation in the envelope as SOAP does, REST uses HTTP verbs such as GET, POST, PUT, PATCH, and DELETE to instruct the server on what should be done. Instead of sending function calls, it deals with resources just as HTTP does with the more-known HTML pages or images that are served through the web; it just happens that these resources are functionally oriented, such as a customer or a contract. Each of these business entities has URLs just like a web page or resource has. Their representation can be in HTML but is more suited to XML or JSON, the latter of which is also lighter than its predecessor XML. Hypermedia, format negotiation, and headers are also used for the equivalent interop function. Authorization is simply left to the equivalent feature in any HTTP call, using Basic Authentication, Bearer tokens, and so on. In short, REST stripped the web-based interop to the bone and eliminated every ounce of fat to focus on the pure and complete use of existing standards. REST does not actually need anything else beyond existing standards such as HTTP, JSON or XML, and Unicode. In this way, it is more a practice than a new protocol.

And it worked… It actually worked so well that commentators on the internet did not hesitate to talk about SOA 2.0, or even *SOA made right*. Some introduced new architectural terms such as *Web-Oriented Application* to separate this approach from the original SOA. The best proof of success for REST is that no editor has built on its fame to try and impose a proprietary implementation: REST works well because it does not add anything but reduces any software layer to nil since everything already exists, and engineers only have to use it in the way it was intended.

What we are missing to reach actual service reusability

This is where we are at the moment of writing this book, and there is absolutely no doubt that the situation will keep evolving, but we have reached a point where actual web-based interop, including between two separate entities, is an everyday reality for lots of companies, which is already a huge victory in itself. Sure, we can always go further, but the main path has been paved and the remaining tasks now are mostly about spreading good practices in this way of interoperating rather than imagining a new approach that overcomes any current shortcomings.

In fact, most of the remaining problems are the presence of mediation connectors due to the lack of accepted formats for functional data exchange. If we want to reach the ideal place where global, universal interop will not be an issue anymore but rather a problem of the past, we would need to have an indisputable standard for each of the data streams. This is of course not possible and we are very far from such a satisfying state, but some precise, very widespread, and technically easy forms of exchange are currently covered. For example, authentication and identification are now well implemented by OpenID Connect, SAML, JSON Web Tokens, SCIM, and a few other norms. Sure, there are lots of legacy software and even expert engineers that do not use these, but the general orientation is that they are the future and everyone globally accepts this and works towards these norms, which will become convenience standards in the future, just as ASCII and Unicode are for text binary representations. A few other domains are covered, or at least have nice, fully-featured norms that could solve the problem, such as CMIS for electronic document exchanges or BPMN 2.0 for workflow modeling.

But the vast majority of exchanges are not covered by an indisputable standard and legions of connectors are still developed to establish correspondence between applications. This is a major waste of resources in global IT today, as these mediation connectors do not add any additional value to customers and end users. But the reality is that crafting a standard takes a lot of time, as we have seen in *Chapter 2*. Let's try to focus on the positive, though: the movement is now active and the situation is getting better every year, with a strong interop foundation where the technical bits are now considered solved. Only the semantics and functional interop remain to be handled. This will be the subject of the next chapter but, before talking about this, we need to come back to the very notion of *service* and explain how a good service should be defined. We will then use these principles to draft the first services for our demonstration system using the architecture principles shown in the previous chapter and applying them to the demo application that has been shown previously and that we will develop in greater detail throughout the rest of the book.

Characteristics of a service

Service is such a blurry designation that a complete section will be necessary to give a good sense—rather than a single definition—of this concept.

As a service explained

The expression *as a service* is used in many formulations: **SaaS** for **Software as a Service**, **PaaS** for **Platform as a Service**, **CaaS** for **Containers as a Service**, and so on. Have you ever considered why such different things use this common denomination? This in itself gives maybe the best definition of what a service is: something that benefits from the advantages of something else without having to deal with the usually-associated externalities. A hotel room is a service because you benefit from a bed and a roof without needing to buy and maintain a house, or even clean the room. SaaS is a service because you can use the software (manipulating its interface, storing data and retrieving it, realizing complex computations, and exporting the results) without having to install the software, buy a long-term license, operate it, install new versions, and so on. IaaS is a service because it offers what you expect from infrastructure (CPU power, RAM, I/O, storage, network bandwidth, and use) without you needing to worry about the hardware aspect of buying servers, operating them, renting some room, sorting electricity and cooling, securing them physically, renewing the hardware when there is a failure, and so on.

This explanation of the *as a service* expression was necessary because the word *service* by itself is very generic and one may be a bit lost as to why we talk about *service-oriented architecture*, then web services in the sense of SOAP web services, then services in the context of the web, and so on. When we talk about service in this book, we really mean service as a software function that is proposed to a user without them having to work on its implementation: the user does not have to know which platform is used, where the servers are, and so on. They only have to know the minimal information possible, namely a URL and the contract defining the exchange grammar, to interoperate with the service.

Does this remind you of something? Depending on only the functional definition of something, without any software-associated constraint? This is something that has already been exposed in the book, in particular where we talked about the four-layer CIGREF map model:

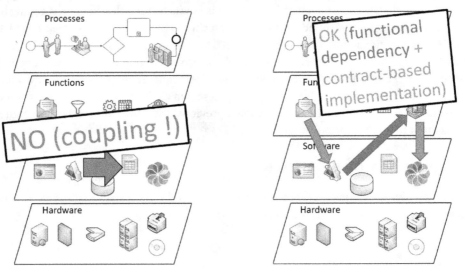

Figure 8.4 – Decoupling illustrated with the CIGREF map

When talking about providing a *function as a service*, one can view it as having something in the second layer from the top (the Business Capability Map) without having to worry about how it is implemented in the third and fourth layers (the technical ones).

Getting rid of middleware altogether

The nice advantage of the *as-a-service* approach is that it allows us to get rid of the middleware altogether. Indeed, what we really want to avoid is the direct, point-to-point interop that causes a lot of coupling, as shown in the following diagram:

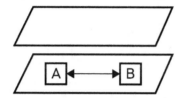

Figure 8.5 – Point-to-point interoperation

But the middleware, while introducing an indirection layer, poses two problems. The first one is that it introduces an additional software complexity, which can be hard to maintain. The second one is that we are still in the software layer of the CIGREF map, and this means that, if done badly (without standardizing the messages), we could very well end up with two steps of coupling instead of simplifying it! The following schema expresses this potential danger:

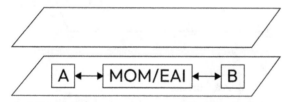

Figure 8.6 – Interoperation through a middleware

ESBs are often presented as a solution to avoid a centralized entity, but the way they actually work still implies the presence—though distributed—of software agents that can cause coupling:

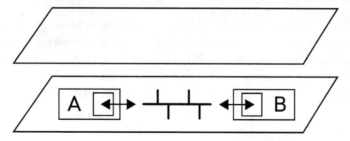

Figure 8.7 – Interoperation with Enterprise Service Bus

One way to avoid this coupling is to standardize the messages from a functional point of view:

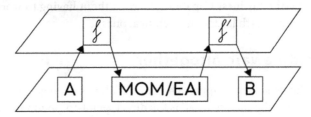

Figure 8.8 – Interoperation with standardized decoupled functions

But if we reach this state where a functional standard has been created, the middleware actually does not need to map data anymore or translate any format, because the f and f ' functions are actually the same (otherwise they would not have been included in a single stream of data). The middleware's sole functions remain routing, authentication, and some other features that can simply be realized by HTTP and do not need any middleware. Thus, the intermediate simply disappears and we reach the ideal situation that was expressed previously:

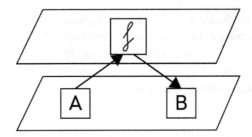

Figure 8.9 – Principle of decoupling by indirection

Here, the only difficulty remaining is a functional one, that of describing the business-related need. Admittedly, this can be a very difficult thing to do, but the main difference is that this is intrinsic complexity that we need to overcome in any case (otherwise the software will simply not work correctly) and not accidental, technical complexity that steps into our design phase and adds unnecessary problems of versioning, maintenance, and so on. This is the essence of decoupling and making it easier for the system to evolve.

Again, even though this is something we should strive to achieve in as many cases as possible and definitely a way to create some low-coupling interop, this kind of interaction is not always easy to realize. MOM and other middleware systems will not be retired any time soon, as they remain a good choice to interoperate complex messages, apply mediation, and ensure robustness of delivery when it is not possible to put in place a complete standardization of messages in the information system.

External interop finally becoming a reality

All of this may sound a bit theoretical, but this approach is what enables us to finally reach the stage where the interop between software A and software B in the preceding diagrams (*Figures 8.5 to 8.9*) does not depend on middleware or other artifacts that get in the way and make it complicated. The best way to show this is to provide a few practical examples.

In a company I worked for in the past, two customers (a regional council and a town) wanted to interop in such a way that, when the regional council added an association to its list, the city would automatically receive the information and store it in its own database, provided that it was the given city of registration. The way this was done necessitated some important preliminary work that had been done by my employer, which was to define a standard format for French associations. Since we knew the subject well, this took only a few days and we proposed this format to the French government for publication in their open source forge as they did not have any existing standard for this. This format was the functional contract between the two customers. They agreed that, whatever changes they might make to their software, the content of the association JSON would always be the following (this extract is highly simplified and translated into English for improved readability):

```json
{
    "name": "Old-time developers of Brittany",
    "registrationNumber": "FR-56-973854763",
    "organizationType": "uri:ORGANIZATIONS:ASSOCIATIONS",
    "creationDate": "2019-01-04T12:00:00Z",
    "representatives": [
        {
            "role": "accountant",
            "lastName": "Gouigoux",
            "firstName": "JP"
        }
    ],
    "legalAddress" : {
        "streetNumber": 282,
        "cityName": "Saint-Nazaire",
        "zipCode": "44600"
    }
}
```

It happened that the regional council was already a customer before this project, so they already were using our moral person referential software based on this format. So, on this side, we only had to customize the event management system to call the second customer callback address whenever the events of creation, modification, or removal of an association happened. This was done with the following grammar:

```json
{
    "webhooks": [
        {
```

```
                "topic": "POST+*/api/organizations",
                "callback": "https://saint-nazaire.fr/referentiel_
associations/modules/index.php?refOrga={registrationNumber},
                "method": "PUT",
                "filter":
"organizationType=='uri:ORGANIZATIONS:ASSOCIATIONS' and
zipCode=='44600'"
            }
            {
                "topic": "PUT+*/api/organizations/{registrationNumber}",
                "callback": "https://saint-nazaire.fr/referentiel_
associations/modules/index.php?refOrga={registrationNumber},
                "method": "PUT",
                "filter":
"organizationType=='uri:ORGANIZATIONS:ASSOCIATIONS' and
zipCode=='44600'"
            }
            {
                "topic": "PATCH+*/api/organizations/{registrationNumber}",
                "callback": "https://saint-nazaire.fr/referentiel_
associations/modules/index.php?refOrga={registrationNumber},
                "method": "PUT",
                "filter":
"organizationType=='uri:ORGANIZATIONS:ASSOCIATIONS' and
zipCode=='44600'"
            }
            {
                "topic": "DELETE+*/api/organizations/
{registrationNumber}",
                "callback": "https://saint-nazaire.
fr/referentiel_associations/modules/index.
php?refOrga={registrationNumber}&setActive=false,
                "method": "PUT",
                "filter":
"organizationType=='uri:ORGANIZATIONS:ASSOCIATIONS' and
zipCode=='44600'"
            }
        ]
}
```

To give a bit of explanation, webhooks are registrations of an external system to events emitted by the given application. In our case, when the regional council actors' referential service received data of a new organization or a change in existing data, through the referential service's API methods, associated events were raised and the aforementioned customization file extract associated these with calls of the provided URL. This URL was exposed by the second customer (the city of Saint-Nazaire) using PHP (but the specific technology doesn't matter). When, for example, we applied POST to a new organization, the callback URL was called with the identifier of the created entity along with the

`PUT` verb. This is also where we introduced the fact that the city was only interested in associations (not all organizations), and in particular, the ones based in their territory, with the `filter` attribute.

The URL implementation was then free to work as it pleased, without any dependence on the emitter. In some operations, the fact that there was an event on a given identifier was enough (for example, to deactivate the association in case of a `DELETE` order in the regional council information system). In some other cases, for example when an association was created, the JSON content—the exact grammar of which was agreed upon between the two participants—would be retrieved through a `GET` operation owing to the identifier obtained in the callback (where there was a use for all information of the association) or simply read in the body of the callback call (as the most important data were sent there, using the same contractual grammar, of course).

This example proved to be a successful experiment, as each of the customers was then free to evolve their systems in the way they wanted, changing technologies or other parameters without their partner even needing to know about it. At some point, the city would be interested in associations outside its own zip-code area and could simply register a new webhook content with the updated filter. This did not impact the emitter of the event, not even in its authorization scheme: if the city had requested to be called for associations outside its department (a French geographical unit between a region and a city), the event would have been sent, but reading the information with the help of the identifier received would simply end up in a `403 Forbidden` HTTP status code. This particular mechanism was something that initially made us decide to never send any data in the callback request in order to simplify authorization mechanisms. But, at some point, it was decided that forcing the called entity to always reply with a `GET` call to obtain the name and basic information of a new association was a waste of bandwidth. Performance was not so much an issue, but simplicity was more important in this context than the risk of authorization mishaps, since this data is public in France and easy to obtain.

Interop made real with standards

The preceding example demonstrated a case where a particular data schema (we call this a *pivotal format*, but we will come back to this in more detail at the end of this chapter and in the next one) had to be devised to exchange data in a free and decoupled manner. But an even better case is where this contract already exists in the industry. This is another practical case I had the pleasure of dealing with, in particular, because the small company I worked for by then forced a much bigger one to comply with our way of working, simply because we used a recognized standard. Let me explain the situation better…

Our flagship application, a kind of ERP, generates PDF documents and other binary files, and these should be stored. For quite some time, those would be stored alongside the database in a network share or, sometimes, in a dedicated server accessed through a UNC link. Electronic document management systems started to become mainstream after a few years and we needed to adapt our application so that it could use these systems to store documents. The natural choice for this was the Content Management Interoperability Services norm, as OASIS published a fully-featured 1.1 version supporting multiple metadata schemas, classification, versioning, and many more functions that we did not even need. It also happened that this was the only standard in use in this functional area, which makes for a very easy architectural decision.

So we ended up using a few operations from the standard (in the first step, we only needed to create documents, add metadata and binary content to them, and then retrieve documents through a query on their metadata content), which took us a few weeks to add to our application. Customers were quite satisfied because a simple customization of the software would make documents appear in their Alfresco or Nuxeo EDM systems, since these applications are natively CMIS 1.1 compatible. But what really demonstrated the importance of such a normative approach was the first time we had to deal with a customer equipped with a proprietary EDM: the editor, a quite large company, with an important footprint in the information system of our common customer, wanted us to make changes to our application in order to support their proprietary web services in order to send documents and metadata. After an initial refusal from us, the situation got a bit tense but we were lucky that the information system owner was a clever person who understood perfectly the value of low coupling. She intelligently asked what the effort would be for one partner if she had to select another supplier for the services this one talked to. The EDM provider stated that they would not have to do anything if our company was replaced by another one. As far as our company was concerned, I explained that—in the reverse hypothesis—we would have to rewrite some parts of the code to adapt to another proprietary protocol. This was enough for the customer, even if she was not a technical expert, to realize that something was wrong with this way of operating and to demand that a standard-based, contractual communication channel was used. Disapproved by the customer, the EDM provider had no choice but to implement, at its own cost, support for the CMIS standard in its product.

This proved a very satisfying experience for many reasons:

- First, I have to admit that replaying David against Goliath was one of the best ego boosts I had in my career.

- Second, we went out of the meeting without having anything to add to our software, since it was already CMIS-ready.

- Third, the customer appreciated our expertise in helping them reach a better, more evolutive, system and not trying to push them into a vendor lock-in situation as the other *partner* did.

- Fourth, the interop project was technically very easy to lead because we would simply provide the partner with a Postman collection of the API calls we needed to work and they were able to validate them from the CMIS norm point of view. There were no "hidden parameters" in the interop calls, everything was explicit and strictly regulated through the OASIS standard. We only had one tweak to add in the case of authentication.

- Finally, even the initially reluctant partner admitted at the end of the project that this approach helped to avoid the ping-pong effect in the project, where both partners reject responsibility for a non-working call to the other, ending up in a global loss of time and the customer not being satisfied. And I am truly convinced that the CMIS support would open new opportunities for their product further down the line.

Keeping complete compatibility

All this sounds like a beautiful dream, with pink unicorns and rainbows everywhere, but having great APIs using international standards and norms does not prevent one last danger in interop. Actually, it is quite the reverse, and the cleaner and more usable an API is, the bigger this danger is. Sounds weird, doesn't it? Welcome to Hyrum's Law (`https://www.hyrumslaw.com/`), which states the following:

With a sufficient number of users of an API,

it does not matter what you promise in the contract:

all observable behaviors of your system

will be depended on by somebody.

The more successful your API gets, the more important forward compatibility becomes as it is impossible to break the uses of many clients. But after all, this is just the flip side of success and not a bad price to pay if your API is the most used in your context, which ensures a large market share and notable income. Hyrum's Law is harsher because even some parts of the API that you have no formal engagement with will become things that get you into trouble. A sudden change in performance, for example, might make it impossible for one of your biggest customers to continue working with your API. Even a smaller, non-contractual, modification may get you into this kind of trouble. You know what? Even removing a bug might make some of your API users unhappy because—in some twisted way—their system depended on this particular behavior to operate. That may sound silly but it occurs more widely than you may imagine. After all, it is very common that some API users consume response attributes by their order instead of their identifier.

To a certain extent, Hyrum's Law can be considered as the API equivalent to the Liskhov substitution principle in object-oriented programming: even if a class can replace another one by implementing the same interface, if its behavior is not the same when the function calls have the same parameter values, then actual compatibility (and thus substitutability) is not achieved.

Managing APIs

Even if this is more of an operational concern, managing a number of APIs, with all the authorization access issues, logging, and possibly invoicing for API consumption, follow-up of versions, and so on can make for a tough challenge. Some dedicated software products exist for this under the common name of *API gateways*. They generally are implemented in the form of reverse proxies that act as a frontal server, hiding the actual API expositions.

Depending on whether you need a very low-coupled system or a very integrated one, you could respectively use systems such as WSO^2 or Ocelot (in the case of an ASP.NET implementation of your API system).

Inversion of dependency for services

If you remember the following schema from the previous chapter, you will recall that a port and adapter pattern is used in order for the satellite modules to depend on the main one that implements the business domain model, even if the calls come from the latter and go to the former:

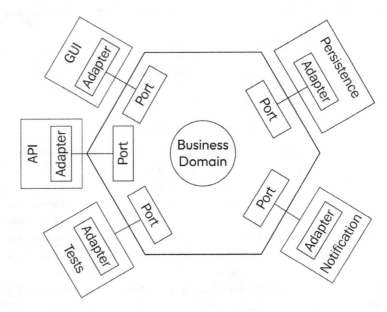

Figure 8.10 – Hexagonal architecture

This is simply the principle of dependency inversion applied to architecture, with the description of a conventional interface being called by one module, without knowing what implementation is used behind this interface. In **Object-Oriented Programming (OOP)** code, this is generally done by object injection.

In service-oriented systems, and in particular, when using web APIs, the indirection level is done by the URL that the caller uses without knowing what is behind it. If having a dependency on this module is not a problem, then the call can be direct. But if the business domain module calls an API, a direct dependency is not a good idea for evolution and a way has got to be found in order to reverse the dependency.

This is generally done by using some kind of callback mechanism, where the domain model module is instructed from the outside (the dependency, in our case) with the URL it should call, possibly in its customization but also in the runtime initialization steps. In the first explanation of the preceding webhooks, this is what happened when the town needed a change of filter on the events the regional council should take into account for informing the town: it would not be normal for the regional council to depend on the town, since the town is the functional requester of the information. This is why the best way for the town to provide the callback URL to the regional council is by registering for the events, possibly through a /subscribe API.

This way, we reach a nice separation of responsibilities, as the regional council is responsible for the following:

- Exposing an API that allows clients to create, modify, and remove organizations from the data referential service's persistence mechanism

- Exposing an API that allows clients (possibly other ones, possible the same ones) to register for events on organizations

- Calling the URL provided by these clients upon registration whenever the event appears in the code

- Applying the filter provided upon registration to only emit requested events

On the other end, the town is responsible for the following:

- Registering on the organization referential for the events it needs to observe

- Providing a URL for callbacks that is reachable, and points to the necessary implementation

When this kind of event-based mechanism is used for every interaction to provide a very low degree of coupling, the jargon term is **Event-Driven Architecture (EDA)**. In its most advanced form, EDA adds lots of very precisely defined responsibilities to allow for the following:

- Different authentication and authorization methods for the registration and emission mechanisms

- Management of robustness of delivery by reapplying the calls if necessary and, if needed, warning an administrator that, after a certain amount of tries, the event has been stored for later emission to certain registered clients

- Handling high volumes of events

- Handling large numbers of registered clients

- Service-level agreement management, among many other features

In its correct implementation, an EDA-based system is the most accomplished outcome of decoupling in software systems, allowing for a completely transparent evolution of the different modules and linear performance. But despite its long theoretical existence, there are very few actual implementations of this.

Now that the notion of *service* has been presented and studied from various points of view, we are going to return to our sample information system and apply this new knowledge to it.

Application to our demonstration system

Now that the notion of *service* should hold any secrets for you, it is time to see some practical applications of what we have covered on our demonstration system to reinforce the takeaways from this chapter. Since we aim at something modern, the choice is quite obvious that the different modules of the example system will interact with each other through REST APIs. As much as possible, we will try to keep the middleware as transparent as we can. We may need some connectors for mediation in some cases, but other than that, applications will talk to centralized APIs that will then be implemented separately (this will be done using the concept of service in the container orchestrator that will be put in place).

Interfaces needing analysis

First, we will start with a hexagonal architecture diagram to list all the business domain models and their dependencies. The C4 approach used in the previous chapter showed that we will need at least three business domains, namely books, authors, and sales.

If we concentrate on books, for example, the dependencies are the persistence mechanism, the authors cache module, the books' GUI system, the books' API controller, and some technical satellites such as logging, and identity and authorization management. This can be schematized as follows:

Figure 8.11 – An example of hexagonal architecture

In terms of the Agile approach, I am not saying that this contains all the interfaces that will be present at the end of our journey together. But in order to keep this exercise as realistic as possible, I am creating the sample information system at the same time as I write the book, in order not to leave anything hidden and so that you can follow the precise method of design that I recommend, and that I of course try to follow myself.

So, now that the first interfaces have been listed, we need to be a bit more precise than just a name. What are they going to do? How will they be designed to provide for clean, future-proof usage? Most importantly, how are these choices going to reflect the business/IT alignment principles that I have pushed forward since the first chapters?

Using norms and standards

Since I have spoken so much about the crucial importance of norms and standards, it would have been a terrible signal not to start with them for the precise definition of the interfaces. And describing more precisely the interfaces we talked about in the previous section is as easy as can be when we use a standard because we simply need to name it (and possibly cite the version of it that will be used) and all the operations, formats, semantics, and other functions of the standard are immediately clearly defined through the documentation of the standard.

For example, let's start with the authentication and identification service. For this particular interface, we will use the OpenID Connect protocol, based on OAuth 2.0 (RFC 6749) and using JSON Web Tokens (RFC 7519), the JWT profile for OAuth 2.0 itself being standardized (RFC 7523). Again, the great thing about norms and standards is that they greatly simplify our work. If I had to describe with the same degree of precision the use of an interface without a standard, this chapter would be extra long. For this service, citing a few RFCs (and of course, in the next chapters, using good implementations of these norms) is enough to make everything explicit.

How about the database interface or, to be more precise, the persistence interface? The decision is to use a NoSQL document-based approach since it sounds the most adapted to the business entities we talked about and the volumetry we want to address. It may not be a very well-known fact about MongoDB, but most protocols used are open standards, and are, in fact, used by many other NoSQL database implementations. If you want to improve on your local MongoDB database, all you need to do to switch to an Atlas service or an Azure CosmosDB instance is to change the connection string, as everything works the same. The MongoDB Wire Protocol Specification is licensed under a Creative Commons *Attribution-NonCommercial-ShareAlike 3.0* license. The BSON format (`https://bsonspec.org/#/specification`) used is documented openly and can be implemented by any software. And the list goes on. In addition to the appropriate adaptation of the software to our needs and the fact it is easy to create a free database, the standardized aspect is the cherry on top that makes MongoDB a sound choice for our sample application.

OK, now on to authorizations! There happens to exist two main norms around software authorization management, namely **eXtensible Access Control Markup Language** (**XACML**) and **Open Policy Agent** (**OPA**). In our sample application, we will use the second one, as it is more modern, less complex, and easier to implement together with our REST API approach. Actually, it may be argued that OPA is to XACML as REST APIs are to SOAP. One important thing: the presence of a norm does not mean you have to implement it fully: you simply use the parts of it that you need. This particular case of authorization management is the right context to explain this: if your access rights are quite easy (the `admin` role has all rights, the `operator` can read and write entities based on a portfolio, and the `reader` role can only read data, for example), then using OPA would not be the right choice, as it would add lots of overhead. Of course, the real question, again, is to take time into account. Of course, by the end of the book, our sample application will be so simple that using OPA would be over-dimensioning. However the goal of this exercise is to show how to work if we aim at a real, industrial, freely evolving information system. And since we operate under the hypothesis that rights management is going to be more complicated, then we will start right away with the adapted interface, which means OPA 1.0 in our case.

The logging feature is a bit of a different situation because this is not something directly functional, but rather a technical feature. However this does not mean that the same approach of standardization should not be used. The only difference is that this level of indirection will not be standardized at the international level as with other norms, but rather locally to the platform. Our sample application being implemented mostly uses .NET Core, so we will use whatever is standard for this technology, and there happens to be a standard global interface in `Microsoft.Extensions.Logging` called `ILogger`, which exists also as a generic class `ILogger<T>`. We will return in the technical chapters to see how to use it and maybe we will even spice it up by using a semantic logging system such as Serilog. But for now, suffice it to say that the logging mechanism will be standardized as well.

It is worth noting that some players in the field currently work towards a first level of standardization, such as Elastic with the ECS specification (see `https://www.elastic.co/guide/en/ecs/current/ecs-reference.html` for details). As Elastic is one of the major publishers of observation platforms and the specification is open source, we can place some hope in the spreading of this as a standard, although only time will tell.

Where do I find norms and standards?

When I teach or consult about business/IT alignment and in particular about this need to refer to norms and standards, the same question always comes up at some point: *How do we search for norms?* I should say that I am really astonished by this question, and for several reasons:

- Finding them is as easy as any internet search and virtually all of them are public, with the need to be as visible as possible in order to achieve their goals, so there is absolutely no technical difficulty in finding them. It shows that most people working in the IT industry (and with the number of people I have trained or taught, I do have significant statistics) are ignorant of the standards of their industry, which is quite annoying. I can understand that not a lot of people know BPMN 2.0, for example, as processes are a specific use case, and not all applications need a workflow engine. But how can some architects not know about OAuth 2.0, since this is used almost everywhere on the internet and almost all software applications need some kind of authentication at some point?

- Even some people outside of the profession know some of the most identified providers of norms and standards, such as ISO or IETF. Even just the term **Request For Comments (RFC)** is understood by many people. Granted, some IT-specific organizations producing norms, such as OASIS, are lesser known. But then again, the **World Wide Web Consortium** (**W3C**) is a very active and recognized institution. So how come people asking this question do not have the reflex to start with these organizations and search them for what they need?

For some customers, I even created a whitepaper at some point with almost a hundred norms and standards used in the business context I was working in at the time (public and government organizations) because this question came back all the time and I wanted to have a quick answer, not only telling them where to find what they needed, but providing them with the answers already found for them. This is for a simple reason: because I found out that the real issue was not that these people did not know "where to find the norm," but because it indicated doubts in their ability to use them. Norms and standards can be a bit intimidating with their hundreds of pages explaining all the possible cases. Even simple RFCs are indeed not easy to read.

But there are other answers to this as well:

- First, finding the right norm and starting to use it does not require you to read the norm specification. In fact, only if you need to implement large parts of it will you gain a benefit from reading it.

- In most cases, you will use components that implement the norm and all you have to do is check that they are recognized, well-established modules.

For example, in order to use OpenID Connect in our sample information system, we will basically need to know nothing about the protocol itself since we will rely on Apache Keycloak, which implements it in a transparent way for us. All we have to deal with is the choice of identity provider and some customizations made easy by the Keycloak GUI.

Even if you have to dive into the details of the norms, most of the time, you will only need to understand a very small portion of them. For example, in our sample application, we will certainly need at some point to implement some kind of support for binary documents for authors' contracts; which means we will of course use CMIS 1.1 since this is the recognized standard for this use case. But as we will only send documents, add binaries and metadata, and query documents in return, we may only use 10% of the whole norm.

Finally, a good norm is normally quite spread out and used internationally already. So, reading the full-blown specification is always an interesting read but let's be honest: the way you will be exposed to the standards in the first steps is simply by mimicking some sample calls that you will find on reference websites and adapt to your needs. Only if you reach a certain level of complexity will it be easier at some point to find the exact nitty-gritty detail of implementation in the full text of the RFC.

Pivotal format for the other interfaces

And for the last part of this subject, the next question that arises is, logically: *What do we do when there is no norm or standard for our context?*

My first reflexive reply to this question is always, *"Are you willing to bet that there is indeed no norm I can show you on this?"* Most of the time, this question goes back to the previous one and just shows that the person asking it is simply not comfortable with norms, or is afraid as they think it is going to be difficult (where in reality, on the contrary, norms free you from all the difficult design aspects).

Because, let's face it, we have norms for virtually everything today. All right, there may be fewer norms in IT than in the mechanical domain. But there are standards for every common feature. You have norms for all generic techniques, norms for every entity used in international data transfers, and for every common human activity including banking, insurance, travel, and so on. You even have an ISO-Gender norm (ISO/CEI 5218) for representing human genders in numeric format.

The second part of the answer concerns what we should do when there is indeed no applicable norm for your context. And the answer to this has already been given a bit earlier in this chapter: you then create what is called a **pivotal format**, which has the same goal of standardization as a real norm, but limited to your own context. Of course, it is always better to aim at something universal. Not only because, you never know, but your format may become a norm if you put enough effort into it and other people have an interest in it (this is the way norms appear: it always starts with the effort of an individual who knows the business domain extremely well and makes the effort to transcribe their knowledge into something technical, which is then agreed upon by other participants as a sound basis for exchanges). But also because aiming at something universal will make your pivotal format as close to a norm as possible, with as many resulting advantages as possible.

And the rule for this is to fall back on existing norms as quickly as possible. Sure, there does not seem to exist an international norm for the concept of authoring (though the Dublin Core `creator` attribute allows us to draw a link between a resource and the person or organization that authored this resource), but since it points to individual persons, lots of other related norms will quickly apply, such as Social Security Numbers for unique identification, ISO 8601 for the date of authoring, and so on. The same applies to books: of course, we may not find the perfect standard to precisely address what we need for our sample application, and in particular its persistence system, but there are nonetheless norms for languages (ISO 639), internationally-recognized standard codes for registered book identification such as **International Standard Book Numbers (ISBNs)**, and standards for virtually everything we will set out to record in the descriptions of the books in our system.

Now, the real question is what to put in the book and author's pivotal format? And this is such a huge question that it will necessitate a chapter on its own. The good news is that the following chapter will explain how to answer this.

Summary

In this chapter, I have used a short historical approach (a detailed one would be a book in itself) to explain what the stakes at play are in service orientation and how this seemingly simple yet hard-to-define word of *service* has been implemented in the past decades. We are definitely not at the end of the story yet, but nowadays, it seems the best approach is to use REST APIs with a middleware, reduced as much as possible through the use of norms and standards. This not only avoids the costly mediation connectors that translate one format to another, since everybody in the interaction talks the same language but also helps us know whether our design is the right one since consortiums and experts have thought a lot about this business domain.

Standardized APIs are what make it easy today to change some parts of important information systems without breaking them. They allow for international banking, much more efficient insurance systems, simplified travel abroad, and many other feats of the industrialized IT world.

We talked about norms, but also compatibility, the evolution of services, how services will be integrated through interfaces, and much more. By the end of this chapter, we came back to our sample application and showed which norms would be used to implement a few of the services it will expose. Now a difficult question remains: when there is no standard format for a business need and we need to create a pivotal format (of course, using norms as much as possible for its inner attributes), how do we determine the content of this format? The best answer I have is to use **Domain-Driven Design (DDD)**. And this is the subject of the next chapter.

9

Exploring Domain-Driven Design and Semantics

The previous chapter ended with a promise to provide a method to deal with a pivotal format and the design of an evolution-ready, functionally correct entity when no standard exists in this precise domain. This is the subject of our present chapter.

In order to reach this objective, a very important prerequisite is to keep thinking in *functional terms*. I know most of you will certainly have a technical background and may wonder when we are finally going to get to the code. Having you wait this long without doing anything technical has been done on purpose and is part of the pedagogical journey provided by the reading of this book. You have to stick to functional and business-related concepts as long as you can because, as soon as you transform this knowledge into software, it gets solidified and way harder to modify afterward. I promise that, as soon as the next chapter, we will start getting our hands dirty with some technical decisions. Then, in a few chapters, we will put some code together to actually show in very concrete terms what all this translates into. But for now, let's stick to business functions only and think about our format without any relationship to anything technical, just like we are taught by clean architecture. This is the main guarantee we have to build the right information system. Actually, if you remember only one thing from this whole book, I would love it to be this practice: think as long as you need to understand your problem from the functional point of view, and only then start thinking about how to deal with it from a technical point of view. Delay implementation as much as possible; think of data, not databases; think of models and business rules, not attributes and methods.

If you do so, you will soon have some questions about the vocabulary used in the business domain. A technical approach has the huge drawback of constraining the approach. But we should at least recognize that it pushes us into being extremely precise, as computers are as dumb as a box of rocks, so they oblige us to be explicit in information designation. Considering semantics and using a method called **domain-driven design** (DDD) will help us to be precise in functional terms, but without depending on anything technical that would hinder our evolution later on; this way, we get the best of both worlds.

Once the principles of this method are understood, again, we will turn back to our long-run example and apply DDD to our sample information system in order to draw its bounded contexts and describe its ubiquitous language (we will soon explain these two important concepts).

Finally, the chapter will explain how all of this applies to clean information systems as we try to design them, and what the links to the concepts of service and API are, which we have exposed in the previous chapter. There, we will talk about the importance of the life cycle analysis of a business entity and discuss a few recent orientations in information systems architecture.

In this chapter, we'll cover the following topics:

- A functional approach to a functional problem
- The importance of semantics
- DDD
- Application to clean information system architecture
- Link to the API and services

A functional approach to a functional problem

As explained in the introduction, it is of utmost importance to use a functional approach to solve the problem of designing a pivotal format. In the four-layer CIGREF map, all layers are a consequence of the one above. Hence, starting with layer 3 (software) without having a correct design of the context studied in layer 2 (business capabilities) is bound to create malfunctioning software. This is made much worse by the fact that once turned into software, the error will be fixed in code and possibly shared through APIs used by many users and machines all over your information system or external ones, which may make it almost impossible to correct the design error.

The vast majority of IT problems come from this lack of business alignment that we have talked about a lot and, right now, we are at the root of the problem: the design of the business entities. When we do not have any dedicated standard to rely on and spare us of a complex reasoning process, filled with risks of misunderstandings that may have important consequences, we will have to pay particular attention to any details, which, in practice, means extensive and guaranteed access to experts of the business domain.

A standard is generally expressed in technical terms in order to be extremely precise and irrefutable, but it represents, in a shared, acknowledged way, a functional concept. For example, RFC 7519 describes what a JSON Web Token is, and what the issuer, subject, expiration time, and all other attributes are for, but it does so in a very constrained way (with a precise definition of **must** and **should**, for example, in the dedicated RFC 2119, which indeed makes it an RFC about how to write other RFCs) and with very concrete translations, such as the exact name of the JSON attribute, which must be `iss`, `sub`, and `exp`, for the information we have cited previously. This way, we can say that a norm both lives in layers 2 and 3 of the CIGREF map and binds them together. This is why norms and standards are so important because they are the concrete actors of business/IT alignment. To give a second example, OpenAPI is also a great illustration of how norms and standards bridge the potential gap between functional and software approaches, by providing a list of all functions that should be accessible through an API on a given business domain, while, at the same time, giving a precise JSON or YAML-based technical description of what this means in the data streams that will be exchanged between servers.

A pivotal format should aim at the same result by joining the functional aspect and the technical representation of it. This is why it is important to describe it with technical means, whatever they are. These technical means can be XML Schema or DTD if you are using SOAP web services, or OpenAPI if you are designing API and their components; they can even be a simple Excel file showing the exact names and structures of the data messages you intend to move around your system. The only important thing is that it is technically written, but not technically restrictive.

The phrase "technically written, but not technically restrictive," may seem paradoxical, so, let me explain this. The technical writing of the norm is important because it ensures preciseness (nobody wants a vague description of something important). This is why a new API has got to be described in an OpenAPI contract. This way, there is no place for argument over how an attribute is written, with capital letters or not, or only the first one; for example, it is written in computerized text in the OpenAPI JSON or YAML, so there is no possible discussion. However, at the same time, attention should be paid to the fact that the pivotal format (like any norm) should never be constrained by any technical issues. We all agree that a norm to represent, say, countries, would not make any sense if the authors of the norm used some Java primitives that would make it difficult to use the norm with another platform. It would be a limited, technical way of doing it, but definitely not a real norm. The same should be applied to your pivotal format design and it should never expose anything from your technical implementation, even if it is expressed in a functional way.

By the way, this is yet another reason for always designing from the functional point of view before thinking of software. If you force yourself during this period to let go of the database choice, for example, you reduce the possibility of creating a pivotal format that is bound to your database orientation. I understand this might sound a bit unrealistic and you might wonder how someone could bind the design of data to a database. Well, the devil lies in details and there are sadly many ways—some subtler than others—to fall into this trap:

- One might express data attributes with types that are only available in some databases and not others. If we are used to talking about VARCHAR(n), for example, we might imply in our data design that there is a limit to the size of an attribute, though it is not justified from the functional point of view. Everyone has seen an application that truncates family names when they are too long, although this creates an incorrect data value.

- The same can happen for date formats. The norm-referenced ISO 8601 (also nicknamed as ISO-Time) makes a clear difference between administrative dates and instants, but most databases do not. If we think in SQL terms, we might miss this essential difference.

- Identifiers can be badly influenced by the well-known standard database mechanisms. The autogeneration of identifiers by SQL databases, based on counters, is quite practical but these identifiers scale very poorly and are one cause for the lack of distribution of such databases. **Globally unique identifiers (GUIDs)** are better and, quite often, used by more modern systems such as NoSQL databases. However, both will definitely be the wrong choice if you need to assign a unique identifier for an entity representing a patient in a health information system since the largely acknowledged (and sometimes legally required) identifier in this particular case is a national security number.

There are actually so many other cases where some technical knowledge can waste the design of a pivotal format that I have personally formed the habit of always designing them by animating groups only made of Product Owners, even going as far as detecting those with a technical background and excluding them from the design group in some cases. I still could badly influence the process, as I have a technical approach, but I generally help design pivotal formats on business domains I do have not much experience with, so, it is easy to play the role of a complete beginner, knowing nothing about the business domain and then focusing only on this comprehension. Also, I know from experience how early technical thinking can have a negative impact, so I always think of what could go wrong because of this.

Sometimes, the coupling can be extremely subtle. For example, let's take a URL such as `https://demoeditor.com/library/books/978-2409002205`. It sounds like a great identifier, since it is based on norms only (URLs, ISBNs for books, and a DNS for the host) and there is apparently nothing else. However, one could argue that prefixing with the (`https://`) scheme is already a hint for how we are going to technically access these functional entities, in this case, through web-based APIs. Luckily, a solution always exists and, in this case, this is through resorting to **URNs** (short for, **Uniform Resource Names**) instead of URLs (both are types of Uniform Resource Identifiers). Our example entity could then be identified extremely precisely but without any reference to technical implementation by using `urn:com:demoeditor:library:books:978-2409002205`.

At this point in the book, you should hopefully be convinced that taking the functional point of view on a problem is always the best option and that the technical aspects should come afterward. This being said we need some method to analyze a problem from a functional point of view only, and this is how semantics can be used.

The importance of semantics

In the previous section, we demonstrated how to use a technically-backed but not technically-coupled approach to define entity formats that will be precise. However, we have not yet touched upon the functional analysis itself and, looking at our example URN, `urn:com:demoeditor:library:books:978-2409002205`, we can spot what needs further analysis in the different parts of the string:

- `urn`: This is the scheme of the URI. It is here just to state that this is a unified resource name.

- `com:demoeditor`: This is the reverse of `demoeditor.com`, the domain name of our sample company. The information is there to serve as a prefix to differentiate the entities from another vendor that would have entities with the same name, and it is reversed in order to keep the information in a logical reading order from the coarsest to the most granular.

- `978-2409002205`: This is a sample ISBN. Again, as soon as we can, and this is essential inside pivotal formats, we turn it back to an existing standard. There are norms for virtually every single piece of information!

- `library` and `books` are the parts of the URN that bear some functional value and we have not yet explained where they come from. Let's say for now that `library` is the domain (the management unit of books and other entities related) and `books` is the chosen name to talk about these resources that `DemoEditor` manages. We will come back later to this.

Semantics is the science of relating meaning to words. In our case, since we have to represent existing real-world concepts in a software-based, virtual world, semantics is what is going to relate the concepts written as text in the IT applications to the concepts we need to manage. Of course, using the right semantics will be important not only so that IT reflects the right business concepts, but also for actors of the IT system to share knowledge based on applications and databases. This is more important than the first argument because, as far as computers are concerned, you could use `x24b72` instead of `books` and they would not mind at all; whereas introducing misunderstanding in the terms used by your information system is bound to create problems at some point.

Let me tell you of an anecdote on this: I was consulting at a company in the information domain and one of the workshops I had with them was about designing a pivotal format around the people who bought the information. Marketing people and salespersons were there and, at some point, their voices started to rise as they disagreed on terms that were used differently. Their argument was about the relationship between a *prospect* and a *customer*. Marketing explained that customers are the best prospects since they already know the company, whereas the salespersons replied that the commercial pipe was quite clear concerning the fact that a cold lead becomes a hot lead, then a prospect, and then a customer if he buys something, leaving—by definition—the status of prospect. In fact, they both were right and something was simply missing in the model: the fact that "customer" and "prospect" are not names of entities, but business rules. If one includes in the model the notion of product proposition, then things get clearer: a customer of a given product is indeed a great prospect for another product in the catalog of the same company, but they are still not a customer for this second product.

Reading this, you might say that this situation was benign and that no harm was done since the discussion cleared out the problem. This would lead to ignoring two things. First, this misunderstanding created some real tensions between marketing and commerce and incomplete future sales reports, which lasted for months before I had the chance to spot the problem in the workshop organized by the CTO of the company. Second, when there is just an oral misunderstanding, this is indeed fine, but the real problem is that this mistake has been solidified into the information system (remember that you should never start working on layer 3 before having a good understanding of the analysis context in layer 2). If this was just a mistake in a single company, that would not be so bad, but even ERP editors

(I will not cite any names) make the same mistake right into their default database models! Several of them sport data tables named `customers` and `suppliers`, which can cause lots of trouble.

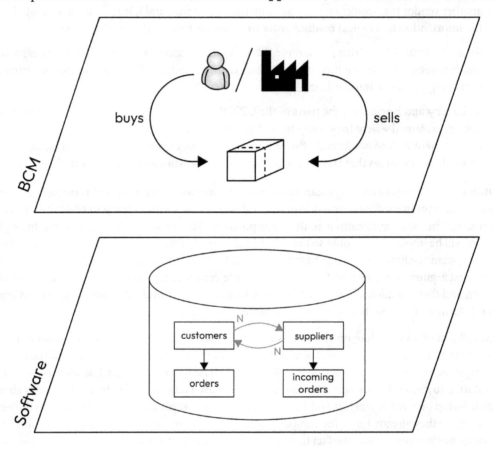

Figure 9.1 – Bad semantics

What happens in this case when a given company you work with is not only one of your customers but also one of your suppliers? This happens very often in the negotiation market, and, generally, people would not make this mistake. However, in this ERP, which I will not cite, the editor clearly did not do the job of understanding all of the markets they wanted to address and tried to propose a generic model that was not adapted to any company where this situation would happen. Of course, how do you think the consultants handled the problem when discovered? You got it: they tried to compensate for the layer 2 problem using layer 3 tricks. In a particular case, I remember, the consultants first started creating a database trigger that, when a customer changed address or bank coordinates, would replicate the modified information into the `suppliers` data table. Then, a few months later since the problem happened, they implemented the same trigger to modify the `customers` data table when a supplier was the modifier and created an infinite loop that crashed the database!

Things would have been so much easier if the data tables had been designed with a single `actors` data table (or `individuals` or `organizations` if you only deal with this kind of actors; again, semantics matters). The notion of a customer would simply arise from a business rule that states that an *actor* is a *customer* if a record exists in the `orders` data table pointing at this actor, and with a value date not older than 18 months. The same would be applicable to a supplier, which would be a business rule stating that an *actor* is a *supplier* if a record exists in the `incoming-orders` data table linked to this actor, or if an entry in the `equipment` data table has a guaranteed owner pointing at this actor.

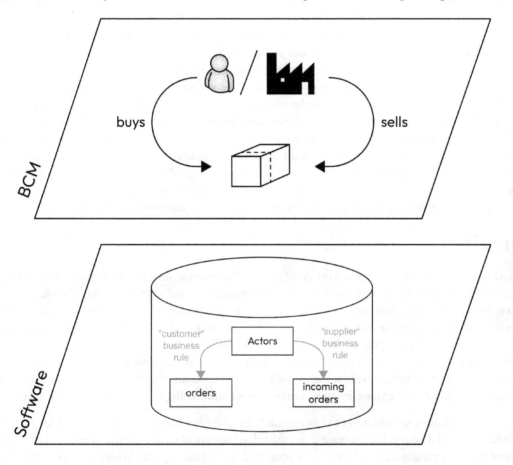

Figure 9.2 – Correct semantics

Those business rules are, of course, purely arbitrary ones, but notice that the entity schema does not change a bit if the rules evolve. This may be the most important thing in this model. If, at some point, marketing decides that the rule should be that the customers list contains only the actors with whom we have done business in the past 12 months instead of 18, what will happen? Here starts the real problems in the bad design, as you will have to create a migration routine to get your customers out of the table and activate the archive procedure. Since you may have orders pending, the risk is to lose a pointer to the right data and so many other things can go awry. With the right model design, on the other hand, what should we do? Well, simply modify the business rule! If it is in code, you can change 18 to 12 and recompile. If you were careful enough beforehand, this business rule is in a custom property somewhere and you will not even have to recompile or deploy anything. Also, if you had a reporting API producing the list of customers, then it is your lucky day: you modify this implementation and, without any other action, the behavior is changed everywhere in your system!

You might think that these examples are too easy and that this approach will not stand the complexity of a real system; it is actually quite the contrary because this approach is based on designing the business complexity in the software model. In the preceding example, for instance, we could very well have a different business definition of the addresses and the owner of the information system could decide that addresses should not be shared between customers and suppliers, or maybe only in some cases. For example, some addresses would be used only for the customers, such as the delivery address. No problem: we would adjust the model by keeping the addresses separate from the actors, and then add the "type" information to them, in order for the delivery address to be pointed at by the actors only when they are customers. We could even add some authorization rules on this to ensure that this address is never even read when the actor is seen as a supplier! Again, a good design would have allowed all this to be smooth but you have to get this clean design. Also, this happens to be one of the hardest bits of your architect job – gathering domain experts and coming up with something close to perfection. Fortunately, methods exist to structure this job. It is time to introduce DDD.

DDD

DDD (please note the last *D* is *not* for *development*, but indeed *design*) is a complete method for functional design that has been created and documented by Eric Evans, in his foundational book, *Domain-Driven Design: Tackling Complexity in the Heart of Software*, released in 2003 and, since then, has been famously known as *the blue book*. This important piece of work has influenced many software designers, despite it being quite a difficult read. Through its hundreds of pages, this book dispenses lots of best practices in modeling data and functional design. It is oriented towards software but everything it says can help, even before the first line of code, and it is a wealth of advice for understanding your business functions before you even start thinking about automating them through IT solutions.

That being said, our goal here is not to talk too much about the book, or to unroll the complete method. You have to read it yourself if you want to get the full advantage of such a seminal work, or watch Eric Evans' excellent presentation at `https://youtu.be/1E6Hxz4yomA`, where the expert explains the essentials of the method, which are as follows:

- Creative collaboration of domain experts and software experts

- Exploration and experimentation

- Emerging models shaping and reshaping the ubiquitous language

- Explicit context boundaries

- Focus on the core domain

What we will now do is show how some tools from the book can be used in order to help our design of a pivotal format, aiming at a nice business/IT alignment. Going back to our sample company, how could we describe what we are doing from a general point of view? One could say we are in the business domain called *book edition*. We need a subdomain for authoring and another one for selling. These two can be considered core domains since this is the bread and butter of our sample company: supervising the writing of books and selling them. There will also be some supporting subdomains such as human resources or accounting: those are not directly implicated in the core value-addition work of the company but are nonetheless absolutely necessary for it to work correctly.

The word "edition" here refers to literature in general but editors and salespeople do not have the same vocabulary for books: the former talk about a *work* and the latter about a *product*. Still, this is a similar entity. Also, they will not use the same attributes. Editors will be very interested in the number of chapters, the progress made in the writing, and other such attributes of a book, which is mostly a work in progress for them (when they go to sales, their job is basically done). On the other hand, a salesperson will check attributes such as the price of the book and maybe even the weight to calculate the transportation fees. Again, there are attributes, however, that are of interest for both roles: the number of pages, the ISBN of the book, the date of publication, and so on.

To solve these apparent paradoxes in the naming and potential difficulties in the separated management of attributes, DDD proposes two concepts.

The first one is the concept of **ubiquitous language**. DDD recognizes that different names can be used in different contexts for the same functional entity, and thus can account for local jargon while keeping a unique name shared between all actors of the information system. In our example, that could be "books," which is something sufficiently significant and still widely accepted to designate what editors call "works" and salespeople call "products." To be perfectly clear, DDD does not recommend finding a single expression for each concept and abandoning all others, but rather deciding on a given expression that will be shared by all actors of the model (hence the ubiquitous qualification). Local jargons are not forbidden, because they are generally useful for fast communication inside a given context, but the *standard* expression should be used every time there is the slightest risk related to misunderstanding.

The second concept introduced by DDD is the **bounded context**, which is the perimeter containing entities and business rules, inside of which the vocabulary is consistent. We talked about this context in which the alternate vocabulary can be used without causing trouble if limited to the actors of this context only; this context is indeed what is called the bounded context. Finding the bounded context in a complete business domain is important because it helps define where the interactions are and, as a consequence, where it is most important to be perfectly clear on the language. Bounded contexts can be aligned to business subdomains, but this is not mandatory. As we will see in the next chapter, the question of the entity life cycle has to be taken into account as well.

To summarize this graphically, see *Figure 9.3* for bounded contexts for our edition domain:

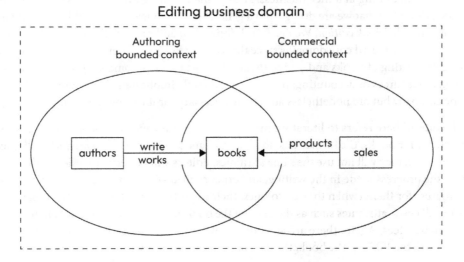

Figure 9.3 – Bounded contexts in Edition DDD

Since we obey an **Agile approach** to the design and development of our sample information system, we will not go further into the design than this very first step for now. Once we have applied this first level of knowledge to the creation of a first version of the data referential (see *Chapter 10*), we will dig a bit deeper as needed. Indeed, trying to cover the whole business domain would take too long and too many pages without adding anything to the understanding of the method. Before we move on, let us refresh our knowledge of **data referential**. Data referential is a service that is dedicated to handling data for a particular functional entity but also metadata, data history, authorization, governance, and many other functions as a complement to the traditional database that only handles persistence. Data referential is the basis for good **master data management**.

Application to clean information system architecture

Now that we are clear on the semantics and the domain decomposition of our business model, we can take a step forward in the design (though technical questions will only be introduced in the next chapter) and start envisioning how these entities are going to be introduced to the IT system. Until now, everything we said could be applied to a non-software-based information system. Starting from this section, we will admit the reality of there not being any such information systems in design anymore and that *every* company is now a software company. Since we are talking about entities, and their pivotal format is considered as designed, the next step is to talk about the way they will be manipulated—and thus stored—by the information system.

Using entities in referential applications

The very first question about storing and manipulating functional entities is about their decomposition. Since complex business attributes may have hundreds of attributes to qualify them, it is of course necessary to at least categorize them, and, if possible, create a tree-like structure to classify them. Entities always have some base attributes that are used by everyone in the information system, and the rest of the data attributes are mostly related to one subdomain, or, at least, one of these domains can be selected for each as the ideal maintainer of the data quality. This decomposition is often used to represent data referential as flowers (see *Figure 9.4*), with the core of the flower containing the shared data and the petals around the core containing the subdomain-related data. As a petal is always attached to the core, *Figure 9.4* shows that peripheral data has no meaning without identifying the core data of an entity. It also states that petals can be independent and that a flower without some of the petals might still be useful to some users. Finally, the metaphor shows that, if the core of the flower is thrown away, the petals go away with it.

The application of this approach to our book entity should be quite obvious, following what we had said before:

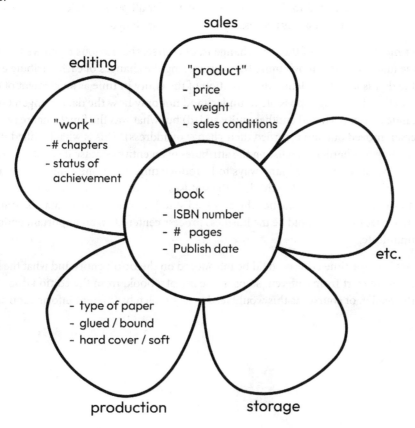

Figure 9.4 – Metaphor of a flower for referential

Though we had only talked about the two main petals before, there might be some others around the flower, such as the one about the physical production of the books, and the one about the storage of the printed units. Again, since this book is about the method and not about designing an IT system for a real book editing company, we will not go into these details; but you should definitely at least get to know all the petals in an entity flower when you design one for real, even if you do not get all the details of each petals in your first analysis.

Managing the life cycle of entities

Also, since we question the design of the storage, it is important to include time in the equation, as was explained in *Chapter 4*. A common mistake once an entity is designed is to think that we need to store and manipulate the data attributes that have appeared during this phase. However, there are many other things around the entity that have an impact on storage. Time is, of course, the first one and an important entity generally needs to have all its states in time persisted and not the last known one. In some cases, versions and branches of entities might have to be handled. Metadata on the entity (who created it, which state it is in, etc.) might be seen as a dedicated petal for history, but it generally is a complete set of data attached to the entity and available for all petals while still not being at the core of the flower, since it is not always necessary to have this metadata.

If we stick to time, traceability of the data change is, of course, the obvious thing we think about, but taking time into account is much more than just storing the changes of each attribute each time they are modified: it is about modeling the evolution of the entity in time as an element of business knowledge as well and making it possible to understand not only how the data changed (addresses an array with index 1 removed and another entity added) but what was the functional reason behind it (e.g., the person moved out and recorded their change of address). This is what is called the **entity life cycle**. Designing it is harder than listing the attributes of an entity because it is not a usual design activity and also because there are many ways to introduce time, each one complementary to the others. For example, it may be used to think of the statuses the entity will go through in its lifetime (created, draft, valid, etc., until it reaches the archived state). However, there may be times when having a design closer to what would be the business process centered on an important entity will be easier to communicate.

Figure 9.5 shows how the time criteria could be introduced on the book entity and what the life cycle would look like if we start from different steps in the life of a book, from the edition domain point of view (not the reader, of course, as this would lead to a completely different information system):

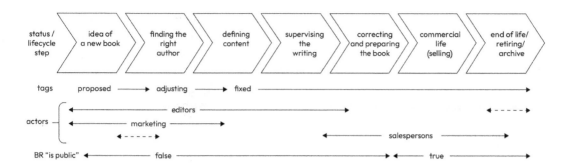

Figure 9.5 – The life cycle of a book

As you can see, the top part of the diagram shows what will happen in the lifetime of the book in the editing company. It always starts from an idea, even if this phase is extremely short, such as the idea coming from a meeting with a candidate author. In this case, the process will directly jump to the second phase. Phases look quite linear here, but that does not mean they have to be. For example, when a second edition is created, the writing and correcting phase will start again, but generally speaking, this diagram helps to envision the entity not only as the sum of data attributes but as a living object in the information system.

The evolution of an entity in time, of course, has an impact on many aspects of its data, but also on the business rules that apply to it. In the diagram, I only showed a few examples of such impact:

- Tags associated with the book in order to categorize them will evolve at first, but soon will be fixed and cannot evolve afterward, as it would create problems if the subject of a book evolved too much once the salespersons have started talking about it on salons, social media or to resellers.

- Actors working on a book will, of course, evolve throughout its life cycle: marketing will create the vision of the book, editors will help the authors create it, and at some point, after the content is reviewed and validated, the main actors will be the salespersons, until the book reaches archiving, when editors will work a bit on the book again.

- An example of a business rule has been provided, though there are always many in an important entity of the information system. In this diagram, I showed that the business rule is *public*, which is false as long as the book is not validated by the reviewers and becomes true after, with the particular case (not shown here) that, once public, a book cannot go back to being private, as people have been informed about it.

The notion of **status** is one that should be analyzed with particular care, as confusion often arises from the fact that it is often understood as data when it in fact is a business rule coming from complex business-related conditions. To give an example on the domain we have chosen as an exercise of application, the *ready to publish* status on a book may seem like the value of a `status` attribute that an editor would change on the `book` entity instance. But it may also be bound to a business rule stating that a book can become *ready to publish* once the following tasks have been completed:

1. Its main editor or two editors have given their vote for it.
2. The author has signed their contract and, in particular, the financial amendments.
3. The printing company has approved the provided files.

Business rules can also cascade on each other. For example, we could authorize payment of an author only if their bank details have been validated for less than three months, which means verifying the bank owning the account, which means, in turn, checking the SWIFT number is correct, and so on.

Finally, all of this is complicated further by the fact that some business rules may be stabilized as data at some point. This happens for performance reasons, where the computations are so long that it becomes acceptable that the result of the calculation is not always up-to-date (which happens when reading the value is more frequent than computing its result anew). There might also be some functional reasons as to why a state evolves against business rules and then gets fixed, without the possibility of going back for regulatory reasons (typically, this is what happens when an entity reaches an `archived` state: its content is then removed from the database and placed in archives, so, returning to an `active` state is not possible, since the data is now only accessible by archivists). In this case, the state overrides the business rule itself (or the business rule starts reading the recorded state and continues calculations if it is not overridden by this state).

Relation between subdomains and time

This notion of a life cycle is also important because it helps define important entities, and thus subdomains, in your information system. For example, a book is definitely a major entity in the domain because, as we have shown, it has a life cycle. Authors also have a life cycle in the information system because, as they are created in it, their contact data will change, they will hopefully write several books, and, at some point, will be erased from the database after a given time of inactivity (under which there is certainly a regulatory business rule, such as the GDPR in Europe, for example). However, tags are not an important entity, as they do not have a life cycle outside of books. Sure, a tag might disappear, but this will only be a result of there being no more books in this category. It is even easier to decide that the addresses of authors are definitely not a major entity, as they will never exist outside of an author, and they will always disappear with their parent entity.

The definition of the entity itself can evolve over time, which is completely natural in an Agile approach where we do not try to have everything conceptualized at first as we know that things will get clearer in time, and we should just prepare for the changes that will be added in the next versions (while still knowing enough of the business to ensure compatibility and smooth evolution of the system). The cutting of subdomains should normally never evolve in time. Additional domains may appear following a change of strategy of the company owning the information system, but there should be something majorly important to justify such a low-level change.

Now that we have seen DDD in action and its relation to a correctly structured information system, we are going to talk more specifically about the consequence of all of this in the design of API contracts.

Link to the API and services

We spent a lot of time talking about the entity from the point of view of its evolution and not with a data focus, but this is done on purpose because we generally take too much time defining the attributes and not enough thinking about the business entity as a whole, living object. Now that this is done, let's use the final section of this chapter to come back to the notion of services and API that we have detailed in the previous chapter.

Including time in an API

One of the first consequences of thinking about the entity as a whole, including its history, is that the writing methods on a corresponding API should not work exactly the same. As reading methods, they are similar to the following:

- `GET on /api/entity`: This is used to read the list of entities
- `GET on /api/entity/{id}`: This is used to read a given entity

However, if you want to take action and be able to access history, you should add some methods such as the following:

- `GET on /api/entity/{id}?valuedate={date}`: This is used to read a given entity state at a given date
- `GET on /api/entity/{id}/history`: This is used to read the full history of a given entity

Changes should also be made to the writing parts of the API, with one method that does not change:

- `POST on /api/entity`: This will always be about creating an entity instance. (Do not forget to follow the standards and send a `201` HTTP status code, together with a *Location* response header containing the URL of the resource just created.)

However, the traditional calls for an API are limited when you think in terms of a complete life cycle:

- `PUT on /api/entity/{id}`: This should not be allowed as it destroys the eventual consistency and the ability to avoid locks, as was explained before.
- `DELETE on /api/entity/{id}`: This should also be adjusted, not in its exposition but in how it works. Most of the time, since the resource is not really removed but only made unavailable by reaching an `archived` or `disabled` status, the equivalent modification call could be used in the same way and be more explicit.

Also, an existing but lesser-known verb should be used in order to act on the state of the entity:

- `PATCH` on `/api/entity/{id}`: This together with a request body content, following RFC 6902 (JSON Patch), should be used to write the data in a progressive, eventually consistent, and lock-free way (optimistic and pessimistic locks have been explained in *Chapter 5*)

- `PATCH` on `/api/entity/{id}?valuedate={date}`: This can also be allowed in some cases where the history of the entity does not strictly follow the flow of orders in the API server, and the value date should be taken into account

We will come back to these definitions in the next chapter, *Master Data Management*, and show the implementation of them.

Aligning an API to subdomains and the consequences

If you use a strict service architecture, all major entities, and thus all business subdomains, should have their dedicated process. However, since we said they should have their own API (minor entities will be under the major entities of the domain they are related to; for example, addresses will be in `/api/authors/{id}/addresses`), which equates to the fact that one API should always have its own process. Also, if you follow the rule of one process in one Docker container, you will have the equivalence of one API to one Docker service (scalability taken into account, as a service is a set of Docker containers of the same image).

> **Note**
>
> Docker is the most known implementation of the principle of containerization for software installation. This technology allows the deployment of instances of self-contained **black boxes** of software, which contain every dependency needed and remain isolated from other instances, while not necessitating a heavy mechanism such as virtualization.

If you consider that all calls should go through an API since they are the guarantee of unique business rules management and source of truth for a given API, then that means that nobody but the API exposition will ever access the application layer. In this case, why bother to separate these two layers? In the next chapter, we will follow this simple rule and implement API business codes directly inside the ASP.NET controllers. If you're wondering about validations and how they can be done as soon as possible, then deserialization will take care of a good part of all this, and preconditions inside the implementation code will do the rest.

We will, of course, keep things separated for everything that is a dependency, such as persistence and logging. However, as far as the business behavior is concerned, everything will be handled in the API code itself, in just one big block. This might seem like something that is not obvious given the principle of clear separation of responsibilities, but this has been done on purpose in this book and in particular in the code associated with it. It does not mean that cutting down into layers, as explained at `https://timdeschryver.dev/blog/treat-your-net-minimal-api-endpoint-as-the-application-layer`, is useless, but simply that the first versions of a sound and evolving information system can very well start with a very simple implementation of a restricted API, leaving it to future versions to evolve to an extended API content and something more sophisticated for its implementation.

API testability

One last thing on the API and its alignment with entities: nothing beats a nice Postman collection for manually testing the content of an API and then using these requests as the basis for a set of automated tests. There sure are other tools for specialized testing purposes but, in my personal experience, I have not yet found something as versatile as Postman for API discovery and testing.

> **Note**
>
> Postman is the reference tool for API testing. A collection is a set of HTTP calls that can be tested manually or in an automatic, sequential way.

If you can gather what your clients, internal team, and partners, external or public, do specifically with your API and integrate their code into your **quality assurance** (**QA**) Postman collections, then this is definitely the best way to ensure non-regression and backward compatibility. Sure, it will not replace unit tests and integration tests, but the former is a development tool, and the latter a QA tool. Everything in between will be beautifully held at the API level, which then becomes your interaction level but unifies with the testing interface of your model since it is aligned with the API. Whatever your level of interop, regression testing is best done at the API level.

If you completely follow the preceding principles, you end up with a perfect alignment of the following:

- One business subdomain
- One major entity
- One API contract (in OpenAPI format)
- One Git repository for the code implementing this API
- One process for the delivery of this code
- One Docker image for the deployment
- One orchestrator service for the running of API calls in a coordinated way
- One Postman collection for the tests of the API

Summary

Here we are, finally reaching the point where we will get into the code! The previous chapters paved the way for a global understanding of most of the constraints of creating an evolution-capable, feature-rich information system. In this chapter, we saw how we should enter details about the data of entities and also their life cycle to create a clean and future ready architecture.

DDD and the method shown previously using semantics will hopefully help you find the best way to structure important entities in your information system. The right schema makes exposing these entities through APIs easier and more loaded with functional value. This approach also allows the evolution of the system in the smoothest possible way, as technical evolutions and functional ones should be separated if the design has been correct. This way, not only is the information system better in its current form but it will also be much easier to evolve.

In the next chapter, we will see how the functional entities that we have designed are going to be implemented in the technical layers. We will not go into the code details right away, but we will start with how the data will be organized in the logical servers, how the entity life cycle that we talked about will be implemented in the software applications that will be put in place, and why master data management and data governance are important for ensuring that these nice functionally correct pivotal formats we designed in this chapter are efficiently exploited.

10
Master Data Management

In the previous chapter, we showed you a method to design information entities in such a way that they do not have any technical coupling, in an effort for the information system containing them to be free to evolve when the business changes. If the data model is a pure reflection of the business represented, it makes it much easier to follow business changes (and change is the only constant) because there won't be some technical constraint in our way forcing us to compromise on the quality of the design, and thus on the performance of the system as a whole.

In this chapter, we will start talking about the implementation of the data model into something concrete (if this can be said about software, which is mostly virtual). It is only in *Chapters 16 to 19* that we will code what we will call for the rest of the book the "*data referential(s)*". For now, we will start some actual software architecture to welcome the data model, persist the associated entities, and so on. There are many responsibilities in the data referential, and the discipline of handling these essential resources in an information system is called **Master Data Management** (**MDM**). At first sight, these responsibilities might look like those you would trust a database with, or even find in a resource-based API. But this chapter should convince you that there are many more things to the model that justify this use of a neologism like "data referential".

In addition to defining the functions of the data referential, MDM is about choosing the right architecture, defining the streams of data in the overall information system, and even putting in place governance of the data, which involves finding who is responsible for what action on the data in order to keep the system in shape. Having clean and available master data may be the single most important factor in the quality of the system. Reporting cannot be done without clean data and most business processes depend on the availability of the data referential. Also, some regulatory reasons, such as accounting or compliance questions, demand high-quality data.

After showing the different types of data referential that you may encounter – or create – in an information system, we will finish this chapter with an overview of possible issues with data, patterns of use, possible evolution of data in time, and some other general topics that will hopefully provide you with up-to-date knowledge about the MDM architecture.

Responsibilities around the data

The concept of the data referential as a unique point of truth for the data entities of a given domain has already been explained globally, but we have not formally described what functional responsibilities are contained in it. This section will explain each of the main responsibilities and features of the referential. Looking at the responsibilities explained in the following subsections, you might ask why we're talking about the data referential instead of simply using the better-known expression of a database, but we will see in the second part of this chapter that a referential is much more than this.

Persistence

Persistence is the responsibility that immediately comes to mind when we talk about managing data. After all, when we trust an information system's data, the very first demand we have is that the computer does not forget it once it has learned about it. This demand is crucial, as even an electricity failure should not have an impact on it. This is why databases were invented and why engineers went such a long way to ensure the safe travel of data between memory and hard disks, both ways.

Persistence may be often reduced to **CRUD** (which stands for **Create, Read, Update, Delete** – the four main operations on data), but this concept is way too limited compared to the features encompassed by the data referential, though it is enough for most of the standard uses of low-importance data in the information system. Since we talk here about primary data used in many places in the information system, some other aspects of persistence have to be taken into account. The first one was talked about at length in *Chapter 4* – namely, time. When one includes time in the MDM equation, storing the so-called "current" state of the data (which, most of the time, is only the last-known or best-known state of the corresponding business reality) suddenly becomes much more complicated and means, at least, storing the different states of data over time, with an indication of time to follow the history of these successive states.

As we explained in *Chapter 5*, a good MDM is a "know it all" system that, instead of states, should store the actual commands modifying the data to enable us to retrace why the state of such an entity has evolved. This means that what will be written in the database is not a state with a date but, ideally, a "delta" command causing a change from one state to another – for example, modifying the zip code in the first address of an author in our sample information system. This way, not only can we reconstitute the state of a business entity at any time in its life cycle but we also avoid the complexity of optimistic/pessimistic locks, transactions, data reconciliation, compensation, and so on.

Metadata is also an important addition to the simple CRUD approach. Indeed, it is of great importance in the manipulation of master data to be able to retrieve and manipulate information linked to the data changes – for example, its author, the IP of the machine where the command came from, the identifier of the interaction that has caused this change, the actual date of the interaction, maybe also a value date if it has been stipulated by the author, and so on. This allows for traceability, which becomes more and more important for the main business entities in an information system. It also provides powerful insights into the data itself. Being able to analyze the history of the data will help you fight fraud (for example, by checking which entity changes its bank coordinates often, or limiting how many representatives of a given company can change in a given period of time). It can also help with some regulatory questions that are becoming more and more common, as we will see a bit later when talking about data deletion.

When talking about persistence, we often think of a given entity (and I, for one, have only been given examples of such atomic manipulations previously mentioned), but the ability to manipulate masses of data is also an important responsibility of the data referential. In most cases, this translates into being able to perform actions in batches, but the consequences are also in terms of performance management and the capacity to handle referential-wide transactions (which are very different from business entity-centered translations, which the data referential should help eliminate).

The question of identifiers

As soon as a business entity unit is created, the question of how to identify it arises, since persistence is the capability of retrieving data that the information system has been given, and this naturally means that a deterministic way must exist to point at this given entity. At the very least, a system-wide identifier should exist to do so. It can take a lot of forms but, for the sake of applicability, we will consider the following as a URI, for example, `https://demoeditor.com/authors/202312-007` or `urn:com:demoeditor:library:books:978-2409002205`. This kind of identifier is supposed to be understood globally, by any module participating in the information system. It acts a bit as the ubiquitous language in Domain-Driven Design but allows pointing at a given entity instead of defining a business concept.

Of course, local identifiers may exist. For example, the book pointed at by `urn:com:demo editor:library:books:978-2409002205` could be stored in a MongoDB database where its technical `ObjectID` would be `22b2e840-27ed-4315-bb33-dff8e95f1709`. This kind of identifier is local to the module it belongs to. Thus, it is generally a bad idea to make it known by other modules, as a change in the implementation could alter the link and make it impossible for them to retrieve the entity they were pointing at.

An entity can also have business identifiers that are not local per se but bear no guarantee of being understood anywhere in the information system. The book generally identified by `urn:com:demoeditor:library:books:978-2409002205` could be retrieved only by its 13-digit ISBN `978-2409002205`; in fact, it is the variable part of the unique system identifier. However, other identifiers exist. For example, the same book can also be retrieved by its 10-digit ISBN, which is `240900220X`. Business identifiers can also be created inside the information system for particular uses. In our sample edition company, one could imagine that a serial number is applied to a book to keep track at the printing station, where batches are used and a single integer might be easier to handle than a full-blown ISBN, without risking any confusion as the workshop only prints books of the sample editor.

Additional technical identifiers are more often encountered, particularly in information systems with legacy software applications. Indeed, those generally insist on having their own identifiers. This way, the accounting system of *DemoEditor* might know the `urn:com:demoeditor:library:books:978-2409002205` book by its local identifier, `BK4648`. The ERP system might have a technical identifier of `00000786` if the book is the 786th product that has been entered into it. And so on. Of course, the dream would be that all software applications are modern and can handle an externally-provided, HTTP-standards-aligned URN. But this is rarely the case and even modern web applications seem to forget that interoperating with other applications means using the URL that they provide indiscriminately.

To provide a good service and account for this reality of information systems, the data referential should provide the capacity to store the business identifiers for the other software modules participating in the system. At the very least, this should be a dictionary of identifiers associated with an entity, with each value pointed at by a key that globally identifies the module in the system. For example, `urn:com:demoeditor:accounting` could be the key that points to `BK4648` and `urn:com:demoeditor:erp` could point to `00000786`. When defining the keys, there is a natural tendency to use the name of the specific software used to implement the function, and it would not matter much because the identifier is indeed specific to this software. But it still remains a good idea to stay generic in order to prepare for any cases. To give just an example, in the fusion of two administrative regions in France, it proved very useful to have such a separation. The two existing software applications for finance management were competing to have a unique market after the merger. It happened that one of the software applications was more customizable than the other and could handle external identifiers, which was part of the decision to keep it as the new unique finance management application. However, since the identifiers used by the abandoned software were prefixed by a vendor mark and the key for the software that stayed was not generic but used its name, there were some strange associations of identifiers such as `urn:fr:region:VENDOR1=VENDOR2-KEY`. Since the two brands were well-known competing companies in France and the merger of the two administrative regions caused lots of team modification and change management, this additional confusion quickly became an irritant, with people not even able to tell which software they should use to manipulate financial data. In the end, switching to a generic key such as `urn:fr:region:FINANCE` really helped, even if this sounded like a little technical move.

I will finish this review of identifiers with a very special case, which is the change of business identifier. Identifiers are, by essence, stable since they should be a deterministic way to point at an entity in the information system. A documented case of a change of global identifier is when a social security number is designated to a person who has not been born yet, typically because surgery is necessary on a fetus. As the first digit in French social security numbers uses the ISO gender equality standard to specify the gender of the owner, it may happen that instead of using 1 (for male) or 2 (for female), a social security number starts with 0 (for unknown). The identifier is then changed to a new one after the birth of the individual since the first number is then known (or maybe unknown in some other conditions – in this case, the norm specifies the number should be 9 for an undetermined gender). This is admittedly a very special case that provokes the change of the global, system-wide identifier.

However, the architecture of the system has to be able to handle *any* existing business case (which does not mean there cannot be some manual adjustment for these cases) to be considered "aligned."

The single entity reading responsibility

Persistence really means nothing if data stored somewhere cannot be retrieved afterward for subsequent use. This is why reading data is the second responsibility of the data referential that we will study. This section details the different kinds of read operations and, contrarily to persisting the data, they are actually very diverse in their forms.

The first reading act we naturally think of is the retrieval of a unique entity, directly using its identifier. In API terms, this sums up as calling a `GET` operation on the URL that has been sent back in the response under the `Location` header when creating the entity. Or at least, this sends the latest known state of the data because parameters can be added to specify which time version of the data should be retrieved. This normally raises the question of how to get the state of data since we said we would store changes, not states. The response there can be simple or complex, depending on the level of detail we go into. If we radically apply the "*premature optimization is the root of all evil*" principle made popular by Donald Knuth, then it is enough to specify that states can be deduced from changes by applying them to the previous state and consider this recursion uses the initial state of the date, which is an empty set of attributes designated by a unique identifier.

I know very well that most technically minded people (and thus at least 99% of you reading this book) will always think a step further and ponder the huge performance problem the data referential would have to deal with if each `GET` operation caused the iterative application of hundreds of patches to an entity in order to find its state at some point of its life cycle. The very least we would do would be to cache the calculated states to improve on this. But when you think about it, the vast majority of read operations ask for the best-so-far state of the entity, which is the latest known state. So, to improve storage while still keeping good performance, caching the last known state of the entities is the right choice.

But there are, of course, some exceptions and, as has been many times explained in this book, business-justified exceptions have to be taken into account – not only because it is the goal of the alignment but mostly because these exceptions are generally great challenges on the data model and if it can accommodate them while staying simple, it means this design is mature and has a much greater chance of correctness and, hence, stability. One such exception can be when the data is often read using a `date` parameter value. In this case, improving the performance might mean storing all calculated states, but this uses lots of storage and wastes most of it, as not all states will be called in time. A good compromise might be to store only a state calculated every 20, 50, or 100 changes. This way, we can start from an existing state all the time and quickly calculate the specified state because we need only apply a few limited patches to the data. Depending on business constraints, some states that are more often used than others can be the milestones that are kept in the cache. For example, in financial applications, it is generally interesting to keep the value just before and just after the change of fiscal year.

Another detail that has got to be taken into account is the optional possibility of inserting modifications in the life cycle of the entity. I understand how this may sound weird to "rewrite the history" and insert changes with potential impacts on the following ones, but there are some cases where this makes sense. For example, I have seen this happen in accounting systems when errors have been made and calculation rules were reapplied to find the correct result, inserting correcting operations at the time the initial error arose. Again, this is a rare case and it should be conditioned by strict authorization rules, but the situation has to be cited for the sake of exhaustivity.

Other kinds of reading responsibilities

There are cases when the unique system-wide identifier of a business entity is not known or has been forgotten (which means not stored outside its original referential) and, in this case, the responsibility of searching the entities corresponding to given criteria has to be used. This responsibility is often called **querying data**. Based on the criteria specified in the request, the operation will return a set of results, which can be an empty set or one that contains corresponding data. There can be cases when the query attributes are such that the results will always contain zero or one entity – for example, because the constraint used is a unique business identifier. But there can also be cases where the results are particularly numerous, and an additional responsibility called **pagination** will be quite useful to reduce bandwidth consumption.

Pagination can be active (the client specifies which page of data they want) but also passive (the server restricts the amount of data and provides a means for the client to request the next page of data). A standard way to implement the first approach is to use the $skip and $top attributes, as specified in the **OASIS** (short for, **Organization for the Advancement of Structured Information Standards**) standard called **Open Data Protocol (OData)**. This standard also includes a grammar for the possible values of the $filter attribute, which is used to specify the constraints reducing the query results that have been cited previously, when talking about performance in retrieving data. This book is not the place to explain the richness of this standard, which is sadly not used as often as it should be. Most API implementers indeed chose to use their attribute names, without realizing that they recreate functions (such as pagination offset, for example) that have been done so many times that they are completely normalized. Lack of interest in standards, and a "not invented here" syndrome that many developers suffer, are dragging our whole industry back. But enough ranting about this: a complete chapter has been dedicated to the importance of norms and standards, so we will just close the subject by taking you to the study of the OData standard, or in this case, the GraphQL syntax as well, since these two approaches can be seen as competing (though they are complementary one to the other, and a great API exposes both protocols).

Another type of reading responsibility is reporting: this can sometimes be implemented directly by the data referential but this is quite rare, as reporting is often done by crossing data coming from several business domains. Even if there are only a few of the reporting needs that demand such an external, shared-responsibility implementation, then it is better to handle all data for reporting to this entity. Depending on the technology you use, this may be a data warehouse, an OLAP cube, a data lake, or any other application. Again, the implementation does not really matter: as long as you are clean on the interfaces, you may change them any time you like with limited impact on the system.

In the case of reporting, these interfaces can be solicited with different time-based approaches:

- A synchronous, on-demand, call is always possible but generally not used for performance reasons, at least in complex reports (this is the "pull" mode). Indeed, if the reporting system needs to wait for all sources to answer and then only calculates the aggregations on its side, the results are, of course, as fresh as possible, but they may take minutes to come and this is often not acceptable by users.

- The asynchronous, regular read of data, is the most commonly used pattern. Here, data is collected at a given frequency (once a day or more often, sometimes down to once an hour), generally by an ETL, and sent to the data warehousing system where it is aggregated and prepared for reporting. This way, reports are sent to users quicker (sometimes, they are even produced and made available directly upon data retrieval). The counterpart is that the data is not as fresh as possible, and moving the cursor to a quicker sending of data increases the consumption of resources. Optimizations are possible – for example, by reducing the transfer to only new or updated data – but this only goes some way into improving the minimal time needed to update the whole data warehouse. The greatest technical drawback of this approach is that most of the calculations are reproduced even if the source data has not changed, which is a waste of resources.

- To go on further in the "push" approach, it is possible to use webhooks to register data refresh to an event of source data change. This way, the calculations are reproduced only when the data has changed, and the moment is as close as possible to the interaction that has changed the data, which means the reports are very fresh most of the time. Dealing with large amounts of events is a challenge, but grouping the changes into minimum packages (or with a maximum freshness time constraint) can help.

- A very modern but technically demanding approach is to mix these "push" and "calculate on demand" strategies by using a system with queues of messages containing data changes and a dedicated architecture to apply fine-grained computations on each of these messages as needed. Such implementations of a big data approach include Kafka architectures or Apache Spark clusters. The goal here is not to detail these approaches but just to explain that they will collect all events at the source and then smartly calculate the consequences in aggregated data (the smartness being in the fact that they know the consequences and calculate only what is needed and they can balance these computations on many machines of a cluster and grouping the result in the end). They can even go as far as producing the final reports on aggregated data and making them available to end users, achieving a complete "push" paradigm.

These four approaches are symbolically represented in the following schema:

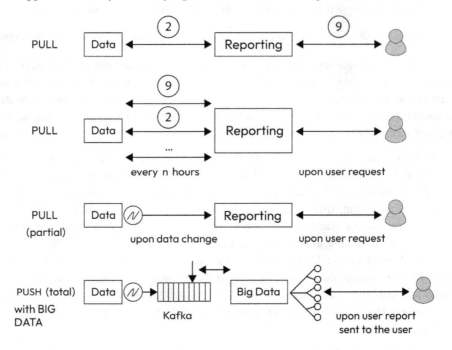

Figure 10.1 – Modes of reporting

To be exhaustive on these additional reading responsibilities, indexing is another function that is used to accelerate data (and some simple aggregates) reading. It does not go as far in data transformation as big data and the preceding reporting approaches, but can already prepare a few aggregates (such as sums, local joins, and so on) and make them available through simple protocols as raw data. Indexing engines such as SOLR or Elasticsearch are generally used to accompany the data referential on the speed of data retrieval. In this case, the data referential itself concentrates on data consistency and validation rules and then handles reference data to the indexing system to make it available in quick-read operations.

The complex art of deleting data

If deltas are stored instead of states, there is not much difference between POST, PUT, and PATCH operations on a resource, as they all translate into a change in the state of the entity, the particular case of a resource creation being a change from something completely empty. But as far as the DELETE operation is concerned, we are in a different situation. Indeed, we could blindly apply the same principle and consider that DELETE removes all attributes of an entity and brings it back to its initial state, but that would not be exactly true as the entity still keeps an identifier (otherwise, one would not be able to delete it). This means it is not in the same state as when it did not exist and there is no way to go back to this situation.

The best way to handle the situation is generally to use a particular attribute of the date stating that it is not active anymore. When using a `status` attribute to keep the value of the calculated position in the life cycle of an entity, this attribute may be used with a value such as `archived` to realize a similar operation. This is the way the data referential can store the fact that the data has been deleted without actually suppressing data (which is incompatible with what has been said previously about the data referential and its responsibility for history persistence). Of course, this creates a bit of complexity in the referential because it has to take this into account in every operation it allows. For example, reading some piece of data that is inactive should react as if the data did not exist (with a `404` result, in the case of API access), unless in the exceptional case that the user accessing the referential has an `archive` role and can read erased data. Other questions naturally arise then, such as the possibility of reactivating data and continuing its life cycle (hint: it is generally a bad idea, as many business rules are not thought to handle this very peculiar case).

But let's stop this digression here and come back to the initial idea of data conservation even after a deletion command has been issued. The functional rationale behind this is mainly regulatory, such as traceability, but also prohibiting data erasing for other purposes, such as forensics after a cyber-attack. An interesting fact is that some regulations also exist that specify when data should indeed be erased for real (and not simply rendered inactive). For example, the European GDPR states that personal data should not be kept longer than some legally defined periods, depending on the processes they are associated with. In the case of personal data collected for marketing reasons (with the consent of the user, of course), the delay is generally a year. After this time, without renewal of storage consent, the data shall be erased from the information system that has collected it. That means the actual removal of the data everywhere it may be (which includes backups).

Relation to specialized links

As always, the devil lies in details, and links can become a problem when dealing with data. Imagine we use a link between a book and an `author` entity. The simplest expression of such an RFC link is the following:

```
{
    „isbn13": „978-2409002205",
    „title": [
        {
            „lang": „fr-FR",
            «value»: «Open Data - Consommation, traitement, analyse et
visualisation de la donnée publique»
        }
    ],
    "additionalIdentifiers": [
        {
            "key": "urn:com:demoeditor:accounting",
            "value": "BK4648"
```

```
        }
    ],
    "links": [
        {
            "rel": "self",
            "href": "https://demoeditor.com/library/
books/978-2409002205"
        },
        {
            "rel": "author",
            "href": "https://demoeditor.com/authors/202312-007",
            "title": "JP Gouigoux"
        }
    ]
}
```

The links are often inherited from specialized links – in our case, a specialized author link that could contain additional important information in its schema, for example, extract restricted to the portion of JSON that has changed, for readability purposes:

```
{
    "rel": "author",
    "href": "https://demoeditor.com/authors/202312-007",
    "title": "JP Gouigoux",
    "authorMainContactPhone": "+33 787 787 787"
}
```

Having additional information in links is useful when you know this is a piece of information that will frequently be used when manipulating the link, as it avoids an additional roundtrip to the other API to find this information. Of course, there should be a right balance, and including the phone number here is questionable because it can be considered volatile data, not changing frequently but on some specific occasions in the mass of authors of an editor's database. The consequence is that all links should – in this case – be updated, which accounts for quite a large amount of work. When you know it is a piece of data that does not change (the author's name, for example, does not change very often) or even data that should never be changed for regulatory reasons (an approved version, for example, should not be modified even if further versions appear), there is no such problem.

This is the first issue that should be taken care of with links. The second one is more subtle: since the title attribute (which is not an extended one added through inheritance but exists in the standard RFC link definition) has been used to store the common designation of the author, as expected from the definition of this attribute in the RFC, deleting an author will end up with their name still existing in the book's data referential through these links. This may be interesting for archiving reasons (even if we do not deal with this author anymore, for example, even though they have died, the books are still in their name). However, in some other regulatory contexts, this can be a tough problem: if we go back to the example of the European GDPR "right to be forgotten" for personal data, that means that when the author is deleted from the database, we should also go over all the books they authored and replace the `title` contents with something like `N/A (GDPR)`. This is how `DELETE` operations can work under specific functional circumstances!

So-called secondary features

Though we might think we have covered all the responsibilities of the data referential since we passed on the four letters of the CRUD acronym, the spectrum of a good application is much larger. To be thorough, we should talk about all the functions that are generally called "secondary," even though they are critical and – in some cases – equally important as the persistence of the data itself.

The first one of these additional features is **security**. There should be no doubt about the importance of this one anymore, but if it is necessary to convince anyone, let's just stress the fact that the four criteria commonly used in security categorization are all about data:

- **Availability**: The data should be available to authorized persons, which means denial of service (among others) has to be treated. Though unavailable data is a good way to prevent leakage or unauthorized access, it remains the primary criterion, as the whole idea is to provide a service in a solid way. Availability also means that a simple mishap should not get the whole system offline.

- **Integrity**: The data should not be tampered with by anyone and its correctness should be guaranteed – the consequence being that all functions underlying the service have to be secured as well (database, network, source code, etc.).

- **Confidentiality**: This is the counterpart of the first criterion, as access should be forbidden to non-authorized requesters. It is the basis for authorization management systems (more on this in the next chapter).

- **Traceability**: This criterion is a more recent one but becomes more and more important with regulations on IT systems; it states that the modification and use of data should be stored in a log that cannot be tampered with, allowing it to retrieve what happened back in time. Traceability is most important after an attack has happened to understand where the vulnerability was and what the attackers have done.

Performance and **robustness** are also so-called secondary features that have a high importance in MDM. They are very much linked to the first criterion (availability). Indeed, the robustness of the software underpins its capacity to answer requests in time with great confidence, and performance is a quality associated with the availability of the data. After all, if someone gets a response to their request for data after 5 minutes, they would not think of the service as being available, though it could be qualified as such since the data indeed arrived… at some point. Rapid availability of data has often been a drive to move existing "manual" information systems to a software-oriented approach.

Dealing with these features is the subject of many books, so we will just leave it there for now, since those are indeed responsibilities expected from the data referential.

Metadata and special kinds of data

Finally, the data referential should handle data and also metadata. Metadata is all the information that sits around the data entities themselves and allows for a good comprehension of them. This provides some additional richness to the data, but please be aware that metadata should have a different life cycle from the data itself. For example, storing information about the history of data is not metadata, though it can abide by the definition of metadata just given. As has been exposed many times now, the data referential keeps track of every change in the entities it hosts. So, information about who changed what at what times is data and not metadata for a complete and correct data referential. In the same way, dates of changes, indicators of modification, or reading frequency can be directly deduced from the series of operations in the data referential, so they are not metadata either.

A good example of metadata is the units associated with numeric data. Having a number in a named attribute of the entity is often not enough. Sure, the attribute can have a name that describes its content and also the units (examples would be `populationInMillions`, `lengthInMillimeters`, or `nbDaysBackupRotation`), but that does not make it any easier to manipulate the values and, in addition, that makes for longer names, which can be a bit cumbersome when the unit sounds obvious. Having metadata somewhere in the schema of the referential that states that *this* attribute of *this* entity uses *this* unit is a better way to communicate the handling of the data, and can also help in some modern database engines to directly calculate formulas between attributes that are not on the same scale of units, and even provide some warning when the formulas are not safe in terms of units definition, such as adding meters and seconds. These new servers generally use a standard definition of units that includes the powers in **MKS** form (short for **meters, kilograms, and seconds**); this makes it possible to express almost any kind of scientific unit. For example, the Newton unit of force can be described as $kg.m.s^{-2}$. When the unit metadata is well implemented, it also allows us to specify a multiplier and a name associated with the unit. For example, the kN unit is associated with $M^1K^1S^{-2}$ as seen previously, but with a multiplier of $10^{\wedge 3}$ and the name `kiloNewton`.

Geographical attributes are another good example of metadata addition to the usual data in a database. Generally, longitude and latitude were expressed as double-precision numbers in `lon` and `lat` attributes, but this did not account for the kind of world-map projection (which can create some discrepancies in the number) and would not prevent silly computations such as adding the two numbers. With database or geographical servers able to understand the metadata added to the coordinates data, it is now possible to calculate distance, transpose coordinates from one projection system to another, and so on.

Metadata is the long-forgotten cousin of data. Apart from CMIS, the standard for electronic document management systems, where they enjoy first-order citizenship (supporting groups of metadata implemented in schemas that can be applied to the documents, used in the queries when searching, and sometimes even versioned independently of the documents supporting them), there are not that many standards that formalize them. The evolution of this depends entirely on engineers who are interested in doing their jobs in a professional and clean manner. As long as "quick and dirty" tricks are used in the software programming and the structuring of information systems, metadata will continue to be set aside. When people – hopefully after reading this book and some others advising in the same quality and long-term approach – decide that the burden of the coupling is too high and they have to address the problem by modernizing their information system, metadata use should naturally rise, making it in time as standard and usual as any other practice.

Now that we know how the data referential should be defined, we will dive into how this can be provided by a software system.

The different kinds of data referential applications

We will not talk about technical aspects in this section (this is the role of the following one, called *Some architectural choices*) but about architectural strategies to structure the data persistence.

In the previous chapter, the metaphor of the flower was introduced to show how data can be organized inside an entity. We will follow this idea to represent how persistence can be implemented in the data referential that manages instances of such an entity. Before we dive into the main architectures, please remember that the main criteria of choice should always remain functional, which, in the case of data, means that the life cycle in your system is what will drive you principally into this or that architectural choice. Also keep in mind that the *people* aspect of data management is as important as the *technical* aspect; governance, designation of people responsible, and good communication about which team owns which pieces of data are essential to the correct use of data in your organization.

Centralized architecture

The centralized (or "unique") referential is the simplest one (as shown in *Figure 10.2*) that everybody first thinks of and that solves so many problems in the information system when it can be applied: it consists of having a single storage mechanism for every bit of data concerning a given type of entity (including, of course, history, metadata, and so on). This way, all services working in the system know that, when needing to read or write something, they have to address their request to a single repository service, as the whole "flower" is in one well-known place.

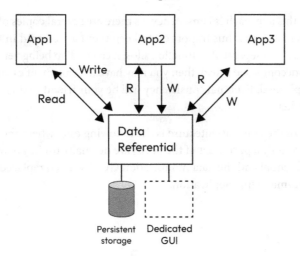

Figure 10.2 – Centralized data referential architecture

The great thing about this approach is that it simplifies the work for everyone in the information system. Of course, this constitutes a **single point of failure** (**SPOF**) and, if the implementing application is down, all applications needing this referential information will be impacted. But this is just a technical problem, with many battle-proven solutions such as active/active synchronization of the database, scaling of the application server, redundancy of hardware, and so on. By now, you should also be convinced that the functional aspects are always more important to take into account than the technical ones. As technicians, we tend to focus on low-occurrence problems such as hardware failure or a locked transaction, whereas the immensely greater problems in information systems nowadays are duplicates of data, poor cleanliness of the inputs, and other commonly observed issues that urgently need to be addressed. The SPOF might be more important in the people organization: a centralized data referential might mean that a single team or even a single person is in charge of the management of this set of data, and some drawbacks are always possible with too much centralization (feedback not taken into account, the relative obscurity of the changes, etc.).

Clone architecture

One way to address this SPOF limitation is to locally copy some of the data that is needed by important applications. In this case, some applications will keep part of the "flower" in their own persistence system, and it is their choice to manage how fresh the data should be compared to the central referential, which remains the global single version of the truth.

When the data is initially scattered around an information system, it can be a first step toward cleaning it, by obeying centralized business rules while still keeping the data as it was stored. The advantage is that, for legacy applications, nothing changes: they still consume the data locally, so all reading functions work as before. With some effort, writing commands can even be kept in the software – for example, by using database triggers that will implement a return of data to the unique referential. Most of the time, though, and particularly if the application is composable and has a unique graphical interface to create entities, it is easier to plug the referential GUI into this application instead of the legacy form.

The main difficulty with this approach is consistency: as there are several copies of data in the system, discrepancies can happen and it is thus important to keep them as reduced in time and impact as possible. If applications are well separated in function silos, it can end up being very easy, but if the way the application has been decomposed is bad, then you may have to implement distributed transactions, which can be quite complicated. Eventual consistency will be your friend in this situation, but it may not be applicable everywhere.

The most efficient form of the clone architecture is the following one, where the cloning of the data (only part of the flower, as only a partial set of the petals are normally useful) is synchronously based on events in the data referential and the data modification GUI has been replaced by the one coming from the centralized data-managing application:

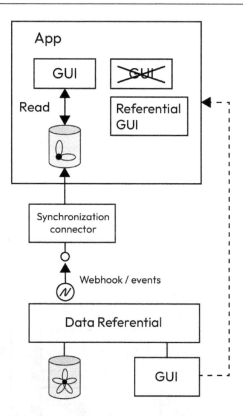

Figure 10.3 – Cloned data referential, the most efficient form

An option in this form is to add a synchronization mechanism for all the data, which compensates at night for messages of data change that could be skipped during the day due to network micro-failures or such low-frequency but still existing incidents if one does not want to put a full-blown **message-oriented middleware (MOM)** to work for this simple stream.

An alternative to the first form is when the synchronization connector uses an asynchronous, typically time-based mechanism to keep the clone database similar to the referential information. The best approach in this case is to call the data referential APIs, as they give the best quality of information:

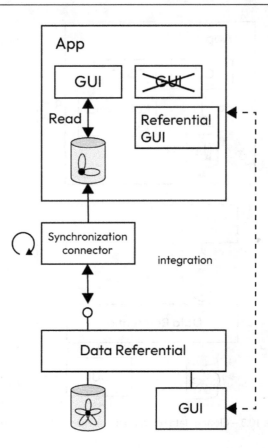

Figure 10.4 – Cloned data referential, with asynchronous alternative

An often-seen alternative (but I really do not recommend it) is to have an ETL perform the synchronization, as shown in *Figure 10.5*. This is often seen in companies that have invested lots of money in ETL to keep data in sync with their system and use this tool for everything. When there is an API (and every good data referential should have one), it is better to not couple ourselves directly on the data. Sadly, lots of companies still have this kind of stream in place, starting their own "spaghetti dish" of an information system, with all responsibilities and streams of data entangled and not clearly defined (see *Chapter 1* for more explanation on this).

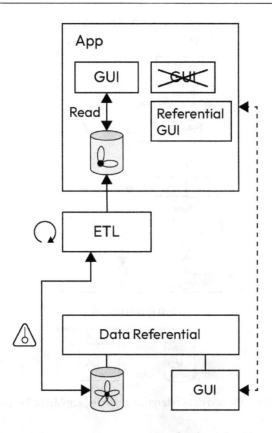

Figure 10.5 – Cloned data referential, using an ETL (not recommended)

As explained previously, some implementations cannot be changed and have to rely on their legacy GUI. In this case, the only possible approach is to rely on specific triggers on the database to get the creation and modification commands and send them as requests to the MDM application:

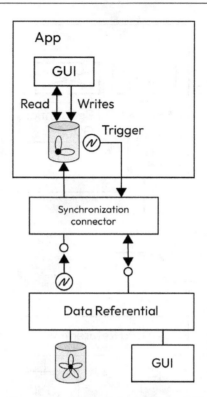

Figure 10.6 – Cloned data referential, with legacy GUI still in place

The difficulty in this approach is when the data changes in the data referential due to some business rules, as the change cannot be sent back to the GUI. Indeed, most applications will keep the state of the data when they have submitted the change to their server. Even for the rare applications that listen to the returned data by their back office, the difficulty is that the complete roundtrip will not be finished before this reading, and the "updated" data will only be the latest in the local database, but not the latest that will come back moments later from the webhook callback. When stuck in this situation, it is best to explain to the users that this is a temporary situation before reaching the centralized referential architecture and that they can refresh their GUI a bit later to see the effects of their change. Even better, learn how to use the new centralized referential, which will always give them the freshest information, at the price of using two graphical interfaces instead of one (which is not such a high price when those are web applications that can be opened in two browser tabs).

> **Important note**
>
> In *Chapter 8*, we briefly talked about enterprise integration patterns. They are the ideal bricks to construct the synchronization connectors that we talked about previously, particularly if a **message-oriented middleware** (**MOM**) solution is put in place during the project of information system reorganization/data referential structuring.

Consolidated and distributed architectures

This type of referential consists of exposing, from a central point of view (an API, generally), data that is actually placed into different parts of the information system. Generally, the core of the flower and some petals are in the data referential dedicated persistence. But for other petals, persistence can stay in the business applications they are associated with because it is considered they know the content of these petals better. In the most collaborative form of this approach, the referential exposes the full data for every actor of the information system and shares ownership of the petals:

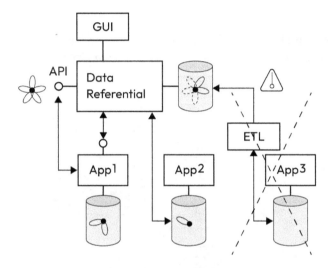

Figure 10.7 – Consolidated referential architecture

The data referential can produce an entire flower of data through, and expose it in, its API but that means it has to consume the different petals it does not own from the business applications (keeping a local cache of these petals is a choice of implementation based on freshness, rate of change, and performance but does not change the ownership of the data). To expose the whole flower with fresh content, the data referential needs to have access to its own database, and also to the business applications data (or, again, the cache it may keep locally).

Also, some applications, such as App2 in *Figure 10.7*, may not need anything other than the petal they own (notice that, of course, everyone has the core of the flower, by definition). Some applications, such as App1, may need some additional petals, and in this case, they have to call the data referential API to obtain this data.

Another difference has been made in *Figure 10.7* to show that the data referential may use a business application API to obtain the data (best case) or may resort to direct access to the database of the business application, which causes more coupling but is sometimes the only way to go. The alternative shown on the right is dangerous and should not be applied: in this case, App3 is not talked to but this is not the main problem. The actual issue is that using an ETL to feed the referential database should never be done, as this shortcuts the business and validation rules inside the data referential. No application should ever touch the referential database but the referential application itself. In fact, this rule is so important that, when deploying on-premises, it is a good practice to hide, obfuscate, deny access, or use any other possible way to prevent anyone from directly accessing a referential database. The results are already bad enough when this is a "normal" database, with its trail of coupling and other bad consequences; doing so on such an important database is the recipe for problems.

When the data referential exposes all the data possible on an entity (the complete "flower"), the architecture is also called "consolidated." It is possible, in some cases, that some bits of data are only useful by the owner application and will not be of any use to anyone else. In this case, the term "consolidated" is not appropriate as some data is – willingly – not available, and the referential should be considered "distributed" only. Such a situation would be schematized as follows:

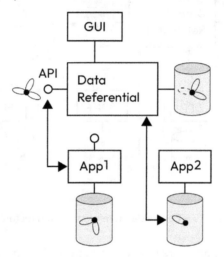

Figure 10.8 – Distributed referential architecture

The main difficulty of a distributed referential architecture is to maintain performance. Optimizations are of course possible, such as the cache mechanism we talked about or the parallelism of calls to the different business applications when no cache is used, but all of these technical additions come with a price that should not be underestimated, particularly when we know that the situation is temporary and that the goal is a centralized architecture. It often happens that a "temporary" situation, supposedly cheaper and made as a stepping stone to the next architecture, actually costs as much as directly putting in place the target architecture. Most of the time, the decision comes from the fact that the difficulties of the target vision are well known, but the ones associated with the intermediate step are less envisioned, mostly because these unstable situations are numerous and thus not as well documented as the final architecture.

Let me give you an example of how hard it can be to set up an intermediate distributed system, by talking about the pagination of data. When calling the data referential API with a `$top=10` query attribute, if the referential is distributed and consolidates data from two business applications, it will have to make two requests to the application, but the limiting thing is that, depending on the order of the data requested by the `$order` attribute, there may be zero data coming from one source and 10 pieces from the other one, or the other way around, or any situation between these two extremes. This means that the gateway in charge of merging the data will have to take 10 lines from one application and 10 lines from the other, then re-apply an ordering algorithm on the 20 lines, finally sending the first 10 to the requesting client and discarding the following 10 lines.

Do not think it would be easier to use a local cache, as you would have to implement the query mechanism on it in addition to the ordering algorithm just talked about. Imagine if this has to be done with more applications! With 5 business applications, you already cache 50 lines in order to actually use only 10, which is an 80% waste of resources. You may think of pre-querying the applications in order to know which will provide data out of the filtered values, but that means you should already query one application and then adjust the counting querying to the other ones, maybe to realize that the optimization will not reduce the number of queries but only the number of lines retrieved. The choice of a pivot application may be difficult in itself for a resulting improvement that may be weak since we deal with reduced sets of data anyway (this is the goal of paginating the requests). Wait! We have not talked yet about the worst part of it: when paginating for the 10th page of data (between 90 and 100, if we stay on 10-line pages), you will not be able to simply call 10 lines from each of the 5 applications, because there may be one application that will account for almost all the lines in the order applied since the beginning of the pagination, and some others will provide nothing in the same range. This means that you may very well have the first result coming from an application only when calling the 10th page! You now see it coming, don't you? Yes, we will have to query the 5 applications for 100 lines to extract the 10 lines corresponding to the 90 to 100 range of the aggregated data, which means a huge waste of 98%… and, the cherry on this sad cake is that, if an application does not support dynamic range, you will have to query it several times in order to compose the complete range of data needed. Sure, it may be possible with some implementations to keep cursors on the database queries in the state, but that means that your application is now stateful, and this will account for some other technical limitations in terms of scalability. Well, the only thing that will save us there is that, generally, the users will stop at the second or third page of data, refining their `$filter` attribute to reach quicker results.

Consistency problems also exist, but they are a bit easier to deal with as long as the cutting of data follows a functionally logical order. This is generally the case because the distribution of data is done in business applications, so the risk that they have duplicate data (apart from the core of the flower, of course, which is always shared) is normally very low.

Other types of referential architectures

A "virtual" data referential is a particular case of a "distributed" referential where the central part simply holds no data by itself, and thus has no persistance, relying on the surrounding business applications databases. Schematically, this is the following state:

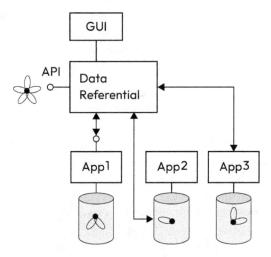

Figure 10.9 – Virtual referential architecture

Other, more exotic, referential architectures exist but it does not sound really useful to expose them here. For those of you who are curious, the French government-issued document called *Cadre Commun d'Architecture des Référentiels* (common framework for referential architectures, freely available on the internet and in the French language) should not be a limitation, as the different possibilities are shown using mainly diagrams.

Now the architecture patterns have been shown, we can talk about the implementation itself, including what technical choices should be made and how when creating the data referential.

Some architectural choices

One of the first is, of course, the database. By the way, I should even say the persistence mechanism because a database is a very well-known persistence mechanism, but there are others, as we will see at the end of this section. Some other technical considerations will have to be dealt with – in particular, on the streams of data.

This section will also be an opportunity for a little rant about the dogmas in IT, and how they delay the long-awaited industrialization of information systems. Lots of technical choices remain based on the knowledge of the teams available rather than on the adequateness of the functional problem at hand. This is not to say that competencies should not be taken into account, but training should sometimes be forced on technical people who have not changed their way of thinking for decades and may hinder your information system development because they simply apply the wrong tool to the problem. You have likely heard the proverb "When all you have is a hammer, all problems look like nails." If you have this kind of person in your team, a manager's job is to open their eyes through training, whether it be internal, external, formal, or not.

Tabular versus NoSQL

One of the very first decisions one has to make when implementing the data referential is the kind of database paradigm to use. Should it be tabular or document-oriented? SQL or NoSQL? Knowing that the natural shape of 99% of business entities is a document structure with many levels, like a tree of attributes and arrays with varying depth, the obvious choice if you want to reach business / IT alignment should be a NoSQL database, adapted to the shape of your data: document-based NoSQL if you manage business entities, or graph-based NoSQL if you manipulate data entities linked to other entities by many typed relationships, causing a network of entities that can be traversed by many paths, and so on.

If one really applies business/IT alignment and looks for a persistence mechanism that closely mimics the shape of their data, SQL tabular databases should be used for business entities that are naturally tabular… which is almost never! Sure, there are cases, just like there are some for key-value pair lists in the NoSQL domain, but they are very scarce. In fact, it looks like the main reason SQL is still largely used for the data referential is simply history. And this is a justified reason when dealing with legacy software… After all, if it has worked for years this way, you are better off not touching it. But the real problem is when a new data referential, designed during a project of information system modernization, also uses a non-efficient approach.

Why do I say *non-efficient*? The history of computer science and databases should be invoked in order to explain why… In the old times of data storage, when spin disks were used with random-access controllers, data was not randomized in the magnetic disks but placed in sequences of blocks (preferably on the outermost lines of the hard disk, as the linear speed was higher, providing for quicker reads). In order to quickly access the right block, database engines would force the size of a line of data in order to quickly jump to the next, knowing the total length of each line of data in advance. This is why old types of strings in the database required a fixed length, by the way. This is also why the data has to be stored in tabular blocks, and structured data decomposed into many tables where lines are related to each other by keys, as this was the only way to calculate the next block index.

These assumptions came with a high price, though: since data was tabular, the only way to store multiple values for an attribute of an entity was to create another table in the database and join the two lines of data. The consequence of this was that complicated mechanisms were necessary to handle global consistency, such as transactions. In turn, transactions made it necessary to create the concepts of pessimistic and optimistic locks, then manage isolation levels for transactions (as the only fully **ACID** ones, which are the serializable transactions, have a dramatic impact on performance), then deadlock management and so many other complicated things.

When you think about it and realize that hard disk controllers have been providing randomized access for decades (and the very concept of a spinning disk does not exist in SSD), it is hard to understand why the consequences of this remain so pervasive today. One of the reasons is the change management, as nobody likes changing. But if there is a job where one should adapt and embrace change, that should definitely be a developer. I can also understand that SQL is still used in workshops where people only know this as a persistence technique. It is much better to start an important work with maybe not the best tool but one that is well known by the whole team, and I would not advise starting with a complex technology that nobody knows. But in this particular case of not using NoSQL for a business entity data referential, there would be two problems:

- First, this would be a training problem, as these technologies have been here for more than a decade now, and returns on experience are perfectly established, with trustworthy operators.

- Second, there are actually few technologies as easy as document-based NoSQL. Take MongoDB, for example – writing a full-fledged JSON entity into a MongoDB-compatible database is as simple as follows (example in C#):

```
MongoDBConnection conn = new
MongoDBConnection(ConnectionString);
conn.Insert("Actors", "{ 'lastName': 'Gouigoux', 'firstName':
'Jean-Philippe', 'addresses': [ { 'city': 'Vannes', 'zipCode':
'56000' } ] }");
```

The equivalent with an SQL-based tabular **RDBMS** (short for, **Relational Database Management System**) is the following:

```
SQLConnection conn = new SQLConnection(ConnectionString);
SQLTransaction transac = new SQLTransaction(conn);
try {
    transac.Begin();
    SQLCommand comm = new SQLCommand(conn, "INSERT INTO ACTORS
(lastName, firstName) VALUES (@lastName, @firstName)");
    Comm.Parameters.Add(new SQLParameter("@lastName",
"Gouigoux"));
    Comm.Parameters.Add(new SQLParameter("@firstName", "Jean-
Philippe"));
    string idActor = Comm.ExecuteGetId();
    comm = new SQLCommand(conn, "INSERT INTO ADRESSES (id, city,
zipcode) VALUES (@id, @city, @zipcode)");
    Comm.Parameters.Add(new SQLParameter("@id", idActor));
    Comm.Parameters.Add(new SQLParameter("@city", "Vannes"));
    Comm.Parameters.Add(new SQLParameter("@zipcode", "56000"));
    Comm.Execute();
    transac.Commit();
} catch (Exception ex) {
    transac.Rollback();
```

```
        throw new ApplicationException("Transaction was cancelled",
    ex);
    }
```

And I am not even talking about the **Data Definition Language** (DDL) commands to create the tables and columns, which would add many lines. MongoDB does not need any, as it is schemaless and collections are created as objects are added.

Again, there are cases where SQL is needed. Reporting tools are very numerous using this grammar and it is good practice to expose SQL endpoint to access data, as it eases its consumption. Big data tools and even NoSQL databases have SQL endpoints. This is valuable as there are lots of people who are competent in using this way of interrogating data and computing complex aggregations. However, choosing a tabular database to store structured data just in order to be able to use a well-known query language is a problem, as it will cause lots of unwanted complexity. In your next data referential, please consider using NoSQL, as you will gain a lot of time with it. And if you know this kind of project will arrive next on your project portfolio, start getting training for your team. Only a few days are required to understand everything that is needed to be proficient with document-based NoSQL servers such as MongoDB, and they are extremely well adapted to storing business entities.

CQRS and event sourcing

While we are at it, you may also want to ditch your old data stream architectures where reading and writing are handled by the same process. After all, these two sets of operations are so different in their frequency (most **Line Of Business** (LOB) applications have 80% of reads and 20% of writes), functions (no locks necessary for reading, consistency needed for writing), and performance (low importance for unique writing, very important for massive queries) that it sounds logical to separate them.

This is what **Command and Query Responsibility Segregation** (CQRS) is about: it separates the storage system receiving the commands for altering or creating the data from the system ready to answer queries on the same data. Event sourcing is closely associated with this architectural approach as it stores a series of business events generated by writing commands and lets the queries use this storage to get the aggregated results they need in a highly scalable way, thus allowing performance on large data.

In some way, CQRS could be thought of as a type of referential architecture between the distributed and the clone approaches. It does not separate data between applications with a criterion that is on the data itself, but rather on the kind of operation that is going to be performed on it (mainly, write or different kinds of reads). At the same time, the prepared reading servers can be considered as clones of the *single version of the truth* data. As their number can rise without any limit since the single version of the truth is in the main persistence, the performance can always be adjusted, however complex the queries and with high volumes as well.

Again, this is not the place to discuss these subjects in detail but they had to be cited in a chapter about data referential and MDM, as they are the indisputable best approach to implementing high-volume solutions.

One more step beyond databases – storing in memory

Let's come back to our discussion about the origin of the tabular database system and even a bit before. Why do we actually need database and storage systems? Mostly because hard disks can store more data than only RAM, and because databases would not fit in small amounts of RAM. Thus, it is necessary to have a system that is good at quickly putting data on disk (in order to keep it safe in case of hardware failure, the database first writes in the log files) and good at retrieving some parts of data from the disk and putting them back into memory for application use (this is the SQL part and, in particular, the role of the `SELECT` and `WHERE` keywords).

Of course, this was a major problem when computers had 640 kilobytes of RAM and databases would need a few megabytes. But how about today? Sure, there are huge databases, but we commonly have databases with a few gigabytes only. And how about server RAM? Well, it is very common to have servers with tens of gigabytes, and it is easy to acquire online servers with 192 GB RAM. In this case, why is there a need for manipulating data in and out of the disks? Sure, SSD disks are some kind of memory, but they are still slower than RAM. Also, there is indeed this persistence under hardware failure that has to be taken care of. But how about the manipulation of data itself? Would the manipulation of queries into RAM not go much quicker?

In fact, it does and there is a rarely-used and scarcely-known technique called "object prevalence" that acts as an in-memory database. We are not talking about files stored in a RAM disk or a high-speed SSD, but having the data used directly in the object-oriented model of your application. How can we be sure not to lose any data if there is a hardware failure, you might ask? Well, exactly as a database does: by keeping a disk-based log of all the commands sent to the system. The difference then is that the reference model for manipulating data and extracting, filtering, and aggregating results is not on some tabular writing on the disk that has to be accompanied with indexes in order to improve performance, but directly in the RAM, and in a binary format that is the one directly used by your application, which means nothing can go faster. By doing so, requests in SQL are replaced by code in your language of choice – for example, C# with LINQ queries.

It is quite astonishing that object prevalence has never reached a wider audience but all the people I know who used it were convinced of its high value. Personally, when I need to implement a data referential that is limited in volume but has one of the following requirements, I always go for this technology:

- High performance required
- Very complex queries that would be hard to write in SQL
- A data model that evolves often

One of the best data referential implementations that I have participated in was on a project calculating advanced financial simulations and optimizing them with genetic algorithms; the performance boost was huge and the ability to write extremely complex cases of data manipulations made the whole project a clear win for the customer, who was surprised in the first test drives by the sheer velocity of the simulations – the old platform this one replaced provided results in minutes, whereas the new one responded in no more than a few seconds.

Another example of a successful implementation was in the handling of low-moving data such as country codes. In this particular example, people were not feeling great with the in-memory approach, even though data is safe in logs on the disk (and we even had a backup, as a third set of data for improved reassurance). So, testing this quite innovative approach with some data that they could easily feed back into the data referential made it more comfortable for a first try of the technology. The test went well, but the customer did not expand it to other data. Sadly, I do not know more examples of uses of this technology, which is a bit sad as the potential was huge.

Though this example may not be the best as this technology did not hit it off, the message still remains: in order to respect business/IT alignment, which is the best way to ensure long-term evolution, always favor a technology that closely fits your business needs and data shape.

In the last section of the book, we are going to talk again about time and how it influences what we do with data referential, in our case.

Patterns of data evolution in time

In *Chapter 4*, we studied the importance of time management in information systems, and one of the major impacts of time handling is on data. Data handled in MDM systems must be taken into account with the time factor, and we talked abundantly about data history management. But the very act of MDM should also be done according to time.

Data governance

Data governance is the act of establishing functional responsibilities around the management of data referential. Who is responsible for which reference data? Who can manipulate and clean the data? Who decides the evolution of the model? How are impacted teams and applications informed about changes? What business rules should be followed when manipulating the data? When should data be erased or archived? Those all are governance questions and they are always related to time. In particular, the responses have to be reviewed at regular periods, just like business processes, in order for the data to remain under control.

Data governance is mostly handled in the second layer of the Cigref map, which is the business capability map and usually contains a zone dedicated to reference data management. This is where you should draw the different data referentials, and store the detailed definition of the entities that are stored, along with the versions to prove compatibility between them or document incompatible changes. Here, you should also find at least the name and contact of two of the main data governance roles:

- **The data owner**: This person is ultimately responsible for the quality and usability of the data inside the information system. They define all the business rules around the data: how it must be manipulated, who can access it, in which conditions, and so on.

- **The data steward**: Under the delegation of the data owner, this person is responsible for the daily maintenance of the data. Following data manipulation rules issued by the data owner, they clean data and ensure its availability and integrity, as well as the respect for authorization rules.

One of the obvious consequences of having data governance is that there is a clear responsibility for a given data referential. Having shared responsibility for a referential is a problem because there can be competing needs that evolve in an uncontrolled evolution of the entity's format or the services provided. In the worst case scenario, the IT team does not know who to consider as the decider and implements both demands, making the data referential progressively harder to use and not fit for its purpose. Having no responsibility is even worse because, as the implementation belongs to the IT team, the technical people become, by default, the owners of the data, which may be the worst move ever as they do not have the best knowledge of the business stakes associated with the data. Sure, they basically know what the data is about (after all, we all know in a company what a customer is or a product) but again, the devil lies in the details, and when the IT team is in charge of defining data, no one should act surprised that organizations only support one address, or that there is no distinction between a product and an article. Such mistakes would never be made by a specialist in the subject, and we all know how destructive a bad entity definition can be. So leaving such business-driven decisions to the IT team because nobody wants to take ownership is a risky move and everyone should be warned about this.

Progressive implementation of a unique referential

When presenting the distributed and consolidated data referential architectures, it has been stated that, sometimes, these intermediary steps toward a centralized referential (which, most of the time, is the ultimate goal) can cost as much as directly going to the target state because of hidden efforts or lesser-known shortcomings. On the contrary, there are times when directly addressing the final vision is impossible, and convergence toward this should be done in several progressive steps. This might be because the information system is so coupled that a violent move may destroy it; most of the time, the problem is with the human capacity to embrace change, and a progressive approach has to be followed for the organization itself to be able to adjust.

This has been the case for me in many situations where I consulted for companies who, in order to successfully manage their merger or acquisition of another company, needed to apply a merging program to the two information systems, incorporating them into a single system. This kind of thing generally takes years in big organizations (the quickest that I have ever witnessed was done in less than 18 months, but all flags were green, which rarely happens). As you will see in the following sections, these plans need numerous steps to be realized.

For privacy reasons, I will show a mix of two progressive transformations that I designed for a public customer (a fusion of two regional councils in France) and an agriculture cooperative that was born out of the merger of two giant entities in the West of France. Both of them needed to address the MDM of the individuals and legal entities that their information systems deal with (customers, agents, prospects, farmers, students, etc.). In order to simplify the diagrams, I will consider the starting point to be where the two entities each had a consolidated data referential, with some applications showing a clone referential pattern. This often happens when there are many applications needing referential data: the most important are directly plugged into the highest-level data referential application, and the secondary applications are simply cloning what happens in their leading business application. In the following schema, I have also highly reduced the number of applications, again, for simplification reasons. I have not drawn the relationships between them and with other software in the information system, as they were mostly ERPs with much interoperability.

Step 1 – same infrastructure but no link

All this being said, the first step can be schematized as this:

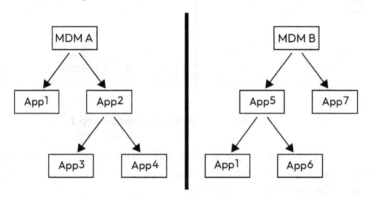

Figure 10.10 – Fusion of two MDM systems – step 1

The two companies have completely separate MDM systems, hence data referential for their "actors," if this is the name we should use to describe the entities at play. Notice that most applications are different in each case, except for `App1`, which is a common ERP between the two companies (this does not mean it will be compatible, as versions may differ and customization definitely will, but this can make a good candidate to put things in common at some point). The very first step is, of course, to connect the two internal networks, even if everything that will be shown next could very well be applied by only communicating through the Internet.

Step 2 – providing a common interface

The second step was to provide all users in the new fusion entity with an API to read actors:

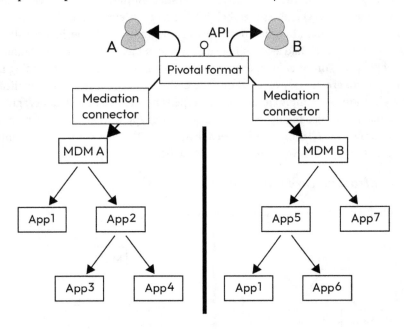

Figure 10.11 – Fusion of two MDM systems – step 2

Notice how symmetrical the diagram is: choosing a neutral pivotal format was of utmost importance, as using the proprietary format of one of the companies would have been a clear disadvantage to the other one (which would have to change all its connectors and mapping code) and this would have caused human problems, as tensions are always exacerbated during company fusions, particularly when they were previously competitors. We thus spent a lot of time crafting a nice pivotal format for the users, using the best data representations coming from both sides. At this step, not only is reading the sole operation available but no company can read the other's data! You might wonder how useful this step is since the goal is to reach a unique MDM system for both companies and, for now, it does not change anything. In fact, it is indeed harder for no functional effect, but preparing a common pivotal format is the basis for adequately sharing data. Also, it provides a way for all new software functions that would be created during the fusion process to read actors in a standardized, fusion-ready way. This means we will not have to come back to these new applications, and this is much-appreciated news when you know hundreds of applications have to be dealt with in the whole project. Finally, it started the work on the mediation connectors (there again, this is the kind of thing that is best implemented in Apache Camel, or another flavor of enterprise integration patterns), which was an important piece of work, better started early in the project.

Step 3 – merging data from the secondary source with the primary

From now on, we will only represent the difference in streams from the point of view of company A, but the opposite is always true. The next step was to start obtaining some data from one information system and transferring it to the other. Again, this was very progressive: it was only done for the reading operations of the data for now and, as shown in the following diagram, the data was first read on the system of the person initiating the request, and then only completed with data "from the other side of the barrier." At any time, the data from the originating side would win, except if the date of modification clearly showed that the data coming from the other information system was fresher.

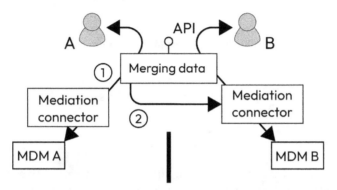

Figure 10.12 – Fusion of two MDM systems – step 3

For this previous step to work, it was necessary to find a way to look for similar actors, for example, with their VAT number or other business identifiers.

Step 4 – storing identifiers from the other source

Since this is a complex operation to realize, once the correspondence was found, the technical identifier from one side was stored in the other, and vice versa, which will allow for quicker access next time. This was the first time the system would write in the MDM system, but this was limited to storing the identifier from the other side:

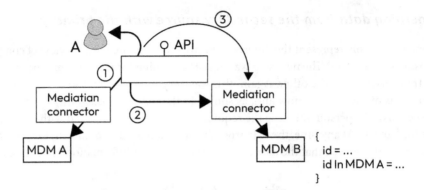

Figure 10.13 – Fusion of two MDM systems – step 4

However, this opened up an entirely new approach to sharing data because, once the *write* authorizations were provided and the "external" identifier known, each side was able to share information with the other side.

Step 5 – sending information to the other side

Every time there was a change in the actor on one side, the other was informed. The receiving information system was free to deal with this information at its own pace, maybe doing nothing with it the first time, but then choosing which pieces of data were interesting and storing them, and so on. At this point, keeping the origin of the data change was necessary in order to not start a loop of information, sending back to the initial information system the event that the data changed because of its initial event. The diagram – once again represented only from A to B for simplicity reasons – was as follows:

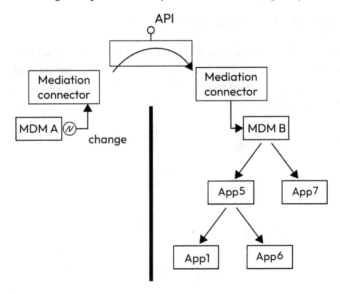

Figure 10.14 – Fusion of two MDM systems – step 5

Now, since the initial write was started and information systems (and people) were starting to trust each other better, the next step was to generalize the modification of data.

Step 6 – centralizing the writing of data and extending the perimeter

This means that both sides started to use the centralized API in writing and the implementation of this API was to push the data on both sides, in order for each information system to know about the latest data. Again, using the data depended on whether the receiving end knew the actor (or should record it) but, in some cases, data was simply ignored, for example, when this was about a change in a supplier that was only used in the other company. As for prospects, though, the data was shared because the commercial approach started to get unified between the two parts of the slowly emerging fusion company.

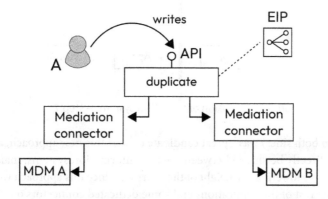

Figure 10.15 – Fusion of two MDM systems – step 6

The enterprise integration pattern used in the MOM implementation was a "duplicate message" one, sending the data pushed by the initial request in two similar messages to the mediation routes and waiting for both acknowledgment messages to come back in order to emit its own acknowledgment along the route it was called by, effectively creating a robust delivery of the change both sides.

Step 7 – access unified

This was the time when the old data referential started to act only as gatekeepers for the messages, checking that they were related to their side of the information system. But, since actors were now largely shared, this was not such an important feature, so some applications started to register their actors' messages directly to the top data referential:

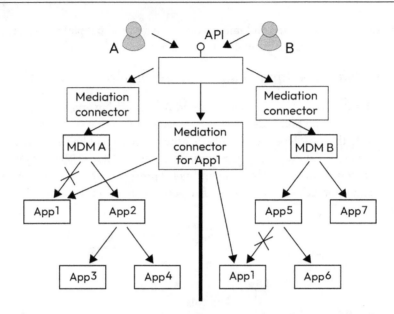

Figure 10.16 – Fusion of two MDM systems – step 7

App1 (an ERP used on both sides) was a great candidate to start this new approach, as the mediation connector to it could directly be shared between the two information systems, making for the first common deployment, thus lowering the height of the "barrier." Since this approach worked quite well, it was a kickstart for the rest of the applications and some dedicated connectors quickly appeared on the other application, which was easy since the common pivotal format had then evolved and was easier than the previous ones, also covering more business cases.

Step 8 – eliminating unnecessary calls

The situation rapidly evolved to something schematized as this, where the old MDM system basically had nothing more to do since all data was coming from the new centralized one:

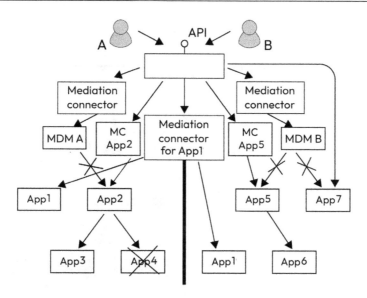

Figure 10.17 – Fusion of two MDM systems – step 8

In addition, some applications, such as App7, had time to evolve and were able to directly take some JSON-representing actors without resorting to a mediation connector. Also, some applications started to be used in common between the two organizations (which clearly appeared more and more like becoming a single organization at this point), and App4 disappeared in favor of the common use of App6.

Step 9 – removing unnecessary intermediate applications

Some "low strategy" applications remained under the control of business applications such as App3, but this was not a problem as their parent application was now under the main data referential and would deal with the change of format for them. These applications did not see any change in the system, which was great for their users, who were not impacted at all by the otherwise major change. The resulting information system started to look like the following:

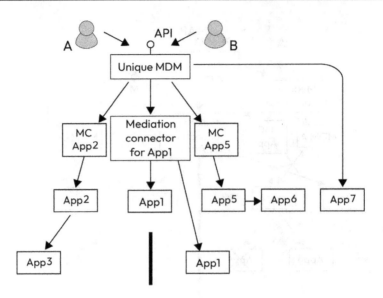

Figure 10.18 – Fusion of two MDM systems – step 9

Since App6 was used by all teams, the barrier between the two formerly separated companies went one more step down, reaching a point where it was not a problem, as it only divided some secondary business applications used in some corner cases by a dedicated team on activities that were not part of the fusion process. There was now a unique centralized MDM system, with a few important applications acting as local referential for actors on which secondary applications would clone some parts of data. This took many years in total but the objectives were reached: merging the actors used by both sides and doing this in such a progressive way that the business was never affected by technical choices.

Keeping an eye on dogma and useless complexity

I hope I have convinced you in this chapter (and in this book in general) to keep a critical eye on technologies and practices for which their use sounds obvious. As with SQL databases for data referential storage, or hard disk-based data manipulation, there are lots of preconceived approaches in the development itself that do not fit the problem very well when you think purely in business/IT alignment terms.

Just one example is data validation. In most programming languages, validators are associated with fields or properties of data entities, through attributes, for example, in C#. This approach is, in my opinion, very wrong and has proved several times in my practice to be a real pain as one will almost always find a particular case where these validating attributes are not correct. In the case of business identifiers, product owners would sometimes insist on the fact that no entity should ever be created without such a value, and then, within a year or so, realize that there is this particular case where the identifier is not known yet and we should still have the entity in the system. This can, for example, be the case with a medical patient database where the product owner will assure you that an entity without a social security number would make no sense as it is absolutely mandatory before even considering providing medication to them… After insisting on putting a strict NOT NULL validator on this for data quality reasons, the same person may come back a few months later, when the database is in production and a major impact change would have a huge cost, telling you that they forgot the particular case of a newborn that should be given drugs but they do not have a social security number yet.

In this particular example, I have personally taken the habit of never describing any entity attribute as mandatory, as only the context of its use makes it mandatory or not. And it is so easy to add a business rule or a form behavior blocking the null value that it really is not a problem to not put it on the entity itself. On the other hand, sorting out the mess when this mandatory characteristic has been implemented down the lowest levels of your information system is such a pain and a cause of errors that it is, in my opinion, never justified to call a field "mandatory" (except the one exception for a technical identifier, as there is otherwise no way to univocally retrieve an entity once created).

Important note

I really like it when I read articles such as https://jonhilton.net/why-use-blazor-edit-forms/, where the author questions there being "too much magic" in the technology exposed. Generally, there is, indeed, and such critical eyes are the best reads one can have on a given technology, rather than the numerous blog articles that simply explain how to use a function without digging into when it is useful and when it is actually more of a danger than a real advantage. This article really has a great point of view on validation included in forms and data definitions.

The same goes for cardinalities as for the identifiers cited previously, by the way: if you do not have the absolute, definitive, and fully responsible engagement of your product owner that an attribute should be with a zero or one cardinality, always make it as an array with N cardinality. What is the worst that could happen? The arrays always being filled with only one item? Well, that does not really matter, does it? A developer complaining that, on all these occasions, they must type deliveryAddresses[0] instead of deliveryAddress? Show them how to create a property in the language they used and it will be sorted out. As far as the GUI is concerned, we will simply display a single piece of information for as long as we do not have a use case corresponding to handling several values in the array. Only when this new business case appears, where we need to handle several pieces of data, will we adjust the GUI with a list instead of a single text zone, for example. But the great thing about this approach is that this will be done smoothly, as the previously unique data will simply become the first of the list, and much more importantly, all the clients of the API will remain compatible and will not be broken by this new use. They will even be able to continue using only the first piece of data in the list as long as they do not want to use the other ones and stick to the old behavior. Since all clients and the server can advance at their own pace on the business change, we know we have a low coupling.

This extends to many other technical approaches that are supposed to help the business but, in the end, can hinder it. To name just a last example, most of the technical approaches to data robustness actually go against the business concepts. The outbox pattern (`https://microservices.io/patterns/data/transactional-outbox.html`), for example, should only be used when eventual consistency is not an option. But when you know that even banks have always used eventual consistency (and will definitely go on doing so in the future), that limits the usefulness of such techniques quite a lot. Of course, understanding the business in depth is less fun than using the latest technology or pattern that will drop your rate of transaction errors to a bare minimum. But it is the only way to win in the long term.

So, once again, because this is such an important message, **think of the business functions first and then find the technology that adapts to it**. In order to do so, the easiest way is to imagine what would happen in the real world between business actors if there were no computers involved.

Summary

In this chapter, the principles of MDM have been applied and implementation techniques exposed, not only from the architectural point of view but also with technical choices that may prove useful when constructing a data referential. The main behaviors of such server applications have been covered and their evolution in time has been described with a few examples. This should make you quite knowledgeable about how to implement your own data referential.

We will come back to the subject of MDM in *Chapter 15*, where we will go down to the lowest level of implementation, with actual lines of code and the design and development of two data referential implementations in C#, in order to deal with authors and with books, respectively. This will be the final piece where we will join the principles of service management and APIs learned about in *Chapter 8*, the domain-driven design of the entities shown in *Chapter 9*, and the architecture approaches described in the present chapter.

But before we reach this point, we will study the two other parts of an ideal information system, just like we did here for the MDM part. The next chapter is about business process modeling and how we can use **BPMN** (short for, **Business Process Modeling Notation**) and BPMN engines in order to implement business processes inside our information systems. Some other subjects such as middleware, no-code/low-code approaches, and orchestration versus choreography will be exposed in the next chapter as well.

11

Business Processes and Low Code

Business processes are all over information systems, and when talking to CEOs, they generally inform the way they see IT: as a way of automating the business processes of the company, providing reliability, repeatability, and – in the best systems – visibility of what happens that brings value to customers, be they internal or external. Business processes are at the heart of the company's information system because each activity is generally ported by an instance of a process. When the company is certified with ISO 9001, each of the processes in the certified perimeter is precisely documented and its use by each of the actors concerned can be verified in practice. The processes thus structure the activity of the company.

This chapter details how a clean architecture should behave from the business process point of view by explaining the notions of business process modeling, business activity monitoring, and business process mining, and then by showing how business processes can be used in an IT system. Low-code and no-code approaches are discussed, as well as BPMN-2.0-based approaches. Throughout the chapter, we will provide examples to relate the practice to our demonstration information system and make the processes used in IT more concrete. Finally, another approach to business processes through service choreography will also be detailed.

In this chapter, we'll cover the following topics:

- Business processes and the BPMN approach
- Business process software-based execution
- Other associated practices
- Other approaches to business process implementation
- Should I use BPMN in my information system?

If you remember well, *Chapter 5* introduced the idea of a utopic perfect information system that would be made of only three modules. **Master Data Management** (**MDM**) was the first one and it has been studied in detail in the previous chapter. Now, we will analyze what the second module, namely BPM, is all about. In the next chapter, we will end this with a thorough explanation of a BRMS. As you will see, the business process approach is not currently very widely adopted in information systems, certainly even less than the culture of data referential services. The norms and standards are there for the taking, but very little implementation can be observed. This is an important point to consider, as what will be shown in this chapter is more of an ideal (at least for now) than a recommendation to orient existing information systems. Only time will tell whether IT gets structured around this solid approach or whether the costs will remain too high for widespread use in the IT industry.

Business processes and the BPMN approach

In this first section, we are going to explain in more detail what business processes are and how we can model them with a software approach, using, in particular, a norm called BPMN. Before talking about processes in the IT world, it is indeed interesting to come back to the definition of a process from a purely functional point of view, as we always do in this book about business/IT alignment.

What is a business process?

A **business process** is a coordinated set of human and automated actions realized to reach an objective. The term "*process*" is often replaceable by the almost equivalent "workflow", which better expresses the fact that the actions (or "tasks") are actual work that is being done by a human actor or a piece of software and that they are realized in an organized stream (the "flow") to achieve the business goal that the process aims at.

As explained in the introduction, business processes are everywhere in an organization, as an "enterprise" is – by definition – a group of people with a means to achieve a goal that cannot be reached alone. Processes are the ways an enterprise achieves these goals. There is generally a main strategic goal that explains that several processes are, in fact, necessary. For example, the strategic objective of a company may be to become the world leader in the editing and publication of software books. Its strategy could be, for example, to cover all possible subjects in great detail, by employing many different expert authors. Working out how this is going to be realized requires several smaller, operational objectives. In our example, this means a good recruitment process, since finding the right experts for all the software subjects will require a well-organized approach. Another operational process will be the follow-ups of writing, enlisting editors, correctors, proofreaders, and so on.

This kind of process is the first one we generally think of because it is directly oriented toward the goal, which here is to produce and sell books. There are two other kinds of business processes, though:

- **Support processes** are all business workflows necessary so that the company goes on working, without being directly related to the goal of the company. In profit-oriented companies, paying employees' salaries is not the strategic objective; it is absolutely necessary to keep the company and the processes running, but it is not the reason why the company was created. These processes that are not established as goals of the company but are necessary for it to realize these goals are called supporting processes.

- **Piloting processes** are the ones that deal with the governance and analysis of the other workflows. One such workflow is the analysis of the activity indicators of the company. Another process that can be classified as "piloting" is the management of quality, which covers all operational processes and has the goal of continuously improving how efficient they are. Governance or piloting processes, just like support processes, are not directly operational. The difference with piloting processes is that they sit on top of all other processes, whereas support processes are dependencies for the operational processes.

Granularity in the processes

As we just saw, there often is a main, high-level, strategic objective for an organization, and several processes are necessary to realize the different lower-level objectives necessary to reach the high-level goal. When grouping the processes at such a large granularity, we often talk about macro-processes, because they are very general. They are easy to define as such because their objective is not a concrete deliverable outcome but a general idea of what a company does. For example, one could talk about "commerce" and "production" as macro-processes, because their outcomes are very general, respectively getting money from sales and producing goods or services. It is hard to say how we could really achieve this in detail.

When talking about business processes, in contrast to macro-processes, the outcome is quantifiable. For example, producing a car is a business process because we can count how many cars leave the factory in a week. Writing software is another example of a business process because the outcome is the release of a piece of software, together with the ways to exploit it (documentation, setup software, etc.). The tasks that compose the business processes are associated with a type of actor, such as "assembling the engine," "writing the summary of the book," or "producing a commercial quote." This is how they differ from macro-process where a single part of the process can necessitate many different profiles, such as "product marketing" or "invoicing". Units inside a macro-process can be business processes. In our editing example, the macro-process of "producing books" needs a business process to recruit authors, another one to supervise their writing, and yet another one to correct the book.

The same idea of level-based decomposition and granularity of processes can be observed one level down, by decomposing the different items of a business process into detailed steps, which themselves constitute another process, this time a fine-grained one, often called a "procedure". This time, the procedure not only states the tasks that have got to take place for each actor, but the precise operation that each of them has to carry out to realize a given task of the business process. For example, inside the business process of "selling a book", there may be a task called "sending an invoice". A detailed procedure for this task would be, for example, composed of the following "elements of procedure":

1. Every month, list all the customers that have ordered books.

2. For each customer, gather all books and quantities that have been sent from the stock database.

3. Check that the packages have indeed been sent.

4. Verify the discount agreement with this customer.

5. Calculate the total amount due for the books sent.

6. Subtract possible credits that the customer may have due to book returns or warranties.

7. Enter all this data into the "invoice" template.

8. Print two copies and store one at the accounting office.

9. Send out the other copy to the customer using its billing address.

This last example may seem a little bit old-fashioned in the virtual world we live in today, filled with Customer Relationship Management (CRM) and Enterprise Resource Planning (ERP) systems, selling only e-books and ordering/invoicing online. The reason why this example is used is two-fold. First, as has been explained previously, it is always interesting in business/IT alignment to consider a problem by removing anything related to IT. This allows us to concentrate only on the functional problem and understand it in its most intricate details before thinking about implementing it technically. This way, software-based hypotheses, which can lead to coupling, are left out of the scope, at least in the first analysis.

The second reason for showing such an old procedure is to illustrate how the software implementation of business processes has led us to almost forget them. There are good chances that when reading this list of steps, you thought, "No one does this manually anymore." And you would be right: all these operations are mostly done by ERPs and dedicated invoicing software applications nowadays. But… there has to be at least one person who knows about these steps, and this is the person who will design these software applications! As this is what this book is about, it is important – once again – to start from a pure and correct understanding of the business aspects before trying to implement them in an information system.

Also, establishing this detailed procedure before automating it will allow you to gain some insights from business experts. For example, someone from accounting will tell you that you forgot about multiple VAT rates you will have to deal with if you want to sell internationally. Another person will add that credits should not be taken into account if the customer has a debt towards you. Yet another colleague may argue that, in some cases, the payer for the order might be a different legal person than the one who should receive the invoice. And so on…

The very principle of processes containing some other processes as detailed explanations of a single task is one of the important concepts in the business process modeling approach and we will see in this chapter how this can be detailed in a formalized way. The three main levels are macro-processes, business processes, and procedure but, depending on the context, some other levels may appear.

Limits of the processes

If you have been exposed to business processes in an organization as an actor in a given task, there are strong chance that you have a bad opinion about them. Business processes have suffered from a bad reputation because of many badly-led implementations. There are many ways of missing the point with processes and doing more harm than good, but let's start with the way that works. If you want to improve your organization with process management, the very first rule is that the processes should always reflect what happens in reality.

This might sound obvious at the beginning of the process design: most process designers will start by looking at what happens in the organization to draw up the process. Yet, there still are some organization leaders who think they know better how operational people work and will create a process that is not based on reality. This, of course, leads to useless processes, and this is why Gemba is one of the important concepts in the Lean method, designating the place where the value is created. In industrial organizations, that means going to the factory floor to understand what really happens.

Another fallacy is to believe that, once the process is well established, improvements will flow from optimizing its representation. This is a much-observed sequence:

1. A process analyst observes the work of an operational team.
2. The process is drawn up and correctly reflects the actual work.
3. The process analyst detects a possible optimization in the process.
4. An improved version of the process is designed.
5. The team goes on working with the existing process and no improvement is observed in reality.

This is simply another case where one forgets that the process should always reflect what happens in reality. The process analysis might find some place for improvement but the only way to achieve improvement is when the operational team takes this into account and discerns – by itself and within its own organization – how to alter its way of working to avoid the problem. Once this is done (and most of the time, the solution found by the team will be different from the one imagined by the solution analyst), the process should be updated to account for the modification done by the operational team and to keep on reflecting it.

The worst that can happen is when a business analyst has hierarchical power over the operational team and tries to force them into following a process that comes out of pure analysis and does not come from operational observation. Except purely randomly, there is no chance this process will have a positive effect and improve the way that people actually work. What will happen is the contrary: working with an unadapted process will lower the operational team's morale and increase the likelihood of people going around the process, or even failing to realize the task by finding a flaw in the process and voluntarily acting on it to show how bad the process is. Sounds crazy, doesn't it? Yet it happens every day in many companies, simply because people use processes in a bad way, thinking that theoretically knowing them can lead to an improvement "on paper." Processes get a bad reputation in this kind of situation, where they are thought to be more important or more correct than the actors themselves.

Again, processes can only be a representation of what happens in a real, concrete organization. They may be a great tool to uncover bottlenecks, design solutions, and even in some cases simulate them. But the only reality always comes from the factory floor and processes can never be more than a useful representation of people's actual work.

Business process modeling

The previous section may make you think that processes are a bad tool, and indeed, they often are. But that does not mean that they cannot be correctly used and their advantages, when doing so, are numerous. First, they are a great visual way to communicate around coordinated work for a team. Just like a Kanban board is a visual way to share a common view of the advancement of a project, a well-formulated process is a great way to share a common understanding of how a team works together. When a team comes together around a process description, there is almost never a case where it doesn't lead to an interesting optimization, whether this is by better sharing the information ("I did not know you were the person doing this task; next time, I will inform you directly of this situation that might affect your step in the process") or by proposing different ways of doing things ("What if I passed the information directly to the actor after your job? Since they are not dependent on your output, they could start right away and the total cycle time would be reduced").

"Visualization," "drawn," "visual way": all these terms clearly indicate that a process should be a graphical reality and, guess what? I am sure you have already drawn many processes in your life without even knowing you did so. How about this simple diagram?

Figure 11.1 – An extremely simple example of a process diagram

This is already a process diagram, even if an admittedly very simple one: it contains two tasks; they are coordinated (the arrow shows that the second task should be done after the first one is complete); and they are done in order to reach an objective, namely getting paid for books.

Business Process Modeling or **Business Process Management** (you will find both decompositions used for the **BPM** acronym) are about formalizing such processes in a way that any organization process can be described in detail and that the process descriptions can be used for more than just graphical representations, which means, for example, clear communication about the tasks of every actor, change impact analysis, or process optimization, and so on. When talking about formalization, you should normally have the reflex by now to think of a norm or standard that would help this. The good news is that these exist; the bad news is that they were so numerous that it took almost two decades in order to reach a point where a single standard is complete and widely accepted as the reference for BPMN.

The history of BPM standards

There have been so many approaches in software text-based representation of processes that the evolution of those tentative standards and their cooperation, competition, and crossovers can be represented as complex chronological diagrams that are almost impossible to display on a single page. You will easily find those diagrams with an internet search, but since all of this dates back ten years or so, there is simply no use in reproducing it here. What might possibly still be useful is to trace the big milestones in this work:

- In 2000, the WfMC consortium created WPDL 1.0, having started the design in 1997.

- A few years later, it adopted the then-new XML approach and created Wf-XML 1.0, followed by a few other versions.

- WPDL itself evolved into an XML-based grammar called XPDL, which was also developed into later versions by WfMC, reaching 2.2 in 2009. This made for an awkward situation where two standards were proposed by the same organization.

- Meanwhile, another consortium called BPMI created BPMN around the same time as WPDL was released in the early 2000s. **BPMN** stands for **Business Process Modeling Notation**. This standard itself reached 1.0 in 2004 and is about representing any process.

- At the same time, IBM was working on WSFL, which evolved under a joint effort with Microsoft and BEA into BPEL4WS in 2002. **BPEL** stands for **Business Process Execution Language**, and takes a bit of a different approach than BPMN, as it insists on the execution of the process and not its representation. BPEL4WS targets web services as the means of execution of the processes.

- OMG is another consortium that is famous for its definition of **Unified Modeling Language** (**UML**). This consortium took care of the evolution of BPMN by replacing BPMI in 2006 in this work and releasing BPMN 1.1 in 2008. BPMN exchanged concepts with XPDL, rendering the latter less useful in time.

- OASIS, another well-known consortium, used the same approach for hosting the works on BPEL4WS 1.1 and supervised the transformation into WS-BPEL 2.0 in 2007. OASIS had an older standard called ebXML, which was integrated into WS-BPEL 2.0.

- The lack of support for human activity gave birth to BPEL4People in order to complement WS-BPEL 2.0.

- In 2010, OMG released BPMN 2.0, which effectively grouped most of the concepts of the existing standards for business process representation into one XML-based grammar.

The BPMN 2.0 standard

While XPDL continued to evolve for some years after the birth of BPMN 2.0, and WS-BPEL 2.0 is still used but only for process-driven execution of web services, BPMN 2.0 is generally considered the go-to standard for process representation nowadays. Its versatile approach makes it able to model virtually any human or machine process from any kind of organization, thus making it possible to apply all operations in formats such as visual representation, of course, but also process optimization, transformation, and monitoring with dedicated tools. Execution is also possible, since the format is very generic, which makes BPMN 2.0 a serious contender for even specialized standards such as WS-BPEL 2.0. Since the latter is also coupled to the web service stack, which is largely considered outdated in favor of REST API approaches, it looks very much like BPMN 2.0 is the standard to learn if you need to use software-based processes.

You will find plenty of resources on the internet to learn about BPMN 2.0 and how to design processes with this standard. If you need a starting place, there is a great poster containing – in a single image – all concepts of BPMN 2.0 and explaining them all including their relationships to each other, which can be found at `http://bpmb.de/index.php/BPMNPoster`. Nothing can be clearer and more concise than this graphical sheet but I will nonetheless provide a few quick explanatory notes about the main concepts of BPMN 2.0 as follows, in order to make it more accessible to follow the examples in the rest of this chapter.

Let's start with the simplest BPMN diagram possible:

Figure 11.2 – Simplest BPMN diagram

It contains a start event, a task, and an end event. The events have been labeled with text, but this is not mandatory, as their representation is enough to distinguish them. The task needs some text, though, and the convention is to always use a verb in its imperative form to describe the task.

The text representation of this process is the following (as output by the Camundi design tool – available online at `https://demo.bpmn.io/` – that I have used for most of this book's figures):

```
<?xml version="1.0" encoding="UTF-8"?>
<bpmn:definitions xmlns:xsi="http://www.w3.org/2001/XMLSchema-
instance" xmlns:bpmn="http://www.omg.org/spec/BPMN/20100524/
MODEL" xmlns:bpmndi="http://www.omg.org/spec/BPMN/20100524/DI"
xmlns:dc="http://www.omg.org/spec/DD/20100524/DC" xmlns:di="http://
www.omg.org/spec/DD/20100524/DI" id="Definitions_14d2537"
targetNamespace="http://bpmn.io/schema/bpmn" exporter="bpmn-js
(https://demo.bpmn.io)" exporterVersion="16.3.0">
```

```
  <bpmn:process id="Process_0bj1gnx" isExecutable="false">
    <bpmn:startEvent id="StartEvent_1psg9fg" name="Start">
      <bpmn:outgoing>Flow_0xih2cf</bpmn:outgoing>
    </bpmn:startEvent>
    <bpmn:task id="Activity_1tbqy2q" name="Do something">
      <bpmn:incoming>Flow_0xih2cf</bpmn:incoming>
      <bpmn:outgoing>Flow_0v43nfz</bpmn:outgoing>
    </bpmn:task>
    <bpmn:sequenceFlow id="Flow_0xih2cf"
sourceRef="StartEvent_1psg9fg" targetRef="Activity_1tbqy2q" />
    <bpmn:endEvent id="Event_08ckjbl" name="End">
      <bpmn:incoming>Flow_0v43nfz</bpmn:incoming>
    </bpmn:endEvent>
    <bpmn:sequenceFlow id="Flow_0v43nfz" sourceRef="Activity_1tbqy2q"
targetRef="Event_08ckjbl" />
  </bpmn:process>
  <bpmndi:BPMNDiagram id="BPMNDiagram_1">
    <bpmndi:BPMNPlane id="BPMNPlane_1" bpmnElement="Process_0bj1gnx">
      <bpmndi:BPMNShape id="_BPMNShape_StartEvent_2"
bpmnElement="StartEvent_1psg9fg">
        <dc:Bounds x="156" y="82" width="36" height="36" />
        <bpmndi:BPMNLabel>
          <dc:Bounds x="162" y="125" width="24" height="14" />
        </bpmndi:BPMNLabel>
      </bpmndi:BPMNShape>
      <bpmndi:BPMNShape id="Activity_1tbqy2q_di"
bpmnElement="Activity_1tbqy2q">
        <dc:Bounds x="250" y="60" width="100" height="80" />
        <bpmndi:BPMNLabel />
      </bpmndi:BPMNShape>
      <bpmndi:BPMNShape id="Event_08ckjbl_di"
bpmnElement="Event_08ckjbl">
        <dc:Bounds x="412" y="82" width="36" height="36" />
        <bpmndi:BPMNLabel>
          <dc:Bounds x="421" y="125" width="19" height="14" />
        </bpmndi:BPMNLabel>
      </bpmndi:BPMNShape>
      <bpmndi:BPMNEdge id="Flow_0xih2cf_di"
bpmnElement="Flow_0xih2cf">
        <di:waypoint x="192" y="100" />
        <di:waypoint x="250" y="100" />
      </bpmndi:BPMNEdge>
      <bpmndi:BPMNEdge id="Flow_0v43nfz_di"
bpmnElement="Flow_0v43nfz">
        <di:waypoint x="350" y="100" />
```

```
        <di:waypoint x="412" y="100" />
      </bpmndi:BPMNEdge>
    </bpmndi:BPMNPlane>
  </bpmndi:BPMNDiagram>
</bpmn:definitions>
```

This may appear quite complex because the XML-based BPMN grammar is quite verbose, but it is the smallest possible file, as it represents the process above, consisting of only one task.

Note that only the first part of the file is an actual BPMN standard representation, as can be seen by the use of bpmn: prefixes, associated in the header to the http://www.omg.org/spec/BPMN/20100524/MODEL namespace. The rest of the file, where the tags are prefixed with bpmndi, corresponds to proprietary additions from Camundi in order to overload the graphical positioning of the elements.

Without going into too many details, one can notice in the first part of the XML representation the following:

- As said previously, the representation clearly states what an event is and what a task is, even precisely what kind of event is used

- All entities receive a unique identifier, which makes it possible to link them

- Flows (which correspond to the arrows in the visual diagram) are also represented

- There is volunteer duplication of information between the incoming/outcoming attributes and the sourceRef/targetRef attributes

There is an important notion in BPMN, namely that an activity may be a task but also a sub-process itself made of several activities. This allows us to implement the different levels of granularity we talked about above. It is thus possible, in the BPMN standard, to represent a business process where an activity is described in a very generic manner like the following:

Figure 11.3 – Representation of a collapsed sub-process

And, if the tool supports it, the diagram can be simply expanded into the full definition of the content, which provides the following representation:

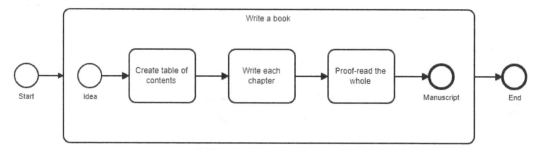

Figure 11.4 – Expanded representation of a sub-process

In the BPMN standards, tasks can be decorated with an icon that specifies how they will be operated:

Figure 11.5 – Different kinds of tasks

Events can also be specialized to account for time-based, message-driven, or other kinds of events:

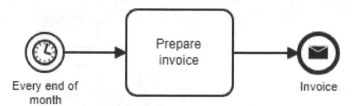

Figure 11.6 – Example of specialized events

If several actors are needed for a given process, they are drawn with what resemble – and as described by the official jargon – swimming lanes in a pool:

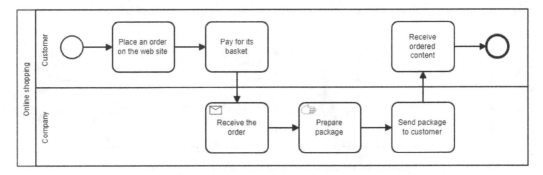

Figure 11.7 – Using lanes in a process

The last essential concept to know in BPMN is the one of gateways. Gateways can derive the stream of a process depending on conditions, and also duplicate the sequences in a given portion of the workflow. Two main types of gateways are exemplified in the following diagram:

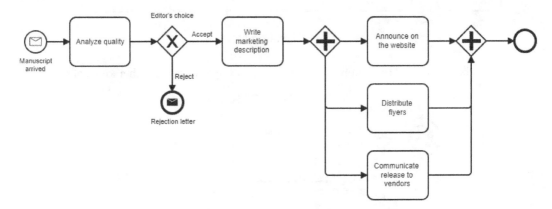

Figure 11.8 – Two main types of gateways in BPMN

The first type of gateway shown on the left-hand side is an exclusive gateway (symbol **X**), which means only one path can be used (in our example, the manuscript can be accepted or rejected by the editor). The second type, using a + symbol, is a parallel gateway, used to execute multiple tasks before joining and continuing the process when all of them are finished (in our example, when all marketing operations have been realized, the process reaches its end event).

There are many other things to know about BPMN, but this book is not the place to become proficient in using this standard format, so I will just stop with these very basic concepts that will be used afterward and advise you to get deeper into BPMN if it will help in the modeling of your organization's business processes. If you doubt at some point that BPMN could correctly represent your activity, remember that 20 years of experts in consortiums have finally achieved the creation of a global standard that is thought to be able to formalize just about any possible assembly of human and computerized tasks. Some may be difficult to design but this is always because of a lack of knowledge of the standard. With a bit of practice, you will be able to model anything with BPMN, and this will, of course, bring a lot of value to your information system design activity, because this means that all functional activity in it will be detailed and formalized, which will highly favor the much-sought alignment of IT.

Business processes software-based execution

As explained, using the BPMN standard already has great value in itself: the simple fact that you use a formalized way to represent your business processes will provide you with great insight and give rise to questions that you might have not thought about but that could become important issues once your information system is built if you have not taken them into account. But an additional value of BPMN is the capacity, once modeled, to execute your processes with the help of software, since the formalism of the representation makes it possible for machines to interpret the processes and even automatically execute instances of them.

In this section, we will explain the principles behind BPMN and give a few examples related to our example information system to give a better understanding of these principles. Then, we will explain how BPMN diagrams are decomposed along roles and what kind of software we can use to model and run them. Finally, I will propose a small explanation as to why BPMN is not more used in the software industry.

Principles

The principles of business process execution are very simple: a piece of software called a BPM engine reads the XML-based BPMN-compliant representation of a business process and can start as many "instances" of this process as you wish. Once started, an instance will roughly have the following behavior:

1. The instance is saved on disk or to a database, and instances can be read and modified at the different steps of their development, which correspond to the advancement in the tasks that constitute the processes.

2. Each instance of a given process is completely separated from the other instances, though they execute the same process definition. How the process is executed in the instances may radically differ depending on how gateways are passed.

3. Once started, an instance follows the process as it was designed at its moment of creation. If the process design evolves afterwards, all running instances will go on with the old version, in order to keep workflow consistency.

4. Every task in the flow is "executed" by the engine. The actual execution of the step depends on its type and on how the engine is configured to handle it:

    -   When a service task is reached, the execution is supposed to be automatic. An API can be called or a connector to an application, and so on.

    -   When a manual task is reached, the engine warns the user that something needs to be done by them. This can be done by a notification through email or any other channel. When the user has finished the task, they are usually invited to inform the BPM engine about this, so that the execution of the instance of the process can go on.

    -   When a user task is reached, the user is also informed but since the action expected is to fill in a form or at least realize something on a machine, a pointer to the required form can be provided in order to speed up the action.

5. If the engine encounters a gateway, it will react differently depending on its type:

    -   If this is a parallel gateway, all subsequent tasks will be run and the engine will take care of waiting for all paths to finish before running the rest of the process.

    -   If this is an exclusive gateway, the decision engine (we will come back to this in more detail in the next chapter) will be activated and the execution of the business rules will define which branch should be taken.

6. On some advanced engines, notifications can be designed in order for stale process instances to emit a warning to a functional administrator, in order for them to get things done.

7. When the end event is reached, the process instance is considered as finished and archived. It is not possible to execute anything in it, but it is kept for statistics or traceability reasons.

Application to our example information system

The best way to comprehend a technology is with examples, and this is why we have followed an example of information system design from the beginning of the book. Let's come back to DemoEditor and take a look at what business processes could be designed – and maybe even automated.

The first example will show how processes accumulate data throughout their execution. After all, processes in IT systems most of the time are about creating or gathering data. A book is a piece of data, sales are data even if their main goal is to bring money to the company, and so on. A process can be seen as a series of tasks that create (or not) data. By the end of the process, enough data has been created or retrieved to obtain the objective of the process, at least in its instance. In the following example, we request information from an editor and ask an author to complete this information because the goal of the process is to publish complete information about a new author:

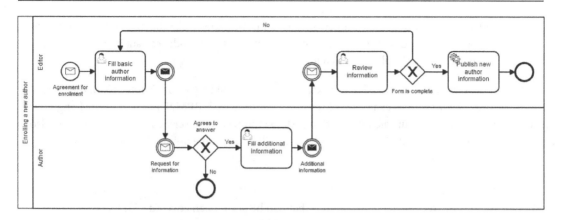

Figure 11.9 – Example of BPMN for author enrolling

The process should be self-explainable, so we are not going to give any details. A second example of a business process of DemoEditor is the following:

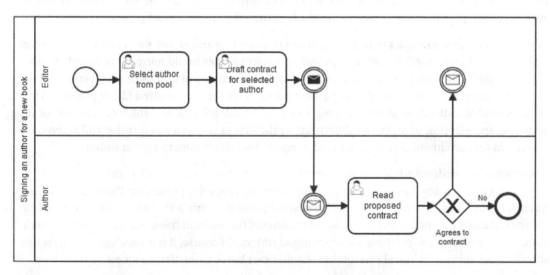

Figure 11.10 – Example of BPMN for contract signing

This time, the collection of data may not be as obvious as in the previous process diagram, but we still can think of the process this way:

- The first task collects a piece of data, namely the identity of the selected author

- The second one creates some data, because the contract draft will be a document, and thus constitutes electronic data

- The third task may not produce functional data, but the simple fact that the author downloads the contract draft to possibly sign it is a signal nonetheless and produces data in the information system (even if very simple, such as a log)

How these processes will be executed in the example software will be part of *Chapter 17*. In the present chapter, we only show them as examples of what can be done with BPMN and why a BPM engine is one of the three parts of the utopic information system architecture described in *Chapter 5*. They also are going to be used in the rest of the present chapter to illustrate some points about process execution.

Link to users

Now is the good time to come back to a notion that had hitherto been covered a bit too swiftly. When talking about how complex diagrams featuring multiple participants are cut into "swimlanes" in a "pool" that the process represents, the notion of "actor" was introduced to explain that each of these lanes had to be associated (and, by convention, named after) an "actor." It may sound like we are talking about a user, but an actor is more generic than this and should rather be associated with a group of users (what some would call a *profile*, even if this word is normally used to designate cohesive sets of authorizations, as in the "role" semantics in the **Role-Based Access Control** paradigm).

In the two preceding examples of BPMN diagrams (*Figures 11.9* and *11.10*), the actors were `Editor` and `Author`. In the user directory, groups with equivalent names would normally be found. A good BPMN engine always includes a user directory supporting groups of users – or even better, can be plugged into a central enterprise directory, using for example the standardized LDAP protocol. This allows a level of indirection as when talking about the functional process, which specific author or editor we are referring to makes no difference. In the first example, a given editor will receive an agreement for enrollment and will then send a request for information to a given author.

It is important to understand this notion correctly and not to couple it too much to the user's directory. Associating a BPMN actor/swimlane with a group is the most logical approach, but the coupling should not be too hard. For example, if it is decided at some point that only a few senior editors can start a contract, the process engine should be able to implement this without being dependent on the user's directory to create a new group for these privileged editors. Of course, it is more elegant if this is the situation, but again, a technical prerequisite should never block a functional request.

The moment when a process is instantiated is also important from the point of view of the actors' selection. The choice of which precise user will correspond to the generic definition of an actor in a "swimlane" will generally be realized upon instantiation of the process in the BPM engine. Most engines will show a dialog asking for who precisely in their user directory will be associated with this or that actor/lane in the process pool. Some even allow the definition of default user affectation or rules to select users depending on other elements of context. For example, we could have a `DemoEditor` process that would automatically select the editor associated with an author when the latter sends its manuscript for review.

Available software for BPMN edition and execution

In such a book that tries to remain generic and push forward standards instead of implementations, we are not likely to find examples of vendors, but the uses of BPM engines in the industry are so scarce that I thought it could be useful to cite a few of them. Of course, I only will list those vendors who play by the rules and make the effort to support the BPMN 2.0 standard, instead of trying to tempt their customers into vendor lock-in by providing sweet but proprietary features. Here are a few BPMN engines (some of them including a graphical editor):

- Bonitasoft (`https://www.bonitasoft.com/library/the-ultimate-guide-to-bpmn2`)

- Activiti (`https://www.activiti.org/userguide/#bpmn20`)

- Bizagi (`https://www.bizagi.com/en/platform/standards`)

- Kogito (`https://docs.kogito.kie.org/latest/html_single/#ref-kogito-app-examples_kogito-creating-running`)

The market always changes, and new versions and features appear; this is why I am not going to recommend or compare these solutions. There are also certainly many others that I simply have not been exposed to. This list is there simply to serve as a starting point.

Why so little use of BPMN in the industry?

As we saw, the norms and standards have existed for quite a long time, and the BPMN engines are there, up and ready for global use. Yet, the overall use of BPMN in the industry is very small. Apart from some very large companies, and only within dedicated perimeters, the use of BPMN-based automated processes is very scarce, although there is no problem in terms of the price or complexity of the norm. In fact, lots of implementations are free and the norm is quite easy to learn, even for a non-technical person. Actually, it may even be one of the easiest norms to understand for business-oriented people, as it represents their everyday job (at least when used properly). So why do so few use it?

One of the reasons may be that automating a process does not bear so much value in many cases. Indeed, setting up a BPM engine is quite a complex task and is only worth it when this investment is compensated either by an important complexity of the process (BPMN drawn up by business specialists and executed "blindly" by the software) or by frequent changes in the process definition (which makes it interesting to delegate execution to a generic engine, as this will allow the rest of the software to remain stable). And when you think about it, lots of business processes are not such frequently moving objects. The process of sending invoices does not change every month and all regulation-associated processes tend to be quite stable. Even operational processes do not change much more often than new versions of applications are released.

Another reason is that designing a process inside a BPMN file tends to add rigidity to it. In lots of organizations, processes are not so clear and depend a lot on how humans execute them, sometimes not respecting a common way of doing so, and most of the time finding a creative way to the process objective, which is – in the end – the only thing that really matters. Sure, this might upset some managers who love the feeling of control that processes can provide. But this is also one of the drawbacks of processes that was explained previously in this chapter: when they tend to replace human choice, they not only bring down team morale but also reduce productivity while trying to increase it.

Or maybe the reason for BPM engines' low use is just that BPMN remains a bit complicated. Of course, the basics of the BPMN 2.0 standard are very easy to understand and apply (a box for each task, a lane for each actor, and arrows between tasks to show the stream of activities) but the rest of the norm can require much more intellectual engagement.

All in all, BPMN 2.0 is not used as widely as it should be because it could really bring more value to the IT industry if it was better known and widespread. This is why we need to try other solutions, and this is what the last section of this chapter is about showcasing some alternatives that may be used as a lighter replacement of BPM, hopefully giving it a new boost, even though not based on the BPMN 2.0 standard. But before that, we are going to digress just a little bit on other operations that can be performed using BPM – process execution is definitely the most advanced one, but not the only one that can deliver operational value.

Other associated practices

Automatic execution of processes is the ideal, but BPM comes with loads of other advantages, some of them much easier to obtain. This section outlines some of these possibilities.

Business activity monitoring

Business Activity Monitoring (BAM) is the use of BPMN process representation to extract statistics about the sequence flows in instances of business processes, and, of course, to gain some insights into the activity represented by the process. The following are some examples of such statistics-based questions:

- Which tasks take the longest in the process?

- How much time does the process take on average?

- Does this average time vary in a regular time-related pattern? Are there seasons where the process is quicker/slower?

- What are the cycle time and lead time in the process?

- What changes in execution times have been associated with the deployment of a new version of the process?

- Has automation of a certain task indeed brought productivity improvement to the whole?

Apart from the promise of automated execution (which brings repeatability and uniformity), BAM is the most sought-after feature that brings managers to BPMN. The reason behind this is that managers need indicators to understand how their business behaves. And what better indicators could they have than metrics directly borne by their business processes?

The implementation of BAM is actually quite simple if you already use a monitoring system. In this case, it is only the adding logs on every input and every output of the tasks, and then using your aggregation mechanism to derive the desired statistics. In addition, showing these values on a graphical representation of the process is a must in order to make them easy to understand. Time-based statistics can be interesting, but sometimes, only knowing how many times a given task has been executed can bring knowledge to the process.

For example, imagine we add counters to our second example of the DemoEditor business process:

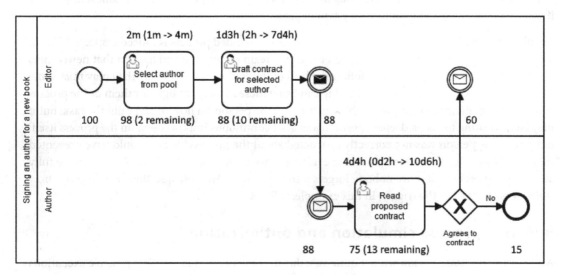

Figure 11.11 – Example of BAM

The counters are shown at the bottom of the tasks and should be read as follows:

- **100** instances of the process have been started

- **98** have passed the first task and **2** remain in this state where the author has not been selected

- **88** have passed the second task and **10** remain at the state of contract drafting (which accounts for the **98** from the previous task)

- **88** contracts have been sent to authors, and all messages have been acknowledged

- **75** contracts have been downloaded by authors; **13** authors have received the message but not processed it

- Out of these **75**, **15** have rejected the contract and **60** of them have signed and sent the approval message

The time-based statistics are at the top of the task. In this example, they only target the tasks and not the transitions. The format shows the average time first, followed by the minimum -> maximum range between parentheses. This kind of statistics could allow us to count how many contracts have been sent over the time interval; how many have been signed and returned; how long it takes on average to draft a contract versus how long it takes for the author to sign it; whether reminders should be used to speed up the process or there would be a waste of time because authors sign quickly their contract; and so on.

BAM is also helpful for finding bottlenecks in the processes. Sometimes, where the process stops is quite easy to find, but in large organizations where roles are scattered, authorizations can be delegated and many steps are involved, some in different services with other managers and where politics could apply, so finding the reason why the whole process has suddenly degraded can be daunting if you do not use BAM. This is just like finding a bug in a distributed cloud application can be almost impossible if you do not have a good monitoring system in place.

Finally, BAM can be used to find out whether the recommended process is indeed respected. When a process is put in place (of course, together with the team using it), it can happen that newcomers are not told about it and they try to follow the lead of the seniors in their field. They may miss some steps of the process and effective monitoring can bring these mishaps to light for them or the process owner. Remediation then simply involves instructing the person on how to carry out the task, but it may be interesting to take a deeper approach and start continuous improvement on the process itself: how come the person was not correctly instructed about the process? What should have prevented this task from being forgotten? Since it is evidently easy to do without, shouldn't this task be fully automated in order to be sure nobody forgets it in the future? All these questions will improve how the process works but the trigger, in this example, is BAM.

Business process simulation and optimization

Another use of BPM, less known but quite valuable in some cases, is to simulate possible executions of processes in order to find a good balance of material resources, tools, and people, and optimize the whole. Imagine for example the following business process:

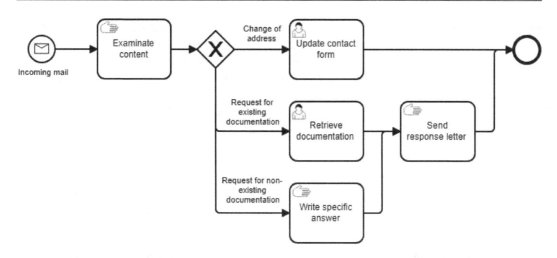

Figure 11.12 – Base for business process optimization

Having this diagram may prove particularly helpful if you have to deal with masses of incoming letters and need to find a way to balance the people doing the five tasks. Let's start with the overly simplified hypothesis that you have ten people in your team, each is able to perform any of the tasks, and each task takes as much time as any other. You may think a good repartition of people would be two people per task... but think again! The **Send response letter** task is necessarily called twice as much as the two preceding tasks. You also do not know what percentage of mail will be going to the **Update contact form** task: if none goes there, then you will gain some time sharing the two persons initially affected by this stage on two other sub-teams.

Keeping the aforementioned hypotheses, once you have a statistical repartition of the types of mail, you may be able to calculate how you should partition people on the different tasks in order to optimize the whole process with a simple calculator. If you start to introduce more realistic behavior, like the fact that not all tasks take the same time, you will most certainly need a spreadsheet or at least some brain power.

Now, imagine that the shares of the incoming mail change seasonally (for example, changes of addresses being more frequent at the beginning of the year and in September); that all tasks have a given average time, but some with a much higher spread (meaning that they can vary significantly around the average time); that some people may be able to handle some tasks and not others, depending on their competencies; that you have to account for the probability of a person being sick or for the fact that people take holidays, though not everyone at the same time due to internal team rules... All of this makes for such a complex system that there is no way you will be able to determine an optimal repartition of people for the different tasks. Luckily, the BPMN representation may help you get an optimal result simply by simulating different team organizations and thousands of process instances, then determining the total duration and choosing the best configuration depending on your criteria.

Large populations-based optimization applications already exist (for example, with genetic-like algorithms or the Monte-Carlo approach) but they all need something to quickly simulate how the system responds: this is where a good BPMN engine will help because it can execute purely automated tasks that simulate the time taken pseudo-randomly. The optimization engine will thus be able to simulate a large number of situations, send them to the BPMN engine for virtual execution, collect the outcome, and finally converge on an optimal solution.

Business process mining

Finally, business process mining should be quickly explained, even if it is not a use of BPMN, but rather an activity that can be the source of a business process. Business process mining (not abbreviated in order to avoid confusion with business process management) consists of determining a business process by analyzing other data that generally comes from software, typically logs.

For example, a process mining system could use logs appearing on the website, together with history tables of invoicing and stocks/expedition metrics, in order to determine a standard process that corresponds to the "normal" act of buying on an e-commerce shop.

There are lots of other applications of BPMN, but we should not deviate too much from our goal, which is to show how business process management can help in reaching business/IT alignment. We have seen how process automation is one of the uses of BPMN that brings the most value, but also sadly how the investment can be quite high, and – for this reason – the industrial uses of such an approach do not frequently arise. Some alternative approaches may help in becoming more process-oriented while keeping the required investment low, or even in some cases without any additional investment than the usual business application.

Other approaches to business process implementation

In this section, we will consider all of the methods that offer an alternative to BPMN engines executing business processes. We will discover which are the more common/more modern ones and will compare their effectiveness. In the end, everything depends on the context, but knowing the specificities should hopefully help you know when to apply this or that approach.

Process in the GUI

You might have not realized but, if you have ever created a software **Graphical User Interface (GUI)**, chances are you have implemented a process without even realizing it. All wizards, for example, are processes since they chain screens in order to provide a way to add data in a sequential order. The most complex ones allow choices, which are the perfect equivalent of gateways in BPMN. They also have a beginning and an end, like any process. Wizards are very similar to processes, but when we take the simple definition of a business process as a series of human and automated tasks organized to reach an objective, then any GUI is indeed a process.

Every GUI allows human interactions that are always the start of a simple process. This process will collect data through forms, and then run some "service" tasks by calling a backend in order to execute some commands. Like wizards, the behavior of the GUI will change depending on business rules or values specified in forms (again, like gateways) and the end of the process will generally be signaled by a toast notification, a dialog box, or simply the fact that the GUI waits for another interaction.

You could argue that a process in the GUI is really a human process where the user follows a process within the software. However, a good GUI leads the user into a particular way of using it that tends to mimic the business process. This is obvious in the case of a wizard, but it also happens in a GUI that has been designed with UX competency. Of course, in the simplest interfaces like a command line, a very large part – if not all – of the process execution is in the hands of the user. But the simple fact that arguments are named after data collected by the BPMN process is already a help in respecting the represented workflow.

Process in higher-level APIs

While we are at simple solutions to the orchestration question, having dedicated APIs that implement an organized sequence of calls to other, simpler, APIs is also a documented approach. In fact, this is a well-known API structure, organizing them in three layers, each building on the one below:

- The first level is the CRUD APIs, which are used to manipulate and read a single business entity. This is the kind of API that we talked about in the previous chapter when explaining the concept of MDM and showing how it could be implemented with REST APIs.

- The second level is about APIs that compose several first-level API calls in order to realize a complex operation in the system. For example, one such API could be exposed as `/api/contract` and its `POST` verb implementation could call `GET` on `/api/authors` to verify the author is already registered, then make a `POST` call to the same API if this is not the case. After that, the code would call the service dedicated to proposing a contract amount, then finally reach the `/api/pdf-fusion` service in order to retrieve the address of the document created and sent to the electronic document management system (of course, using the CMIS standard). At any time where a failure happens, this implementation would have rules to know what it should do and, in the worst case, would provide notifications to a human to clean up the remaining situations that are too complex for the computer to resolve.

- The third level is used to adjust APIs to their callers. This time, they do not necessarily compose several calls but rather add some arguments to the request, adjust and filter some content to the response, and so on. These tertiary APIs are typically used to provide a "backend for a frontend" and, for example, adjust the default pagination and sets of attributes retrieved for a mobile application.

When using an API gateway to include general features such as authentication, authorization, rate limiting, and counting accesses for invoices, the server used can be considered as another level, but it does not sit on top of the three levels described above. Since it can be used to expose any level, a better representation is to show it on the side of the three levels of business-oriented APIs, providing a technical coating for them:

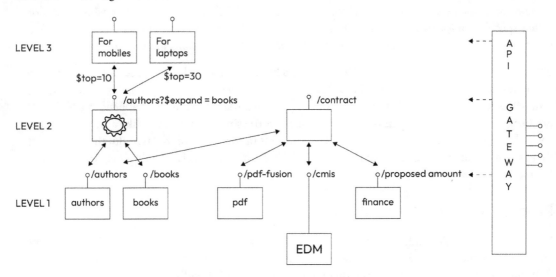

Figure 11.13 – The three levels of APIs

As you can see from the examples chosen in this schema, APIs on different levels do not necessarily use different expositions. For example, it could be decided that the API exposing the authors together with their past books uses the /api/authors route just like the CRUD API exposition of the authors alone does. One rationale for this would be to respect the Open Data Protocol standard, and in particular, the $expand grammar that is well adapted to this case. The API would nonetheless be a second-level one.

One of the advantages of this approach to implementing a business process inside an API is that it greatly respects the principle of single responsibility. A drawback is that the implementation will either be in code, which is more difficult to evolve, or with a BPMN engine but, in this case, this creates some coupling to a technology. There of course is a bit of coupling when choosing a BPMN engine, but using it in many API implementations definitely increases this level of dependency on it.

Now that we have shown two "simple" ways to implement business processes in software, let's analyze some more complex ones. We will start with dedicated middleware servers, which we previously talked about in the context of service-oriented architecture.

Process in a MOM/ESB

In *Chapter 8*, the notion of middleware was introduced and a few well-known implementations were presented, including **Message-Oriented Middleware** (**MOM**) and **Enterprise Service Bus** (**ESB**). These can of course be used to implement processes and orchestrate the different messages or service calls that will practically realize the tasks of a given business process. Though MOMs and ESBs do not understand BPMN, the use of **Enterprise Integration Patterns** (**EIPs**) can be enough to make a business process concrete, then executed and monitored in the middleware.

You might look at this last sentence, as I recommend introducing business functions inside the middleware, while I stated in *Chapter 10* that all business rules should always be inside the MDM service associated with the entity bearing it. Is there a possible paradox here? In fact, there is not if you make use of the Single Responsibility Principle. When talking about business functions that are the responsibility of a given service, it is important that the services are clearly made responsible for each function. For example, if an accounting service needs the net price for a book and the book service only contains the raw price without VAT, it is important not to take the easy path of applying the VAT rate inside the middleware simply because it is quicker and will avoid a change and the deployment of a new version of the book MDM service. The books and all their attributes are clearly under the responsibility of the books referential service, another attribute has to be added to the list of what it exposes (which does not harm backward compatibility, so should not have any impact if clients are correctly written).

On the other hand, let's say that when an editor decides a book should be discontinued, the command should delete the reference from the CRM and from the commercial website, but also set the status of the book to `archived` in the books referential service. Clearly, the action will be sent to the book referential service first, but who should decide about the other actions? We could ask the data referential service to send a message to the CRM and the website, but that would make a clear coupling with these applications. Neither the CRM nor the website should be in control of this interaction since none of them is supposed to be responsible for what happens at the book entity level. The responsibility of this grouped set of commands is not unique, and thus hard to give to a single service.

How the middleware solves this is by providing another application that is responsible for these "orchestration" tasks. It sits on top of messages and has only the responsibility of handling them and routing them to whatever service needs them. Note that the middleware application does not know anything about the content of the message; it simply takes care of the fact that a `DELETE` operation on `/api/books` should be sent to the book MDM service, the CRM, and the website. What they do with it is none of its business. Of course, there are details that should be sorted out. For example, how should the middleware react if one of the services sends an error? Is it its responsibility to cancel the transaction and ask other services to roll back what they have done? These questions will be addressed a bit further on in this section but, for now, keep in mind the old proverb "dumb pipes, smart endpoints": the middleware should never embed any business rules other than pure orchestration, which is the simple dispatch of messages and nothing else.

Low-code/no-code approaches to processes

You have almost certainly heard about the low-code/no-code movement in the industry lately. The idea behind these approaches is that specialized platforms can make it possible for non-developers to create **Line-Of-Business** (**LOB**) software applications by removing most or all code and proposing visual editors to create forms, workflows, data structures, and so on. In a way, they contain everything that is necessary to create a complete system, just like the utopic system we talked about does. The difference resides in the fact that they do so with graphical editors and the user never has to type a line of text-based code (or almost, in the low-code approach, as opposed to the no-code approach that demands that not a single line of code is typed).

Lots of polemics have surrounded these approaches and their associated platforms, some presenting them as a revolution allowing the appearance of the "citizen developer", and others explaining that the code logic and algorithms are still there, simply in a non-textual form. For them, these platforms are not much more than the next avatar of an old promise that went over code generation, fourth-generation frameworks, previous approaches of graphical integrated development platforms… and of course, all the applications that have been created with the most versatile tool ever, which is Excel. As an architect, I try to step away from these opinions and stay focused on the kind of value that such tools can bring.

In particular, better tooling may be an answer to the difficulty expressed above on the limited use of BPMN engines in the industry. Just like MOMs, BPMN engines are quite complex systems and require some setup, maintenance, and expertise. With its focus on the simplicity of use by non-developers (I was about to write "non-technical persons," but that would be far-fetched, as you definitely require a technically oriented mind to use them), perhaps the low-code/no-code tools could be a way to provide an easy-to-use orchestration that would allow the business process execution approach to become more widely used?

This could happen in two different ways, depending on the kind of tools one uses. The first family of tools is the easiest one to use for process automation: namely, platforms that are data-driven. As the ideal information system clearly separates MDM and business process management (and business rules management), that may sound a bit weird, but the concept of reification will help get out of this separation. The word "*reification*" means the making of something into a concrete entity, typically a relationship between two entities. In the case of processes, they can be seen as a series of tasks that contribute to acquiring data, but we could also apply reification and consider that an instance of the process itself is data, after all. This is how data-driven no-code systems, such as Airtable for example, deal with business processes: they simply store data about the process in their data structures, each line corresponding to an instance of a process. In addition, this is made easier by the fact that most simple processes target mainly one business entity, which means that the process and the associated entity can simply be turned into the same entity, managed by the MDM that Airtable and the like actually are.

For example, let's model a process of onboarding for Human Resources. The target entity is an employee which concerns the onboarding process. We will thus simply create a data structure for these onboarded employees, and complete it with data that is more process-oriented, typically the date of entry (beginning of the process), the date of full integration (end of the onboarding process), the URL to the picture that has been taken of the new employee on their first day at work, maybe a pointer to the signed document of IT charts they were required to approve, and so on. As you can see, data from the processes and data about the employees themselves sometimes have blurred boundaries. For example, the date of joining the company is the start of the onboarding process, but this remains important data for the employee long after the end of the onboarding process, as it is used by HR to calculate how many additional vacation days the person receives (determined by how long they have been with the company).

There is a second family of tools that can be classified as *no code* since they allow graphical-only operations: these are the light orchestration tools such as Zapier, IFTTT, and many others that are similar in their use. These platforms allow us to create simple interactions by binding an event (for example, when a GMail account receives an email with an attachment) to an action (storing the file in a OneDrive account, under a given directory named `Pictures`). The GUI to create these interactions can go a bit further, for example by allowing an intermediate task that will filter files based on their extensions and stop if the attachment detected does not end with `.png` or `.jpg`, but this will generally be the most sophisticated use you can have. This limitation is compensated for by providing lots of connectors to third-party platforms. I cited Google Mail and Microsoft OneDrive in the example, but there are hundreds or thousands of editors that have made their applications accessible with these tools. Exposing a public API is generally a prerequisite to doing so, and we will see shortly that webhooks can also be very useful here.

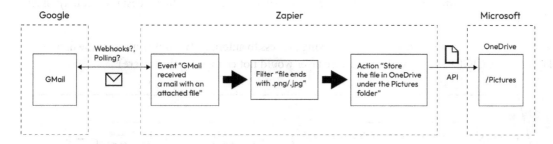

Figure 11.14 – Zapier example

Some platforms such as Microsoft Power Apps are more sophisticated while keeping the same approach of associating business events with actions. Simply, they make it easier to add intermediate filters, duplicate messages, and so on. They could be considered as an implementation of EIPs in their functional approach but, since they do not respect the names of the patterns, they are not eligible for this. An advantage they have, though, is that EIPs' implementations are written in Java or with a technical **Domain Specialized Language (DSL)**, both of which are code, necessitating true developers to be involved. Let's not think that the replacement of this text by a visual diagram editor radically changes the skills necessary to implement "flows," as they are sometimes called: one needs a developer-oriented mindset to be able to set up Microsoft Power Apps workflows correctly. In this way, this kind of tool is really *low code* rather than *no code*, since part of them, such as the complex attributes mapping functions, involve a programming language.

To end this section, simply know that low-code/no-code tools can be great tools to implement an MDM, but also BPM and BRMS. It is easy to create an information system with them but beware: the technical coupling to the platform can be very high. If your goal is to design an industrial-grade, long-term evolving information system, the most important aspect will always be business/IT alignment, and the technical coupling may make you miss some important things and degrade your system's performance. Still, they can be great tools to prototype the orchestration or the entity's definition. And if you keep a clear separation between your services behind contract-first APIs, these low-ceremony tools can be absolutely great as an implementation of the "dumb pipes," while API implementations are the "smart endpoints."

Choreography instead of orchestration

For now, we have only talked about orchestration when considering executing a business process in a software approach: in every implementation exposed, something (a middleware, a BPMN engine, or a low-code platform) was at the center of the game, receiving messages and looking at events from one side, and sending commands to services on the other side. What happens in this case when this central router fails? Since it is a **Single Point of Failure** (**SPOF**), the whole system halts, which is of course a problem. Some people will argue that ESBs have distributed brokers and that a network failure can be handled by the broadcasting approach, delivering the message even in the case of technical incidents. But the functional logic remains centralized and, if a route is badly designed, it may affect the whole system.

To illustrate this, imagine you have the following process to automate (we only represent the first three layers of the CIGREF map, as the hardware layer would not change anything here):

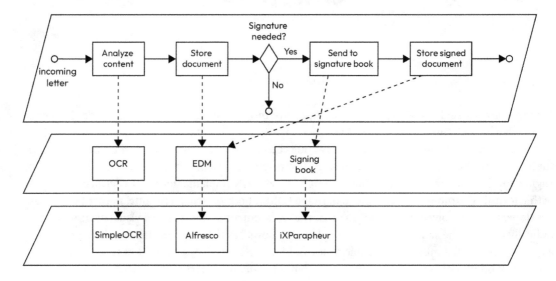

Figure 11.15 – Example of a process to be automated

After analyzing the content of an incoming letter that has been OCR-ized by SimpleOCR, the binary document is stored in the EDM (for example, a Community version of Alfresco). If a signature is required, the signing book (implemented perhaps by the iXParapheur software) is called and, finally, the signed document is also sent to Alfresco to be stored alongside the unsigned version.

If we use the orchestration approach, a BPMN engine, such as Kogito, would be added to the software layer (and its function in the BCM) and the file corresponding to the business process would be in the same layer, as this is a software artifact. Kogito would then call all the functions needed when an instance of the process is run:

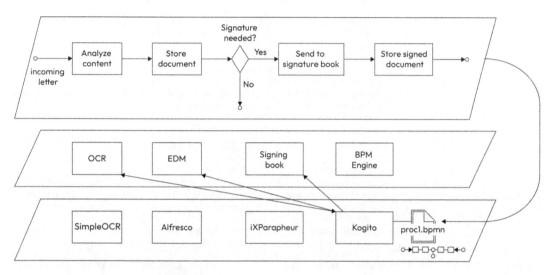

Figure 11.16 – Automation by orchestration

It can easily be seen that if the **Kogito** server suffers a failure or cannot read the proc1.bpmn file, there is a serious problem and the whole process will go down. This is how harmful an SPOF is to an organization. But it could be even worse in terms of evolution and SPOF: imagine that, instead of selecting best-of-breed applications for the different functions in the BCM, we had chosen the "fully integrated" approach, with SharePoint used for storing the file (**Docs** in the following diagram), its OCR features, and also for the workflows (**WFW** in the following diagram) it features. The result would be the highly-coupled one below:

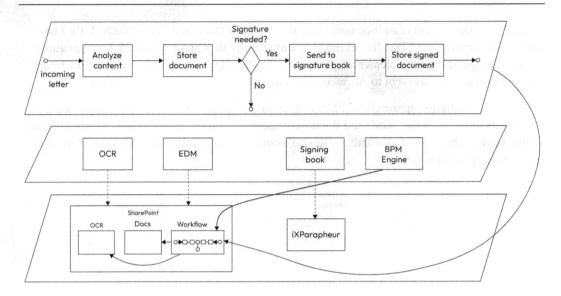

Figure 11.17 – Higher coupling with integrated orchestration

In this case, if there is a failure in SharePoint, not only will the process be down, but also all the features implemented by this server. And since they are most likely used by many other business processes in your organization, the SPOF now carries more risk than ever. Of course, Microsoft has a very robust implementation of SharePoint 365, but you may lose internet access. And if you think you will be better off running SharePoint locally, think again because you will never reach the level of robustness Microsoft can provide for its own solution, however gifted your admins are. So how can we get rid of this SPOF problem for the process execution?

One radical answer to this problem is to simply eliminate all kinds of centralized authority and keep only what is absolutely necessary for the messages to flow between the services, which is a network connection. That may sound quite harsh but, after all, if we really want dumb pipes, how much dumber can they be than simple TCP/IP packets? HTTP and in particular, HTTPS, will add some welcomed low-level features such as streaming encryption and receipt acknowledgment, but they will remain completely agnostic from the business point of view, which is what we define as "dumb".

Eliminating any centralized orchestration is realized by setting up what is called "choreography." In the orchestration approach, just like in musical orchestras, there is a leader who directs the rhythm and tone of all instruments from a physically centralized position. In choreography, a group of dancers does not follow a single leader, but each adjusts by observing their neighbors, just like in flocks of birds or fish. For example, when moving left, a dancer will stare at the feet of their left neighbor; then, when going right, they will synchronize with the dancer on the other side. In these groups, there is no "main dancer" but only a group that is used to working together. By the way, this means that there still is some kind of leader: when learning their choreography, a choreographer will explain the expected moves to the group of dancers, show how they can synchronize, and so on. Practice will then make it so that, once dancing is "in production," the team will no longer need a choreographer. This goes the same in IT choreography: you need an architect to tell each service what signal they should listen to and what their reaction should be. But once set up, the system is on its own and the architect can simply monitor that everything is going as expected.

In our example, implementing this choreographic approach for our process would be done as follows:

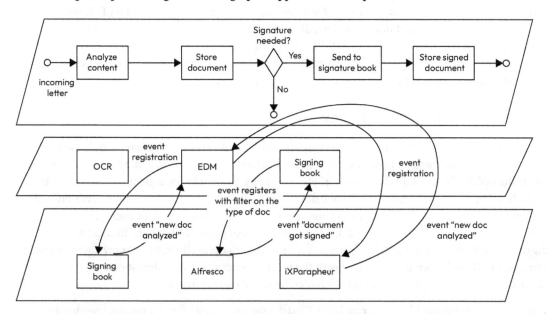

Figure 11.18 – Low-coupling choreography

There would simply not be any additional software because each application would look out for events on the other applications. As a consequence, there would be no possible SPOF. In order to implement the process, the "registrations" would be the following:

- The EDM would wait for a signal from the OCR stating a new document has been analyzed. Upon this signal, the EDM would store this document.

- The signing book would wait for a signal that a new document was stored and would use the business rule to filter documents that need to be signed. This could be done in two ways:

  - The signing book could get all the documents and decide by itself depending on the metadata whether the document needs to be signed.

  - Even better, if the EDM application supports this, it could tell this application to only notify it for specific documents by registering with a "filter expression" that the EDM could apply. This would reduce bandwidth and improve performance, as only documents actually requiring a signature would be notified to the signing book for handling.

- The EDM would also wait for signals from the signing book software (just like it would for signals coming from the OCR platform) and store the signed document when such an event happens.

The great thing about not having an SPOF is that all parts not concerned by a failure will continue to work in the system. For example, if, for one reason or another, the signing book software is not available (let's say it is a SaaS application and your internet connection is down), the rest of the process will work fine: documents that do not need any signature will simply be sent to the EDM and stored; the problem will only be that documents needing to be signed would not be presented for signature (but they would still be stored in the EDM in their original form). We will see a bit further in this section that we could even put in place a "safety net" so that events are not lost and documents to be signed eventually reach the signing book once it is back online.

Implementing choreography

One of the most logical and least-coupled ways to implement this kind of choreographic approach is to use webhooks. We talked about webhooks in *Chapter 8* already and saw they are a great way to invert the order of service. In the context of choreography, webhooks are a great way to eliminate all centralized orchestration, since the responsibilities are shared between two components only: the emitter and the receiver. The emitter can store requests for callbacks on certain events and takes care of sending a message to these callback URLs when the business event happens in their service. On the other hand, the receiver needs to register for the emitter's events it wants to know about, provide a callback URL and listen to it, and deal with the message once it arrives. Almost everything (we will see later that a few things *are* missing) is handled by those two participants.

Then why do platforms like Zapier or IFTTT exist, you might ask? Simply because webhooks and business events are not standardized yet. OpenAPI version 3.0.2 supports webhook definitions, but it is not complete yet and very few editors support it for now. And it will definitely take a long time to define some standards for technical events, let alone business-oriented ones. What Zapier and the like bring to the game is a centralized market of proprietary connectors to hundreds of LOB applications, and this is why they are still in the game. Their value can also be in simple flows that could be implemented with only a webhook plugged into an API, because using them in the middle provides you with monitoring, error detection, automatic retries, and notifications of durable failures, among others.

But in pure theory, it would be possible to implement your **Event-Driven Architecture** (**EDA**) (this is the accepted jargon to define this generalized approach) simply by registering webhooks on all the services in your information system, telling them to call back APIs exposed by others. This would however require that all messages be standardized and a global convention on defining all available business events to exist.

As always, the devil lies in the details, and everything would work fine in such an ideal EDA system until a packet is lost by the dumb pipes, or a network card goes down. Since everything is synchronous and in memory, such a technical incident would result in a functional loss, with potential business impacts, low or catastrophic, depending on what has been lost. This is of course something that cannot be tolerated in an industry-grade system and the reason why, for important data streams, it is a good idea to keep some kind of middleware, such as a message-queuing system. The principle is to keep the pipes dumb, though, and this is where it gets difficult because, once a distributed broker is set up, it is hard to restrain its use to just orchestration. There may be good reasons to do so because this approach can be more adapted to other contexts than the choreography-based one.

Queuing systems will allow for the robust delivery of messages. If you also need the ability to go back in time and do some event sourcing (such as for CQRS, in order to implement Big Data complex calculation, or even to ease eventual consistency), then you may go as far as deploying specialized distributed systems such as Apache Kafka. Again, this is quite a heavy machinery to pull up, so be particularly careful with the balance between losing messages from time to time (and we really are talking about infrequent events, as modern networks are much more resilient than their ancestors) and paying for all the extra cost of a middleware (as a rule of thumb, you can consider that an average-sized middleware will cost you one Full-Time Equivalent). Remember in particular that even huge players in e-commerce accept the loss of consistency and implement many strategies to reduce the consequences (limited time to retain a basket, stock management and acceptance of reservation only if there is enough stock, improved reimbursement if the lock of the product failed, etc.).

Again, remember that functional alignment is key. When thinking about how dangerous it is to miss an event, do not take a technical approach that will immediately draw you into firing up the big guns; think only in functional terms, imagining an information system without any piece of software. How could you compensate for this? Maybe one way would be that the next time you receive an order event, you check with the emitter that all orders that have been passed since the indicated date and time; normally, there should only be the one that triggered the event but if you missed the previous one, you are now informed about it. Another way to implement consistency could be to double the webhook approach with an automated job that will query all new orders every five minutes and verify they have indeed been treated as they should. If you really want to take it to the next level, you could even set up a self-healing system that clones all important data from its neighboring services and only reacts to time-based and interaction-based events to perform its duties, while always keeping and communicating a value date on its operations.

When I talk about this functional-first approach with software architects, they generally tend to reply that a good transactional system will take care of consistency, allowing us to get rid of these functional complications. Or that Apache Kafka is best for this kind of problem and will be the solution – sometimes without having a documented estimation of the cost of such solutions. Though this can be acceptable at times (again, it all depends on the context), getting the deepest possible understanding of the business always brings value to a software architect. One should also remember that, even though the technical approach seems to cover perfectly the difficulties, there is always a way to fail as well, and this should be taken into account (this danger is high when these technologies are SPOFs). On the other hand, when you have a clear understanding of how functionally consistent your system has to be, failure is included in the debate and thus cannot cause surprises anymore. **The functional approach is the only way to solve a problem as a whole.**

If you are interested in digging into this deeper, a good starting point is to understand how sagas work (quickly said, sagas are a way to re-create transactions when you have separated your persistence into several databases, as the MDM and SRP propose to do in order to reduce coupling). The great article at `https://microservices.io/patterns/data/saga.html` shows how to implement them with orchestration and choreography. Please note, though, that in both cases a MOM is necessary, so we still return to the same conclusion: since transactions are a technical solution, they cannot fully cover a functional problem; if you really want a global solution, you have to provide a fully functional solution. In this case, this is about finding business rules for eventual consistency and implementing them. As you are probably used to doing by now, the best way to find this functional solution is to imagine an office without any computer: how would you go about ensuring consistency over a complex workflow, for example, a business case that requires two people working in sequence? Most simply, you would alternate synchronous calls with asynchronous callbacks: "Here is the file, call me when you are done." And when the callback is made, this would trigger passing the business case to the next person, with the same request. When the second one tells the initiator they have finished this part of the job, the whole process can be considered as complete. If there is a stall at any position, the initiator can request the status of the job. If the request has been lost at some point, it can be sent again. When one of the executing agents expresses they are done with their unit of work, the initiator may be used to send them another one, which is great to avoid overloading the agents and building up buffers, which are bad for performance at scale.

Should I use BPMN in my information system?

The length of this chapter may indicate that I am truly passionate about using processes for software information system integration. To be fully transparent, I have long been against processes as I had been mostly exposed to their bad aspects: constraining people into working in a way that has been decided away from the people actually doing the work, being rigid as a working process tends to block any innovation since "it has always worked this way", and so on. Following two **Massive Open Online Courses** (**MOOCs**) from France Université Numérique (French Digital University) named CARTOPRO (mapping business processes) and PILOPRO (using business processes to pilot an organization) completely changed my mind by showing me the power of BPM when used correctly, which means the process is designed by the team using it (the BPMN expert is only there to help them with the BPMN standard and ask the right questions) and continuous improvement is at the base of the whole process strategy. In fact, I went as far as continuing the two MOOCs with an additional digital diploma from the University Jean Moulin (Lyon 3) in France, for which my thesis work was on Agile method process representation (a challenge, since the very first recommendation in the Agile Manifesto is "people over processes").

Why do I tell this personal information? Simply to stress even more the fact that, even if it might surprise you very much, I generally do not recommend using BPMN engines in your architecture of an information system. I know that may sound weird after all I have said and demonstrated here, and particularly since BPM is part of this ideal information system that I have tried to show how to reach since the beginning of the book. But after many attempts at using this approach in production,

I can now honestly say that the tools and – more generally – the understanding of BPM are, in most organizations, not developed enough for a true BPM approach to be worthy.

Hear me well: I am not saying that the approach does not have value at all. If you have a given business process that is complex or moving frequently, the investment may be worth it. But this is a very particular case when you will have to meet the following criteria:

- This process is core to your business and you know it will last for years, or even be so important that it will stay for as long as the company is around

- You know that the tools are not completely stable and you are prepared to change the implementation of a BPMN engine in the middle of the project if needed, considering all costs and consequences

- You have the right expertise for BPM modeling and the maintenance of the BPMN engine, and you are aware that the chances are you will never be able to group this in a single person

- Management understands that turning the workflow of the team into a process-driven approach will require training for everyone and a process of change management that will be long and complex

If you check all these boxes, you have a lot of work in front of you but you are in for a treat at the end of this project as, once the investment is done, the returns are phenomenal: modifying the implementation of a process by changing a few files when the strategy of the company needs to adjust is really the cherry on top of a well-aligned information system. You will need to go through decoupling, appropriate separation of responsibilities, the long evolution of legacy systems, and all the aforementioned BPM hurdles, but once you get there, the information system will not only be the spine of your organization… it will be its main asset.

Summary

In this chapter, we have covered the second part of the utopic information system, namely **business process management**, after talking about MDM. Though the BPMN 2.0 standard is not much used in the standard LOB systems, it is definitely a mature norm, and the tooling is quite complete for edition and runtime. Sadly, the use of BPMN 2.0 is still not soaring, which is a pity as it can truly help adapted information systems to evolve smoothly. Maybe low-code/no-code approaches will achieve a better result at making functional changes easier and less dependent on IT people; only time will tell.

In the coming chapter, we will cover the third and final part of the utopic information system, which is **business rules management**. We will show what this expression covers, how it can be integrated with the two other responsibilities, provide examples based on our demonstration information system scenario and, of course, discuss how such a function should be implemented with which software applications and observing which general recommendations.

12

Externalization
of Business Rules

After the two preceding chapters detailing the Master Data Management and Business Process Management parts of a utopic information system, this chapter will end with the third and last part of such an ideal system, which is the **Business Rules Management System (BRMS)**. We have already discussed briefly business rules in the previous chapters because the data referential may contain some validation rules that are associated with a given business entity, and the business process may also embed some business rules to orient the workflow and decide which branch of the process should be executed, depending on the context. But in the perfectly ideal system that we envision, a centralized system should be responsible for all the business rules, and that is the subject of this chapter.

We will start by explaining in more detail what a BRMS is and what implementing such a solution requires in terms of business rules management, deployment, and architecture of the stream of data in the system. Then, we will show the first example of business rules management using a standard called **DMN** (short for, **Decision Model and Notation**), inside a business process.

Unlike the previous two, we will end this chapter (and the series of three chapters on the different parts of an ideal information system) without providing application examples of our sample information system. The reason behind this is that authorization management is one of the best examples of business rules management, but the subject is so complex that it needs a complete chapter to understand it.

Business Rules Management Systems

A **Business Rules Management System** (we will abbreviate into **BRMS** from now on) is a piece of software that deals with computations and decisions that can be applied to data, in order to output results that have a higher business value. There are lots of concepts in that definition, and we are going to explain them one at a time.

How does a BRMS handle business rules?

A business rule can, for example, calculate the total price of an order line, using the tax-excluding price of an article, the number of articles, and the applicable tax rate. Another example of its application would be to decide whether a piece of document created in an invoicing process should be given an electronic signature or not. In this case, the business rules output a Boolean value, stating whether the result is true or false. Business rules can call each other. In the previous example, we may have to decide how the document will be presented to someone for signature, who will sign if the initial signee is considered absent after several notifications, how many such notifications will be sent and on which channels, and so on – all these are business rules.

As its name suggests, a BRMS manages business rules. But what this entails is not so obvious. In one way, you could consider that a BRMS is the MDM of business rules – it can store them, together with their old versions. It can allow some people to read them, some to write them, or reject others that have no authorization whatsoever on certain business rules. It can group and categorize business rules in order to specify their research. All this is done by an MDM on its referenced business entity, but a BRMS has one more responsibility that an MDM does not have, which is the execution of the business rule. Indeed, a business rule takes input to calculate an output, and it is the main responsibility of a BRMS to do so.

However, responsibility does not mean the BRMS executes everything. Most of the time, it will delegate the enforcement of the rule, as it does not own the data that is addressed by a business rule or the service that executes a business action in a way defined by the output of the rule. That may sound counter-intuitive, but a BRMS can even delegate responsibility for the execution of the rule (which is to calculate the output of the rule from its input). This is the case, for example, where an MDM service validates its incoming data with a rule coming from the BRMS. Some local cache of the rules expressions is also possible since it does not introduce lots of coupling. Nonetheless, the responsibility of the validation rule remains in the BRMS, since if somebody changes the rule in the BRMS editor, then it will apply (maybe after a small delay, if the cache is not immediately invalidated for performance reasons) to all servers using this rule, among which is the MDM of our example.

To summarize, the primary responsibilities of a BRMS are to store, expose, and execute business rules. It is responsible for the right execution of the rules and, thus, either executes them internally or trusts other applications to execute the rules that it provides them. This is often the case, as external applications are the ones that have access to the input data necessary to execute the rule. Also, they will be – most of the time – the ones that use the output of the rule to adjust their behavior accordingly.

Additional characteristics of a BRMS

We often talk about the "secondary responsibilities" of a service for features that are not absolutely necessary but still remain important. In the case of BRMS, there are several such responsibilities.

First and foremost, a BRMS should have high performance, both in execution time and in its capacity to withstand a high volume of requests. Indeed, rule execution is one of the few cases where a cache is difficult to apply. When retrieving an image from a URL, there is a great chance that it is not going to change from one call to another one a few seconds after; thus, it is really worth keeping a cache, as this will avoid a network roundtrip and server request handling and drastically improve performance. This is not the case for business rules, as their main function is to be calculated from varying input and provide an output that depends on them.

Sure, the rule expression can be cached (and rules may not change that frequently), but when you do so, the caller has to be able to execute the rule from its textual expression itself, which may be overly complex and necessitate a rule-execution engine. If the rule is shared among many services, many instances of the engine will have to be kept in synchronization with the BRMS, so it is not efficient. So, we come back to the engine being in only one place, which is the BRMS server itself.

In this case, the engine's moving parts may be cached, or at least kept in RAM, which will provide a quick execution. However, caching the results depending on the input is, most of the time, not efficient, as there are so many possible values. To use our preceding example, there is absolutely no need to cache the result of a calculation of an invoice line total price, as there is almost no chance that another call will come back quickly for the same article, with the same quantity and tax rate. If you add that some rules may be based on ever-changing data (such as stock market values), it can become absolutely impossible to arrange some way of caching. So, we basically fall back to the need for an engine that can output values as quickly as possible. This requirement should of course be supported in case of high volumes. As business data (as opposed to reporting data) tends to be volatile, business rules will be called very often.

The other "secondary" feature of a good BRMS is robustness. When they are used in the industry (which is not very often because they are complex applications), it is because they are a very important part of the business processes. For instance, a BRMS is used by insurance companies to calculate risk, or by mobile phone companies to calculate the price to be paid from data about the conversations (the duration, numbers called, time of the day, etc.) and the contract (a discount for certain numbers, prepaid amount per month, consumption in the month, etc.). Because of the cost of a BRMS, they are generally used for core business functions, where important decisions (in our examples, accepting a contract and sending the correct invoice to a customer) are made based on their output. The robustness of calculations is, thus, an important aspect because no one would work with a system that can miscalculate from time to time.

For the same reason, traceability is generally an important feature of a BRMS. It can, of course, delegate this to calling services because a BRMS mostly works for other services. But even if the responsibility is shared, there should be logs that record whether a rule has been applied to certain context data, providing output that clarifies why a certain rule was made. Even if logs are better suited to the calling application, it is a good idea that the version of the BRMS set of rules is kept somewhere and that the rule engine versions are immutable. This allows you, if necessary, to go back in time, re-execute business rules calculation on the then-used version of the BRMS engine, and understand why an output value was wrong.

Finally, as explained previously, a BRMS is often used in conjunction with other services and is useless by itself. Its low-level characteristics make its integration and good capacity for interoperation of paramount importance. An implementation should normally come with at least some APIs and, if possible, SDKs for as many languages as possible, making it easy to interact with all possible software applications.

Actual use of BRMS

As was hinted previously, the actual use of BRMS in the field is very low. The cost of implementation is such that only a few very particular cases of business rule execution are actually worth deploying a dedicated server for. Also, as we saw, the externalization of a rule comes with a high toll on performance, as either the application that knows the data has to send it to the BRMS and wait for the output to follow its flow, or it has to dynamically execute a business rule expression sent by the BRMS, and maybe cached internally. In this case, the speed of execution is still lower than when the rule is compiled into an application. Sure, the coupling between the application and the rule is then maximal, there is no centralized sharing of the rule, and the many uses can diverge. However, the performance issue can be so significant that these reasons are not as important.

Also, let's not underestimate habitual factors as well – since developers have spent most of their careers taking business rules from use cases and translating them into code put in an application, it is an effort to change this way of thinking, extract the business rule, and place it somewhere else. And with what results? A large performance drop and code that is more difficult to read and maintain. This means that the business rule should **really** change very often for externalization to become an option, but this is rarely the case. In the previous example with the computation of a VAT-net amount, how often does the tax rate change? If, like in France, this is more in terms of years or even decades, that means that there is absolutely no problem whatsoever in implementing the rule right in the code, since there will be many versions of the application released before the next change of VAT rate. Of course, that does not mean that the tax rate should be hardcoded in many places. But a careful developer will place it in a `CommonBusinessValues` class, in a `public static readonly` member, and everything will be fine and ready for an update.

This means that, indeed, in 99.99% of the cases, business rules will be concretely implemented, like in the following C# example, by means of code:

```
public decimal GetPrice(decimal unitPrice, int quantity, decimal
taxRate)
{
    return (unitPrice * quantity) * (1 + taxRate);
}
```

Also, lots of other business rules will be scattered all over the code:

```
if (Document.Type == DocumentTypes.INVOICE)
{
```

```
      SendForDigitalSignature(Document);
}
```

As a side note, it is better to use string values or even dedicated code structures instead of enumerations for this kind of value, as this eases evolution.

In fact, there are so many business rules everywhere in a code base that it is difficult to spot all of them. But this is not what is most important. The real challenge is for the architect/product owner/developer to know, when creating the application, which ones should be externalized, which ones should be centralized, and which should be simply left in the code, even in duplicates, because they will never change. Beware, though, that some things that are thought to never change sometimes evolve over time! For example, you could say that the rule about net price will always be stable; net price will always be the tax-free price multiplied by one plus the tax rate. Well, yes, until the government decides to apply multiple tax rates that apply differently to the sub-parts of a product. And if you have in your Product Information Management software some articles that are composed of a hardware part and an installation service, for example, you may end up with the first part taxed at 5.5% and the second part taxed at 20%. If the calculation has been written in a centralized function, this is not so bad. But if it has been duplicated in hundreds of places in the code (which can be the case with business rules that everyone assumes are constant and immutable), you will face some difficulty, not only because it will take ages for the change to be realized, but also because the one instance you forget will most likely be the one your most important customer uses.

In short, externalizing a business rule in a dedicated BRMS is 99.99% of the time overkill, and the cost is not justified. But you can go a very long way simply by putting the business rule in a function. And most of the time, the only difficulty there is simply realizing you are implementing one!

Examples of a BRMS

Let's say that you are indeed in this very particular 0.01% case where you actually could gain business value from implementing a dedicated BRMS. You would, thus, need a piece of software to do this for you, since, as you can imagine from the required secondary features, this kind of server is quite a complex piece of code. At the time of writing, there were only two serious contenders for BRMS servers – Drools (open source) and **Operational Decision Manager** (**ODM**).

The most used open source BRMS is Drools (`https://www.drools.org/`). It contains a core engine to calculate rules (including some functions such as rule chaining), which is sometimes called an inference engine, as it infers results from a data context and a set of rules. It also contains an application to create and manipulate rules (with a web editor). Drools is written in Java and can be interoperated with other platforms, but not natively.

ODM, from IBM, is a proprietary decision management system that was created to extract the important business rules from legacy COBOL code, in an effort to modernize the information systems on the z/OS platform. Although it can manipulate rules, it is mostly organized around the concepts of decisions on events.

As you can see, the landscape is far from being as complicated as in other fields of IT – for example, Big Data, where a single book could not even describe all the software applications, platforms, and servers that are available, mostly all doing the same things while pretending to be radically different from their competitors. This has the merit of clarity – if you need to implement a BRMS in your information system and you want to reduce costs, Drools will be your first choice.

Of course, there are some lesser-known alternatives. Lots of BPMN engines implement their own language for workflow decisions. Windows Workflow Foundation did so, but it is not supported anymore. PowerApps has some expression capability that can be used for business rule execution, but it can only be mutualized, so it is not a real BRMS system. Another solution, although it involves additional work, is to implement your own BRMS. If you do not need advanced features, you can build one quite quickly if you use an existing expression execution engine. Lots of scripting languages are available, and you can even use C# inside C# with expression trees, dynamic code generation, and other advanced, but still accessible, features.

In short, you have a choice of software, even if it is not as plethoric as in some other fields of IT. However, as you are certainly used to now, business/IT alignment is about reducing coupling, so the choice of software implementation is normally not such an important subject (in the sense it could harm the application evolution) as long as there is a standard norm, widely accepted specification, or even just an organization-wide pivotal format that can serve as a level of indirection between functional dependency and technical implementation. And the great news is that there is a standard for business rules, which is **Decision Modeling Notation** (**DMN**) 1.0. This will be our topic of discussion in the next section of this chapter.

The DMN standard

DMN is a standard that defines decision trees and decision tables, which are the two main concepts concerning business rule implementation. In the upcoming sections, we will show how it works and how useful it can be.

The origin and principle of DMN

The DMN standard is in version 1.0 and was published by the **OMG** (short for, **Object Management Group**) in September 2015. At the time of writing, the latest validated version is numbered 1.3 and was published in February 2021. Version 1.5 has existed since June 2023, but it is considered a beta version currently. Thus, we will discuss only version 1.3.

Note that OMG is also the consortium behind the BPMN 2.0 standard, which works in conjunction with the DMN standard. As expressed by OMG as soon as the first version was launched (`https://www.omg.org/spec/DMN/1.0/About-DMN`): "*DMN notation is designed to be useable alongside the standard BPMN business process notation.*" And there is a type of task that exists in BPMN that directly relates to business rules:

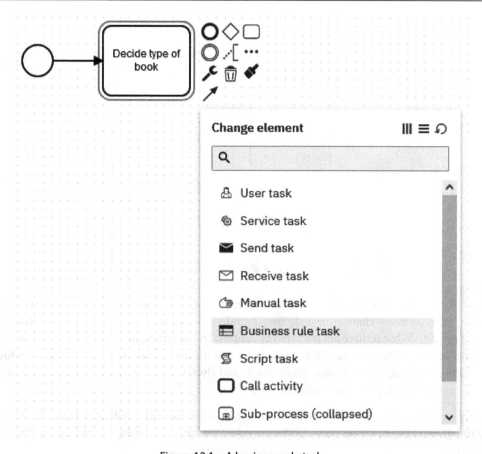

Figure 12.1 – A business rule task

The idea behind this type of task is that there are complex business rules that decide how a BPMN business process should behave (mostly, which path in the gateways should be taken) and that a way to handle such a decision should be made possible. Indeed, imagine a (not so) complicated process like the following:

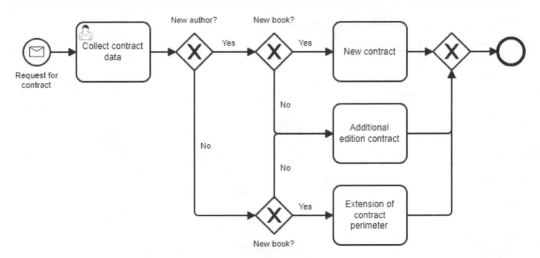

Figure 12.2 – An example of a process with several rules

For now, it is not too bad because there are only three types of contracts. But this is typically the kind of scenario that has many chances to scale (never underestimate the creativity of salespersons and marketing people). What if there are ten types of contract in the future, maybe with a third criterion to take into account? The process will become more and more complex and, soon, illegible, which would be a big problem, since business processes should always remain a helpful tool for teams and, specifically, not something that makes it more complicated for them to work.

DMN proposes a solution, which is to externalize the decision rules in a dedicated place, in order to free up the design of the process itself. In the previous example, we would externalize the decision table like this:

Hit policy: Unique ⌄

	When Author string	And Book string ➕	Then Contract string ➕	Annotations
1	"New"	"New"	"New"	|
2	"New"	"AddEdition"	"AdditionalEdition"	
3	"Existing"	"New"	"PerimeterExtension"	
4	"Existing"	"AddEdition"	"AdditionalEdition"	
+	-	-		

Figure 12.3 – A decision table

This would allow us to draw the process in a much simpler way, as follows (note the icon in the second task, which corresponds to the `Business rule` type):

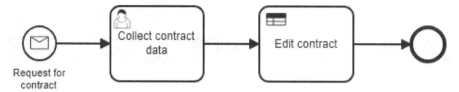

Figure 12.4 – A simplified BPMN process

> **Note**
>
> Sadly, since there is no standardization of how data is collected in the different tasks of a BPMN process, there can be no standardized way either to call a DMN model. But it's worth keeping informed about any updates to the norms at `https://www.omg.org/dmn/` since this is bound to change at some point in the future.

The best part is that now that this logic has been decoupled from the business process itself, we could evolve to a much more complex definition of the type of contract, such as the following, without having to change anything on the process itself:

Type of contract	Hit policy: First ⌄			
When Author _string_	**And** Author age _string_	**And** Book _string_	**Then** Contract ➕ _string_	Annotations
1	"New"	<18	"New"	"JuniorWriter"	This contract needs parents signature
2	"New"	-	"New"	"New"	
3	-	-	"AddEdition"	"AdditionalEdition"	
4	"Existing"	-	"New"	"PerimeterExtension"	
+					

Figure 12.5 – An extended decision table

This time, we also take into account the age of the author to issue some special contract that must be signed by the author's parents. This is done by using a very simple expression here (**<18**), but the FEEL expression language allows for much more sophisticated expressions (if you want to delve into more on this subject, `https://kiegroup.github.io/dmn-feel-handbook/#dmn-feel-handbook` is a great starting point). You also may have spotted that the **Hit policy** dropdown at the top of the preceding screenshot now is set on the **First** mode (the first rule that matches in the table), whereas the previous screenshot showed the **Unique** value – this allows us to specify a condition on the author's age in the first line of the table and simply let the second line to contain no value. I also cleaned the decision table for `AdditionalEdition`, since the result was common for any author as long as the book is a new edition of an existing one.

Having these tables to externalize potentially complex rules is already a great advantage, but DMN also comes with a graphical way to represent the decision process itself:

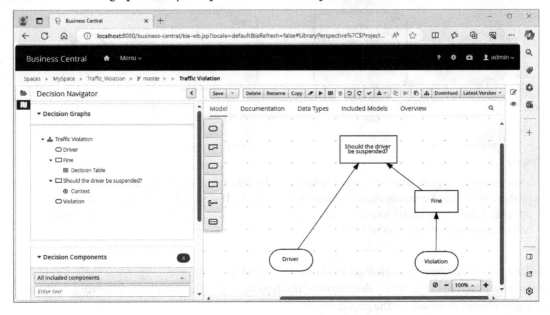

Figure 12.6 – An example decision graph

In our example, the diagram is very simple, since we use only two inputs (author and book information) in order to create a decision (the type of contract), possibly using a "knowledge source," which would be our referential of contracts, although we did not relate to any such use in the simple example previously. However, these diagrams could be much more advanced and show hierarchical decisions if necessary. We could imagine that the type of contract decided on is then itself used to decide the content of a customized contract, depending on the regions of diffusion of the work, and that sales statistics are used to decide on a proposed amount for the contract:

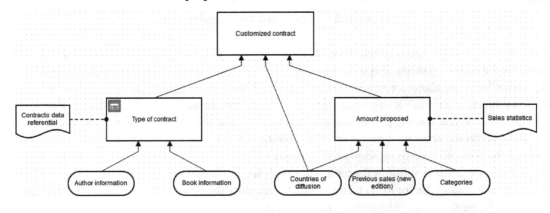

Figure 12.7 – An extended decision graph

To give an idea of the XML structure of a DML file, here is the (shortened) content corresponding to this first example above, where you will easily recognize the first part with the decision rules (starting at `<decision>`) and the second part corresponding to the diagram (starting at `<dmndi:DMNDI>`):

```xml
<?xml version="1.0" encoding="UTF-8"?>
<definitions xmlns="https://www.omg.org/spec/DMN/20191111/
MODEL/" xmlns:biodi="http://bpmn.io/schema/dmn/biodi/2.0"
xmlns:dmndi="https://www.omg.org/spec/DMN/20191111/DMNDI/"
xmlns:dc="http://www.omg.org/spec/DMN/20180521/DC/" xmlns:di="http://
www.omg.org/spec/DMN/20180521/DI/" id="definitions_065qkmh"
name="definitions" namespace="http://camunda.org/schema/1.0/dmn"
exporter="dmn-js (https://demo.bpmn.io/dmn)" exporterVersion="15.0.0">
  <decision id="decision_1u2xbtg" name="Type of contract">
    <informationRequirement id="InformationRequirement_1i0e44v">
      <requiredInput href="#InputData_0wi3jz6" />
    </informationRequirement>
    <informationRequirement id="InformationRequirement_0g0syf3">
      <requiredInput href="#InputData_1jz546j" />
    </informationRequirement>
    <decisionTable id="decisionTable_0cwlzw4"
biodi:annotationsWidth="400">
      <input id="input1" label="Author">
        <inputExpression id="inputExpression1" typeRef="string">
          <text></text>
        </inputExpression>
      </input>
      <input id="InputClause_1hfsajf" label="Book">
        <inputExpression id="LiteralExpression_00wz5lk"
typeRef="string">
          <text></text>
        </inputExpression>
      </input>
      <output id="output1" label="Contract" name="" typeRef="string"
/>
      <rule id="DecisionRule_05kn45x">
        <inputEntry id="UnaryTests_19ou6i4">
          <text>"New"</text>
        </inputEntry>
        <inputEntry id="UnaryTests_0188vr8">
          <text>"New"</text>
        </inputEntry>
        <outputEntry id="LiteralExpression_05irfs8">
          <text>"New"</text>
        </outputEntry>
      </rule>
      <!-- Some rules removed -->
```

```
      <rule id="DecisionRule_1sg8k57">
        <inputEntry id="UnaryTests_1yjyvpp">
          <text>"Existing"</text>
        </inputEntry>
        <inputEntry id="UnaryTests_0qo517n">
          <text>"AddEdition"</text>
        </inputEntry>
        <outputEntry id="LiteralExpression_0ymlni0">
          <text>"AdditionalEdition"</text>
        </outputEntry>
      </rule>
    </decisionTable>
  </decision>
  <inputData id="InputData_0wi3jz6" name="Author history" />
  <inputData id="InputData_1jz546j" name="Books history for the
author" />
  <dmndi:DMNDI>
    <dmndi:DMNDiagram id="DMNDiagram_1r90cap">
      <dmndi:DMNShape id="DMNShape_15dfipm"
dmnElementRef="decision_1u2xbtg">
        <dc:Bounds height="80" width="180" x="330" y="200" />
      </dmndi:DMNShape>
      <dmndi:DMNShape id="DMNShape_14d6htu"
dmnElementRef="InputData_0wi3jz6">
        <dc:Bounds height="45" width="125" x="257" y="337" />
      </dmndi:DMNShape>
      <dmndi:DMNEdge id="DMNEdge_1a3apwq"
dmnElementRef="InformationRequirement_1i0e44v">
        <di:waypoint x=»320» y=»337» />
        <di:waypoint x=»390» y=»300» />
        <di:waypoint x=»390» y=»280» />
      </dmndi:DMNEdge>
      <dmndi:DMNShape id=»DMNShape_0s4bzo1»
dmnElementRef=»InputData_1jz546j»>
        <dc:Bounds height="45" width="125" x="457" y="337" />
      </dmndi:DMNShape>
      <dmndi:DMNEdge id="DMNEdge_0ng7t96"
dmnElementRef="InformationRequirement_0g0syf3">
        <di:waypoint x="520" y="337" />
        <di:waypoint x="450" y="300" />
        <di:waypoint x="450" y="280" />
      </dmndi:DMNEdge>
    </dmndi:DMNDiagram>
  </dmndi:DMNDI>
</definitions>
```

All the graphs shown previously were designed using the great tool provided by Camunda at `https://demo.bpmn.io/dmn`. Now that you have an introductory knowledge of what DMN is about, let's see how we can put the standard to work.

Implementations

The Business Rules Execution System landscape is quite small. The go-to implementation of DMN has been and remains the Java open source project called Drools. Drools is a BRMS that supports its own rules language but also DMN, and since DMN is the standard, all servers using Drools are based on it. You can use Drools directly in your Java applications, or even with some bridges to other platforms. In particular, there has been a Drools .NET implementation, and some projects such as `https://github.com/adamecr/Common.DMN.Engine` can help with that, but the maintenance of such projects is questionable, and I'd rather show you another way that – in my opinion – is more suitable to what we are trying to achieve, which is an aligned and adaptable information system.

To do so, we will get closer to a service-oriented architecture by using a BRMS server that exposes the business rule runtime through REST APIs. Sure, the performance will not be as strong as with an embedded library but remember that, first, "premature optimization is the root of all evil" and, second, that most calls of the business rules are not executed with high frequency (and when they need to be, we will show at the end of the chapter how we can adapt). Kogito has already been cited in the previous chapter, but we did not show a complete example of BPMN with it because, as was explained, this would be overkill for most cases, especially our sample `DemoEditor` information system. What is interesting is that Kogito also supports DMN, and that is why we are going to use it here – or, rather, use JBPM, which is the product that Kogito is based on.

In fact, Kogito is the cloud-native derivative of JBPM, a product maintained under the JBoss umbrella. Since we are not going to deploy in the cloud, instead keeping a Docker-based deployment of our applications to satisfy either SaaS or on-premise conditions, we will simply use JBPM in the following example. Still, keep in mind for your needs that Kogito may be a better alternative, particularly since it offers some functions that could be compared to a light MDM, exposing entities directly by REST APIs generated dynamically. If you want to go in this direction and see how a fully integrated cloud-oriented approach suits you, you can start with Docker images of Kogito, available at `https://github.com/kiegroup/kogito-images`.

The JBoss JBPM server that we are going to exploit is an all-in-one application, providing a frontend and a backend to design and operate BPMN workflows with DMN-based business rules. It works with Maven projects containing some Java code for the unit tests, and possibly for the exposition of the entities, but it can operate with simple standard files in DMN for our example.

In the next section, we will explain how to operate a sample business rule engine in JBPM 7.74, using the Drools engine and a DMN definition of two business decisions with several parameters. For more information about how this works precisely, head over to `https://docs.jboss.org/drools/release/7.74.1.Final/drools-docs/html_single/`. The reason why we use a sample provided by JBoss is that designing from scratch an example on the subject of `DemoEditor` would take up a whole chapter. In addition, it would be a completely artificial exercise, since a DMN rule engine, just like a BPMN engine in the previous chapter, would simply be overkill for our functional needs. It is essential that I respect the principal rule I have been repeating over and over since the beginning of this book, which is that the technical aspects should be completely defined by the functional needs. Although I – like most of us who are passionate about technology – would love to integrate a full-blown Kogito server in our sample information system, the truth is that it would not be suitable for our needs. The implementation of the business workflows and most business rules will simply be in dedicated .NET services. Only a particular case of business rules would be treated with a dedicated external service that strongly resembles a BRMS, namely the authorization rules. But I am anticipating the last section of this chapter, and for now, we are going to show how we can take advantage of a DMN-based BRMS in a business/IT-aligned context, using JBPM.

An example of DMN use

Simple exercises such as the one that follows are really where Docker stands out because it will save us the hassle of installing Java and Maven, getting the right dependencies, updating versions, and so on. Provided you have Docker installed on your machine (and if this is not the case, you really should, as this tool is now part of your basic toolset, just like a web browser and a text editor), you can simply enter the following commands:

```
docker run -d --name jbpm-console -p 8080:8080 quay.io/kiegroup/
business-central-workbench-showcase:latest
docker run -d --name jbpm-bre -p 8180:8080 --link jbpm-console:kie-wb
quay.io/kiegroup/kie-server-showcase:latest
```

Note that the `latest` tag, at the time of writing, was the `7.74.1.Final` version. It is generally recommended to use the `latest` tag as much as possible, but if you ever encounter a functional problem replaying the example, give it a try using this precise version, even if it is not the latest anymore. The first Docker command will start a container based on the image containing everything that is needed to design, build, test, and deploy projects, including the BPMN and DMN assets. This is where we will manipulate the DMN model. If you want to get some more information about this image, the reference page is `https://quay.io/repository/kiegroup/business-central-workbench-showcase`. The second Docker command runs a container onto which the project will be deployed and that will act as a separate business rule execution engine, or a simple runtime if you prefer to think of it that way. The reference page for this second image is `https://quay.io/repository/kiegroup/kie-server-showcase`.

Once everything has started (you should allow for some time – up to one minute – for the completion of the startup routine), you can access the console by navigating to `http://localhost:8080/business-central`, where you will be able to connect with the default credentials, `admin/admin` (the documentation cited previously provides other credentials for users with different authorization profiles, as well as how to set up production-ready authorization).

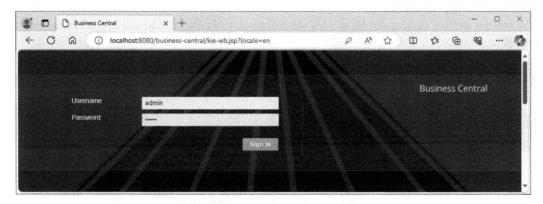

Figure 12.8 – The JPBM login page

Once connected, you will be presented with the welcome page interface, which you can return to at any time by clicking on **Business Central**, or the home icon in the top-left part of the screen.

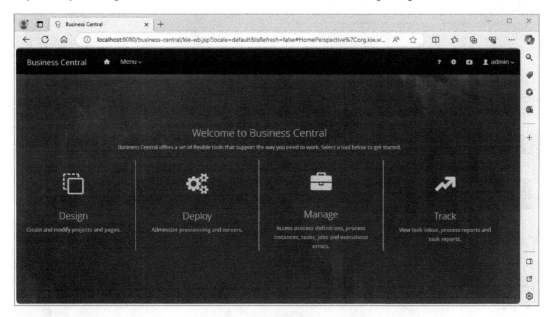

Figure 12.9 – The JPBM welcome page

In the **Design** section, click on **Projects**. This will bring you to an interface in which you can manage your JBPM projects:

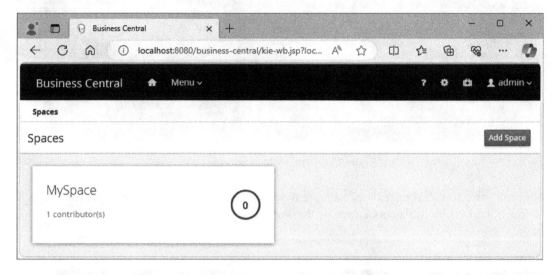

Figure 12.10 – A list of JPBM spaces

Spaces are used to organize work and separate groups of projects from one another. In this simple tryout of the technology, simply select the existing **MySpace** space.

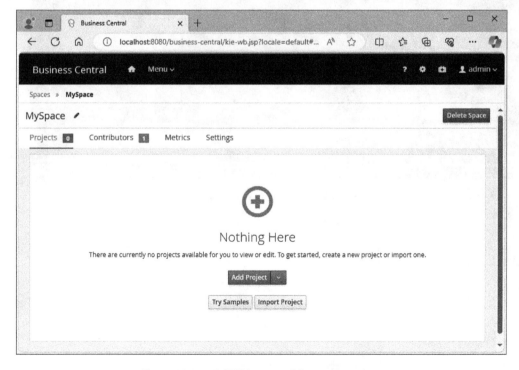

Figure 12.11 – A JPBM space without any project

The space just created is, of course, empty for now. We are going to use one of the embedded examples to illustrate how JBPMN works and what we are particularly interested in right now, namely the DMN rules engine. To do so, click on **Try Samples**, which will bring you to the following interface:

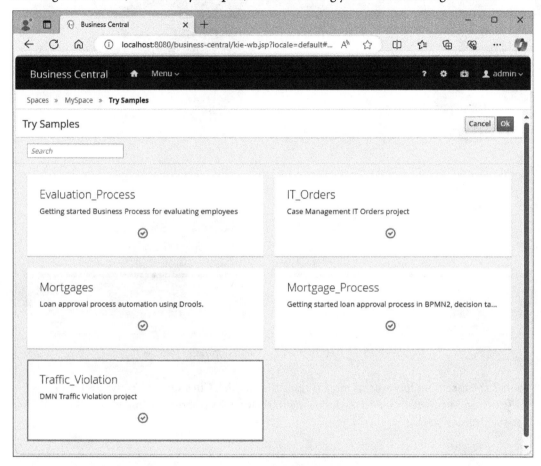

Figure 12.12 – Choosing sample projects

There, select the **Traffic_Violation** sample project and click on **OK**. You should receive a message stating that the project has been correctly imported, and you will land on a page that shows the assets contained by the sample project:

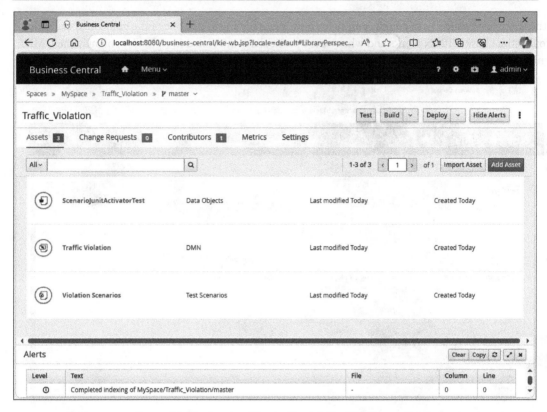

Figure 12.13 – Assets of the traffic violation JBPM sample

Of course, the asset that interests us most is the DMN model. Click on the **Traffic Violation** asset to analyze it, and you will be led to the following interface, which shows the main part of the DMN model, the decision graph:

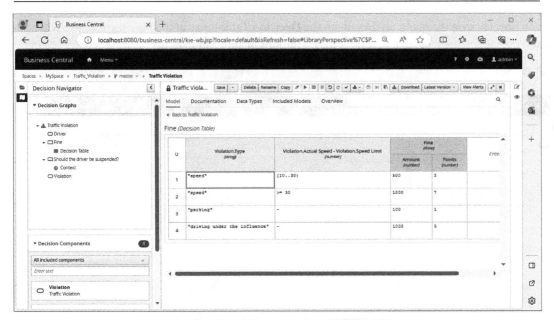

Figure 12.14 – A sample DMN decision graph

If you have a driver's license, the understanding of this example should be self-explanatory – the violation of the speed limit provides data to calculate the associated fine. Then, depending on the fine and additional context on the driver, another decision is taken, regarding whether a suspension of the driver's license should be invoked.

If you now click on the menu on the left, on the **Decision Table** entry of the **Fine** section, you will be shown the following table that describes the conditions of application of the decision:

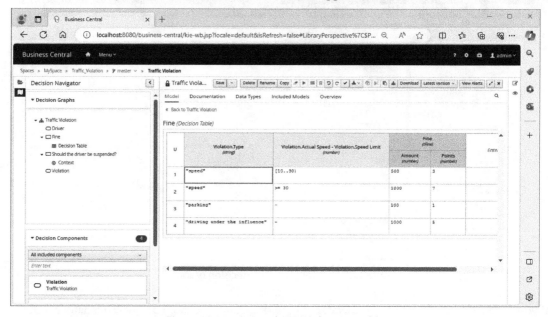

Figure 12.15 – A sample DMN decision table

Now, go back to the project by using the breadcrumbs menu at the top, and then click **Deploy** in the top menu that appears:

Figure 12.16 – The JBPM build and deploy menu

After a bit of time, you should see a message stating that the build is successful, and then a second one, like the one shown in the following figure, that explains that everything is now ready to exploit the decision engine:

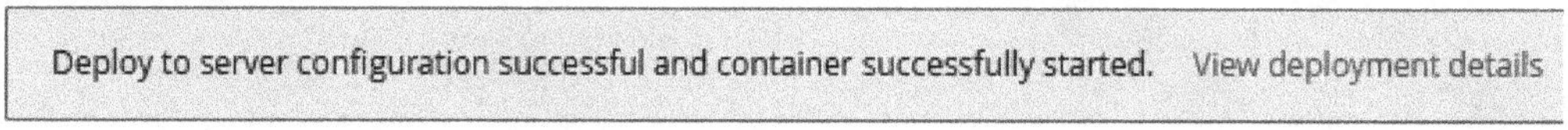

Deploy to server configuration successful and container successfully started.   View deployment details

Figure 12.17 – The JBPM deployment success notification

If you care to take a look at the deployment results, you can activate the **Menu/Execution Servers** command and watch how servers are configured and deployment units are organized on them. You can then start and stop execution servers from this console, or even remove deployments:

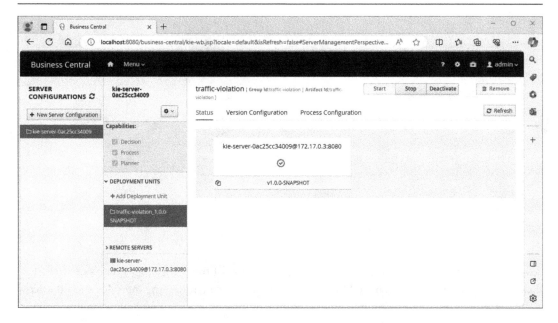

Figure 12.18 – The JBPM servers management interface

Since everything is now set up and deployed, we are able to exploit the business rules.

Calling the business rule runtime

Checking how effective the engine is, is simply a matter of calling a REST API that has been exposed dynamically for us to consume. In order to do so, and since the engine is (logically) exposed through a POST verb, we need a tool a bit more advanced than a simple web browser, such as Postman. To access the API, you will have to use the port number that was associated with the second Docker container we ran – in our example, 8180. The rest of the URL is composed like this:

- /kie-server corresponds to the application server of the rules execution engine (or **BRE** for **Business Rules Execution**)

- /services/rest indicates that we will be accessing the REST APIs

- /server/containers is linked to the fact that BRE servers are exposed through containers, each deployment unit being separate from the others

- /traffic-violation_1.0.0-SNAPSHOT is the identity of the project that we have chosen to deploy in this unit

- /dmn corresponds to the resource we are interested in this project, namely the decision management system

The content of the body should be adjusted to raw/json and contain the following data:

```json
{
    "model-namespace": "https://github.com/kiegroup/drools/kie-
dmn/_60B01F4D-E407-43F7-848E-258723B5FAC8",
    "dmn-context": {
        "Driver": {
            "Points": 15
        },
        "Violation": {
            "Type": "speed",
            "Actual Speed": 135,
            "Speed Limit": 100
        }
    }
}
```

model-namespace corresponds to the unique identifier of the project, and dmn-context indicates the values that should be fed to the rules engine for execution. The interface should look like the following:

Figure 12.19 – A sample Postman call

In order for this to work, you need to go to the **Authorization** tab, choose the **Basic authentication** mode, and instruct Postman to use `kieserver` as the username and `kieserver1!` as the password (these are the default values that, of course, would change if working in production):

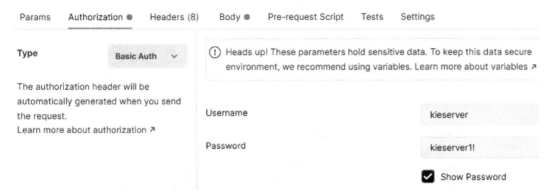

Figure 12.20 – Postman authentication settings

Finally, after sending the message to the server, the complete response is the following:

```
{
  "type" : "SUCCESS",
  "msg" : "OK from container 'traffic-violation_1.0.0-SNAPSHOT'",
  "result" : {
    "dmn-evaluation-result" : {
      "messages" : [ ],
      "model-namespace" : "https://github.com/kiegroup/drools/kie-
dmn/_A4BCA8B8-CF08-433F-93B2-A2598F19ECFF",
      "model-name" : "Traffic Violation",
      "decision-name" : [ ],
      "dmn-context" : {
        "Violation" : {
          "Type" : "speed",
          "Speed Limit" : 100,
          "Actual Speed" : 115
        },
        "Driver" : {
          "Points" : 15
        },
        "Fine" : {
          "Points" : 3,
          "Amount" : 500
        },
        "Should the driver be suspended?" : "No"
      },
```

```
      "decision-results" : {
        "_4055D956-1C47-479C-B3F4-BAEB61F1C929" : {
          "messages" : [ ],
          "decision-id" : "_4055D956-1C47-479C-B3F4-BAEB61F1C929",
          "decision-name" : "Fine",
          "result" : {
            "Points" : 3,
            "Amount" : 500
          },
          "status" : "SUCCEEDED"
        },
        "_8A408366-D8E9-4626-ABF3-5F69AA01F880" : {
          "messages" : [ ],
          "decision-id" : "_8A408366-D8E9-4626-ABF3-5F69AA01F880",
          "decision-name" : "Should the driver be suspended?",
          "result" : "No",
          "status" : "SUCCEEDED"
        }
      }
    }
  }
}
```

What interests us in particular is how dmn-context has been completed, with the results of the decision. In our case, the fine will be 3 points and 500 units of money, and the result of the driver's license suspension decision will be negative. But change Actual Speed to 135 in the body of the request, send it again, and watch the impact on the results:

```
      "Fine" : {
        "Points" : 7,
        "Amount" : 1000
      },
      "Should the driver be suspended?" : "Yes"
```

The engine is, thus, ready to be used in any information system that can handle REST APIs (which is any platform on earth, except for a few very exotic exceptions). Note there is also everything you need inside the JBPM platform to perform tests on the decisions that have been built. A test with Postman was preferred because it is closer to how the BRE would be exploited by another application, but if you click on the Violation Scenarios asset, you will be brought to this nice interface, where you can execute preliminary tests in order to make sure everything works fine before deployment:

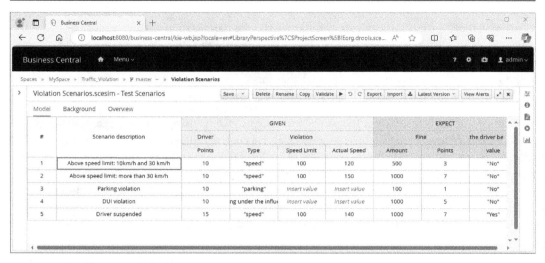

Figure 12.21 – A JBPM-integrated automatic test interface

Also, if you want to understand better how to create your own project (which is outside of the scope of this book, where we focus only on how to use existing projects to correctly structure an information system), the best starting point is the code source of the example used previously, which can be found at `https://github.com/kiegroup/kie-wb-playground/tree/main/traffic-violation`. You can also follow the detailed instructions on how to build this project from the console, as explained at `https://docs.jboss.org/drools/release/7.51.0.Final/drools-docs/html_single/#dmn-gs-new-project-creating-proc_getting-started-decision-services`.

Finally, we will complete this example with a very simple (thanks to Docker) clean-out procedure (be aware that this will erase all data associated with the exercise):

```
docker rm -fv jbpm-console
docker rm -fv jbpm-bre
```

Everything should be reverted to the state the test machine was in before creating this example, which leaves us with the conclusion of this chapter.

Summary

In this chapter, we showed what a Business Rules Management System does, how useful it can be in an information system, and how we can implement one, starting with a functional example and demonstrating afterward another example relating to authorizations, which are one of the most used sets of business rules in software applications.

Just like BPMN engines, BRMS engines are not used very often. In fact, business rules are – in the great majority of cases – implemented in code expressions or compiled into applications. This is absolutely normal because a BRMS represents an important investment, and implementing such complex applications really needs a strong business case, where business rules change very frequently, they are associated with high regulatory or marketing constraints (such as the necessity to trace all business rules and their changes), there is capacity to simulate the effects of new versions of sets of business rules, and so on. It is clear, then, that this approach is currently limited to very rare contexts. Things may, of course, change in the future, with the longed-for industrialization of information system designs, but currently, BPMNs and BRMSs are efforts that are almost always overkill.

And since two of the three parts of the ideal system are not worth using by most organizations, that means that this system remains utopic. Moreover, even a centralized MDM approach is complicated. The MDM practices per se are applicable to every business domain, so there is no problem with data referential – they are not very complicated to set up, as we will see in practice in *Chapter 16* and the upcoming chapters and they bring lots of business value and advantages. However, the ideal system aims for a generic MDM, dynamically adjusting to every entity in the business context of an application. This additional sophistication is also out of context for now, although static code generation for data referential is becoming a viable option, as will be shown at the end of *Chapter 19*.

In addition, we have shown that the three responsibilities of an ideal information system are, ultimately, quite interrelated with each other:

- MDM uses business rules in its validation of data

- A BRMS needs data from MDM to apply business rules and decide their output value

- A BPMN serves mainly as a collector of data to feed MDM, while also consuming data from MDM

- A BPMN also uses business rules to know where to go in the different gateways (and sometimes to calculate some additional data during a given task)

All this proves that, technically, this assembly of three generic servers for MDM, BPM, and BRMS is not so feasible, and neither achieves a perfect decoupling. So, why did we bother in *Chapter 5* and the last three chapters to discuss such an ideal system? Again, the answer lies in the business/IT alignment. The ideal system is not something that can be realized in practice in information systems today (and certainly for at least a few more decades), but it has the great advantage of forcing an architect to think in terms of three generic, always applicable, functional responsibilities. Even if you use a unique software application, knowing how to separate the data management, the business rules management, and the business process execution provides a great step toward decoupling your information system (which is not achieved at all with *n*-tier architecture, for example). As you will see in the upcoming chapters, constructing an information system with these principles in mind will help us achieve a very complex goal, which is to be able to modify important functional rules and behaviors very easily, in most cases without any significant impact on the implementation.

In the next chapter, as explained in the introduction, we will show a particular use case – yet very important application – of business rules management, which is using rules to determine and enforce authorization in a software application. Although we have shown a few examples in this chapter, the most complete description of how to use a BRMS will happen in the next chapter, by applying dedicated authorization management policies to our good old sample information system.

13
Externalization of Authorization

The previous chapter was about business rules management in general. In this chapter, we will analyze a particular case of authorization management, since the rights and privileges of users are one of the most common uses of business rules that you can find in many applications. Since there exist two standards for authorization management (as already explored in *Chapter 8*) we will quickly explain the first and more complete standard, namely **XACML** (short for, **eXtensible Access Control Markup Language**) because it helps understand how it relates to the **Single Responsibility Principle (SRP)**; then, we will create a more complete example with the new, lighter, standard, which is **OPA** (short for, **Open Policy Agent**).

We will then end this chapter (and the series of four chapters on the different parts of an ideal information system) by reflecting on how to implement this authorization in practice, which will open the way to the analysis and the implementation of the information system for DemoEditor, which has accompanied us so far, illustrating with examples the concepts studied, and it will, of course, also serve as a practical example of the implementations of what we learned in the previous chapter.

In this chapter, we'll cover the following topics:

- A BRMS and authorization management
- Applying authorization to our same information system

A BRMS and authorization management

As I quickly mentioned in the previous chapter, there is a functional domain in the DemoEditor sample information system where an externalized business rules engine would be interesting, and this domain is one of authorization. Before explaining the need to clarify the semantics of the "rights" business domain, examine the main paradigms to implement authorization in software applications, and also explain one of the standards associated with this function, which decomposes very well the different responsibilities it entails.

The semantics of identity and authorization management

As explained in *Chapter 9*, semantics is the foundation of all things in architecture, and we will clarify the terms we use for certain concepts in order to not incorrectly define the business domain model. Thus, it is important to clearly define the different subdomains of **Identity and Authorization Management (IAM)** and how we name things inside of them. Let's start with the concept associated with **identification** (who you are) and **authentication** (how you can prove your identity):

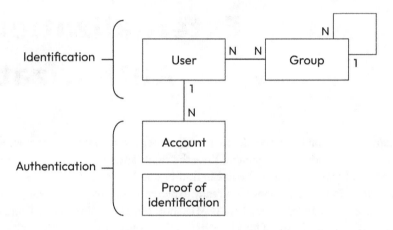

Figure 13.1 – Identification and authentication semantics

A first – and very important – point is that authorization should depend only on your identity (and, of course, some elements of context, but we will come to that later) and never on how you prove your identity. At least, this is how we are going to work on the information system for now. Of course, we may in the future have to take into account that some authentication methods are safer than others and that some applications may request a strong form of multi-factor authentication to open certain features. But this use case will be handled with the addition of attributes to identification to account for this. After all, even in this situation, an application does not need to know exactly what you authenticated but, rather, how strong a trust it can have in the identity that it is provided with.

There are already some cases like this in the standard identity profiles associated with OAuth; for example, in addition to the `email` attribute, the contact profile can provide an `email_validated` attribute that specifies that the identity provider has verified that the identified user indeed has control of a certain email address. This is a way of augmenting trust in the identification without the identity consumer knowing anything about *how* the email has been verified. We are not going to dig deeper into authentication, as this is a hugely sophisticated domain, and what we want to model precisely is the authorization domain. For now, let's just remember that a given user can be authenticated by different accounts/ways to prove its identity.

The important aspect of what will follow is that users can belong to groups, which ultimately will bring them some commonalities in rights management. These groups can be formed in a hierarchical tree in order to ease complex management. Bear in mind that we are still in the identification domain, so belonging to a group does not directly give you certain rights. Groups are simply part of your identity, as are any other attributes such as `lastname` or `firstname`, to give examples from the OpenID Connect/JWT/OAuth standards.

Let's now discuss the other half of IAM, which is **authorization** management. The main semantics are as follows:

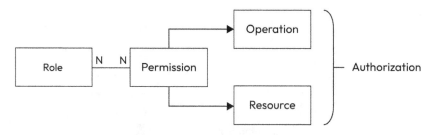

Figure 13.2 – Authorization semantics

The preceding diagram is, of course, just an example, and you may have your vocabulary for the terms used within it. But this is precisely the goal of such a semantics analysis; I know that some people use the word *"profiles"* to describe the groups of people from the identification domain, that some use the word *"group"* to discuss authorization groups, and that some others replace *"role"* with *"profile."* But there are also people using other vocabulary, and the important thing is not who is right; as long as there is no established standard, everyone is. The important thing is to be able to understand univocally what we talk about. In this book, a group will be an entity that organizes sets of users that are similar in their identity, while a role will be a set of authorizations that are often used together.

Let's explain in just a bit more detail the concept of permission, which is defined by pointing to a resource and an operation (or several, if this is easier in your model). For example, removing a book from the data referential service may be something that only some editors have the right to do; we would then design the corresponding permission by pointing to the `book` resource and the `DELETE` verb. The use of the REST-based vocabulary is, of course, intentional – first, it makes it more precise to explain what we mean; second, it allows for a precise alignment of what will happen in the software. In this case, this permission will be associated with the possibility of sending a `DELETE` verb to the `/api/books` API, and it is thus implemented without any possible confusion in the books data referential service.

Of course, some permissions are linked – a senior editor will not only be able to delete books but also create, modify, and read them. This is where roles come into play – grouping together many permissions that make sense together. This is where semantics is also important. Naming the editor role is a difficult choice because we will tend to use the word `editors` for two things that are fundamentally different: the group to which all users belong when they **are** editors and also the role that contains the permissions traditionally assigned to them. A better choice is to keep `editors` for the identification group and use a name such as `book-editor` for the role.

Semantics is important in another area – since there are several applications in an information system and each of them (at least the data referential service) deals with specific resources, it is important to specify the main resource in the name of the role; otherwise, they would get confused with each other. By the way, this is how we will group the two previous schemas, showing this multiplicity of "rights management targets" against the unicity of identification concerns:

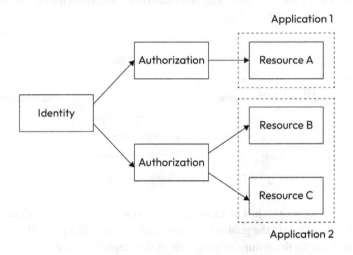

Figure 13.3 – Each application takes care of its authorizations

Before we go into more detail about what is in the **Authorization** boxes, let's make a useful digression on the way IAM is handled in many applications and how it should be used to obtain a neat business/IT alignment.

A digression on IAM implementation

In most existing information systems, identification is still handled directly by many applications, leading to the well-known antipattern represented here:

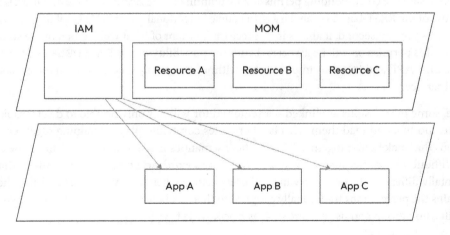

Figure 13.4 – An antipattern where IAM is in many applications

These multiple implementations of a unique feature are one of the most observed misalignment patterns in existing information systems. It leads not only to the duplication of accounts, making it more difficult to manage access rights, but also to the duplication of different passwords, which is a pain for users and quickly causes security issues because lots of them will use similar passwords across their line of business applications, making a password breach suddenly more impactful because the attack surface is increased.

One method that is often used by companies to compensate for this difficulty is to automate the "newcomer" process and implement some kind of tool that will automatically create accounts in every application of the information system. Unless you only have a legacy application and no intention of modernizing your system (for example, because the activity will be closed in a few years), this is always the worst move that can be done, as it tends to crystallize the problem – since you have added another (potentially costly) component to the system, you will be even less keen on changing it again. The following diagram shows the second antipattern in this approach:

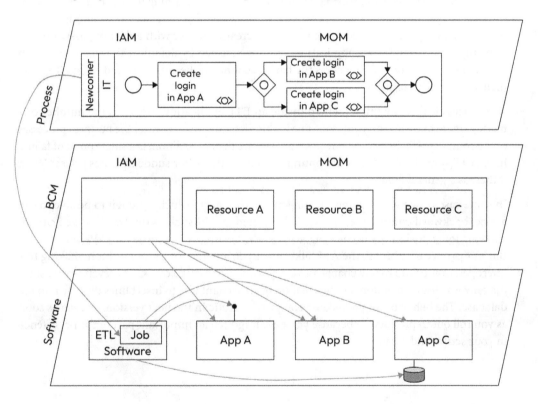

Figure 13.5 – The process for a newcomer in its coupled version

This diagram shows all the additional problems:

- The process is designed in the upper functional layer, but it cannot be modified by the business persons, since its execution is based on a job executed by an **ETL** application and, thus, can only be modified by technicians, which creates some time coupling (a change of regulation will be applied not when the business needs it but when the IT department can get around to in its many projects).

- Talking about the many things the IT has to do, did you notice that the only actor in the BPMN is **IT**? This is logical since all tasks have been designed to be automated and IT is considered responsible for managing the users inside the software, simply because they are the ones who have installed it or know how to access the APIs. This is a very common problem; instead of having functional administrators taking full responsibility for their applications, they rely entirely on IT for this. Although this can be considered normal for technical tasks, this is a problem in this case because trusting IT to add users and determining their default permissions can be a recipe for regulatory disaster. After all, how could you be mad at an intern who has dealt with an urgent ticket from accounting by creating a user with a default password, not knowing that, in this legacy application, users are created by default with full rights, which allows the newcomer user to access the bank accounts of the company and empty them on their very first day in the job?

- The process is directly implemented inside an ETL application, which is the number one misalignment antipattern. If you continue in this direction, very soon, all the business processes of the company will depend on one piece of software that, in addition, is a single point of failure in your IT system. What if it is discontinued? What if the editor suddenly raises prices? What if there is a general failure?

- In some cases, the person doing the implementation may be lucky enough to be able to call a nice, backward compatible, and well-documented API such as on `Application A`, allowing for some kind of decoupling, or even the possibility to expose this API in the BCM. But in `Application B`, the API talks directly to a library of the application, making this interoperation brittle to any change of version. In `Application C`, it is even worse, since the only way found to automate the creation of an account was to insert lines directly into the database. The behavior might become completely erratic in the next version, or even as soon as you roll out in production because you have forgotten an important part of the persistence in your script, and so on.

The preceding approach tends to embed in a system this antipattern, where each application takes care of its own identification and even authentication, whereas it should only handle authorization (this antipattern has to stay there, since the application handles the resources, and the permissions apply to these). Instead of this, the right move would be to progressively adopt the following correct pattern:

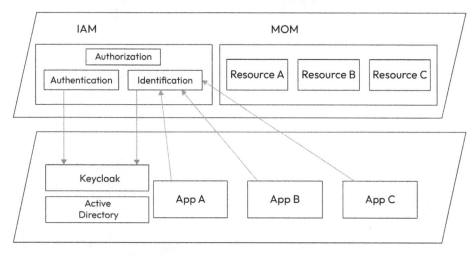

Figure 13.6 – A correct map of IAM responsibility

In this case, the identification and authentication responsibilities are implemented by a dedicated piece of software (in our example, an Apache Keycloak IAM server, plugged into a Microsoft AD user directory) and all applications still take care of authorization on the resource they respectively manage, but they point to this unique identification feature that they need to apply the right permissions (again, without knowing anything about the authentication process). Of course, this would not be done in one day; you need to progressively equip your information system with an application that supports externalized authentication/identification. Nowadays, almost all modern enterprise-grade applications do so, and if they are browser-based, it is even possible in some cases to handle these responsibilities with a frontend protecting them if needed. And since you will likely always keep some legacy applications, you will certainly end up with a "middle of the journey" information system such as the following, which is already much better and easier to handle:

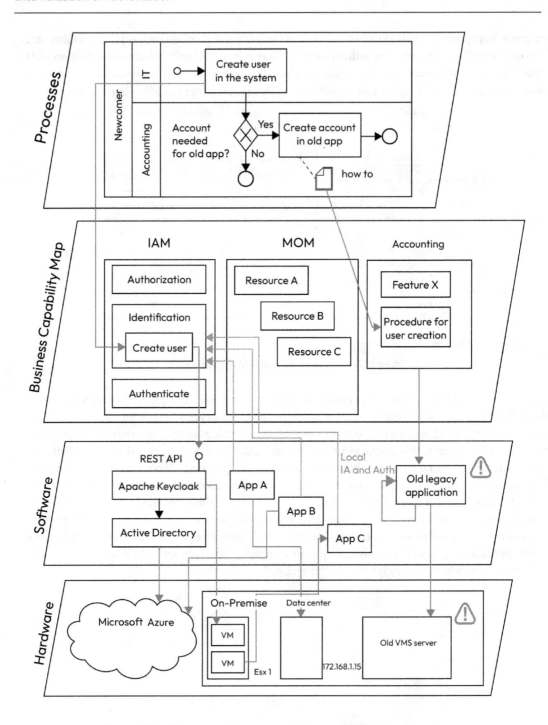

Figure 13.7 – The process for a newcomer in a perfectly aligned version

Do not be put out by the added complexity to the diagram; it is simply more complete because I have added more details – in particular, the hardware layer, which had not been shown before. In this part of the information system, many advantages can be seen on the right of the diagram, but we will discuss them now in more detail:

- The implementation of the process can now be dedicated to any tool, and it will not have any coupling to the technical stack (except for the call to the Apache Keycloak API to add a global user, but it is extremely rare, as this can be based on the LDIF standard and a change of software would not be visible by the process users).

- If the process had to be modified – for example, by adding another step for a legacy application that had been forgotten in the first version – it could be done by decision-makers alone. In the new version, this additional task would work like the existing one for the legacy accounting system – when a user-based task is completed, an email would be sent to the functional administrator of the application, together with a link to the procedure to add the requested user. When done, this person would click on a link in the email received to signal that the task is done, which would also close the process.

- The first task dedicated to the IT department would still be manual, as there would be a form to fill in (the one from Apache Keycloak or – as represented here – a form provided by the BPMN engine that would call the API from Keycloak associated with the `Create user` function of the BCM). If the API from Keycloak follows the LDIF standard, it could be considered as the standardized unique point associated with the function in the information system, making it easier to replace Keycloak with another software if needed.

- In addition, Keycloak acts as an indirection layer to the actual user's directory. If this had to change to another directory, or even use identity federation and several directories, this would be transparent for any user of the API associated with the `Create user` function.

Of course, the problem of the legacy application would not completely disappear, but at least, in this configuration, the legacy impact is progressively reduced and the right functions are ready for the new and more modern applications to work in the way they should. Also, the legacy application is isolated into a silo and will be easier to discard in the future. In this example, we could start by removing the task from the process, and then suppress the old application with its locally coupled identification and authentication features. Finally, we have to verify that the old server with an unsupported or exotic, hard-to-maintain operating system does not serve any other software role in the system.

Role-based access control and attribute-based access control models

After that rather long – but hopefully useful – digression into IAM implementations, we will return to where we were before, which is the fact that, in a good information system, identification and authentication features are unique for all applications, but the authorization feature is duplicated for each resource. Indeed, only the application that handles the resources knows how to handle the permissions on them. In our example with the book data referential service, we saw that a role called book-edition would make sense. But what about in an archiving system? Chances are we would find roles such as archivist or readonly-verifier in there, but book-edition would make no sense.

This is not to say that we could not find common role names between applications; on the contrary – similar names should be considered carefully because they do not mean the same thing. This is why it is so dangerous, even though it is frequently done, to name roles administrator. Of course, everyone understands what this means –users with this role can perform every operation in the software. But, specifically, the definition of "everything" can differ from one software to another. If you add to this situation a group called administrators inside your users' directory, which is supposed to mean that the users in this group should have full permissions in every application, the confusion is increased.

I personally recommend restricting this situation to domain-administrator and arranging for your IT department to never be a functional administrator of an application, only of the machines they are installed on (which does not prevent them from indirectly seeing or manipulating data, but this is another problem that should be dealt with by contractual standards and full traceability of administrative actions).

To account for this, a better representation of the preceding diagram would be the following one:

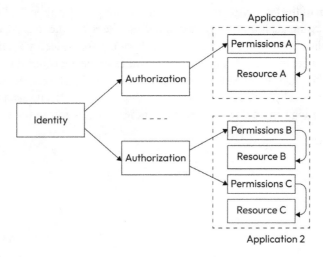

Figure 13.8 – Affecting authorizations on permissions rather than resources

The left-hand side is not as detailed in the preceding diagram, but this is what we wanted. Since we stated that authorizations should be based on identity, how could we do this in practice? One of the easiest and most commonly used ways is the following:

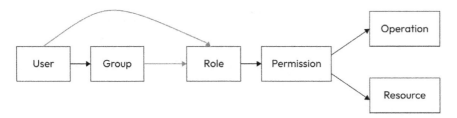

Figure 13.9 – A pure role-based access control approach

When roles are associated (or "mapped") to groups or directly to users, the paradigm of rights management is called **Role-Based Access Control (RBAC)**. The main advantage of this approach is that it is very simple to implement. Since the person who administrated rights only sees the role, the diagram could even be represented like this from their point of view:

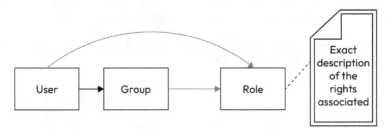

Figure 13.10 – A documented RBAC approach

This eases the work of developers as well because, as long as they respect the contractual text-based definition of the rights associated with the role, they can choose whatever implementation method for the role they prefer, or even a mix of them:

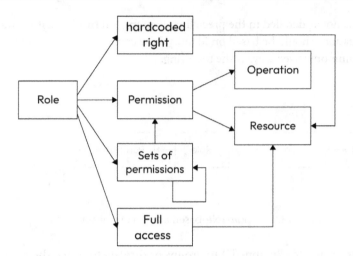

Figure 13.11 – Other possible role implementations in RBAC

The textual definition of the role may cause some trouble, due to the text's imprecise nature and the potential for knowledge to become outdated over time, it is subject to approximation, particularly if the editor role has a high personnel turnover and/or does not document clearly its software features.

Since pure RBAC is quite restrictive, applications often allow for the direct mapping of granular permissions to users or groups, as schematized here:

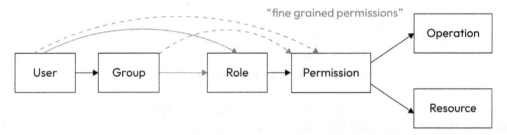

Figure 13.12 – Fine-grained permission as an improvement on RBAC

This extends the possibilities, but it also makes it much more difficult for functional administrators to keep track of the rights given to different users if these cases become more than just exceptions. As the number of users increases, the use of groups and roles becomes more and more important. The temptation to delegate some rights administration responsibility increases as well, but it is essential to implement this with rigid rules and train people carefully, as it can quickly become a mess, where users with the same job title end up with different rights, depending on who has given them these rights. Even worse, some users end up with full permissions on the software because the new functional administrator does not understand precisely how the rights management system works. This is yet another reason to not give this responsibility to IT, however tempting this may be, because they would have control of the technical part of the application.

Another, more sophisticated, way to extend the RBAC features is to shift to what is called **Attribute-Based Access Control (ABAC)**. In this rights management paradigm, rules are set that link attributes from the identification to attributes of the resources:

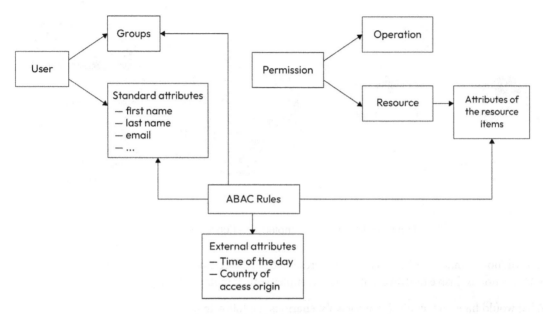

Figure 13.13 – An ABAC approach

This allows us, for example, to overcome the limitations that would happen with RBAC in our sample DemoEditor information system, if authors simply added a book-edition role. Indeed, this role would either give them the right to read and write books, **including those from other authors**, or give them the right to read the books without being able to modify them. What we really want is that authors and editors can manipulate the data on **their** books (the ones they write or supervise) but have very limited rights or none on other books. With RBAC, this is not possible, as the role is related to the books resource but not specific books.

This is a job for ABAC, and the attributes it would use are the following:

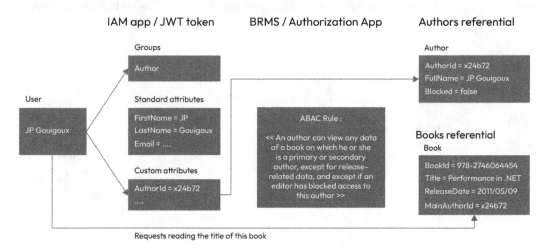

Figure 13.14 – An ABAC implementation with a BRMS

You will notice that permissions are still represented – and we could include roles as well – because ABAC is not exclusive of RBAC but, rather, complements it in its forthcoming.

What would happen technically in such a scenario is the following:

- An application would call the GET verb on /api/books/978-2409002205.

- This request would be accompanied by a bearer-based authentication header.

- The JWT token would include the custom attribute providing the author internal identifier (or another way to go would be to base the association to the author on the email or another standard attribute).

- Upon reception of this request, the books referential service application would call the authorization central API, providing it with everything it knows about the request – the incoming JWT-born identity, the attributes of the book requested, and so on.

- The authorization app would find the rule that applies to the situation – in this case, GET on a book.

- It would first check that the incoming user has author_id and that this is the ID associated with one of the authors of the given book (looking at the book_mainauthor_id attribute and – if necessary – the book_secondaryauthors_ids array of attributes).

- It would then check that the initial request to the book referential service does not contain something such as $expand=release-information, since this data will not be seen by the author.

- It would realize it needs to check that the author has not been blocked and would call a `GET` request to `/api/authors/x24b72`. This would be done with a privileged account with full read rights, as we consider that the BRMS has a justified "right to know" due to its function in the system.

- An alternative to this would be for the books referential service to provide an extended view of the book, just as if there had been a call to `/api/books/978-2409002205?$expand=authors`.

- For most advanced authorization systems, these three checks would be done in parallel to save time.

- If everything is correct, the BRMS will send a `200 OK` HTTP response to the call from the books referential service.

- The book referential service would then grant the requested access.

Of course, if anything goes wrong in these steps, the request will be refused with a `403 Forbidden` status code. This could happen if the rules are not respected, but also if the BRMS system does not respond in time. This behavior is expected, as the so-called "graceful degradation" would imply, for security reasons, that the system does not take any risk to disclose date data or allow any operation if it is not sure it is allowed. This means that the authorization is another SPOF in the system and should be operated corresponding to this requested level of service.

I hesitate to discuss **ReBAC** (**Relationship-Based Access Control**), which looks like a nice complement to the RBAC and ABAC paradigms but, at the time of writing, has not yet reached a mature enough state. In a nutshell, the principle of ReBAC is to manage authorizations based on links between entities; hence, it has a strong link to DDD. For example, this approach allows you to easily give writing permission to an author on their books while keeping the books of other authors with read-only permissions. This can, of course, also be done with ABAC, but ReBAC makes it a little bit simpler, by basing its functioning on relationships instead of simply attributes. To read a bit more about ReBAC, you can start at `https://en.wikipedia.org/wiki/Relationship-based_access_control` and then check what OSO states about this mode at `https://www.osohq.com/academy/relationship-based-access-control-rebac`.

OpenFGA (`https://openfga.dev/`) is also a project that is worth looking at if you need a clean external authorization management system that is ReBAC-capable. Although still nascent, the project has already been referenced as a Cloud Native Computing Foundation project. If you want to check out what it could do for your authorization needs, one of the best ways to start is to tweak the samples provided in the sandbox (`https://play.fga.dev/sandbox`).

The XACML approach

Now that the different organizations of rights management have been discussed, we will start discussing a bit more about implementation and, by now, you will certainly have started wondering what kind of norms and standards are available to us. Since we have discussed the different steps to realize an ABAC implementation, it would be interesting to study one of the most complete specifications and explain how it would fit into these ABAC steps.

XACML (eXtensible Access Control Markup Language) specifies how access control can be executed and administered. It is one of the most advanced ways to handle authorization and establishes five different responsibilities to do so:

- The policy administration point is where rules are defined
- The policy retrieval point is where they are stored
- The policy decision point is the engine that decides which decision should be taken
- The policy information point is where additional attributes that are necessary for rule evaluation are gathered
- The policy enforcement point is the place where the result of the decision is applied

How these five responsibilities are spread across one or many applications defines how sophisticated a system will be. In the most simple approach, all five responsibilities can be implemented inside the data referential service that ultimately has to apply the enforcement point (since the data referential service is the one who owns the data, this cannot be externalized). In this mode, the data referential service not only stores the data but also stores the rules, executes them, and decides what it should do depending on the outcome. The only instance in which a responsibility could still be considered as external is if the referential service needs some external data, but it could very well store that as well. In this case, the responsibilities are affected like this:

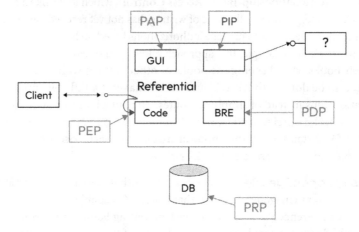

Figure 13.15 – All authorization responsibilities integrated into an application

By contrast, this is how we could spread the responsibilities in the previous organization of responsibilities we discussed previously:

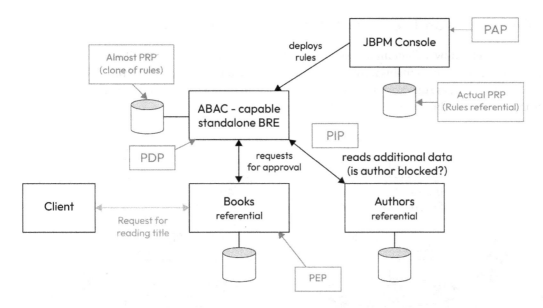

Figure 13.16 – Authorization responsibilities completely spread across dedicated services

In this very clean (but, of course, more expensive to set up) approach, each responsibility is completely separated, and the BRMS and data referential service work together in order to orchestrate them:

1. Before any first interaction, a functional user connects to the PAP and designs the rules (just like what was done in the example previously with the DMN use).

2. These rules are stored in the associated database, which is the PRP.

3. The books referential service receives the initial request. It cannot make decisions on its own and delegates the PDP.

4. It communicates to the deployed BRE the context of the call, in order to get a decision from it.

5. The PDP needs to retrieve the rules in order to process them. It could call the PRP, but luckily, it has a local clone in our case, where we made the hypothesis that the JBPM server has been used and the console deployed a standalone runtime container for the rules execution.

6. The PDP may also need some additional information that it could collect through the PIP, which retrieves the blocked status for the author.

7. The PDP sends the result of its rules decision engine back to the books referential service.

8. As with the PEP, the books referential service uses the decision sent by the PDP to allow (or not) access to its data and possibly respond to the HTTP response that was presented.

Before we show you a practical example of how to set this up, let me make another digression, this time on how services should be separated.

A digression on the granularity of microservices

First of all, let's draw a diagram for a less sophisticated and more common situation, where each data referential service contains its own PRP and PDP in addition to the PEP. In this case, the PAP is generally minimal, as rules are integrated into the code and do not allow for easy management, which means that PRP is simply the code base itself.

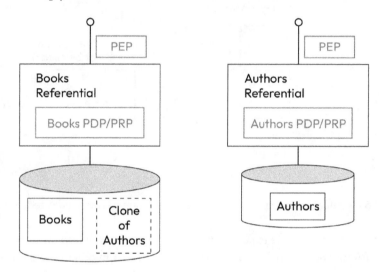

Figure 13.17 – The problem of authorization management when data is cloned

Can you spot the potential problem? The books referential service does not hold the author's PDP/PRP, which is logical, since it is not responsible for it. However, it still stores a clone of the author's data in order to quickly respond to API calls such as /api/books/978-2409002205?$expand=authors. This means that, since it does not know how to filter this kind of data, it might create a breach of confidential data if care is not taken. In a four-layer diagram, this problem can be seen from a strange misalignment that appears this way:

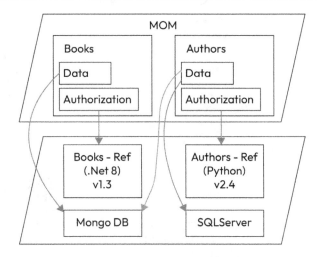

Figure 13.18 – A representation of the authorization antipattern in the four-layer diagram

This kind of misalignment arises from the fact that authorization has been trusted by the application that stores the data. This way, since the data is duplicated, there are in fact two potentially different ways to apply authorization to the same data! This situation can also happen when we externalize data in a BRMS because the runtime and the PAP may not be synchronized, but the advantages in this case are much more important than the actual drawbacks. Indeed, the decoupling between the JBPM console and the BRE runtime container bears lots of added value – the console is a complex server whereas the runtime container is very light; it is much better to separate them because errors are prone to happen in the first, whereas the second one should have an excellent level of service. When the console is used to deploy a standalone server, it can then crash without this being a problem. The runtime, conversely, can be made extremely robust, since it is stripped of almost every bit of code that is not immediately necessary to execute the functions. The fact that the console deploys versions of the rulesets makes it possible to create as many runtime servers as needed for performance reasons (thus, you also avoid the SPOF problem, since this service is required by many others and should be extremely stable) without any risk of a lack of consistency, which would be a big problem (imagine explaining to your customers that their rights of access to their tenant data vary on every new request).

Still, this does not mean that all responsibilities should always be added to as many services and different processes as possible. Of course, it may be useful, but, as is often the case in information architecture, the most important thing is to strike the right balance. There have been so many pointless discussions on the internet over which is best between microservices and monolith applications that you can almost surmise the quality of an article just by looking at its title. Of course, the only correct response to what is best is, "*It depends*," and any competent software architect knows that this is not a "one-size-fits-all" situation.

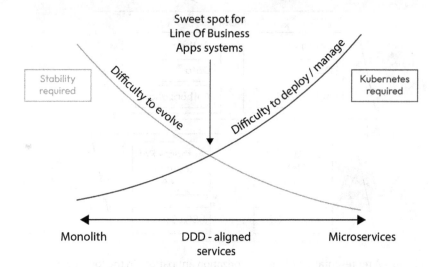

Figure 13.19 – Service granularity advantages and drawbacks

I realize I have been saying this every chapter or so, but it's worth repeating – what should be a priority in the granularity of services for business functions? If you know that the authorization rules change very rarely and having to wait for a new release is not a problem, then implement what should be a priority in the granularity of services for business functions directly in the code of the associated referential service; you will get the best performance, and you only will have to deal with the problem of securing cloned data if you have some. If it is an issue, consider calling the other referential service if there is any doubt; it will also be a way to refresh part of your cloned data. Conversely, if you can foresee that authorization rules are going to change frequently or there are external circumstances, such as regulation, then consider progressively extracting responsibilities from your data referential service. Foreseeing this kind of thing is admittedly a fine line between over-anticipating and adopting too much of a DRY approach, but this is where judgment, long-time expertise, and having suffered from many previous experiences come in handy.

Applying authorization to our sample information system

XACML was explained previously but is quite a complex mechanism to put in place. Also, there is not a reference implementation of the protocol, although several products exist such as WSO², Balana, Axiomatics, or products from AT&T. Although these all have their place in big information systems such as banks or insurance, they would be oversized for the small information system that we have decided to simulate in our example, so we are going to use something lighter and closer to the main internet protocols.

The Open Policy Agent alternative

Open Policy Agent is a project that is supported by the Cloud Native Computing Foundation and that proposes a nice decoupling between grammar to describe policies. In short, OPA is to XACML what REST is to SOAP – a lightweight alternative that takes on 80% of the job with 20% of the complexity. Instead of installing a full-blown XACML server to show an example of externalizing the authorization responsibility, we are going to use Docker to customize an authorization engine.

OPA uses a declarative language named Rego to describe the policies that should be applied to data to make decisions. It can then execute these policies to provide JSON results that can be exploited in other services, or another part of the code if you use the OPA implementation as a component.

Technically, what will happen is that a request like the following will be sent to OPA, and it will respond with whether the requested access should be authorized or not:

```
{
    "input": {
        "user": "jpgou",
        "operation": "all",
        "resource": "books.content",
        "book": "978-2409002205"
    }
}
```

In this example, the user jpgou requests full access to the content petal of the book, identified under the ISBN number 978-2409002205 in the system. If this is granted by the OPA server, it will respond with something like this:

```
{
    "result": {
        "allow": true
    }
}
```

Before diving into the technology again, we need to be precise about what we want to achieve from a functional point of view.

The functional needs of DemoEditor

Let's return to our DemoEditor sample and describe what should be done from the authorization's point of view. Of course, in a publishing company, authors have permission to provide the content of books and adjust it as they will, but they should never be able to read the content of another author's book, in order to avoid plagiarism or even intellectual property theft. Since there are editors who take care of authors, it is logical that they can at least read the content of the books from the authors they manage. Salespersons, on the other hand, do not have any editing responsibility, so they may know some information about the book to prepare sales and orders but have no reason to know anything about the editing process.

In this short description of the stakes at play in the rights management of `DemoEditor`, it is quite clear that pure RBAC will not be enough, and we will have to resort to ABAC to complete RBAC since there are rules based on the attributes of the book, namely who is the author, and even other information such as the link between authors and their editors. RBAC is simply not enough because an author has more rights on their own books than on the ones from a different author, although they both will benefit from the `author` authorization profile.

As will be explained in more detail in the following section, we will also add a few rules, such as the fact that salespersons can only see the book once it has reached a certain status, or another one allowing us to block the rights for an author that does not respect the editing contract and should be denied permissions. To do so, we will use the same metaphor that we used in *Chapter 9*, where designating the different categories of data for books is like placing them in the petals of a flower, the core of which contains the most important, entity-defining data, such as the ISBN number and the title of the book. Although it may be tempting to define these petals based on the authorization rules, it is important to keep in mind they have to be drawn from functional constraints, and authorization management is one of them, but still only one of them.

Creating the authorization policies

Starting with the definition of the authorization policies will allow us to do two things simultaneously:

- Explain what kind of authorization behavior we intend to put in place and how the data referential service for the books should work
- Explore some of the `Rego` syntax and what is at play when externalizing authorizations with such a mechanism

When writing the `policy.rego` file (just an arbitrary name) for the authorization management of the books data referential service, we need to start with a package name, which allows us to separate rules from different groups when they are executed in the same engine. The beginning of the file also contains some instructions to import specific keywords and functions (OPA supports plugins and extensions of the grammar to ease its use) or prepare data (which we will come back to further on in the chapter):

```
package app.abac

import future.keywords
import data.org_chart
```

The body of the file then generally starts with a pattern where a main authorization attribute, which we will call `allow`, is broken down into several finer-grained decision policies. What we want to achieve is an authorization engine that, when exposed to a type of access, will send as a result whether this access should be granted or not. We will come back to this part later when demonstrating how to apply the rules engine, but for now, let's continue with the policy-defining file and show how the behavior we discussed will be implemented:

```
default allow := false
allow if {
    permission_associated_to_role
    no_author_blocking
    no_status_blocking
    authors_on_books_they_write
    editors_on_books_from_authors_they_manage
}
```

In order to implement security best practices, access is forbidden by default. This allows for what is called "graceful degradation" – if there is a problem in the authorization subsystem, it will default to the safest situation. In our case, the safest approach is to not allow access, since a lack of availability is, of course, less of a problem than disclosing data to a person who should not have been able to see it, with all the business consequences such an event could have. This is what the first line in the preceding code is about – setting to `false` the default value of the `allow` attribute.

The second operation states that, in order to make `allow` become `true`, we need to pass five different decisions, each of them needing to be evaluated to `true`. These decisions are, of course, named in such a way they will be easy to understand and debug (setting authorizations correctly is somewhat of a challenge, but this will almost never be 100% correct at the first attempt). The rest of the file will basically be about detailing these five main decisions, but before we declare how they work, we need to prepare some data. Indeed, as we will explain in the next section, we need some referential service data in order for the engine to make decisions. For example, since we stated that editors should have access to books from the authors they coach, there will be an obvious need for the engine to know about the links between authors and editors. Some other information, such as the status attribute of the book, will also be useful. All this data will mostly come from the data referential service to the engine, but it will be basic data, and we may glean some information from it before actually using the whole set of data to infer decisions.

One such piece of information is the roles the current user owns. As stated previously, we will need some bits of RBAC, even if it is not enough. That means the user will be linked to some roles, some of them directly and some of them through the user belonging to identification groups. The following grammar expresses precisely this:

```
user_groups contains group if {
    some group in data.directory[input.user].groups
}
user_group_roles contains role if {
    some group in user_groups
    some role in data.group_mappings[group].roles
}
```

```
user_direct_roles contains role if {
    some role in data.user_mappings[input.user].roles
}
user_roles := user_group_roles | user_direct_roles
```

Groups can be found in a piece of data called `directory`, by looking at the entry in the list designated by the value of the input variable called `user`. Once this user is found in the list, the `groups` attribute will provide the list of identification groups. These will then be used to retrieve the roles associated with the groups, leveraging a collection called `group_mappings`. The same logic will be applied to a collection that sends the roles directly applied to a user, and the two role lists will simply be merged by the last operation shown in the preceding code.

We will also need information about the author potentially associated with the user. This is something I have not explained fully yet, briefly mentioning the fact that an author uses a user to access the information system from `DemoEditor` even if they are not really part of an organization, or at least not employees of it. This means that, first, access will have to be provided to them (we will come back to how to do this when implementing the associated function). This also means that there should be a way somehow to associate these two entities in the information system. A method that happens quite often is to use a verified `email` attribute to link them together. For the moment, we will just consider that the user's information is contained in the `author` entity. The rule to retrieve the association is quite easy to write – it simply loops over the list of authors, and if the `user` associated with an author is the one the request for access relates to, then the author is the one we are looking for:

```
user_author contains author if {
    some author in data.authors
    author.user == input.user
}
```

In fact, we should refer to authors rather than just an author because we know there will functionally be only one, but technically, we will use a list even if the name of the variable remains in the singular form, `user_author`.

The same applies to the book that the request talks about, as we need to retrieve its ID from the list of data to be able to make some decisions from rules on book attributes:

```
book contains b if {
    some b in data.books[input.book]
}
```

In the case of an author, we also have to retrieve the list of books they are the writer of because some rules apply to this as well:

```
author_books contains book if {
    some author in user_author
    some b in data.books
```

```
        b.editing.author == author.id
}
```

Now that all the necessary data is collected, we can start discussing the rules themselves, taking the five units of rules separately and breaking them down even more. The first rule that applies is that permissions should be owned by the user associated with the request. It will not be enough to grant access, but this is nonetheless a necessary constraint. In order to know whether the user should be allowed access to the resource they requested, all accesses provided by roles should be studied. If one corresponds to the type of resource and the operation requested, then it is a match, and the permission is applied:

```
permission_associated_to_role if {
    some access in user_accesses_by_roles
    access.type == input.resource
    access.operation == input.operation
}
```

The following rule makes it happen that if someone gets a right to `books.content`, `books.sales`, or `books.editing` (one of the petals of the flower corresponding to the data referential service), then they automatically obtain a right to the core of the flower, which is logical, since having access to some data without being able to link it to a given entity would not prove to be very useful:

```
permission_associated_to_role if {
    some access in user_accesses_by_roles
    "books" == input.resource
    access.operation == input.operation
}
```

Since we have two rules with the same name (`permission_associated_to_role`) instead of two rules with different names inside the same group, there is a big difference in processing, as it means that rules are considered to be separate by an "or" operator (and not all needs to be true for the result to be true, such as what was set up previously for `allow`). Also, we are even going to add a third case where this part of the policy should be granted, namely when the access provided contains `all` as an operation. In this case, whether the requested operation is `read`, `write`, or any other value, it will be granted (at least based on this criteria):

```
permission_associated_to_role if {
    some access in user_accesses_by_roles
    access.type == input.resource
    access.operation == "all"
}
```

Now, the question should be, how is `user_accesses_by_roles` calculated? Well, this time, it is a bit more complicated, with a sub-decision that walks through some tree-like hierarchy of profiles and their associated accesses, contained in the `roles` entity of the data provided. We will return to the definition in the next section, but for now, it is important to know that we will use a hierarchy to set managers on top, and then salespersons and editors below, and authors under their editors. The interesting bit in this approach will be how easy it is to make it so that the role above receives all the permissions of the one below in the tree. After all, if a salesperson has the right to write the sales values, it is logical that their manager has at least the same rights. The syntax is harder to read, but do not worry about this, as the OPA documentation is well-written, and there are many examples available for even the most convoluted rules:

```
user_accesses_by_roles contains access if {
    some role in user_roles
    some access in permissions[role]
}
roles_graph[entity_name] := edges {
    data.roles[entity_name]
    edges := {neighbor | data.roles[neighbor].parent == entity_name}
}
permissions[entity_name] := access {
    data.roles[entity_name]
    reachable := graph.reachable(roles_graph, {entity_name})
    access := {item | reachable[k]; item := data.roles[k].access[_]}
}
```

When dealing with the rule stating that an author can only have rights on books they are authors of, we need to apply a little trick. As usual, we start by setting access to false, in order to respect security best practices. And we will set it to `true` if we can follow the link of authoring when in the case of a book writer, and also simply in the case where the user is not a book writer. That may sound too relaxed a constraint, but remember that this is not the complete result. In this case, this is just the piece of decision that is about the link between an author and their books; if we are dealing with an editor, this rule simply does not apply, but some others will, and all of them need to align in order for the final, summarizing decision to be positive. The result is the following:

```
default authors_on_books_they_write := false
authors_on_books_they_write if {
    some role in user_roles
    role != "book-writer"
}
authors_on_books_they_write if {
    some role in user_roles
    role == "book-writer"
    some author in user_author
    some b in data.books
    b.editing.author == author.id
    b.id == input.book
}
```

By now, you should start to be better accustomed to the `Rego` syntax, but the third part of the global authorization scheme still requires some thinking, as it needs you to traverse the whole organizational chart in order to retrieve the link between editors and "their" authors, since we need to apply the rule that an editor only has rights on books from the authors they manage:

```
default editors_on_books_from_authors_they_manage := false
editors_on_books_from_authors_they_manage if {
    some role in user_roles
    role != "book-edition"
}
book_author contains b.editing.author if {
    some b in data.books
    b.id == input.book
}
editors_on_books_from_authors_they_manage if {
    some role in user_roles
    role == "book-edition"
    some author in book_author
    some b in data.books
    b.editing.author == author
    b.id == input.book
    user_hierarchy_ok
}
foundpath = path {
    [path, _] := walk(org_chart)
    some author in book_author
    path[_] == author
}
user_hierarchy_ok if {
    some user in foundpath
    user == input.user
}
```

The fourth part of the global decision is simpler – it considers that salespersons cannot see a book if it is not in a published or archived status. Again, in order to account for the "or" approach, we need to calculate twice the `readable_for_sales` attribute, initially set to `false` for security reasons, respectively for the two values of the status allowing access to the salespersons:

```
default no_status_blocking := false
no_status_blocking if {
    some role in user_roles
    role != "book-sales"
}
default readable_for_sales := false
readable_for_sales if {
    book.status == "published"
```

```
}
readable_for_sales if {
    book.status == "archived"
}
no_status_blocking if {
    some role in user_roles
    role == "book-sales"
    readable_for_sales
}
```

The fifth and final part of the decision is even simpler, and we will not explain the code, simply the rule – if an author has been blocked, they cannot have any access to any book:

```
default no_author_blocking := false
no_author_blocking if {
    some role in user_roles
    role != "book-writer"
}
no_author_blocking if {
    some role in user_roles
    role == "book-writer"
    some user in user_author
    user.restriction == "none"
}
```

Once syntax is complete, we need a second type of information to make a decision. This is what decision data is about.

Adding some data in order to make decisions

We had hinted in the previous section about the fact that data should be provided (and even inferred from other data) in order for the rules engine to be able to make a decision. Next, we will show what kinds of information we should put in place for our example to work. First, we need the definition of roles (remember that a role is a set of rights, each composed of a resource type and an operation type):

```
"roles": {
    "book-direction": { "access": [] },
    "book-sales": { "parent": "book-direction", "access": [{
"operation": "all", "type": "books.sales" }] },
    "book-edition": { "parent": "book-direction", "access": [{
"operation": "all", "type": "books.editing" }] },
    "book-writer": { "parent": "book-edition", "access": [{
"operation": "read", "type": "books.editing" }, { "operation": "all",
"type": "books.content" }] }
}
```

The preceding JSON content also defines the notion of `parent`, which creates a tree of roles, with, for example, `book-sales` and `book-edition` placed under `book-direction`, which means that a director will automatically receive all the permissions granted by default to a salesperson *and* the ones granted to editors, in addition of course to the ones described directly on the role itself.

Some data about books will be sent in order to apply some specific rules that need this. In the following example, I have shown a list because I tested different combinations. In actual use, we could simply send the data associated with the only book we request OPA for to decide on its access, and nothing more, in order to preserve performance. Here is the associated data:

```
"books": {
    "978-2409002205": { "id": "978-2409002205", "title": "Performance
in .NET", "editing": { "author": "00024", "status": "published" }},
    "978-0000123456": { "id": "978-0000123456", "title": ".NET 8 and
Blazor", "editing": { "author": "00025", "status": "draft" }}
}
```

Note that the preceding code is not expressed in a JSON array but as a structure. The same is true for data about authors:

```
"authors": {
    "00024": { "id": "00024", "firstName": "Jean-Philippe",
"lastName": "Gouigoux", "user": "jpgou", "restriction": "none" },
    "00025": { "id": "00025", "firstName": "Nicolas", "lastName":
"Couseur", "user": "nicou" }}
```

The organizational chart allows us to define who is the big boss (`frfra`), define which salespersons and editors are below him (three persons in my example), and finally, to place two authors below the editor, codenamed `mnfra`:

```
"org_chart": {
    "frfra": {
        "frvel": {},
        "cemol": {},
        "mnfra": {
            "00024": {},
            "00025": {}
        }
    }
}
```

We then simulate what would be sent by a user directory – for example, the groups that each user belongs to. This is somewhat artificial, as we would normally extract this for the JWT token that has been passed through identification, but this is what we will do when hitting the difficulty in code. For now, we will stay quite symbolic with the following tree:

```
"directory": {
    "frfra": {"groups": ["board"]},
    "frvel": {"groups": ["commerce", "marketing"]},
    "cemol": {"groups": ["commerce"]},
    "mnfra": {"groups": ["editors", "quality"]},
    "jpgou": {"groups": ["authors"]},
    "nicou": {"groups": ["authors"]}
}
```

Of course, now we have the groups, we need the mappings to link them to roles in a true RBAC approach:

```
"group_mappings": {
    "board": { "roles": ["book-direction"] },
    "commerce": { "roles": ["book-sales"] },
    "editors": { "roles": ["book-edition"] },
    "authors": { "roles": ["book-writer"] }
}
```

And since we have decided to be as complete as possible, we will allow – beyond pure RBAC – the possibility to also state some direct association between a user and a role, without the group intermediary:

```
"user_mappings": {
    "frvel": { "roles": ["book-edition"] }
}
```

Now that everything is set for the server to output some results (rules and data), we can go to the next step, which is setting up a real OPA server, feeding it with these two files, and trying out some decisions.

Deploying a Docker-based OPA server

It is so easy to deploy with Docker that we would be asking for trouble not using it to test OPA. The command line to run the server is extremely simple:

```
docker run -d -p 8181:8181 --name opa openpolicyagent/opa run --server
--addr :8181
```

Once the server is up, we will start with a call to push policy definitions in it:

```
curl --no-progress-meter -X PUT http://localhost:8181/v1/policies/app/
abac --data-binary @policy.rego
```

Another call is used to send the data:

```
curl --no-progress-meter -X PUT http://localhost:8181/v1/data --data-
binary @data.json
```

Finally, we are able to test the OPA server with a simple request that is shown in the following code:

```
{
    "input": {
        "user": "jpgou",
        "operation": "all",
        "resource": "books.content",
        "book": "978-2409002205"
    }
}
```

When called with the following command to send this text content to a POST-using API, the OPA server will send a result in JSON, the rest of the command taking care of retrieving only the part of the result we are interested in:

```
curl --no-progress-meter -X POST http://localhost:8181/v1/data/app/
abac --data-binary @input.json | jq -r '.result | .allow'
```

If executed as is, this normally sends true, meaning that this context of the request is granted by the OPA server. If you get rid of the last part of the command and display the whole response, you will get something like the following, which is really useful for debugging, since it shows the values for all intermediate decisions:

```
{
  "result": {
    "allow": true,
    "author_books": [
      [
        "978-2409002205",
        "Performance in .NET",
        {
          "author": "00024",
          "status": "published"
        }
      ]
    ],
```

```
"authors_on_books_they_write": true,
"book": [
  "978-2409002205",
  "Performance in .NET",
  {
    "author": "00024",
    "status": "published"
  }
],
"editors_on_books_from_authors_they_manage": true,
"foundpath": [
  "frfra",
  "mnfra",
  "jpgou"
],
"no_author_blocking": true,
"no_status_blocking": true,
"permission_associated_to_role": true,
"permissions": {
  "book-direction": [
    {
      "operation": "all",
      "type": "books.content"
    },
    {
      "operation": "all",
      "type": "books.editing"
    },
    {
      "operation": "all",
      "type": "books.sales"
    },
    {
      "operation": "read",
      "type": "books.editing"
    }
  ],
  "book-edition": [
    {
      "operation": "all",
      "type": "books.content"
    },
    {
```

```
        "operation": "all",
        "type": "books.editing"
      },
      {
        "operation": "read",
        "type": "books.editing"
      }
    ],
    "book-sales": [
      {
        "operation": "all",
        "type": "books.sales"
      }
    ],
    "book-writer": [
      {
        "operation": "all",
        "type": "books.content"
      },
      {
        "operation": "read",
        "type": "books.editing"
      }
    ]
  },
  "readable_for_sales": false,
  "roles_graph": {
    "book-direction": [
      "book-edition",
      "book-sales"
    ],
    "book-edition": [
      "book-writer"
    ],
    "book-sales": [],
    "book-writer": []
  },
  "user_accesses_by_roles": [
    {
      "operation": "all",
      "type": "books.content"
    },
    {
```

```
        "operation": "read",
        "type": "books.editing"
      }
    ],
    "user_author": [
      {
        "firstName": "Jean-Philippe",
        "id": "00024",
        "lastName": "Gouigoux",
        "restriction": "none",
        "user": "jpgou"
      }
    ],
    "user_direct_roles": [],
    "user_group_roles": [
      "book-writer"
    ],
    "user_groups": [
      "authors"
    ],
    "user_hierarchy_ok": true,
    "user_roles": [
      "book-writer"
    ]
  }
}
```

Testing the authorizations

These sample authorizations are not very complicated, but the level of complexity is enough to be hard to handle manually. There are many specific cases that pose questions. For example, if I told you that a manager requests access to the sales data of a book that has not been published yet, what do you think would happen? And, more importantly, what do you think should happen?

Also, the Rego syntax has a steep learning curve. Writing the rules presented previously took me a few hours, if not a day since I am not a specialist, and I am not sure they work exactly the way I think they do.

This is where having good testers is of utmost importance, and their ability to define a testing campaign, find all the corner cases, discuss them with the product owners/customers, and so on will be a great help. Such a test campaign will be created using a Gherkin syntax (see the following sample scenarios). If you use a tool such as SpecFlow, you can create many of these scenarios and test them automatically so that every modification to the rules grammar does not break anything. Once your complete set of tests is ready, you will obtain a report on whether all the series of tests have passed, ultimately reassuring you that all modes you have thought of are correct.

In order to install SpecFlow in Visual Studio, follow the instructions at `https://docs.specflow.org/projects/getting-started/en/latest/index.html`. Then, you will need to create a project of type `SpecFlow Project`. The result will be some classes with examples of how to use SpecFlow, and we are going to adapt them to our specific needs, which is to test the authorization rules we have set up in OPA. We will use xUnit as the underlying test framework here, but you can, of course, modify this to your preferences:

Create a new SpecFlow project

Framework

.NET 6.0

Test Framework

xUnit

☑ Add FluentAssertions library

Figure 13.20 – Creating a SpecFlow project

The structure of the project created will be based on a sample called `Calculator`, and the very first action is to change the names to fit our own purpose, which is to test OPA:

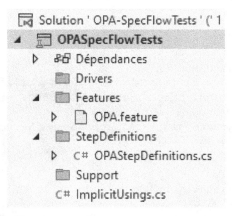

Figure 13.21 – The SpecFlow project structure

The OPA.feature content is, in the first step, modified to the following Gherkin content:

```
Feature: OPA

Scenario: An author has all rights to the content of their book
    Given book number 978-2409002205 with author id 00024 is in
workInProgress status
    And user jpgou belongs to group authors
    And organizational chart is
{"frfra":{"frvel":{},"cemol":{},"mnfra":{"00024":{},"00025":{}}}}
    And user jpgou is associated with author 00024 who has a level of
restriction none
    When the user jpgou requests all access to the books.content petal
of the book number 978-2409002205
    Then access should be accepted


Scenario: An author has no rights to the content of the book from
another author
    Given book number 978-2409002205 with author id 00024 is in
workInProgress status
    And user jpgou belongs to group authors
    And organizational chart is
{"frfra":{"frvel":{},"cemol":{},"mnfra":{"00024":{},"00025":{}}}}
    And user jpgou is associated with author 00024 who has a level of
restriction none
    When the user nicou requests read access to the books.content
petal of the book number 978-2409002205
    Then access should be refused


Scenario: An author that has been blocked has no rights, even on their
own books
    Given book number 978-2409002205 with author id 00024 is in
workInProgress status
    And user jpgou belongs to group authors
    And organizational chart is
{"frfra":{"frvel":{},"cemol":{},"mnfra":{"00024":{},"00025":{}}}}
    And user jpgou is associated with author 00024 who has a level of
restriction blocked
    When the user jpgou requests all access to the books.content petal
of the book number 978-2409002205
    Then access should be refused


Scenario: An editor has all rights to the content of the books from
the authors they manage
    Given book number 978-2409002205 with author id 00024 is in
workInProgress status
    And user jpgou belongs to group authors
```

```
    And user mnfra belongs to group editors
    And organizational chart is
{"frfra":{"frvel":{},"cemol":{},"mnfra":{"00024":{},"00025":{}}}}
    And user jpgou is associated with author 00024 who has a level of
restriction none
    When user mnfra requests all access to the books.content petal of
the book number 978-2409002205
    Then access should be accepted


Scenario: An editor has no rights to the content of the books from the
authors they do not manage
    Given book number 978-2409002205 with author id 00024 is in
workInProgress status
    And user jpgou belongs to group authors
    And user mnfra belongs to group editors
    And organizational chart is
{"frfra":{"frvel":{},"cemol":{},"mnfra":{"nicou":{}}}}
    And user jpgou is associated with author 00024 who has a level of
restriction none
    When user mnfra requests all access to the books.content petal of
the book number 978-2409002205
    Then access should be refused


Scenario: Refusing salesperson access to work-in-progress book
    Given book number 978-2409002205 with author id 00024 is in
workInProgress status
    And user frvel belongs to the group commerce
    And organizational chart is
{"frfra":{"frvel":{},"cemol":{},"mnfra":{"00024":{},"00025":{}}}}
    When the user frvel requests read access to the books.content
petal of the book number 978-2409002205
    Then access should be refused
```

This syntax should be easy to read, even for non-developers; the idea of behavior-driven development is that functional people are able to express their requirements in such a language, called Gherkin (which has many more sophisticated features than we show here for simplicity reasons). In order for this Gherkin syntax to be transformed into an automated xUnit test, we need to create a correspondence between the lines in the scenario and the C# functions that implement this part of the test. This is done in the OPAStepDefinitions.cs file. For example, for the Given and And keywords (which are of the same notion), the corresponding functions will be as follows:

```
[Given("book number (.*) with author id (.*) is in (.*) status")]
public void AddBookWithStatus(string number, string authorId, string
status)
{
    _books.Add(new Book() { Number = number, AuthorId = authorId,
Status = status });
```

```
}

[Given("user (.*) belongs to group (.*)")]
public void AddUserWithGroup(string login, string group)
{
    _users.Add(new User() { Login = login, Group = group });
}

[Given("user (.*) is associated to author (.*) who has level of
restriction (.*)")]
public void AddAuthor(string login, string authorId, string
restrictionLevel)
{
    _authors.Add(new Author() { Login = login, Id = authorId,
Restriction = restrictionLevel });
}

[Given("organizational chart is (.*)")]
public void SetOrganizationChart(string orgChart)
{
    _orgChart = orgChart;
}
```

In the initialization part of the class containing this function, we will, of course, have a member to store the books (and other lists for other entities needed for the test scenarios):

```
private static HttpClient _client;
private static List<Author> _authors;
private static List<Book> _books;
private static List<User> _users;
private static string _orgChart;
private static string _result;
```

The corresponding classes contain all that is needed to vary the context of the rules. As you can see, the first name and last name of the authors have not been integrated into the model, since we have a strong assurance that they cannot impact the output of the rules engine:

```
public class Author
{
    public string Id { get; set; }
    public string Login { get; set; }
    public string Restriction { get; set; }
}

public class Book
```

```
{
    public string Number { get; set; }
    public string Status { get; set; }
    public string AuthorId { get; set; }
}

public class User
{
    public string Login { get; set; }
    public string Group { get; set; }
}
```

Some methods will be used to initiate the values for each scenario, and also for the entire feature:

```
[BeforeFeature]
public static void Initialize()
{
    _client = new HttpClient();
    _client.BaseAddress = new Uri("http://localhost:8181/v1/");
}


[BeforeScenario]
public static void InitializeScenario()
{
    _authors = new List<Author>();
    _books = new List<Book>();
    _users = new List<User>();
}
```

This will allow us, upon calling on the function associated with the When keyword, to realize the call to the so-called System Under Test (what we want to validate is the OPA server, which should have been started and customized with the Rego content and will listen on port 8181 in our setup):

```
[When("user (.*) request (.*) access to the (.*) petal of the book
number (.*)")]
public void ExecuteRequest(string login, string access, string
perimeter, string bookNumber)
{
    StringBuilder sb = new StringBuilder();
    sb.AppendLine("{");
    sb.AppendLine("    \"roles\": {");
    sb.AppendLine("        \"book-direction\": { \"access\": []},");
    sb.AppendLine("        \"book-sales\": { \"parent\": \"book-
direction\", \"access\": [{ \"operation\": \"all\", \"type\": \"books.
sales\" }]},");
```

```
    sb.AppendLine("            \"book-edition\": { \"parent\": \"book-
direction\", \"access\": [{ \"operation\": \"all\", \"type\": \"books.
editing\" }]},");
    sb.AppendLine("            \"book-writer\": { \"parent\": \"book-
edition\", \"access\": [{ \"operation\": \"read\", \"type\": \"books.
editing\" }, { \"operation\": \"all\", \"type\": \"books.content\"
}]}");
    sb.AppendLine("        },");
    sb.AppendLine("        \"books\": {");
    for (int i=0; i<_books.Count; i++)
    {
        Book b = _books[i];
        sb.Append("            \"" + b.Number + "\": { \"id\": \"" +
b.Number + "\", \"title\": \"***NORMALLY NO IMPACT ON RULES***\",
\"editing\": { \"author\": \"" + b.AuthorId + "\", \"status\": \"" +
b.Status + "\" }}");
        if (i < _books.Count - 1) sb.AppendLine(","); else
sb.AppendLine();
    }
    sb.AppendLine("        },");
    sb.AppendLine("        \"authors\": {");
    for (int i = 0; i < _authors.Count; i++)
    {
        Author a = _authors[i];
        sb.AppendLine("            \"" + a.Id + "\": { \"id\": \"" +
a.Id + "\", \"firstName\": \"***NORMALLY NO IMPACT ON RULES***\",
\"lastName\": \"***NORMALLY NO IMPACT ON RULES***\", \"user\": \"" +
a.Login + "\", \"restriction\": \"" + a.Restriction + "\" }");
        if (i < _authors.Count - 1) sb.AppendLine(","); else
sb.AppendLine();
    }
    sb.AppendLine("        },");
    sb.AppendLine("        \"org_chart\": " + _orgChart + ",");
    sb.AppendLine("        \"directory\": {");
    for (int i = 0; i < _users.Count; i++)
    {
        User u = _users[i];
        sb.AppendLine("            \"" + u.Login + "\": {\"groups\": [\""
+ u.Group + "\"]}");
        if (i < _users.Count - 1) sb.AppendLine(","); else
sb.AppendLine();
    }
    sb.AppendLine("        },");
    sb.AppendLine("        \"group_mappings\": {");
    sb.AppendLine("            \"board\": { \"roles\": [\"book-
direction\"] },");
```

```csharp
    sb.AppendLine("            \"commerce\": { \"roles\": [\"book-sales\"]
},");
    sb.AppendLine("            \"editors\": { \"roles\": [\"book-
edition\"] },");
    sb.AppendLine("            \"authors\": { \"roles\": [\"book-writer\"]
}");
    sb.AppendLine("        },");
    sb.AppendLine("      \"user_mappings\": {");
    sb.AppendLine("      }");
    sb.AppendLine("}");

    var response = _client.PutAsync("data", new StringContent(sb.
ToString(), Encoding.UTF8, "application/json")).Result;

    string input = "{ \"input\": { \"user\": \"" + login + "\","
        + " \"operation\": \"" + access + "\","
        + " \"resource\": \"" + perimeter + "\","
        + " \"book\": \"" + bookNumber + "\" } }";

    response = _client.PostAsync("data/app/abac", new
StringContent(input, Encoding.UTF8, "application/json")).Result;
    if (response != null)
    {
        _result = response.Content.ReadAsStringAsync().Result;
    }
}
}
```

The final part of the test execution is carried out by the method associated with the Then keyword, which is the one running the asserts in order to simulate an automated test:

```csharp
[Then("access should be (.*)")]
public void ValidateExpectedResult(string expectedResult)
{
    JsonTextReader reader = new JsonTextReader(new StringReader(_
result));
    reader.Read(); // Get first element
    reader.Read(); // Read result attribute
    reader.Read(); // Get element for result
    reader.Read(); // Read allow attribute
    bool? actual = reader.ReadAsBoolean(); // Get boolean value for
allow attribute
    if (actual is null)
        throw new ApplicationException("Unable to find result");
```

```
    bool? expected = null;
    if (expectedResult == "refused") expected = false;
    if (expectedResult == "accepted") expected = true;
    if (expected is null)
        throw new ApplicationException("Unable to find expected
value");

    Assert.Equal(expected, actual);
}
```

You now can display the Test Explorer by accessing it from the menu or using the *Ctrl + E + T* shortcut. The tests might not display at first, and you may have to run the solution generation for them to appear. Once they are displayed, you can run the scenarios one by one or simultaneously, and if everything works fine, they should confirm the rules work as intended and display everything with ticks on circles:

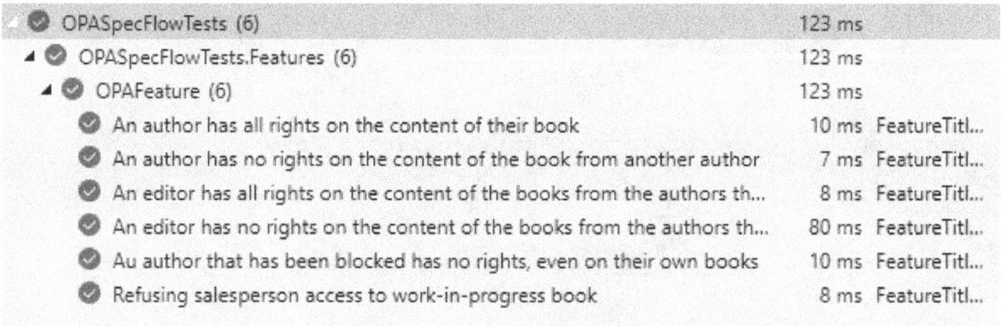

Figure 13.22 – The results of the SpecFlow tests

Six scenarios are not a lot for such a complex set of policies, and in the real world, a few dozen such scenarios would be welcome to form such a powerful harness that everyone would be convinced the system works perfectly as expected. But again, since this is not the main subject of the book, we will leave the automated tests of authorization rules here. By the way, I showed automated BDD tests created with SpecFlow because this is the framework I am used to, but there are alternatives that may be more suitable, depending on your needs and context. The important thing is not whether you use SpecFlow, Postman, or any other method but that such important rules as authorizations should be verified carefully.

Challenges with OPA

OPA is a great approach to authorization rule implementation, but it still offers some challenges.

First of all, there is the complexity of writing rules, as previously discussed. Although this is quite logical as we try to fit some complex functional algorithms into just a few keywords, it can really be limiting for those trying to adopt OPA and the Rego syntax, only to be held up by many incorrect attempts at writing the right rule.

I personally experienced this, and to be perfectly honest, I still do not really understand how the following rule works:

```
permissions[entity_name] := access {
    data.roles[entity_name]
    reachable := graph.reachable(roles_graph, {entity_name})
    access := {item | reachable[k]; item := data.roles[k].access[_]}
}
```

I know that it does because I have tested it, and I can see the point about walking through a tree and picking some data of the path, but the additional recurrent valuation of `access` combined with the use of the `_` keyword and the `reachable` function just make it too hard for me to write this on my own, without referring to some examples that others have written. It may be a lack of practice, but I have tried my fair share of exotic languages over almost four decades of programming, and I still think `Rego` might be one of the most complicated logics I have encountered. After a few attempts at using OpenFGA, it may be simpler to provide an equivalent authorization rule, but I cannot make a commitment to this, since I have not yet used this technology in a production-ready module.

Luckily, some documentation such as `https://www.openpolicyagent.org/docs/latest/policy-reference/` shows advanced examples, and I also found great advanced tips at `https://www.fugue.co/blog/5-tips-for-using-the-rego-language-for-open-policy-agent-opa`, while links such as `https://medium.com/@agarwalshubhi17/rego-cheat-sheet-5e25faa6eee8` provided some clear explanation on how these complex syntaxes work.

Another challenge with OPA might come from the fact that HTTP API calls can cause performance problems if done in volume. And if your authorization rules are complex, chances are that you will be obliged to apply them to business entities one by one. So, how would you handle some calls requesting a list of entities? Calling the API hundreds of times or more is not an option. And what is true for a local Docker container is even more true for a service in the cloud such as OSO (`https://www.osohq.com/`) that proposes a SaaS solution for authorization rules.

Of course, the best approach is still paginating results, which is also good not only for ecological reasons, helping to reduce the strain on resources, but also for ergonomic reasons, by providing users with screens that are less cluttered with data and easier to read and comprehend. However, cases where you need volume may remain, and calling an HTTP server many times is not an elegant option anyway. Luckily, OPA can be accessed directly from your code if you use the Go language, or even as a WebAssembly module, making it possible (although not currently easy) to integrate it at the code level from many platforms.

Here's a final thing to note on authorization management – in this chapter, you have seen a simplified version of the grammar and data that will be applied more realistically in the upcoming chapters. For example, I used simple identifiers instead of URN, some attributes were repeated in order to ease rule execution, and so on. I could have shown the policies in their final form but considered it better to show the work in progress for two reasons:

- Avoiding this additional complexity made it easier to concentrate on the subject of authorization rules

- Showing these adjustments at the precise moment we need to make them will hopefully also make them more understandable, as the situation will show how the simple approach could cause an evolution problem and help explain the change

Summary

In this chapter, we showed what a Business Rules Management System does, how useful it can be in an information system, and how we could implement one, starting with a functional example and then demonstrating another example relating to authorizations, which are one of the most used sets of business rules in software applications.

Just like for BPMN engines, BRMS engines are not used very often. In fact, business rules are – in the great majority of cases – implemented in code expressions or compiled into applications. This is absolutely normal because a BRMS represents an important investment, and implementing such complex applications really needs a strong business case, where business rules change very frequently or are associated with strict regulatory or marketing constraints, such as the necessity to trace all business rules and their changes, the capacity to simulate the effects of new versions of sets of business rules, and so on. We can conclude, then, that this approach is currently limited to very rare contexts. Things may, of course, change in the future, with the industrialization of information system design that we really long for, but at the present time, BPMNs and BRMSs are an effort that is almost always overkill.

And since two of the three parts of the ideal system are not worth using in most of an organization, this means this ideal system is really utopic. Moreover, even a centralized **Master Data Management** (**MDM**) approach is complicated. The MDM practices per se are applicable to every business domain, so there is no problem with a data referential service; they are not very complicated to set up, as we will see in practice in *Chapter 15*, and they bring lots of business value and advantages. However, the ideal system aims for generic MDM, dynamically adjusting to every entity in the business context of an application. This step further is also out of the scope of this book, though static code generation for a data referential service is becoming a viable option, as we will show at the end of *Chapter 15*.

In addition, we have shown that the three responsibilities of an ideal information system are, ultimately, quite entangled with each other:

- MDM uses business rules in its validation of data

- A BRMS needs data from MDM in order to apply the business rules and decide their output value

- A BPMN serves mainly as a collector of data to feed the MDM, while also consuming data from the MDM

- A BPMN also uses business rules in order to know where to go in the different gateways (and, sometimes, to calculate some additional data during a given task)

All this proves that, technically, this assembly of three generic servers for MDM, a BPMN, and a BRMS is not so feasible, and neither achieves a perfect decoupling. So, why did we bother in *Chapter 5* and the last three chapters to talk about such an ideal system? Again, the answer lies in the business/IT alignment. The ideal system is not something that can be realized in practice in information systems today (and certainly not for at least a few more decades), but it has the great advantage of forcing an architect to think in terms of three generic, always applicable, functional responsibilities. Even if you use a unique software application, knowing how to separate the data management, business rules management, and business process execution provides a great step toward decoupling your information system (which is not achieved at all with *n*-tier architecture, for example). As you will see in the coming chapters, constructing an information system with these principles in mind will help us achieve a very complex goal, which is to be able to modify important functional rules and behaviors very easily and, in most cases, without any significant impact on the implementation.

In the next chapter, we will use everything we have learned so far to design the information system of `DemoEditor`. In the following chapters, we will finally get hands-on and implement all the different parts of this information system, using C# and .NET as a programming platform and Docker as deployment architecture.

Part 3: Building a Blueprint Application with .NET

After a theoretical part and one on the architecture principles, we will now dive deep into the technical aspects of the method by implementing some important parts of the sample information system. We will create some ASP.NET services implementing the API contracts and a graphical user interface that uses them and implements some of the business processes. Since some features have been externalized to bring more industrial-grade quality, we will also show how to interact with these modules in a lowly coupled way. Plugging the services into the Apache Keycloak IAM, using standards such as OAuth and JWT, will of course be an important step, but we will also show electronic document management systems in a standard way and talk about many other external services. Finally, the external execution of business processes will be shown, with both orchestration and choreographic paradigms.

This part has the following chapters:

14

Decomposing the Functional Responsibilities

Since *Chapter 6*, we have stepped away from the theory explained in the first chapters and demonstrated how we would turn the principles that we learned into the design of a sample application. Now, we are going to build this sample application, and our goal is that, by the end of the book, you can apply most of the principles learned in a real suite of applications and reproduce a small information system. By using what was explained, this small system will be very easy to evolve, and this is what we will demonstrate in the very last chapters of the book.

I know from a lot of experience in managing developers that most technical readers are eager to finally get some code typed and see what happens "for real" (funny to use this expression for something as virtual and abstract as code), but this is the very last chapter before we dive into the technical installation of dependency services (*Chapter 15*) and code writing for the main services (*Chapter 16* and following).

This chapter explains how the sample application will be decomposed into functional modules aligned to the business, thus following the method explained in the previous part of the book. Since we have talked a lot about `DemoEditor` and used this sample information system to explain many concepts with applied examples, this chapter is mostly about consolidating the many bits of information into one single unit of writing. As such, this chapter will be quite short because it is more of a summary of what we are trying to achieve and how, before diving into the technical implementation in the next chapters.

We'll cover the following topics in this chapter:

- More context about the `DemoEditor` information system
- Decomposing into a four-layer architecture
- Looking for standards for interfaces
- Writing up the specifications for the different modules
- Technical specifications

Some more context about the DemoEditor information system

Since we are going to use this sample information system and its supposed owner company for the rest of the book, we should take a bit of time to introduce what it does, explain how it is organized, what are their requirements, and so on. Reaching business/IT alignment requires the business to be clear, as many architectural and technical choices will flow one way or another from what the business needs. Though the next section might seem anecdotal, these anecdotes are at the core of the method.

Current company organization

The first section of *Chapter 6* was about setting the landscape for DemoEditor, a sample company that edits and publishes books. We went around the different profiles working in the company and the data they manipulate in their daily job. Since this sample company and its information system accompanied us from *Chapter 6* until this one, we analyzed the different business domains associated with a DDD perspective, evaluated which data should be part of data referential services (at least books and authors, but further analysis in this chapter may show that some other entities should be treated as well), and showed how business processes could work in such a company. In this section, we recall this previous information and add some context in order to explain the technical choices that are going to be made in the next section.

We already said that DemoEditor is a small company, with less than 50 employees, mostly composed of editors. A more detailed organizational chart is provided here (this will help us to create an aligned **business capability map** (**BCM**) later):

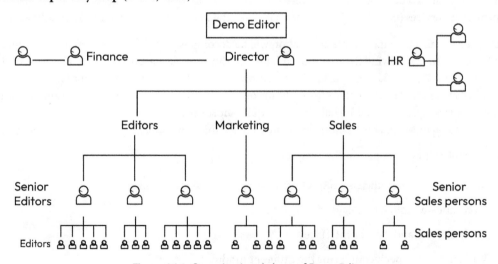

Figure 14.1: Organizational chart of DemoEditor

The current state of important data

DemoEditor does not have much of an information system yet: there is no dedicated IT department and people operate mostly by exchanging Excel workbooks, with the exception of the workbook about authors and the one about books, which are centralized and shared at the company level on the network. Authorizations have not been put in place, as this is difficult to set up in a single Excel file, which causes organizational problems because everyone can make modifications and the data is not clean anymore.

The authors workbook looks like the following:

	A	B	C	D	E	F	G	H	I	J	K
1	Number	First name	Last name	Full name	Phone	Email	Status	Number books	IBAN	BIC	Skills
2	0001	Stacy	Johnson	Dr Stacy Johnson	546 324 567	dr.s.johnson@edu.albertson.gov		1	AT811095027504584340	VOBOIY97M4U	.NET, Java and Python
3	0002	John	Smith	Mr John Smith	+1 789 567 345	john.smith53@gmail.com		1			Architecture, DDD, Tests
4	0003	Joan	Bouchara	Mrs Joan Bouchara	+33 687608093	j.bouchara@free.fr		3	CZ8256023405251438710402	BNIUIZPK	.NET
5	0005	Sebastian	Oconnor	Sebastian O'Connor	0032 765 235 243			1	IE76WFHC46841250788695	XKWQMZ88	Java
6	0006	Ernest	Daren	Mr Ernest Darren	+31 797345293	ernest@daren.com	Blocked	1	AT5007707586516996686	HPWICMWK	Architecture, REST APIs
7	0008	Deborah	Kingston	Ms Deborah Kingston		debbie.kingston@academics.net		1	AT640879301783478105	OFLHKS3IA88	REST API
8	0009	Sebastian	O'Connor	Sebastian O'Connor	+32 765 235 243	s.oconnor@outlook.com		2	EG4669782306282370230137080 32	JTZQWGCIV78	Java

Figure 14.2: DemoEditor's legacy software for authors

You may notice that there is duplicated data. Authors numbered 0005 and 0009 seem to be the same person, Sebastian O'Connor, whose last name had been mistyped the first time he was recorded in the list. His phone number is the same but has been typed in two different formats, where the international indicator has first been prefixed with a double zero and the second time with a + sign. Also, in the first entry, his email had not been recorded. At least this did not create any confusion about which entry contains the right data. As for the number of books, it causes a problem because we do not know whether the number of books has been adjusted with the second "version" of the author, or if the two versions are completely separate and we should add the numbers to reach a total of three books written by Sebastian. Confusion about IBAN bank coordinates is even more worrisome, as this can lead to financial mishaps.

The way DemoEditor stores information about books at the moment is also cumbersome, for many reasons, and you will perhaps be able to spot in the following extract:

	A	B	C	D	E	F	G
1	ISBN	Title	Author	Edition	Summary	Sales	Comments
2	0-9655-9654-0	Programming languages	Dr Stacy Johnson		General book on programming, with examples in many languages for beginners	790	
3	0-4309-3035-6	Domain Driven Design	John Smith		Architecture book on DDD, rather oriented to experts	567	
4	0-1037-8088-2	.NET 6	Joan Bouchara		Reference book on .NET, this one for version 6	3546	Best-seller
5	978-7-8575-5876-2	Rust with examples	Cindy Excelmans		Practical book on Rust	1278	Author is not referenced anymore
6	978-8-6169-8066-4	Designing Java applications	Sebastian O'Connor	1	Complete step-by-step building of applications in Java	345	
7	978-4-7246-4097-1	.NET 7	Joan Bouchara		Reference book on .NET, updated on version 7	8009	
8	978-5-9347-2021-7	REST APIs	Ernest Darren		How to design REST APIs, mostly theoretical, for experts	234	Author has been blocked (false banking information)
9	978-7-9865-6486-3	Programming in COBOL	Erin Doorman		Reedition of an old book, with the original author	125	
10	978-2-6944-4657-9	How to REST	Deborah Kingston		Creation of REST APIs, with examples, for beginners	1034	
11	978-1-7867-6524-6	.NET 8	Joan Bouchara		Reference book on .NET, update for version 8	1789	
12	978-1-6411-8448-9	Designing Java applications, 2nd edition	Sebastian O'Connor	2	Complete step-by-step building of applications in Java, updated in 2020	408	Not a rewrite, more of a completion of first edition

Figure 14.3: DemoEditor's legacy software for books

This time, there is no duplicate (you might think on lines 6 and 12, but these are two different editions of the same title, and consequently considered as two books, with two different ISBNs). The ISBNs, by the way, are a great way to avoid these duplicates because, this time, we can count on this external reference to check whether a book has already been recorded. In the author's workbook, we can imagine that an editor would search for an author by their name before adding it and that Sebastian O'Connor indeed has been duplicated since his last name was first written as OConnor, without the apostrophe.

Talking about authors, this list of books shows only one author for each line of data, and in addition, the value is the full name of the author. This is absolutely fine for book printing, and it is even a good idea to duplicate the data since it should not change if an author changes their name. However, it definitely makes it difficult to find other information about the author by referring to the other Excel workbook. Is the book in line 6 related to the first or second definition of Sebastian O'Connor? This might not seem a problem since the full name is the same, and we are convinced it is simply a mistake. But what about the bank coordinates? Should the royalties for this book be paid on the first account or the second one?

If we go back to the ISBN, we encounter this question: isn't it also a problem for an editor to rely on this outside reference? Since an edition company accompanies authors in creating books, there is a fair amount of time when the book exists but has not received any ISBN. How do you think people handle this situation when they do not have a technical culture? Most of the time, they simply invent numbers or a range of numbers, such as TMP-0-0000-3547-0, with all the associated problems.

A small anecdote here: I once consulted for a newspaper company that had to create a referential service for all the geographic information they handled. One such piece of information, used for paper delivery and also the definition of the different regional editions, was the official code for a city in France. This code is made of five numbers, two being for the "département" (France contains a bit less than 100 of them, roughly numbered between 01 and 97, with some quirks for 2A and 2B replacing old 20) and 3 being numbers used to distinguish the cities inside the department. For example, the code for Vannes, the city I live in, is 56260, 56 being the code for Morbihan, one of the departments of the westernmost region. Generally, there are much fewer than 1,000 cities in a department, which means there is room for additional numbers. This is great since this customer needed to create some "special" numbers to represent areas of delivery of the newspapers where they did not fit exactly with the city's boundaries. For years, this did not cause any problems, since completely new cities simply do not get created in France. However, one day, the French government decided that, in order to increase the efficiency of local services, cities should be encouraged to merge together, thus creating a new entity – and, of course, this involved taking a new number in the list, which clashed with the other use made by the company. Since it was also interested in serving newspapers in the new city area, I let you imagine the mess when two of the same numbers represented two different realities, one being settled in the information system for years (and thus used everywhere in the business processes of article writing, newspaper publishing, local delivery, etc.), and the other one impossible to modify since it was an officially attributed identifier.

Finally, the comment section contains lots of very different information, and there is no way we can handle it in an automated way, so, while it may be useful for the human eye, it is completely useless for the information system in general. Also, the number of sales for the books seems to be in the wrong place there, since they are not an attribute of the book, but rather statistics coming from sales data on books, which is radically different. Chances are that they are not up to date or are made so with an offline process. In short, this way of storing data is a mess, just like for authors, and certainly accounts for a lot of the inefficiency seen in `DemoEditor` by their boss, and the reason why they need a new information system.

What kind of information system is going to be built?

Though the hypothesis is that `DemoEditor` already has a rigidly coupled, hard-to-evolve spaghetti dish of a legacy information system, it would not be proper for a book to set up such a complete information system just to get rid of it a bit after, however progressively we create the steps to do so. In the following chapter, we will thus design and create an information system as if it were from scratch, apart from the fact that we will use the information already existing (typically, the content of the Excel files). This is of course not what the virtual scenario is, but we have discussed several times how to progressively shift from a legacy system to a newer one, in particular when talking about data referential services, the constitution of which is almost always progressive from scattered databases.

Let's imagine, then, that we are the director of `DemoEditor` and have the financial capacity to invest in a well-defined, low-coupling information system while keeping the legacy working. First, this would mean having at least an internal resource dedicated to IT management and considering right from the beginning that the information system will be the backbone of the company. A consequence of this is that IT should not be considered a mere cost center and its manager would be part of the board. Another consequence would be that the information system map should be discussed at the board level, as well as the strategy level, to design and evolve it, with a multi-year planned and budgeted project.

If we imagine ourselves on this board, the very first thing we do is create a map of the information system. Of course, what is better than using a four-layer diagram to do so? This is what we are going to do in the coming section. After that, we will follow and design parts of the maps more precisely, such as a few important business processes that we will implement, or data referential services that should be completely developed to show all aspects of this work. However, we will, of course, not use all the methods we discussed in the previous chapter.

You may wonder why the proposed application does not follow every best practice in architecture that has been explained before. The obvious answer is that lots of different, concurrent approaches have been explained, and putting them all into practice at the same time is not feasible in practice, precisely because they have different recommendations for some points of architecture. Yet, one could have been chosen among the others, demonstrating the principle that the best tool has to be chosen for the given context.

This is precisely what we are going to do, but since the application is just a sample, a very limited one, there is no need for any complex setup with hexagonal architecture, C4 approach, or anything "industrial" – we can already go a long way with a simple but clean separation of concerns using code interfaces and dependency injection, which is provided by default in the .NET skeleton projects that are going to be used. This will be also a way to state that only the goals are important (loose coupling, capacity of evolution, and separation of concerns) rather than any given architecture or framework.

Decomposing into a four-layer architecture

Having a map of the information system is the single most important thing to do when creating an IT service or when working on an existing one. Let's start with a map of the required information system in a top-down approach that will be better suited to detailing what we want to achieve with this example.

Business process management layer

Some business processes of `DemoEditor` have already been illustrated in the previous chapter but, before coming back to this level of detail and providing a reference list of the processes we are going to turn into software, it may be interesting to look at the company from an even higher perspective than macro-processes, and this is what **value stream mapping** (**VSM**) is about. VSM diagrams integrate processes into a global view of the stream of information and goods of the company, in order to ease reflection on production time and value created by the company, often within a network of suppliers and partners.

In our case, such a VSM diagram could be designed:

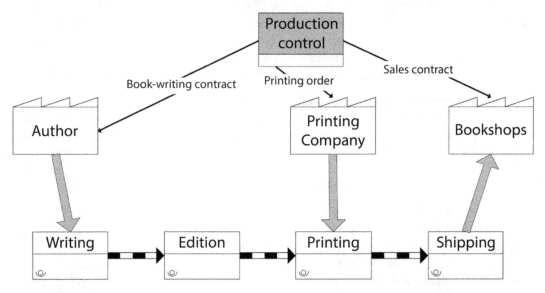

Figure 14.4: VSM of DemoEditor

This type of diagram is useful when the value produced by the company must be followed precisely. When applied to logistics or material production companies, it is generally accompanied by indicators at each step with the cycle time and the lead time, which helps find bottlenecks in the stream and improve efficiency. In our case, the VSM diagram is only useful for showing the different actors and the general streams of data between them. Here is the process:

1. The company appoints an author to write a book.

2. After the writing itself, there are activities that need to be completed in order to finalize the edition.

3. Following this, the work is printed and becomes multiple concrete instances of paper books. This is piloted by the company that uses a supplier for the printing activity.

4. They are finally shipped to the bookshops. Bookshops also are external actors, but their relationship with the company is about sales.

In our case, there is no particular use in showing the CT/LT times, since storage of books is not a primary concern (it may be in the real world, but it would not reflect directly on the information system, which is our subject). The diagram helps us understand the high-altitude work of the `DemoEditor` company. Once we know the general streams, we can focus on the more detailed processes. Let's first show the few processes that have already been shown in the previous chapters. In *Chapter 11*, we talked, for example, about how the actors are enrolled by the editors:

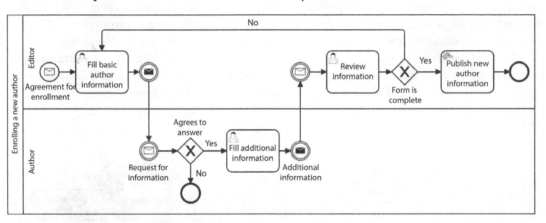

Figure 14.5: Example of BPMN for author enrolling

Another example of the process was provided in the same chapter, about contract signing:

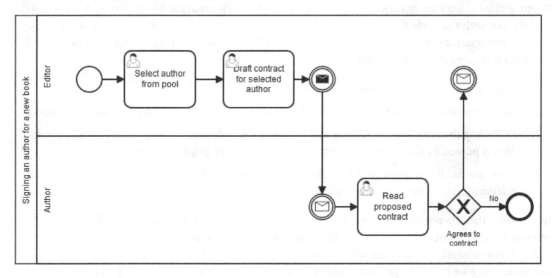

Figure 14.6: Example of BPMN for contract signing

Even if we are not going to create a fully realistic set of processes, these two are, of course, not enough to even partly address the complexity of the modules needed for the information system. We are thus going to document a few additional ones that will guide us in its design. A good idea is to get a global map of the macro-processes before designing the content of each. This way, we continue our top-down approach from the value streams. When listing processes, they are generally decomposed into operational, support, and piloting ones. Here is a first draft of such a macro-process diagram:

Figure 14.7: Macro-processes map for DemoEditor

If we take a closer look at the "**Finding authors**" macro-process, it can be defined in more detail as the following chain of processes:

Figure 14.8: Decomposition of the macro-process of finding authors

The first one of these processes can itself be refined into a micro-process (the closest to a procedure, where each task is detailed in such a way that one can operate it with minimal previous knowledge of the job):

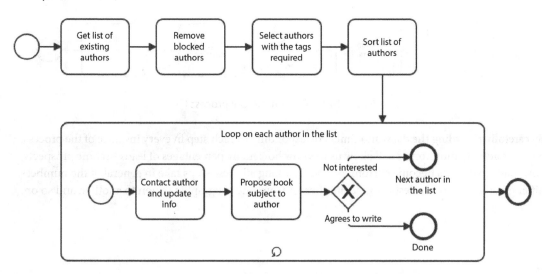

Figure 14.9: The process for finding an author in the existing list

In the way DemoEditor editors work for now, the first tasks of this process would correspond to opening the Excel workbook with the authors (or, more probably, a local copy of it, potentially not up to date), then applying a filter in order to remove the authors that have been blocked (which may be difficult since the Status column contains simple text, which means values can be mistyped), and finally reviewing manually the list of skills, in order to establish who is suitable for writing the next book. When the list is ready and sorted from the best candidate to the least wanted in the list, the process consists of calling every author and proposing the writing of the book to them (while updating the data for the author; this is where the local copy of the Excel workbook becomes a problem since the fresh data will not be shared). This is repeated until an author is found. If this is not the case, the next process in the macro-process, which is to enroll a new author, will have to do the job.

Again, there is no use in making a complete functional analysis here, as our goal is only to show the method and to detail the processes that will bring some specific implementations of services in the software layer of the information system. Still, we are going to draw at least one last process. Indeed, as was explained in *Chapter 6*, the director does not have a clear idea of how long it takes to gather leads, turn prospects into actual customers, and so on. So, one process that is necessary to design along this information system wave of improvement is the one about the commercial pipe:

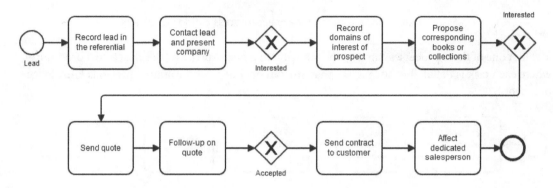

Figure 14.10: The commercial process

By carefully recording the dates and times corresponding to each step in every instance of the process, business activity monitoring will allow us to know how many percentages of leads become prospects, how many prospects become customers, and how long all these steps take in general, if the numbers differ greatly from one salesperson to another, or from one geographic region to another, and so on.

The BCM layer

A possible BCM is represented here:

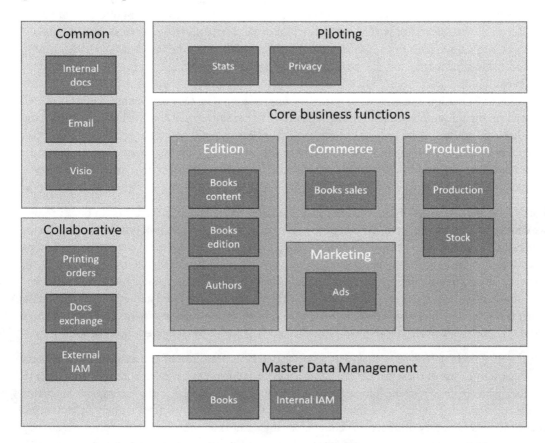

Figure 14.11: BCM for DemoEditor

I use the word *possible* because a BCM is always about choices. For example, the governance of each function has been attributed to the most important direction, but it has been decided that the responsibility around the books should be split, even if the book core management remains a master data management issue, and is located inside the dedicated zone, stating that the core identity of the book is something that concerns the whole company. Notice, by the way, that the BCM business zone is well aligned with the preceding organizational chart.

As far as the authors are concerned, after imagining how such a small company would work, it appeared that the authors are not really a data referential service in the sense of master data management. Sure, there will be a dedicated application and database for them, but when you think about it, only editors really manage their information. Commerce is not concerned at all as they sell books without taking into account who the author is, where he lives, and so on. Marketing may get some information on authors from time to time to contact them and organize book presentations, but they basically only consume this information, just like how accounting (not represented here, for the sake of simplicity) will consume the bank information of the author to pay royalties on their book, but will not know any other information and will never write any on these entities. One of the consequences of this decision to include authors' management in the `Edition` sub-zone is that we have not shown the details of the authors' data; there is the contact information, bank coordinates, areas of expertise, contact information, and so on. However, all of this is included in the `Authors` block. It would be represented in a focused BCM, but remember that it is a trap to want to have everything in a single diagram. Maps are there to help solve a problem and this one is there to express ownership of functions.

I briefly created the areas of expertise to add some interesting data on the authors. In our example, we could consider that knowing the competencies of the different authors is of critical importance for `DemoEditor`, as they have a hard time finding good authors. Tags for domains of expertise would thus be included in the authors' attributes, and the long-term idea would be to have an extranet for them, allowing authors to read and sign their contracts, but also to complete their domains of knowledge. For now, this function is not represented but, should it appear, it would be placed in the `Collaborative` zone, as this is something that would be turned towards other information systems or, more generally speaking, the "outer world" for our sample information system owned by `DemoEditor`.

Also, notice that two functions have been duplicated between internal and external. Documents management is one of these functions, as we are going to need to generate and send contracts to authors, who are known by the information system, but physically outside of the company. Since this needs authentication, it drives the internal/external split of a second function, namely IAM. Indeed, there will be a directory of internal accounts, but the authors are not part of the company, so they should be separated. Since we have a referential service of authors, we could use it to associate with external accounts through validated email addresses, for example. However, we would still need some ways to establish the identity of the incoming requests on external accounts.

As was said in the previous section, printing is outsourced, so `production` is really an internal responsibility, but it is called a secondary function, which is `printing orders`, and requests the actual printing of batches of books. Since this is done by an outside company, this function is placed in the `Collaborative` zone as well.

Finally, some common functions and piloting ones are shown in the BCM, since they will be useful in some parts of the processes that have been shown, or in the scenarios that were discussed in the previous chapters in association with roles in the company.

Software layer

Since this book is about creating a new information system (greenfield projects are rare in the real world, but evolving a legacy information system into something sustainable would make for a complete book on its own), it is quite easy to get a very good alignment between functions and software. However, coupling always lurks, as will be explained here.

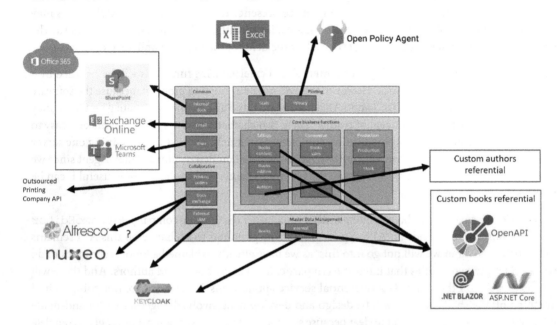

Figure 14.12: BCM with software implementations

Let's just walk around the BCM and explain the software choices (in particular when alignment has not been made one to one), as well as detail the remaining coupling points.

First, the three common functions (mail, videoconference, and internal documents management) will be handled by three applications offered by Microsoft Office 365, which are, respectively, Exchange Online, Teams, and SharePoint. Using several applications from the same suite makes it much easier to create coupling and more difficult to get rid of one while keeping the others, particularly when the editor integrates the solutions very strongly. At the same time, this is what improves the performance and usefulness of the applications. As always, there is some balance to strike and decisions to be made as to what is more important. In this case of common functions, better integration clearly beats harder coupling.

Excel, which is part of Office 365, will also be used for statistics. Storage of the associated files will certainly be done on the cloud. As far as the documents exposed to the outside of the company, we could very well stick with the choice of SharePoint and create a dedicated site, since SharePoint is capable of handling intranets and extranets together. Still, the security of the authors' documents is paramount to a book editor, and the features expected are not the same. Since dedicated functions for book edition and content browsing are expected, the choice is made to choose a different, CMIS 1.1-compatible content management system, with an open-source presence to make it easier to extend. This may be Alfresco, but it could also be Nuxeo, as both are aligned with the OASIS CMIS 1.1 standard. The great thing about such standards is that they allow the delay of the decision and keep options open.

As for the IAM, we do the contrary: both internal and external IAM functions will be handled by a single instance of Apache Keycloak. However, this will not be a coupling problem, because the software handles **OpenID Connect (OIDC)** and SAML2 standards, and also happens to manage multitenancy extremely well, owing to its concept of realms. These two characteristics together will make it easy to handle internal and external authentication completely separately, while still maintaining one server only and consuming fewer resources. Authorizations will be handled by Open Policy Agent since we have extensively described how it works in the previous chapter and explained how useful it can be in our sample case.

Let's get rid of a small detail about printing orders: we consider the printing company that `DemoEditor` uses as a supplier to provide an API to pass orders. The grammar to exchange on this API remains to be constructed but we will not go into this, as we have already explained a lot of techniques to do so based on the two entities that interest us in particular, namely books and authors. And there will be lots of work on these two data referential service applications. Indeed, we are not going to find anything off the shelf, so there will be design and development involved, together with hand-made integration. The choice there is quite clear because so-called generic software almost never gives complete satisfaction. They can be very well adapted to big companies where there is an interest in forcing the entire enterprise to adopt industry-grade, standardized processes. However, for a small company such as `DemoEditor`, going into a full-blown ERP would not only be unadapted but also hugely expensive. Alright, development is not cheap either, and there are good chances that `DemoEditor` does not have any internal person capable of handling these pieces of software. However, we will show in the next chapters that a bit of knowledge in .NET 8 and the ease of use of Blazor for creating web and mobile applications can already carry you a long way. Integration can be made easy with simple webhooks and API calls, without needing any middleware; and there are also no-code / low-code alternatives that can help in this matter.

Hardware layer

As has been said before, the hardware layer is part of the four-layer map, but its drawing is often skipped or at least made with reduced effort, as the main problems of evolution are in the higher layers. Personally, out of the almost hundred organizations I have been in contact with about information systems, there has not been a single case where a profound business problem actually came from the hardware layer itself. It might be *part of* the problem but is almost never the *source* of it.

In the case of our example, there is no need for a dedicated diagram. Indeed, lots of functions will be in the Microsoft Office 365 cloud, and all the other applications can be run in Docker containers. This means that there is almost no adherence to the servers below. We might deploy in Azure Container Instances, maybe with the local Docker Compose. However, that means that it will be easier for anyone to simply deploy the application, which was a requirement associated with the book, in order to not block any reader for a question of complex requirements. If needed, anyone can get a free cloud account with enough resources to test the sample application. If we were to run this as if we were in a real company, we would simply use a commercial account or buy some real machines in a data center and operate them for real, with backups and redundancy. However, this would not change anything, since there would only be more nodes in the Docker cluster and its maintenance done professionally, but without any impact on how we are going to design the rest of the information system.

Looking for standards for interfaces

We already touched on the subject when we talked about the software layer, but choosing norms and standards is of such great importance when designing or modernizing an information system that a dedicated section is not too much.

OIDC, together with OAuth 2.0 and JWT, is the obvious choice for authorization and identification management, as it has become the de-facto standard nowadays. SAML, though still quite used in the industry, definitely lost the battle to OIDC for modern uses such as single-page applications and API management. Still, on the identification function, the **System for Cross-Domain Identity Management (SCIM)** specification could have been used to manage identity information, but it would be overkill for such a small organization. SCIM is only useful when one needs to manage lots of attributes on users. If we were to embed the organizational chart in the IAM, though, it would definitely be with SCIM 2.0 Enterprise User Extension (RFC 7643).

XACML could have been the standard for authorization management, but a lighter approach is more adapted here, which is why Open Policy Agent will be used. In addition, Open Policy Agent makes it easier to adapt to changing rights management policies. Finally, the fact that a light implementation in a Docker container is readily available is a plus for our sample information system.

OASIS CMIS 1.1 does not have any competitors in normalizing electronic document management, so it is almost a non-choice. We still insist on this because lots of EDM and CMS editors do not "play the game" and tend to ignore a standard-based approach to stay in vendor lock-in attitude, which is a pain to the business in general. Arguments are always the same: CMIS is not powerful enough, our APIs will bring you more performance, other editors do not implement CMIS either, and so on. Those are only pretexts, and any good editor of such a solution will be able to provide a CMIS connector if the customers insist on it, at least for a reduced perimeter of the standard. After all, most of the time, we do not need the full perimeter of the norm; however, respecting it for the functions the client needs is the guarantee of being able to evolve freely in your document management.

The **Dublin Core Metadata Initiative (DCMI)** would also be a great source of norms and standards for DemoEditor since it contains the definition of metadata about authoring, the language used in publications, types of literary work, and so on. Though I have not found definitive standards for complete book descriptions, the DCMI attributes can be used to create a pivotal format, and additional standards such as ISBN descriptions and the like will be included.

Of course, as in any pivotal format, low-level standards should be used for every data of certain types: **ISO 8601** (also called **ISO-Time**) for dates, **ISO 3166** for country codes, **RFC 5545** (also named **iCalendar**) for events descriptions, **RFC 6350** (aka **vCard**) for contacts data, and so on.

Finally, **Open Data Protocol 4.0** (normalized by OASIS) will be used (at least partially) in every API definition where we request referential service data. Examples of queries with attributes starting with a $ sign have already been shown in *Chapter 10*, such as $top, $skip, $filter, and $order. In order to be able to easily extract data from the business information system into the piloting Excel workbooks for stats and key performance indicators, it is essential that data is correctly exposed, as it will make it easy to manipulate with the Power Query module of Excel in our case, but also with almost every modern business intelligence tool.

Writing up the specifications for the different modules

Now that we are equipped with the four-layer map of the information system that we want to build and the norms and standards to kickstart entities' technical implementation, we only need to close this chapter with a more detailed design of each of the modules that are going to be built and assembled.

Overall system specifications

Before we dive into the specifications for the different modules that the information system will be made with, a few considerations apply to the overall system. For example, all modules with web exposition will be in the demoeditor.org domain. Most of the time, you will notice that I only talk about relative URLs starting with /api; this is done on purpose, as we are going to do as much as we can to delay the application of absolute URLs, in order to reduce the coupling with the domain exposition as much as possible. When calling URLs from services, we will use relative URLs as well.

The following image recalls what business domains will be implemented in the system, from a global point of view:

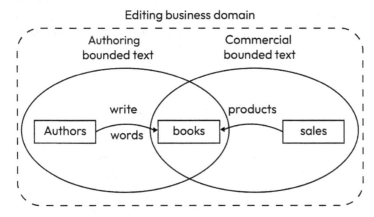

Figure 14.13: DemoEditor's business domains

The book's referential service will be in conjunction of the authoring and commercial bounded contexts. It will have its own application, as the master data management of books is the most essential one in this sample system. A dedicated referential service is thus adapted and will be the basic stone on which we will build the rest. We could call the API books but since there can be secondary objects that we will discover along the way in this business domain, it will be called library, and books will be the name of the entities inside it.

The edition domain will be the one managing the authors and their relationship to books. In particular, contracts will be handled in this domain. This choice is based on the life cycle of each entity: a book can exist without any data on its authors whereas, from the edition point of view, a person without any link to a book (existing or in the project) is simply not an author and should not be present in the database. authors will then be the name of the entity, and there will also be contracts. Please remember that we talk about the functional entity there, and not the paper or electronic document that formalizes the contract itself. A contract is simply an agreement between individuals (private persons) or organizations (moral persons) that is beneficial to both of them and entails rights and duties. In many cultures and national laws, no formal signature is needed to establish a contract, though it is recommended to make one if the matter is important enough to justify this formal approach. In the information system, the contracts entity in the edition API will, of course, refer to binary documents, which will be stored in the EDM application that will be chosen.

Finally, the commerce business domain will contain mostly the sales entities. It will not be detailed because all we are going to do with it is extract some data to generate some statistics in business intelligence scenarios.

The books referential service

As the `books` data referential service is the most important one of the whole `DemoEditor` information system, we will take a bit of time describing all the features that should be in this piece of software. To get a global list of these features, we can refer to the previously shown example of a hexagonal architecture:

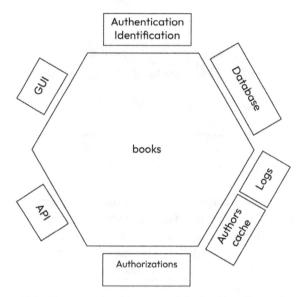

Figure 14.14: Hexagonal architecture of the books referential service

Apart from the traditional safety features around IAM, the books referential service will sport two interfaces: a GUI for human manipulation, and an API for automated interactions. It will be equipped with persistence in the form of a database, in which books will be held, but also a cached list of authors (at least the ones associated with books, otherwise, they have to use inside this referential service). Finally, a log mechanism will be put in place, or rather interfaced with the application.

If we go into a bit more detail on the persistence part of the referential service, another schema that was drawn when we talked about methods can be repeated here, since the example was indeed about how books should be structured as entities. At their core, there will be identifiers, such as an ISBN but also an internal book identifier, which is there to enable the identification of an entity before it has received an ISBN. Attributes that are used by anyone talking about this book, such as the number of pages and published date, will be also available as main attributes of the entity, which generally means they will be available for reading for any authenticated user, whichever role they own in the application.

As for the petals, corresponding to the different works around the domain of books, they are stated in the same schema and will be completed according to the needs during the implementation phase:

Figure 14.15: Flower metaphor for the books referential service

Part of the technical specifications should be about authorizations, even though extracting them in dedicated software such as OPA makes it very easy to adjust afterward. Let's just make the hypothesis for now that petals will be adjusted to the role of the person connected to the referential service while keeping the inheritance of the roles down the hierarchy, just like we showed in the OPA example, which was also about books.

As far as functional content specifications are concerned, the description should ideally go down to creating the OpenAPI syntax. After all, this is a specification of what the software will do. Of course, the first time your Product Owner is in charge of writing the OpenAPI contract, they may be hesitant about getting their hands dirty with something that looks very technical. The job of scrum masters or lead developers/architects is to show them the right GUI or explain that the YAML/JSON format is not that difficult, particularly considering the advantages of having an API contract-first approach directly led by the people who know the business the best.

So, in order to apply this "contract-first" approach, which is at the heart of the method shown in this book, let's head over to `https://editor.swagger.io/` and, starting from the provided **Petstore** example, work our way into an OpenAPI for the books referential service, which would be exposed with URLs such as `https://demoeditor.com/api/library/books/978-2409002205`. When starting, the editor looks like this:

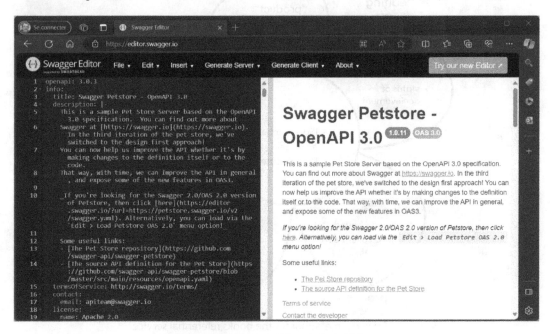

Figure 14.16: Swagger Editor in its initial state

Since the example is quite complete and provides a great description of the expected content of the contract, we quickly came up with something such as the following file, which we will present in small chunks in order to explain the different parts in it. The very first part is a header containing general information about the API (its name and version, license and contact, etc.):

```
openapi: 3.0.3
info:
  title: DemoEditor books referential service API
  description: Contract for books exposition at DemoEditor company
  contact:
    email: api@demoeditor.org
  license:
    name: Apache 2.0
    url: http://www.apache.org/licenses/LICENSE-2.0.html
  version: 1.0.0
servers:
```

```
      - url: https://demoeditor.org/api
tags:
  - name: books
    description: Manipulating books, existing and in project
  - name: params
    description: Setting parameters for the books data referential
service
```

This header is followed by a section containing the path exposed by the API and, for each of them, the HTTP verbs exposed. For example, this is the POST verb, corresponding to a description of the creation of a book, together with the errors that can be produced by such a call (this is also part of the contract, in order to explain the reaction in case of problems):

```
paths:
  /books:
    post:
      tags:
        - books
      summary: Creates a book in the referential service
      description: This operation allows the creation of a new book in
the data referential service.
      operationId: addBook
      requestBody:
        description: Description of a book, with at least the
mandatory attributes
        content:
          application/json:
            schema:
              $ref: '#/components/schemas/Book'
        required: true
      responses:
        '201':
          description: Successful creation
        '403':
          description: Is returned if the operation is forbidden
        '422':
          description: Is returned if the content proposed for
creation is invalid
```

The next section uses a different path because the GET (at least the one retrieving a unique item) and DELETE operations work on a single resource of the list:

```
/books/{id}:
  get:
```

```
(...)
        responses:
          '200':
            description: Successful read
            content:
              application/json:
                schema:
                  $ref: '#/components/schemas/Book'
          '400':
            description: Is returned if an invalid id is provided
          '403':
            description: Is returned if reading operation on the book is
forbidden
          '404':
            description: Is returned when the id provided is valid but
no books exist with this identifier or ISBN number
      delete:
        tags:
          - books
        summary: Deletes a book from the referential service
        description: Rather than actually deleting the book from the
data referential service, this operation actually makes it inactive
and invisible. The data is still there, but can only be retrieved by
high-privilege or technical administrators. This is done in order to
make possible full traceability.
        operationId: deleteBook
        parameters:
          - name: id
            in: path
            description: Identifier (internal or ISBN) of the book to be
deleted
            required: true
            schema:
              type: string
        responses:
(...)
```

In some cases, API operations can be about other resources than the ones corresponding to main business entities. Here is an example of a resource for pure customization, in this case, of the possible values for a book editing status:

```
/params/statuses:
  put:
    tags:
```

```
              - params
        summary: Defines the possible values of status for the books
        description: |-
          This operation allows to customize the possible status key /
value pairs. By default, the statuses will be the following:
            - 1 / Project : when the book is only a project
            - 2 / Draft : when the author has been chosen and a structure
is written
            - 3 / Writing : when the author is writing the book
            - 4 / Review : during the review of the book
            - 5 / Validated : book reviewed, ready for printing and
selling
            - 6 / Archived : no further edition is planned but sales go on
            - 7 / Retired : no further sales, the book is out of the
circuit
(...)
```

The third section of the OpenAPI contract, after the header and the paths, is the one where we define the data structures. For example, we have said that POST would send Book, but we need to be precise about what such a resource is in detail, and this is what the following schemas explain. As Book is the main resource, this definition is quite long – and this is just a reduced example here – but it is worth noting that this grammar contains everything in the book structure, except for the "satellite" types, which have been pushed to the end of this section:

```
components:
  schemas:
    Book:
      required:
        - id
        - title
      type: object
      properties:
        id:
          type: string
        isbn:
          type: string
          description: Official ISBN number of the book, once obtained
          example: 978-2409002205
        title:
          type: string
        numberOfPages:
          type: integer
          format: int16
        publishedAt:
```

```
        type: string
        format: date
```

Notice that each attribute is accompanied by its type and potential format. After those simple types, we arrive at the ones corresponding to more complete structures, in particular, those representing the petals:

```
    editing:
      type: object
      properties:
        numberOfChapters:
          type: integer
          format: int16
        status:
          $ref: '#/components/schemas/Status'
        mainAuthor:
          $ref: '#/components/schemas/AuthorLink'
```

This last attribute refers to a type that is described a bit further down, and that corresponds to a specialized link to another resource from a different domain:

```
    production:
      type: object
      properties:
        typeOfPaper:
          type: string
        assemblyMode:
          type: string
          description: mechanical assemblying mode used to bind
 pages of the book in paper format
          enum:
            - bound
            - glued
  (...)
```

Since `AuthorLink` inherits from `Link`, the latter would typically be common to many schemas, but there is no simple way to include definitions from another file, so they are generally copied, which is a best practice to ease versioning:

```
  Link:
    type: object
    properties:
      href:
```

```
          type: string
          format: url
      rel:
          type: string
      title:
          type: string
AuthorLink:
  allOf:
    - $ref: '#/components/schemas/Link'
    - type: object
      properties:
          authorId:
              type: string
```

For easier reading, again, this technical text can be put in Swagger Editor, or any other such editor, which will provide such an interface:

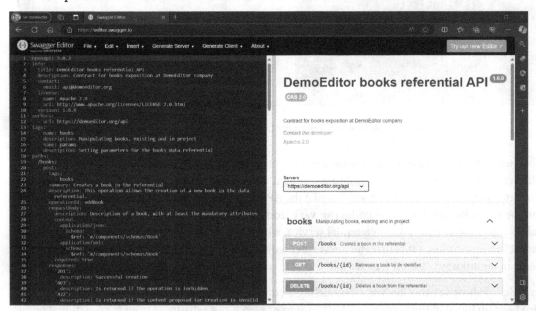

Figure 14.17: Swagger Editor for the books API

It is not our goal to be exhaustive here but to simply show how an OpenAPI specification is a great way to precisely state what we need from a functional point of view while being much more precise than any human language could ever be, however lengthy the descriptions are.

Just to specify a few particularities of the syntax, please note that the `assemblyMode` attribute is an enumeration whereas the type of `status` is a `Status` object, and a dedicated PUT API operation has been created for people using the books data referential service to be able to customize the possible values of statuses:

Figure 14.18: Swagger Editor for books parameters API portion

This has been done because the two attributes receive complex values, but there is a big functional difference. In the case of the assembling mode, it was considered to be a very stable, embedded characteristic of the book. Thus, it was OK that the API should be upgraded if the values were to evolve. On the opposite end, the status codes can vary from time to time, so it was decided to make these values a parameter of the data referential service, which can be modified by a special API operation inside the application. Of course, business rules would still apply, which would prevent books from being in an unknown status, for example.

In order to show yet another way of modeling a complex attribute, the link to the main author has also been included in the preceding YAML example (even if not exhaustive from the functional point of view, it was important that it carried lots of different approaches). You will notice that this time, a dedicated `AuthorLink` type has been created, extending the `Link` entity type that is completely copied from the RFC 5988 Web Linking standard. This is what will allow us to write such links in the book's JSON definitions (extract):

```
{
    "id": "00278",
    "isbn": "978-2409002205",
```

```
    "title": "Performance in .NET",
    "editing": {
        "numberOfChapters": 11,
        "status": { "code": 5, "title": "Validated" },
        "mainAuthor": {
            "href": "/api/authors/authors/00024?version=12",
            "rel": "dc:creator",
            "title": "JP Gouigoux",
            "authorId": "00024"
        }
    }
}
```

Remember the link to the authors should point at them with a given time measure. Indeed, if an author changes their name (typically since they married after writing), the name on the book should remain the one that the author had at the time of the publishing of the book edition. There are several ways to do so. One way is to simply copy the content of the author of the book. This way, the cloned data will not be changed if the author is updated in its referential service. This works but is not very resource-efficient. Another solution is to add a business rule in the book author entry stating that, when it gets the author data, it should automatically add criteria of the date value based on the value of the publication date of the book. Since everyone is calling the books referential service to get this data, they will be ensured consistency – but nothing prevents the client from separating the two pieces of data and losing this consistency. Finally, the best way is to include the business rule right in the link, by pointing `href` to a URL that includes the precise state of the author, either by version number (`https://demoeditor.org/api/authors/authors/00024?version=12`) or by value date (`https://demoeditor.org/api/authors/authors/00024?valueDate=2011-05-09`). This way, not only will the client have fully semantic information, but it will also make it more difficult to lose consistency, as the client would have to modify a part of the URL to do so. Of course, this does not mean that the link cannot contain a few often-used pieces of information about the author, such as its full name, but doing it this way makes it clear that this is a local copy of the data based on a precise instance in time.

Another example where dealing with a stable and immutable version of the data rather than pointing to the "moving" entity is when signing up contracts: it is generally a good idea to record the state of signers, as the address of the person that is going to be taken into account for the contractual relationship should remain stable during the signing process for regulatory reasons (it is a way of enforcing identification when the person has homonyms), even if the person moves out at the same time and "modifies" their address.

The preceding proposed values for the possible status codes and names are only default ones and the feature dedicated to customization would of course be used to implement the previously proposed life cycle for books, which I reproduce here:

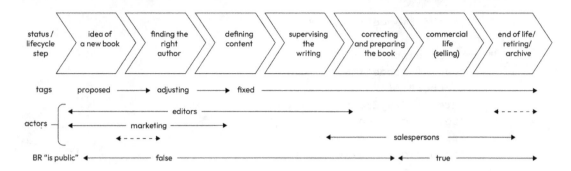

Figure 14.19: The life cycle of a book

In a real production project, we would, of course, go much further in the analysis and specifications of such a crucial entity as the books are for an editor. We stopped here because we talked about everything necessary for the implementation (and, of course, chose attributes that were diverse enough to try and show as many use cases as possible, in order to help you choose the right model for your own needs when designing an API).

In particular, the choice of specific names and attributes is of utmost importance, as explained in *Chapter 9*; this will have an impact everywhere in the information system and processes afterward. In our sample company, we could imagine that designers would be very interested in standards such as `https://schema.org/Book` for modeling the entity, though the fact that there is only one author possible on such an entity might be a problem for an editing company.

The issue with migration

For now, we have only talked about specifications for the data referential services when they will be used as the central version of the truth for their entities, but one of the challenges in any architecture activity is to integrate them into the existing legacy. As stated in the hypothetical `DemoEditor` situation described before, the data referential services must be built while the operations going on in the company, which means the data migration tool needs to be automated. This way, it is easier to test it along the way while the Excel workbooks still remain in use and validate at the same time that the new application is ready. When this is so, the migration will be run one last time and the old Excel file removed in favor of the new data referential service; by then, of course, all processes will have undergone adjustment (if needed) to accommodate such a change.

The main idea in this kind of move is "no Big Bang" – it is sometimes possible to break large pieces of the IT, in particular when the system is so broken that it does not give any satisfaction anymore. Most of the time, though, the difficulties are there but they act more like a thousand little stings that, on the whole, hurt productivity but do not individually justify stopping the incriminated piece of software. In order to show how this can be handled, we explain the different actions, step by step, that would end up changing the data referential service while minimizing the impact on the operations. In the example afterward, we may not show each and every step because it would not bring any added comprehension, but at least they will be explained here. By the way, these steps are very similar to the example of the fusion of two data referential services, which was shown in *Chapter 10*.

The steps (once again, this is an atomic decomposition but, depending on the situation, one may apply several evolution steps at the same time if the situation prescribes it):

1.   In the very first step, the existing system is not touched at all; we simply create the new data referential service on the side of the existing Excel worksheet.

2.   After that, we create a migration tool that takes as input the Excel file and outputs API calls to the new data referential service. Again, the Excel file is not touched, as it is only called in read-only mode. If there is the smallest risk of locking the file, a copy might be used to further reduce the impact potential. This migration tool can be a completely external tool, but also a function embedded in the new data referential service application, or even a macro in the existing Excel file; the location does not really matter for the function and the choice will be technical.

3.   The next step is recurring. As Excel continues to be exploited, with new books, changes, and so on, the migration tool is going to be refined in order to ingest the existing data (with all its quirks, bad namings, duplicated data, and so on) and clean it by code to produce a nice, clean view of the books in the data referential service. Ideally, Postman automated tests will be conducted on the data referential service in order to validate that the data is indeed as expected.

4.   By that time, the new data referential service can be opened to people benefitting from its use, but only in read-only mode, since the old Excel workbook is still where the modifications are done. However, when employees find out that the data does not have any duplicates, that it is better sorted, that the new referential service provides additional filtering features, and so on, there should be a tendency to increase the frequency of data refresh.

5.   At some point, the new data referential service is going to be univocally considered as a better source of data and people will start asking why they cannot modify data in there. The conduct of change is then won, and the rest is going to be pulled by the users. This is when the big switch is going to happen, and the new data referential service will become the official version of the truth in the system. The old Excel file will not disappear yet, because there are potentially many read operations (reporting, statistics, direct readings by processes that cannot transfer to an API call, legacy software pointing directly to the file, etc.), but the relationship will now be inverted and the source of data will be the new application, while the Excel file becomes a read-only copy of it. This is the biggest step because you need to inverse the connection between both applications (which is normally easy because you have mapped the other way around; cleaning bad data is hard, but making it dirty again from clean data is – sadly – very easy) but also shift all the practices of the people using the books data, and make them create and modify books in the new product. This is where a nice interface and new features are particularly useful, in order to fight against the "but we have always used the old way" reflex and the difficulty for people to accept change. If the legacy application is very old, you may even have to prepare a dedicated plan to conduct change.

6. Starting from this step, things are generally easier, as the big changes have been done, and users will enjoy the improvements of the new application for books, retrieving them in an easier way, being able to share URLs on them, perform reporting with a whole world of applications and, generally speaking, integrate books everywhere due to the API exposition. Authorization management should also benefit greatly since everything is exposed in an Excel file. Your job, in this step, is simply to encourage new uses of the repository, which will progressively replace the uses of the old legacy Excel file, even though it is still kept up to date with the new data referential service for now.

7. At some point, users will be transferred to the new data referential service for most of them. Also, old legacy applications based on the Excel file may have upgraded to new ones being able to use API, or at least plug into a file that is generated by them. At the same time, since new features and additional data will be added in the new referential service, the Excel file will miss some data that users find important, and will become more and more outdated. After some delay (this depends very much on the company culture, whether it is very conservative or very eager for change), the Excel file will simply not be used anymore and you will be able to archive it, together with decommissioning the data propagation functions that kept it updated since then.

This migration process has been explained for books but will be the same for any other referential service, such as authors in our example. The only variations are the speed at which it is run. In some cases, users and IT owners may be very careful; working in a complicated environment and the full seven steps can take years. In some other, smaller and more agile companies, several steps can be taken at once and the whole process is covered in a very reduced time. The only important thing is to never force the "Big Bang" approach, going directly from the initial state to *Step 7*, without the possibility of going back, as this is known to create more harm than good.

Authors referential service

As we have dived in depth into the specifications for the books referential service, we are only going to talk about the specificities of the authors as data, since the whole process and technical specification (including migration) has been explained in the case of books. Since the authors referential service is, for now, also an Excel workbook, all the same recommendations apply.

As for the content of the data, the specification of authors should follow as much as possible the norms and standards. In particular, the DCMI will be studied in order to find as many standard attributes as possible for content creators, their naming, and so on. Also, `schema.org` has been identified by the deciders in the company as a good source for standards, particularly since it models the Books / Author / Person model that is the bread and butter of the company. In particular, `https://schema.org/Person` might be a good way to model an author. As stated previously, the difficulty may be that `https://schema.org/Book` only contains one author; maybe the solution is to use `https://schema.org/Organization`, which contains several persons in this case, but the company's deciders were not convinced. Finally, they decided to use as much as possible of the schemas.org descriptions as guidance for the names of attributes, but not to embrace them fully. In particular, the use of RFC links between entities is more expressive and should be kept (as shown in the books referential service specifications).

As a side note on this subject, using a norm not only does not mean that you should use all of its aspects if you do not need them but also that you can add your own additional attributes. In fact, most of the norm implementations allow some way to manage additional attributes. This is a common objection to the use of norms: I saw several teams refusing to use an EDM because the CMIS standard was "too complicated," before realizing that, since they only had to send documents to the EDM for archiving purposes, simply calling the POST function with createDocument and using another call to retrieve the document was enough for their needs and could be done in a few lines of code. The same objection could go for https://schema.org/Book, arguing that the single Author attribute does not account for the business complexity. However, earlier on, in our simple example, we had indeed shown only the mainAuthor attribute, showing implicitly only that there would be a secondaryAuthors array. What prevents us from using the standard grammar for the main author and still adding another attribute for the additional authors? Well, nothing, and this is what will be done in the implementation in *Chapter 16*. By the way, the same goes for datePublished instead of publishedAt. The following shows the adjusted part of the OpenAPI contract for the books:

```
components:
  schemas:
    Book:
      required:
        - id
        - name
      type: object
      properties:
        id:
          type: string
        isbn:
          type: string
          description: Official ISBN number of the book, once obtained
          example: 978-2409002205
        name:
          type: string
        numberOfPages:
          type: integer
          format: int16
        datePublished:
          type: string
          format: date
        editing:
```

```
        type: object
        properties:
          numberOfChapters:
            type: integer
            format: int16
          status:
            $ref: '#/components/schemas/Status'
          author:
            $ref: '#/components/schemas/AuthorLink'
```

The corresponding book sample is adjusted like this:

```
{
  "id": "00278",
  "isbn": "978-2409002205",
  "name": "Performance in .NET",
  "editing": {
      "numberOfChapters": 11,
      "status": { "code": 5, "title": "Validated" },
      "author": {
          "href": "/api/edition/authors/202312-007?version=12",
          "rel": "dc:creator",
          "title": "JP Gouigoux",
          "authorId": "202312-007"
      }
  }
}
```

As for the author that is pointed out by the preceding URL, a GET call would bring something similar to the following, using https://schema.org/Person as a basis for the attributes:

```
{
  "id": "202312-007",
  "givenName": "Jean-Philippe",
  "familyName": "Gouigoux",
  "email": "jp.gouigoux@demoeditor.org",
  "birthDate": "25/12/1974",
  "telephone": "+33 2 97 02 02 02"
}
```

Of course, this will be extended in time with the attributes needed for the correct management of authors. For example, the legacy Excel file shows that there is a need to list fields of competency for the authors (though it has been done in a poorly designed way, listing them in a single text column, which makes it hard to search for them). We could express them like this:

```
{
  "id": "202312-007",
  "givenName": "Jean-Philippe",
  "familyName": "Gouigoux",
  "email": "jp.gouigoux@demoeditor.org",
  "birthDate": "25/12/1974",
  "telephone": "+33 2 97 02 02 02",
  "skills": [
    ".NET",
    "Tests",
    "Docker",
    "Digital transformation"
  ]
}
```

Again, it is important to use standards where they are available. For example, IBAN bank coordinates are used for paying royalties to the authors, as this is an internationally agreed-upon standard.

Identifiers of authors should be something like `https://demoeditor.com/api/edition/authors/202312-007`. As the company's domain name has not been reserved, since `edition` is still unsure as the bounded context for the `Authors` entities and `/api` could disappear if it is decided to use hypermedia and content negotiation to expose web interfaces and JSON content for machine-to-machine links, it is recommended to not make this part of the specification to the developers and make it easy to customize in this end when integrating the functions.

IAM

The legacy information system of `DemoEditor` is extremely rudimentary, as we saw, for example, with the Excel file as a books referential service. This means that IAM was not really deployed. Sure, the company used Office 365, and thus an Azure Active Directory connection. Maybe it even had multi-factor authentication activated and we could consider this as state-of-the-art. However, the book's listing is still an Excel file, positioned on a server file share, with all of the persons with file-writing authorization able to change just about anything they want in it, without any other traceability than the fact they have accessed the file. This is a bit limited compared to the authorization management needs that we explained in *Chapter 13*!

Thus, one of the first specifications that should be handled is to plug the data referential services into the centralized IAM, which is the Office 365 organization. This means in particular that identification and authentication should be delegated to Microsoft Azure Active Directory. But since `DemoEditor` should also handle the identification of people outside its organization, namely the authors, who should have access to their book content on the company's IT, it is essential that the IAM supports multiple user lists or, as the IAM jargon says, directory federation. Azure B2C could be used, but the request from `DemoEditor` and the mere fact that an author has been added to their referential service, with a validated email address, should be enough for the author to validate their entry ticket to the company's system. In fact, this should even be part of the enrollment process in that an author validates their digital identification.

The company's board has also been advised by their technical consultant to beware of vendor lock-in, so the specification for IAM is to include an intermediate local IAM software that will then delegate to the different lists of users. As far as internal users are concerned, the IAM will delegate authentication to Office 365 and establish a single sign-on approach inside the internal IT. As for the authors, it is intended that the presence in the authors referential service will act as identification provided that there is a correspondent on the email address and that the user coming in provides the password generated during the author enrollment process. The technical consultant has informed `DemoEditor` that this will imply a custom development, but the company has agreed, as this is a key feature for them.

When the technical specifications have been done, Apache Keycloak has been elected as the centralized IAM company software, with a delegation to Azure Active Directory for the internal users, and a customized interface to the authors data referential service for the external users. This intermediate approach has convinced the deciders as it will make it possible, in the future, for authors to provide an external authentication such as Microsoft Live ID, Google ID, or the like as proof that the user has rights to the email address. They insisted, though, that their identification attributes should be then completed by those coming from the author's data referential service, as there is some important information needed for authorization (in particular, book IDs) that will never be outside of the editing company's control.

As additional security, the board of `DemoEditor` requested for there to be some kind of manual validation by the editor of an author in order for them to gain access to the information system. This way, in case of any doubt, the editor will be able to contact the author by phone to check whether they created the authorization request. In the future, the company could even become interested in a feature where they proactively create their authors' access, but, for now, it should remain an option in the enrollment process, with validation by the editor.

Authorizations

Of course, we will base our authorization management on OPA and the exercise that has been carried out in *Chapter 13*. Since we have detailed almost completely the policies of rights management during this exercise, there is no need to be more precise in the specification of the authorization management system itself. We will only explain a bit more about how it will be attached to the data referential service.

OPA will be used when access to the books referential service is performed. Dealing with all the performance problems that a simplistic approach would cause with high volumes would be too long for the exercise, so we will simply consider that each book queried from the books data referential service generates a call to the OPA server in order to see whether this access is granted or not. This call will be synchronous and considered safe as both services will be placed in the same network, without direct exposition to the outside. Though security would be tightened in a production-ready example, for example, by establishing a certificate-based trust relationship between the books referential service and the authorization service, this will not be demonstrated here (we have done enough work on functionally aligned design).

Technical specifications

Now that the functional specifications have been drawn (in summary, of course, as this is for exemplification, and not designing a real information system), a few additional details on the technical choices are provided, in order to ease comprehension of the coming chapters.

Explaining how the interface will run the main use scenarios

Though we have shown BPMN engines in *Chapter 12*, we have also explained how complex they are and that they should not be used for simple purposes. In our sample applications, the most complex process will be the enrollment of an author for a new book, and this is not such a complicated and rapidly evolving set of operations that it would justify a BPMN engine or any orchestration means. We will simply use the graphical interfaces and passing of entities between applications through API to realize this operation. The corresponding actions could be described as follows:

1. An editor has an idea for a book and will check the authors referential service to see whether anyone already has the skills to write it. The editor may add a new person to the authors referential service even if they have not written any books for the company; they are then called "prospect authors" and their status attribute reflects this.

2. The editor will also create a book in the books referential service, with a `project` status, and a function that will allow the automatic sending of invites to some authors. This interface will use the `authors` API to retrieve the list, possibly filtering on skills. When applied, the function will use the notification service to send the invitation to authors/prospective authors.

3. At the same time, a "middle office" service will be activated and will propose a vote for a request to the destination of the preceding notifications. The potential author will receive a link that will lead them to a page where they can select a vote for "interested" or another for "not interested." Of course, the request will contain the text provided by the editor to explain what they are looking for in this book project.

4. Each vote received by the middle office service will trigger an action. In the case of a refusal, information will simply be stored for the editor to check out if they want. In case of approval, an email will be sent to the editor (which means the editor requesting the vote should be kept

in the request sent to the middle office service, even if it is not visible to the potential authors), and this will be considered as the end of this part of the process, as the rest is going to be a different one, between the editor and the chosen author alone.

By the way, this validation function (or, as we say in functional jargon, the "middle office") could also be used when the editors validate the request for an external account to be associated with an author in the company's information system. The associated editor would receive a request for validation, verify by other channels that it is justified, and vote for or against this request, leading to different outcomes in the rest of the process.

Integration of the applications in the system

The technical specifications should end up with a schema of the whole information system to show how all modules will interact together, but it happens to be something we had already designed in *Chapter 7*. As a consequence, we only reproduce this container view:

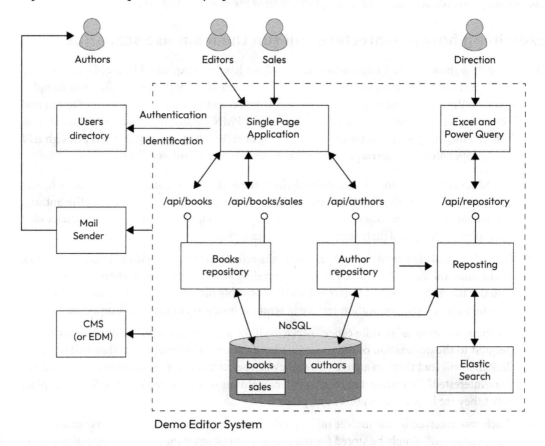

Figure 14.20: C4 container-level diagram

We have mostly talked about the books and authors repositories in this chapter because they are the most important ones in the scenarios of `DemoEditor`. Of course, in a real situation, all services should undergo a review at least as deep as the one that has been carried out for the main data referential services, and those should be analyzed in full, with detailed OpenAPI and discussions about attribute names and so on.

Technical implementation of the applications

The same remarks stand for the technical implementation of each of the applications. In *Chapter 7*, when showing the C4 approach, we had already used our sample application and shown how the books repository would be architected:

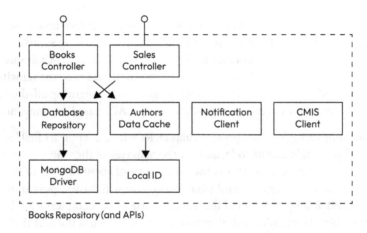

Figure 14.21: C4 Component-level diagram

We will not go further in this chapter, not only because there is no need to be exhaustive for the purpose of the book but also because, in reality, an agile approach would commend neither to go into too many details right from the beginning. The proof is in the pudding and only the first drafts of the code will guide us in the design questions and next version implementation choices.

Again, for simplicity reasons, we will, of course, not show every detail of every application in this book but only where it is useful for learning the method. The source code of the whole application is made available for you if you want to have a complete view of what the system would be like, and run it by yourself on a Docker-compatible infrastructure (the only prerequisite for running this sample).

Summary

In this chapter, we have shown how the decomposition of the functional requirements in a business- or IT-aligned information system works, and, in particular, how the use of norms and standards can have an impact on the design of entities, pushing some modifications and choices that might be a bit different to what was initially thought of.

I hopefully also convinced you of the power of the four-layer map approach, which generally helps us understand much better a complex system. Of course, our sample here is not complicated at all and I could have simply explained it by text. I trust that using the CIGREF approach has given you a better understanding of what we are going to implement and, in particular, of the challenges and important design decisions.

Now that all concepts have been explained and the functional specifications drawn, we are eventually ready to get our hands dirty in the technical bits and (finally) start building something. In the next chapter, we are going to install the IAM application (Apache Keycloak), the EDM software (Nuxeo, with a CMIS connector), and some other utilities. Once this is ready, we will code the remaining modules in the following chapter, which will be arguably much longer than configuring off-the-shelf modules but should also bring more knowledge on design choices in an API-based, business-aligned IT.

If you only remember one thing from the present chapter, let it be the "contract first" approach: note that we have not shown a single line of code, and yet the behavior of the whole applications system should normally be quite clear to you. This is the true power of spending lots of time on the API interfaces: it allows us to better comprehend what the system will do and pose important design questions, and, in the end, it simplifies a lot of the work because all the complexity has been addressed before we solidify anything through technical implementation (which is the part that makes it hard to change once started). The key ability is here: delay the coding as much as possible, and only start coding when the business is absolutely crystal clear: this way, your system will be flexible and all evolutions will be easier.

15
Plugging Standard External Modules

We are in the last third of the book, and you should find some changes in the present chapter since this will be the first one when we turn theory and architecture into concrete, ready-to-run code. This action of putting into practice concepts of architecture and information system decoupling is extremely important for many reasons. Firstly, we all know the devil lies in details and great theories are not worth much as long as they have not been battle-tested with some actual working code. Secondly, some of the concepts that have been presented may not be perfectly clear yet, and seeing them from the implementation point of view may help you understand them better. Lastly, the proposed methodology is still – at least in my personal experience – not seen as standard good practice for engineering (otherwise we would not need so much in the way of consulting services for ready-to-collapse information systems). Showing that this method is not purely theoretical but has practical applications and has been positively enacted in some companies has played a key role in the method's spread.

An important approach in modern information systems is to not reinvent the wheel, and the "not invented here" syndrome is one of the most dangerous for sound IT operations. Much of the sample `DemoEditor` will be made of existing modules, taken off the shelf, and simply parameterized to our needs and integrated into the chosen infrastructure. Of course, there will be some dedicated, custom-developed modules (mostly because there are no adequate master data management generic solutions available right now), but that will be for the next chapter. In the present chapter, we will see how the standardized modules will be put in place, using existing implementation. We will see how modules can be evaluated and how their interaction with the whole system, not just the target application, will impact the choice made between the different implementations, and how a standards-based approach makes it possible to change an implementation late in the project, even while the system is working.

Five functions will be considered for externalization in the `DemoEditor` information system, namely **Identification and Authorization Management (IAM)**, NoSQL persistence, **Electronic Document Management (EDM)**, **Message-Oriented Middleware (MOM)**, and monitoring. They will be implemented with Apache Keycloak, MongoDB, Nuxeo, RabbitMQ, and the Grafana stack, respectively.

In this chapter, we will cover the following topics:

- Externalizing IAM with Keycloak
- Externalizing persistence with MongoDB
- Externalizing document management
- Externalizing queue management with RabbitMQ

Externalizing IAM with Keycloak

The very first thing to externalize is in a modern information system IAM. Except for extremely rare cases, nobody should ever implement their own authentication mechanism as the security stakes are so high and the entry barriers are way too high for most organizations. When a simple password was enough, almost any application had its own authentication mechanism, sometimes with clear-text passwords in the database. That time is now over, and the smallest company needs at least a multi-factor authentication, which necessitates the ability to send SMS, for example, for the confirmation of authentication. This is of course way out of reach for small companies such as `DemoEditor`, and even large companies prefer to turn to security specialists for their needs.

In addition, since `DemoEditor`, like a lot of small and medium-sized enterprises, has an Office 365 offer that comes with user management via Azure AD, it would be a waste not to use it for its internal user authentication (and part of the identification at least). Of course, there is still the need for authenticating external users as well (authors might need access to the information system in order to deposit their manuscripts, for example). This could be done using Azure B2C, which is a service provided by Microsoft, but the managers would rather keep this information on who writes for them secret and have asked the technical consultant to find a way to embed the information in the IAM service.

Apache Keycloak has been chosen due to its good reputation, its open-source nature, the free-of-license-costs model, and its capacity to easily accommodate such situations as described previously, with multiple user directories. The following sections will describe the different aspects of Keycloak and how we are going to install and customize it to our needs.

Realms

Apache Keycloak happens to be a software-multitenant application. This means that, instead of installing many instances of Apache Keycloak on different servers (or with different exposition ports), you can install the server once and make it work through software-configured tenants as though there were several instances of it. This saves resources while keeping a complete separation of tenants from the data point of view.

In Keycloak jargon, these tenants are called **realms**. In our case, one could wonder whether we need two tenants: one for internal users and the other one for external users, but this is not what tenants are made for. We need to use only one tenant because there is only one information system that both internal and external users will address. Inside the only realm, there will, however, be two different ways to access the system: using Office 365 authentication, and using standard username-and-password authentication for the authors.

One of the first things to do inside Keycloak is to create a `DemoEditor` realm. But first, we of course need to start Apache Keycloak. To do so, a simple Docker command is enough:

```
docker run -d -p 8080:8080 -e KEYCLOAK_ADMIN=forster -e KEYCLOAK_
ADMIN_PASSWORD=4AFXbm5vX7YFjNOrMYKK --name iam quay.io/keycloak/
keycloak:23.0.7 start-dev
```

A quick explanation of the options used is given here:

- The `-d` option runs the Keycloak container in the background and gives you back control of the command line, instead of blocking you on the console outputs.

- The `-p 8080:8080` option redirects port `8080` of Keycloak (its standard exposition port for the web GUI) to your port `8080`. If it is not available, please adjust the first value accordingly.

- The `-e` option allows you to send environment variables inside the Docker container. In this case, these variables are for the admin username and password. Please use a normal account name (`admin`, `administrator`, and even `batman` should be banned, as they help an attacker to find the high-privileged account and focus their hacking attention on them) and a robust, machine-generated password.

- The `--name` option enables naming the Docker container, which will make it easier to manipulate afterward.

- `quay.io/keycloak` is the name of the repository, and the rest of the text points to the image tag we want to use. Please update this when new versions come, as IAM may be the single most important module in terms of security and should always be up to date with security patches.

- Finally, `start-dev` enables the development mode in the Keycloak process, which makes it less secure and lowers performance but makes it easier to use for debugging.

> **Note**
>
> As Docker has become such a widely used solution, we assume that you already have it installed on your test infrastructure. If not, please refer to the online documentation on how to install it, and read a basic tutorial to understand the simple commands that will be used in this chapter and the following one.

By default, a single realm is created, and it is called `master`. As its name indicates, this is the tenant that will be used to manage all others. As an important software server, Keycloak must be secured from unauthorized access and thus needs an IAM service. Since this is what it is made for, of course, Keycloak acts as its own IAM, and this is what the `master` realm is for: it is going to manage access to the application itself, and in particular, the ability to manage other realms that will serve as IAM for other services. Initially, the `master` realm has only one user, which is the admin user that we have defined using the `KEYCLOAK_ADMIN` and `KEYCLOAK_ADMIN_PASSWORD` environment variables.

Running a web browser on `http://localhost:8080/admin`, we can connect using the values we have provided for these environment variables:

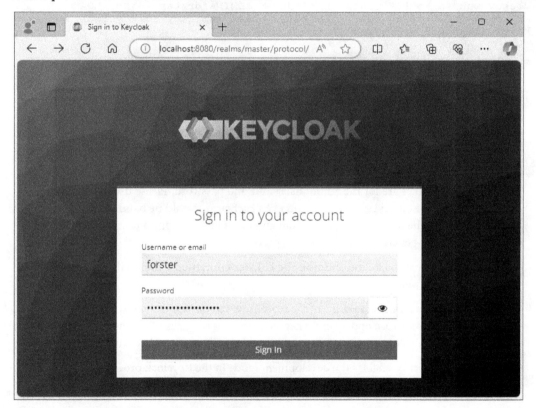

Figure 15.1 – Administrative access to Keycloak

Once inside the interface, the burger menu in the top left of the interface enables you to select realms, as well as to create new ones with the **Create realm** button:

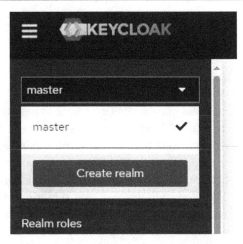

Figure 15.2 – The Create realm button in Keycloak

In our case, we are going to create a realm called `demoeditor` for our sample information system. We will not use a preconfigured JSON file for the settings, as most of the customization will be shown later in order to give you a better understanding of what Keycloak can do for us.

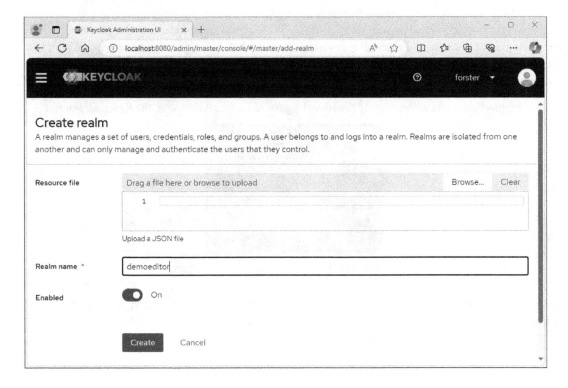

Figure 15.3 – Creating a demoeditor realm in Keycloak

The interface automatically places us in the just-created demoeditor realm, and the burger menu on the left allows us to manage and customize it, which is what we are going to do in the following sections.

Delegation to Office 365 for internal user SSO

As was already explained, the employees of DemoEditor want to be able to log in to the software services using the same login that they use for their Office 365 services, and if possible without having to type the password another time; they want to enjoy the same **Single Sign-On** (**SSO**) experience they do with Office and their connection to Azure AD on their workstation. This is why we are going to plug Keycloak into the DemoEditor Office 365 subscription. Actually, this will rather be my own Office 365 organization, since I did not want to pay for a new subscription as I already benefit from the one associated with my Microsoft Most Valuable Professional title. In a real production case, I would have bought the demoeditor.org domain and associated it with a dedicated subscription. But, as you will see, the associated domain for Azure AD (or rather, Entra ID, since this is the new name of the service) will be jpgouigouxhotmail.onmicrosoft.com.

Please note that this manipulation needs you to be an authorized administrator of an Office 365 organization, so there is a chance you will not be able to follow this in your standard environment. Also, depending on the security constraints in your organization, you may have to go through extended domain validation operations, which are too variable to be explained here. If you encounter any configuration problems, do not worry, as this will not prevent you from following the rest of the exercise: you will simply add internal users outside Azure AD, in the exact same way as we plan to do for external users, by using Keycloak's embedded users directory (see the next section, *Minimal implementation for external users*, for more details).

This configuration starts with selecting the **Identity providers** menu option in the Keycloak interface:

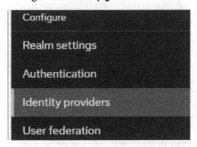

Figure 15.4 – Identity providers option in Keycloak

This enables us to plug Keycloak into external identity providers, whether they use the **OpenID Connect v1.0** or **SAML v2.0** protocol. Well-known providers are also listed:

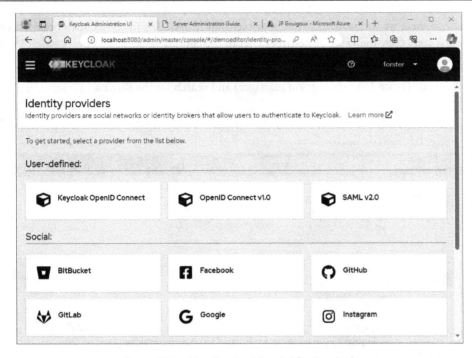

Figure 15.5 – Identity providers list for Keycloak

Going down the list, you will find **Microsoft**. Click on it to reach the following screen:

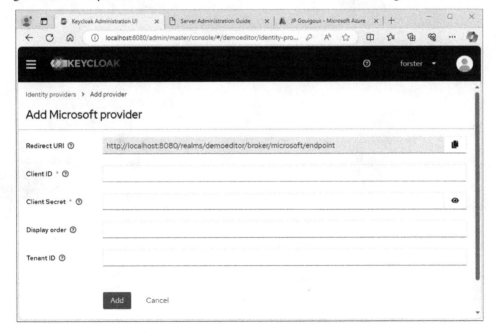

Figure 15.6 – Microsoft identity provider configuration screen

This is (partly) where we will realize the necessary settings to associate the Entra ID service with our IAM. As a first step, copy the value of **Redirect URI** in the clipboard.

We will now switch to the Azure portal for the intermediate part of the setup phase. Connect to your portal (or ask your Office 365 subscription manager) and search for the Entra ID service:

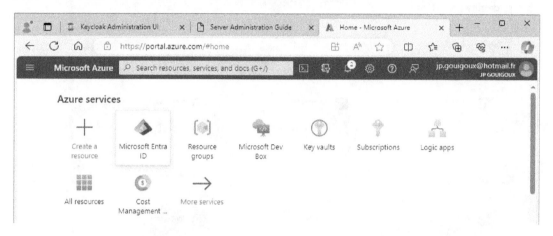

Figure 15.7 – Looking for Entra ID in Azure Services

When selecting this service, you will be brought to an interface with a left-side menu allowing you to create an app registration:

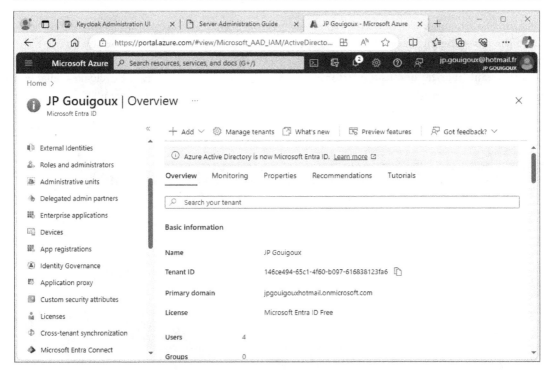

Figure 15.8 – Inside the Entra ID service

Once there, you can click on the + **New registration** menu at the top of the inner interface:

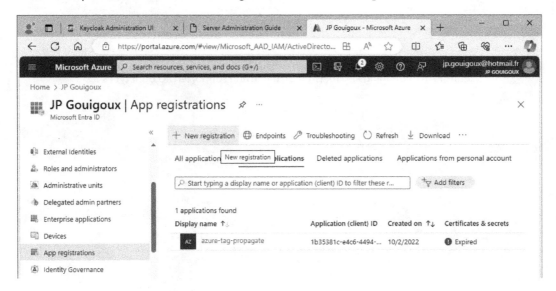

Figure 15.9 – New registration button

Giving an explicit name is a good idea to help easily retrieve application registrations when there are many. The following interface is where you specify it, as well as choose who will be able to use the application. Note that, depending on your settings, you may have to switch **Supported account types** to **Multitenant** (if an authentication error happens in the following, the logs of Keycloak, available through `docker logs iam`, will be of great help in debugging this kind of error):

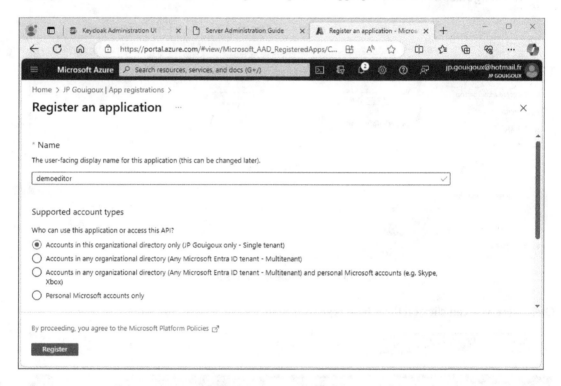

Figure 15.10 – Defining app registration name and supported account types

When you click on the **Register** button, the app registration is created and, moments later, you are brought to the second part of the interface:

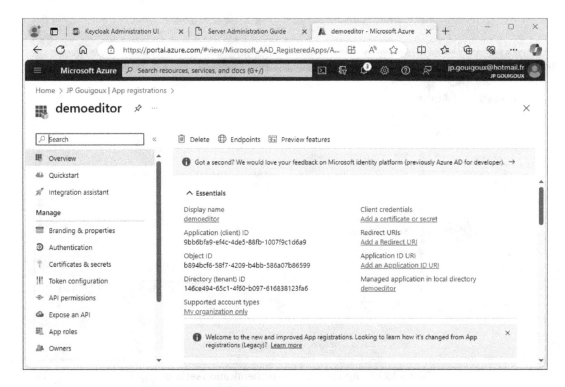

Figure 15.11 – Parameters for the app registration

Clicking on **Add a redirect URI** in the overview or the **Authentication** menu on the left brings you to the following form, where you will be able to click on + **Add a platform**:

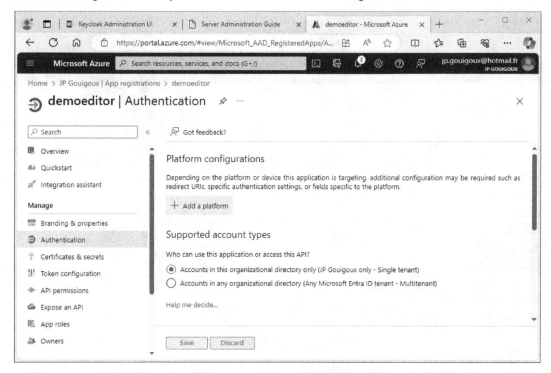

Figure 15.12 – App registration authentication settings

We will add a **Web** application:

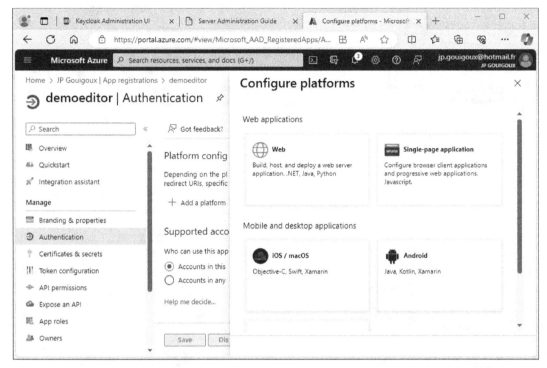

Figure 15.13 – Choosing the platform

The next screen is where we will paste the value of the Keycloak redirect URI (`http://localhost:8080/realms/demoeditor/broker/microsoft/endpoint`) that we had copied to the clipboard previously:

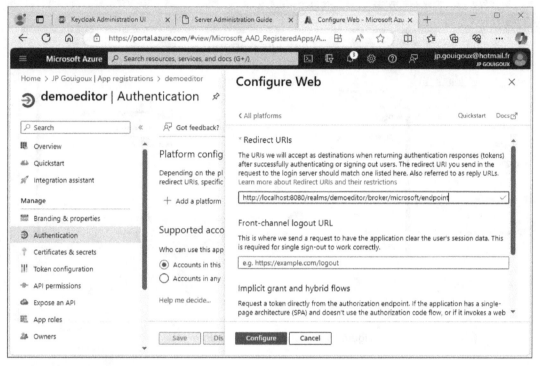

Figure 15.14 – Configuring the web platform

You can click on **Configure** to validate this entry. Now, we need to switch to the **Certificates & secrets** menu to establish a secure exchange channel between Keycloak and Entra ID. To do so, we create a new client secret using + **New client secret**, with an appropriate name:

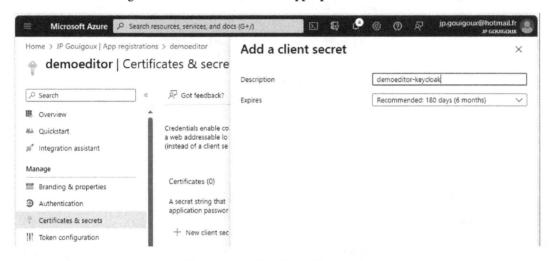

Figure 15.15 – Creating a client secret

Once created, we need to copy the value of the client secret:

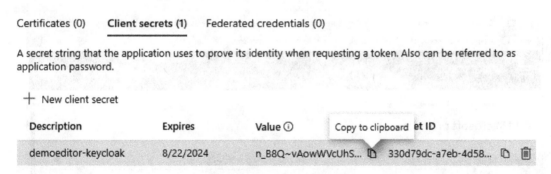

Figure 15.16 – Client secret value

This is what we will paste back into the Keycloak interface, and we will do the same copy-paste operation for the **Client ID** value retrieved in the **Overview** section of the app registration:

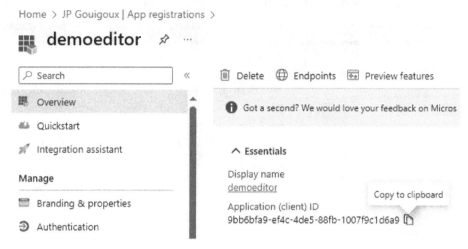

Figure 15.17 – Retrieving the client ID

The wizard in Keycloak should now have all the mandatory settings filled, and clicking on **Add** will normally establish the bridge between Keycloak and Office 365 authentication:

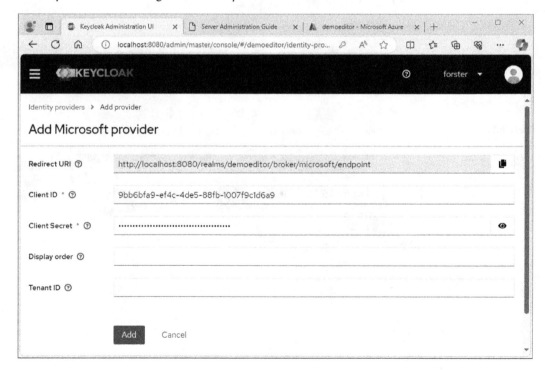

Figure 15.18 – Microsoft provider creation interface, ready to validate

If all goes well, you should receive the following success message:

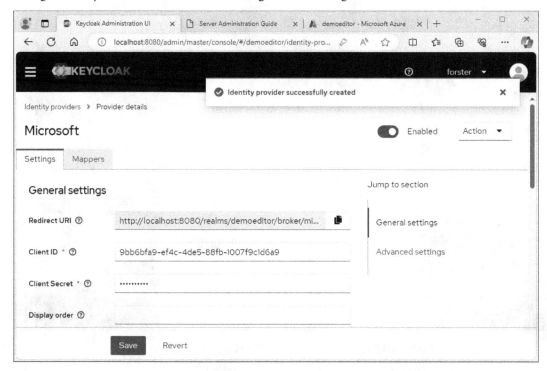

Figure 15.19 – Identity provider successfully created

In order to test that everything works as expected, we will connect on the account interface (where users can view their profile). The best way to avoid login confusion, since we are connected on the Keycloak GUI with our `forster` administrative user, is to open a new web browser interface in private mode and navigate to `http://localhost:8080/realms/demoeditor/account`. There, we can click on the **Sign In** button, which will bring us to the following interface:

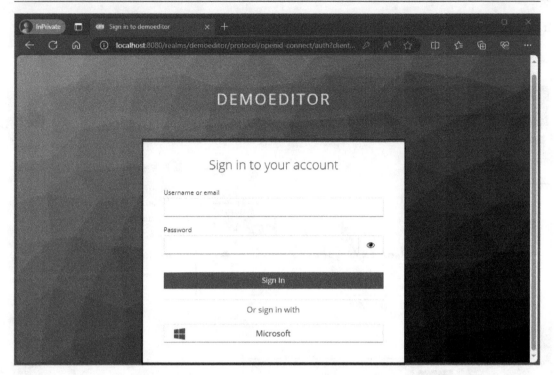

Figure 15.20 – Connection with the Microsoft provider

As expected, we now have the **Microsoft** button at the bottom of the interface, allowing us to log in to the demoeditor realm with a Microsoft account. For my example, I have created a new user in my Entra ID tenant called Test. I can click on the **Microsoft** provider in the screen shown previously, then input my test user email address in the screen that comes up (notice the domain is Microsoft-handled):

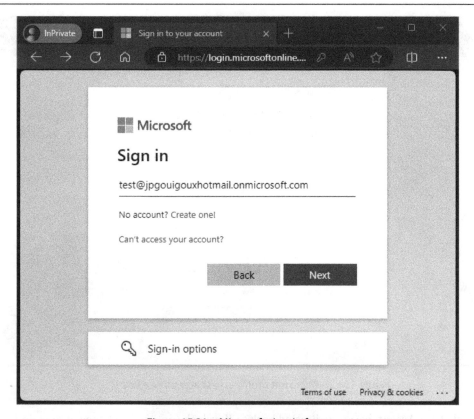

Figure 15.21 – Microsoft sign-in form

After that, we need to enter the password, and – the first time we connect – answer in the affirmative on the following screen, which warns that the application is not an official one from Microsoft and enables us to accept the reading of the user profile:

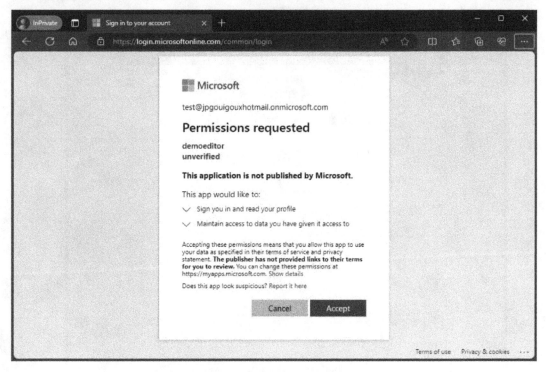

Figure 15.22 – Microsoft profile access acceptance screen

After we click on **Accept**, we are redirected back to Keycloak (notice the change of domain in the following screenshot). The login shown in the title bar on the left contains the name I associated with the `test` user and, if we click in the **Personal info** menu, we see the identity information of this user inside Keycloak (getting this data back to Keycloak was the reason that Microsoft asked for permission):

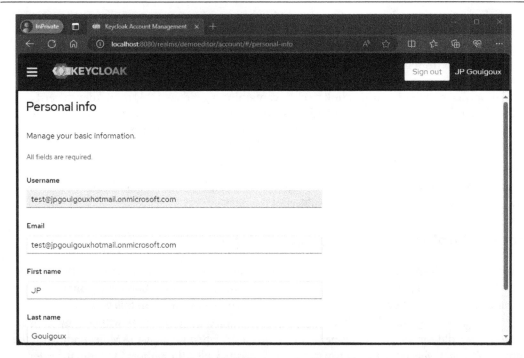

Figure 15.23 – Personal information for the Test user

If we go back to the Keycloak administration console, switch to the `demoeditor` realm, and look at the **Users** menu, we see clearly that the user information now is also local to our IAM tenant:

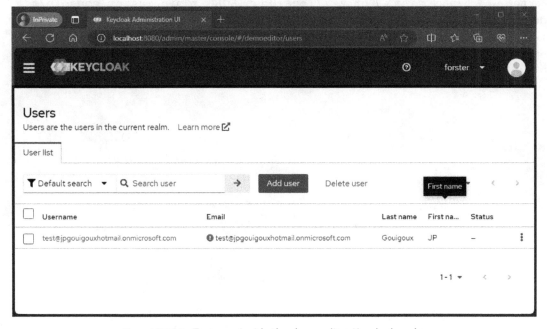

Figure 15.24 – Test user inside the demoeditor Keycloak realm

> **Note**
>
> The alert icon on the left of the email address is not a problem of identification but is only there as a sign from the directory that this address has not yet been validated by an exchange of mail with the owner (identity providers using external addresses generally send an email to the supposed owner of it, with a link to click containing a generated token in order to verify the person indeed has access to this address; since I have not done this verification yet, the Entra ID provider has sent an `email_validated` identification attribute with `false` as its value to Keycloak, in order to be transparent about this fact).

Keycloak is now ready to accommodate the signing-in of employees of `DemoEditor`. But we also need to let the authors in, and this is the subject of the next section.

Minimal implementation of external users

Authors are not really part of the `DemoEditor` company but they still need access to the information system (of course, it's a partial system, but this is the domain for authorization management and not authentication/identification). We thus need to create accounts for them. In this very first approach to the subject, we are going to simply create a few authors directly in Keycloak without any external backend, as we did for the employees before. In *Chapter 23*, when we adjust IAM depending on the feedback of the first version of the information system, we may plug the authors IAM into the external identity providers, provision them automatically using the Keycloak API, or use an extension of Keycloak to look for them in the authors data referential (the third one is the most interesting one, but a bit more complicated to put in place).

For now, we will simply go back to `http://localhost:8080/admin/master/console/#/demoeditor/users`, if needed, and use the GUI provided by the **Add user** button to create a few sample author users:

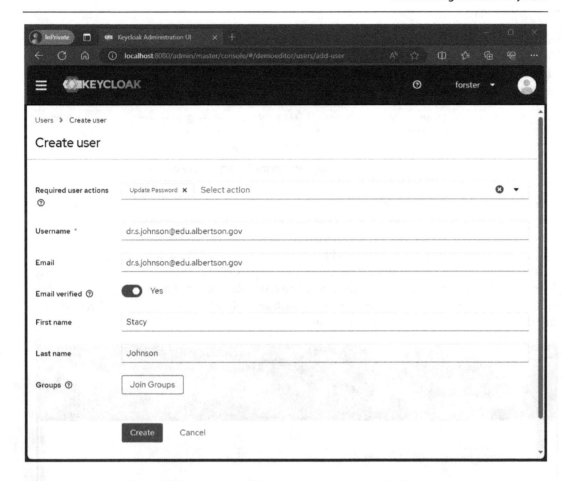

Figure 15.25 – Creating an internal user for an author in Keycloak

Notice that the choice in this case has been to pre-verify the email. Indeed, since this user is from the data we created using the existing authors directory (see the following figure for a refresh on the Excel workbook), we assume that the editors have exchanged enough with the author to treat the email as basically verified. If this is not the case for a given author (for example, one where the editor tells us they are quite new and have not been contacted yet), we could add to `Required user actions` an additional `Verify Email` together with `Update password`.

	A	B	C	D	E	F	G
1	Number	First name	Last name	Full name	Phone	Email	Status
2	0001	Stacy	Johnson	Dr Stacy Johnson	546 324 567	dr.s.johnson@edu.albertson.gov	
3	0002	John	Smith	Mr John Smith	+1 789 567 345	john.smith53@gmail.com	
4	0003	Joan	Bouchara	Mrs Joan Bouchara	+33 687608093	j.bouchara@free.fr	
5	0005	Sebastian	Oconnor	Sebastian O'Connor	0032 765 235 243		
6	0006	Ernest	Daren	Mr Ernest Darren	+31 797345293	ernest@daren.com	Blocked
7	0008	Deborah	Kingston	Ms Deborah Kingston		debbie.kingston@academics.net	
8	0009	Sebastian	O'Connor	Sebastian O'Connor	+32 765 235 243	s.oconnor@outlook.com	

Figure 15.26 – List of existing users in the legacy Excel workbook

Of course, in a real situation, we would not do this by hand but create a small application to inject this data, taking into account that blocked authors should not be added but included in security groups (we have not created any for now in Keycloak but this is an existing – and expected – feature), etc. But a huge advantage of using Keycloak in this kind of situation is that all the difficult, risky stuff (such as creating verification emails, validating reception, forcing password robustness, and so on) is already done by a solid, battle-hardened application. In our case, we will anticipate how we are going to migrate the authors data referential and show the result of such an application on the cleaned referential after duplicate data and errors have been sorted out:

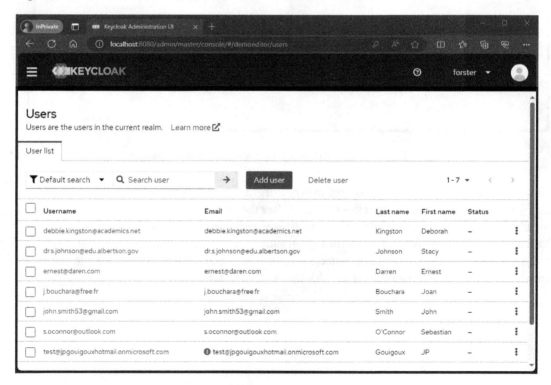

Figure 15.27 – List of users in the demoeditor realm

Once created in its simple form, the user can be modified and completed with many additional attributes, such as roles, groups, consent information, and customized attributes. We can even generate a – temporary or not – password for them in advance, requesting that they change it at first login:

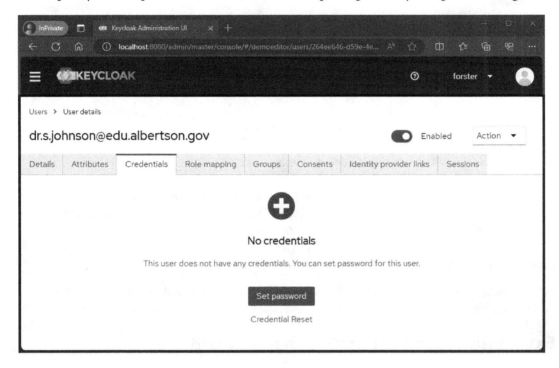

Figure 15.28 – Credentials tab for users in Keycloak

If writing space was not limited, we could also go into detail on how an **authors** group could be added, as well as how to use roles in order to link IAM to authorization, but this is not a book about IAM nor Keycloak, so we will only explain that these features exist and should, of course, be used in a real production system. In the coming chapter, we will show how roles can be used for authorization.

The list of Keycloak features is absolutely huge, but let me show you a final one that will be of great help when we test the authorization features of our business application. When in the user form, one can use the **Action** menu and the **Impersonate** command to act as the user:

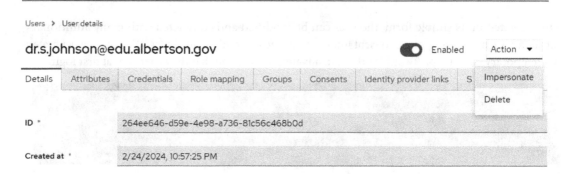

Figure 15.29 – Impersonate menu

This is particularly useful if you need to debug a behavior acting as the user. Of course, this operation is reserved for highly privileged users of Keycloak. In our case, since we have not customized further the `master` realm, the only user for now is the Keycloak administrator user (the one that we have called `forster` in the initialization customization). The great thing about this feature is that it is traced and thus cannot be seen as pure user hijacking. The logs of Keycloak show that the admin has done this impersonation action and we can then trace the actions executed in the name of the user back to the admin that has taken the identification of this user.

IAM is now ready for use in the information system of `DemoEditor`, and we can go to the next big dependency block we will need, namely a good database for our business entity referentials.

Externalizing persistence with MongoDB

Persistence will be dealt with in more depth in the next chapter, but still, you will see the choice of a NoSQL approach is absolutely clear in our situation, since the APIs are going to receive standardized, structured serializations of the business entities. I could use other document-based NoSQL databases, but MongoDB has become quite a de facto standard (the Mongo grammar is implemented by MongoDB but also other databases, such as Azure Cosmos DB), so this is what we are going to set up here.

Choosing between local and online storage

Once the type of database is chosen, the next decision is about its location. We could go all the way toward externalizing the database and choose a free tier in Atlas or Cosmos DB, which are basically MongoDB as a service, but for the sake of simplicity, we will simply start a Docker container with MongoDB inside. Again, once the prerequisite Docker is active, this is done with just one line:

```
docker run -d -p 27017:27017 --name db mongo:7.0.5
```

Again, the choice of version is important and the tag should be adjusted depending on the available version and your versioning policy. You might want to stick to the `latest` tag, but you should be careful about two things. First, once the `latest` version is in your local cache, you need to explicitly use a `docker pull` command to get the new `latest` version of the image, and this is a moving tag. Also, sticking to such a moving target might mean you can suffer stability issues. For a simple exercise, it does not really matter, but in production uses, you are better off validating a major version and following the patches. The preceding choice is done to be strict on the complete version definition, as I want to avoid as much as possible any side effects when installing the dependencies for the exercise.

A little digression about old CPUs

If, like me, you have an old CPU on your computer, you may see the following message when typing `docker logs db`:

```
WARNING: MongoDB 5.0+ requires a CPU with AVX support, and your
current system does not appear to have that!
  see https://jira.mongodb.org/browse/SERVER-54407
  see also https://www.mongodb.com/community/forums/t/mongodb-5-0-cpu-
intel-g4650-compatibility/116610/2
  see also https://github.com/docker-library/mongo/
issues/485#issuecomment-891991814
```

This can be handled in two ways. Of course, you could revert to an older database, by removing the previous container and creating a new one on MongoDB version 4, like this:

```
docker rm -fv db
docker run -d -p 27017:27017 --name db mongo:4.4
```

This option has been chosen for the code available on GitHub to make it as accessible and easy to use as possible. For my tests, I considered it was better to stay on the latest images, so I created a virtual machine in Azure, using a Clear Linux OS, logged in to it with PuTTY, and installed Docker on it (note that, sometimes, the update service might prevent you from doing so; if you prefer not to wait for it to give back the access, you can stop – temporarily, of course – this service by issuing the `sudo systemctl stop swupd-update.service` command):

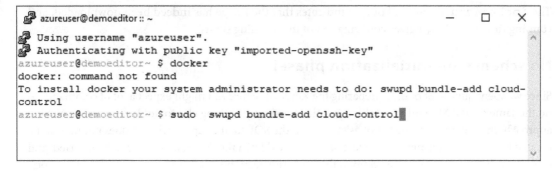

Figure 15.30 – Installing Docker on a Clear Linux OS machine

Once this installation is done, you might want to issue the `sudo usermod -aG docker azureuser` command in order to be able to use the `docker` command line without always prefixing with `sudo`. You will also need to instruct Clear Linux to automatically start the Docker service upon boot, with the `sudo systemctl enable docker.service`. A `sudo reboot` command to verify that all is fine, and you should be able to run `docker version` with success. The machine is now ready to run the MongoDB container with the initial command proposed previously, and the container should be running this time:

```
azureuser@demoeditor :: ~                                        —    □    ×

 Using username "azureuser".
 Authenticating with public key "imported-openssh-key"
azureuser@demoeditor~ $ docker run -d -p 27017:27017 --name db mongo:7.0.5
Unable to find image 'mongo:7.0.5' locally
7.0.5: Pulling from library/mongo
01007420e9b0: Pull complete
bc3bec6a423e: Pull complete
c5db81b694a8: Pull complete
427a1a117df0: Pull complete
dfb180c9e7b5: Pull complete
92e6f08e133c: Pull complete
374f042f3159: Pull complete
73549bb43006: Pull complete
Digest: sha256:5a54d0323fe207d15dc48773a7b9e7e519f83ad94a19c2ddac201d7aae109eb1
Status: Downloaded newer image for mongo:7.0.5
ddafa749902594fd2f93d964e34ec18a65c42c124eebdb847e8cff5c68f02a39
azureuser@demoeditor~ $ docker logs db
{"t":{"$date":"2024-02-25T10:10:13.296+00:00"},"s":"I",  "c":"CONTROL",  "id":23
285,   "ctx":"main","msg":"Automatically disabling TLS 1.0, to force-enable TLS
1.0 specify --sslDisabledProtocols 'none'"}
{"t":{"$date":"2024-02-25T10:10:13.297+00:00"},"s":"I",  "c":"NETWORK",  "id":49
15701, "ctx":"main","msg":"Initialized wire specification","attr":{"spec":{"inco
mingExternalClient":{"minWireVersion":0,"maxWireVersion":21},"incomingInternalCl
```

Figure 15.31 – MongoDB 7.0 running in the virtual machine in Azure

The `docker run` command output indicates that the image has indeed been downloaded, and running `docker logs` gives confirmation of this running status.

No schema, no initialization phase!

Since we spent quite a lot of time preparing the Keycloak setup, you might expect a lot of commands to do the same for the MongoDB database, but there is simply nothing else to do! Indeed, in the NoSQL approach, the schema for data is not as strict as in the SQL tabular approach and does not need to be created before inserting any data. In fact, it is deducted from the data that is going to be inserted, and even the collections will be created upon first use. This section is thus the shortest in the whole book due to the ease of use of MongoDB.

Securing through regular backups

If, like me, you use a local database, the only thing you might want to add to your setup is a backup system. Of course, this is not really useful for an exercise like DemoEditor but it is important to show what should be done in production, or at least a first glimpse of it, because the subject is of course more complicated than the commands we are going to show.

In order to back up a MongoDB database, there is a mongodump command that does the job of creating a single file corresponding to the content of the databases. Since this command is included in the docker image that we have been using to start a container, the easiest thing to do is to execute it directly from the running container. A first attempt shows us the command is indeed there and displays the expected use of it:

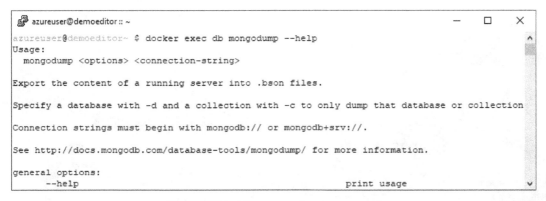

Figure 15.32 – The mongodump command

Once you have read the option, you can issue a command like docker exec db mongodump --gzip --archive=/tmp/dump.bson.gz to get a compressed archive into the file you specified, or even add the -d option to select the database to be exported. In our case, we have not created the databases for the authors and books (they will of course be separated, in order to keep low coupling), so we simply export the whole local server content. Once this is done, the docker cp db:/tmp/dump.bson.gz ~azureuser/mongodump.gz command line will copy the file from the container to the virtual machine it is based on.

We also could have set up a volume so that the export sends the file directly to the underlying virtual machine, and many other options are available. We are not going to dig further here, because this is not our main subject, but the file should be sent to another place (local backups are not very useful, since they allow going back in time on working machines but are lost if the whole VM is destroyed), a cron job should be setup for daily backups, incremental options should be studied, restoration procedures should be added, and so on.

Optionally installing a web console

Even if we do not strictly require it, a GUI is quite useful to get a better understanding of what is going on inside the MongoDB server. You can either install Docker Compass (`https://www.mongodb.com/try/download/compass`) if you have a local database or use a container with a web interface, based on the `mongo-express` image. In all cases, remember that, if you have a cloud-based machine, you are exposed to any attacker on the web, so never expose your MongoDB `27017` port directly on the web: your data will be encrypted with a ransom request within a few days at most, as robots crawl the web for such invitations to crime!

In the case of Compass, you get a Windows/Linux application that helps you connect and manage your servers, send data requests, and perform such operations:

Figure 15.33 – Compass GUI

In the case of the web GUI, it is time to switch our one-shot runnings of Docker command into a Docker Compose approach, with the two services inside, in order to avoid network access problems. You can then stop the existing container with `docker rm -fv db` (note that the `-f` option forces stopping the running container and the `-v` option removes the volume with the data that has not been linked) and add the following code to the `docker-compose.yml` file:

```yaml
version: '3.0'
services:
  mongo:
    image: mongo:7.0.5
  express:
    image: mongo-express:1.0.2
    ports:
      - 80:8081
    environment:
      - ME_CONFIG_BASICAUTH_USERNAME=forster
      - ME_CONFIG_BASICAUTH_PASSWORD=gOU89tRF05
```

Once this is ready, you can set up the global application with `docker compose up -d`, which should bring up a display output similar to the following:

```
azureuser@demoeditor~ $ docker-compose up -d
Creating network "azureuser_default" with the default driver
Creating azureuser_mongo_1    ... done
Creating azureuser_express_1 ... done
```

> **Note**
>
> Depending on your setup, you might use the `docker compose` commands or the `docker-compose` separate application. In the second case on a Clear Linux OS, you will have to install it using `swupd bundle-add docker-compose`.

You can then connect to the IP of the virtual machine you have set up in the cloud (beware that port 80 should be open to have access), using the user and password defined in the preceding YAML file:

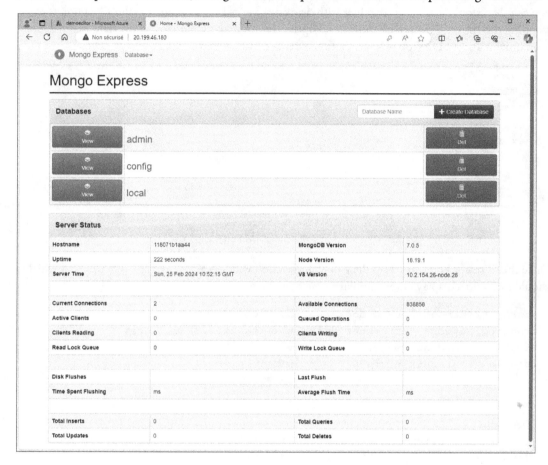

Figure 15.34 – Mongo Express web console

This gives you access to the database standard operations, such as Compass, which will prove useful to debug our manipulations in the next chapter. The section about the persistence is now finished and we can proceed to the next block of dependency.

Externalizing document management

After pure data, binary documents are the second most important objects to keep safe in an information system, and they should be managed in their own application (please never serialize documents into a database: that might be the worst thing you can do as an architect/developer). Such applications are called electronic content management applications. Sometimes, content management systems are also used, but this naming often is associated with web content such as articles and pages. EDMs are usually used for handling PDF, Word, and Excel files, bitmaps associated with your business, and so on. They include metadata management, versioning, the capacity to virtually classify and reorder documents, and many other value-added, document-specific operations. In the following sections, we will explain our choice of an EDM implementation and show how we install it.

Choosing the right standard, not the right implementation

After all our talk about standards and the usefulness of decoupling, it should be crystal clear that we are going to rely on a standard protocol and not on a given implementation of an EDM system. Luckily, there is the CMIS 1.1 protocol, which is perfect for document manipulation. Even if CMIS 1.1 is not used in every EDM system, there is no competing standard so it is an easy choice.

Instead of explaining the CMIS standard in words, it is much easier to make it clear by showing a few examples of API calls corresponding to the most used operations. Let's walk through a few EDM operations we will use in the context of `DemoEditor`:

1. Before anything else, we should retrieve the root folder of the default repository of the EDM system.

2. Then, we should create a folder dedicated to the documents of the company; then we can create some sub-folders for contracts, manuscripts, and so on (we will only show the first example).

3. After that, we will create a document inside the `contracts` folder and upload some PDF content to it.

4. Finally, we can query the EDM system to retrieve the document and download the associated PDF.

We could describe these commands in pure HTTP content or in Postman, but to make it more complete, it would be interesting to start an implementation of the protocol and test the commands at the same time. This way, we will be able to show the answers coming from the server as well as the requests. In these simple calls, and the rest of the book, we will concentrate on very easy operations such as querying documents, creating them, and uploading/downloading content. As a consequence, the 1.0 version of the CMIS standard will be enough. In real-world situations, one would of course take advantage of all the additional functions of version 1.1 (in particular, metadata management with schemas).

Starting up Alfresco

Very sadly, EDM editors do not play nice with standards in general and lots of them provide only proprietary APIs, supposedly because this helps performance and allows access to advanced features, but in fact mostly in order to maintain customers via dependency on their solution through vendor lock-in. Indeed, an EDM system is an important module in an information system, and once thousands of documents are inside it, with dedicated workflows, attributes, and interop, it gets very difficult to get out of this dependency. Standards help to do so, which is in the interest of the community but not of these vendors.

With the buying of Alfresco and Nuxeo by Hyland, Docker images for the latter being now private and the source code for the CMIS connector of the former not delivered by default in the community edition, the CMIS landscape looks grim. But that does not mean that the standard itself is not useful: one always needs an intermediate pivotal format to reduce coupling between the modules of one's information system. If the endpoints support directly the standard, that's great, because it accelerates the plugging. But even if they don't, there is still great value in using the standard as the intermediate, neutral language of interop between the client modules and the server module.

This being said, there still are some easily available CMIS 1.1-compatible applications, and Alfresco in its community version is one of these. Nuxeo in its legacy version 10.10 is also a good choice, but I decided to use Alfresco in this example because a complete deployment exists in the form of a Bitnami-maintained virtual image available on Azure, which makes it easier to quickly set up and exploit (we could also install with Docker a full-blown Alfresco platform as explained in `https://docs.alfresco.com/content-services/community/install/containers/docker-compose/`, but the Bitnami image is much lighter):

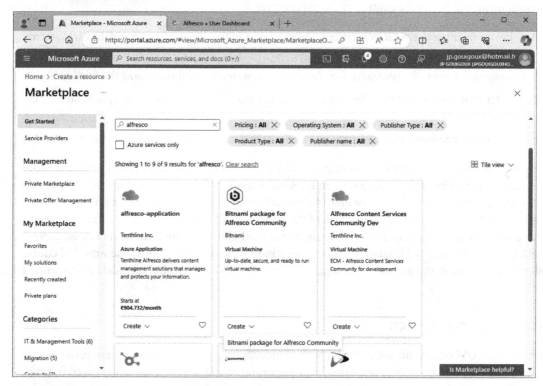

Figure 15.35 – Bitnami image on Azure containing Alfresco

Once set up, you need to connect to the machine to retrieve the credentials in the `/home/bitnami/ bitnami_credentials` file:

Figure 15.36 – SSH access to the Alfresco Bitnami virtual machine

Important Note

There's an alternative way to better way to include Alfresco support. You can read all about it on the book's GitHub repository: `https://github.com/PacktPublishing/ Enterprise-Architecture-with-.NET`.

This can then be used to connect to the web console of the EDM system:

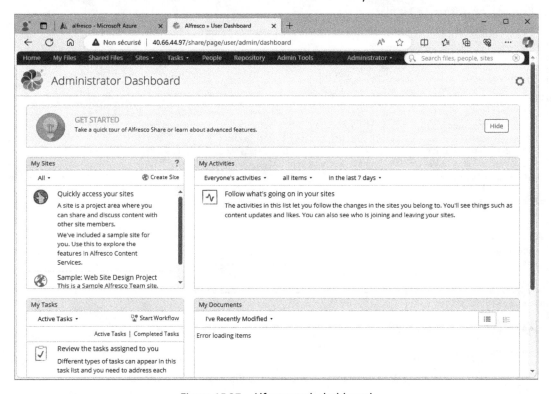

Figure 15.37 – Alfresco web dashboard

Everything should now be done for the setup phase and you can start testing whether Alfresco works as needed for your use, which will mostly be a machine-to-machine interaction.

Alternative for Alfresco setup

If you prefer to execute Alfresco locally because you do not have enough free Azure credits to use the method shown previously, or simply because you want to have a fully standalone application, there is also the possibility to use one of the Docker Compose files provided at `https://github.com/Alfresco/acs-deployment/tree/master/docker-compose`.

A reduced version of the community version is shown here in order to quickly explain the different parts (you will need to use the online version, as I omitted some necessary configuration information here):

```
version: "2"
services:
  alfresco:
    image: docker.io/alfresco/alfresco-content-repository-
community:23.2.1
  transform-core-aio:
```

```
  image: alfresco/alfresco-transform-core-aio:5.1.0
  ports:
    - "8090:8090"
share:
  image: docker.io/alfresco/alfresco-share:23.2.1
postgres:
  image: postgres:14.4
  environment:
    - POSTGRES_PASSWORD=alfresco
    - POSTGRES_USER=alfresco
    - POSTGRES_DB=alfresco
  ports:
    - "5432:5432"
solr6:
  image: docker.io/alfresco/alfresco-search-services:2.0.9.1
  ports:
    - "8083:8983" # Browser port
activemq:
  image: alfresco/alfresco-activemq:5.18-jre17-rockylinux8
content-app:
  image: alfresco/alfresco-content-app:4.4.1
  environment:
    APP_BASE_SHARE_URL: "http://localhost:8080/aca/#/preview/s"
control-center:
  image: quay.io/alfresco/alfresco-control-center:8.4.1
proxy:
  image: alfresco/alfresco-acs-nginx:3.4.2
  ports:
    - "8080:8080"
```

As you can see, Alfresco is not a standalone application and needs a few dependencies to work, among which are a PostgreSQL database, an ActiveMQ MOM, a proxy, and a SOLR index engine. This means that, if you go this way, you can do either of the following:

- Merge the full proposed content with our existing Docker Compose

- Keep it as a separate system and configure some HTTP links by exposing the right ports on your local host machine

I personally recommend the first option because it is more coherent with the approach we have taken so far, where all local services are run in the same set of networks. One important thing, though, is that the example provided uses port 8080 heavily, which is the same as the one we used for Keycloak. It will be way easier to adjust the port of the latter service than the one for the former, so please choose another port for IAM and adjust your samples accordingly.

Finally, a word of advice on the resources required for such a local installation: although the Docker Compose file instigates resource quotas for optimal use of memory, you will find that the whole platform takes a lot of RAM and CPU to run. Use this option only if you have a recent (or old but powerful) machine. Otherwise, I recommend sticking to an online service or simply not putting an EDM system in place: the rest of the exercises do not depend on it, and the coupling should anyway be kept low, as in a real platform where a service should never fail because a dependency does not respond.

Testing calls to the EDM module

Getting back to our list of commands explained previously, we are going to start by reading the information of the default repository of the EDM system. To do so, we simply call a `GET` request on the base URL `http://[your host]/alfresco/api/-default-/public/cmis/versions/1.1/browser/root/`, as shown here:

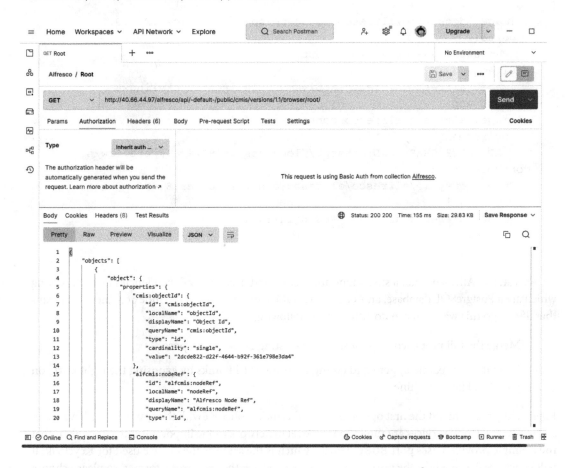

Figure 15.38 – Calling the CMIS endpoint root to get repository information

Note that you need to set up the **Authorization** tab as follows in the collection since we have decided to inherit the authentication of all commands from the collection that gathers them. In my case, the collection is called CMIS and I entered the same credentials as provided by the setup and used for the GUI access:

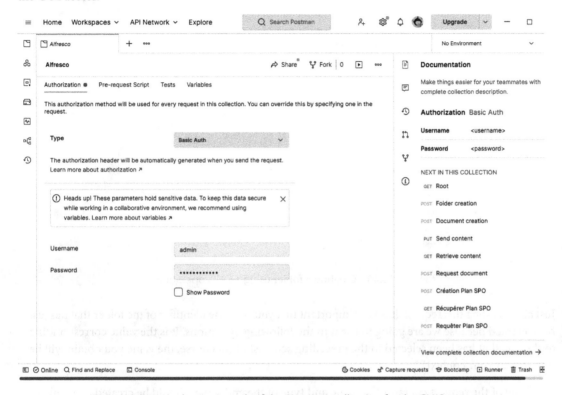

Figure 15.39 – Authorization tab of the Postman collection for the CMIS tests

In order to access the content of the root default repository and create a folder for DemoEditor documents, we need to adjust the URL as follows and, this time, use a POST verb and body content such as { "name": "DemoEditor", "nodeType": "cm:folder" }:

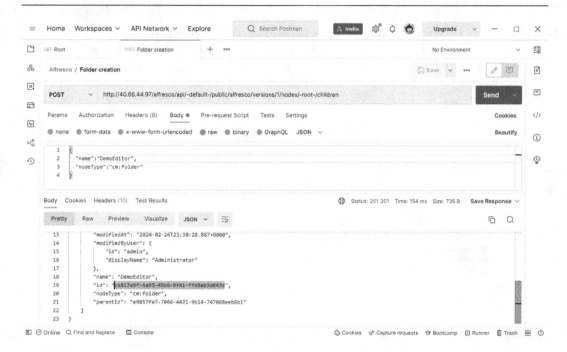

Figure 15.40 – Creating a folder using a CMIS operation

Just after sending this request, it is very important that you copy the identifier of the folder that has just been created because we are going to use it in the following operations. It is the value corresponding to the one that has been selected in the preceding screenshot (of course, the value you obtain will be different in your context).

The body of the request contains the name and type of the entity that should be created as a child of the root folder of the Alfresco repository. One very simple way to check that the folder has indeed been created (but we will show a much more sophisticated one later) is to simply try to send the request again, which will result in an error; the response will state that this would lead to a duplicate folder:

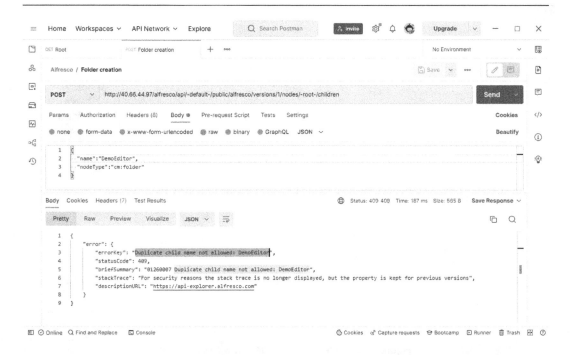

Figure 15.41 – Error message when attempting to create a duplicate folder

Creating a document is done by placing ourselves in the folder that we have created by taking its ID (it is the ID I asked you to copy and store for future use, which was selected in the response of the previous operation) and adding it to the URL. If you forgot to keep this value or lost it, you can find it again by calling `http://localhost:8080/alfresco/api/-default-/public/alfresco/versions/1/nodes/-root-/children` and finding which identifier corresponds to the folder you created (you either named it yourself or used the recommended value of `DemoEditor`). The body again carries a name and a type of entity, but this time the latter is `cm:content` instead of `cm:folder`. Note that the name is arbitrary and does not necessarily correspond to the name of the file we are going to provide as the document content. By the way, you can see that, for now, the content is empty (`sizeInBytes` is 0) even if Alfresco has deduced the type of the document and its MIME type by the extension provided in the name:

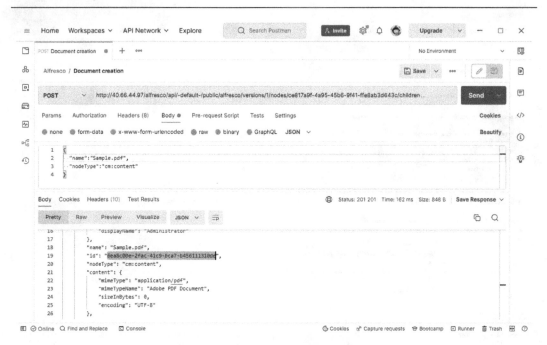

Figure 15.42 – Creating a document with CMIS

Again, we need to copy the identifier of the file because we are going to, as a final operation for this sample of CMIS commands, send binary content in association with the document in the EDM system. This is done by calling `http://[Your host]/alfresco/api/-default-/public/alfresco/versions/1/nodes/[Identifier of the document, copied above]/content`, this time with a `PUT` verb, as this is an idempotent operation:

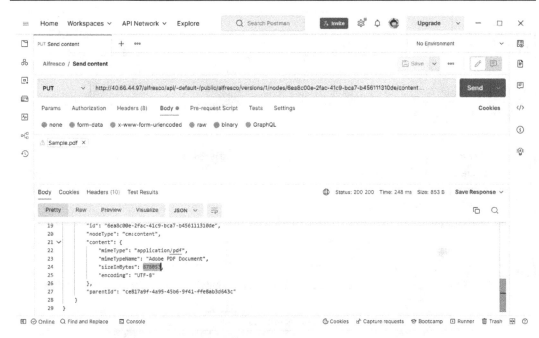

Figure 15.43 – Sending PDF binary content to the document store

The size shows that the file provided in the Postman interface has indeed reached the server, but we can also check this by calling the same URL with a GET verb. Instead of clicking on the **Send** button, simply click on the triangle on the right and select **Send and Download** in order to adjust the behavior of Postman to present you with a binary file instead of some strange text in the **Response** tab.

Using the Apache Chemistry console as an Alfresco CMIS client

Another way to check everything worked fine is to use the reference client from Apache, which is called Workbench and is available in the **Chemistry** project. In its current version (although this is apparently not going to get upgraded in the future), the archive is available at `https://archive.apache.org/dist/chemistry/opencmis/1.1.0/`.

Please note that a Java runtime is necessary for Chemistry to work. If you do not have one set up on your testing machine, please go to `https://jdk.java.net/22/` to download the OpenJDK in version 22, decompress it, and add the resulting `bin` folder to your `PATH` environment variable.

Once the previously cited archive is uncompressed, one can run the Workbench client by double-clicking on `workbench.bat`. The following interface then appears, asking for the credentials and URL of the CMIS server. In our case, we will connect using the browser URL, namely `http://[Your host]/alfresco/api/-default-/public/cmis/versions/1.1/browser/`:

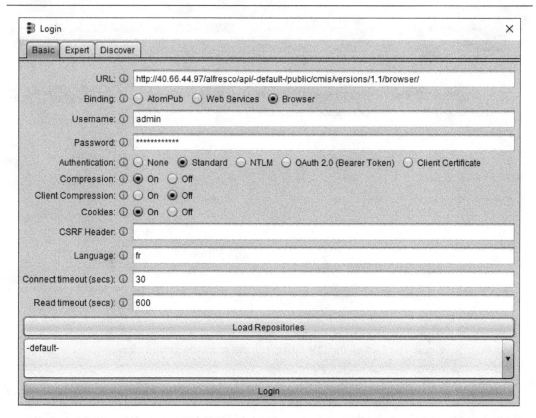

Figure 15.44 – Connecting to Alfresco using the Apache Chemistry Workbench

Once the -default- repository is selected and the rest of the login procedure is finished, the interface shows the DemoEditor folder that we have created:

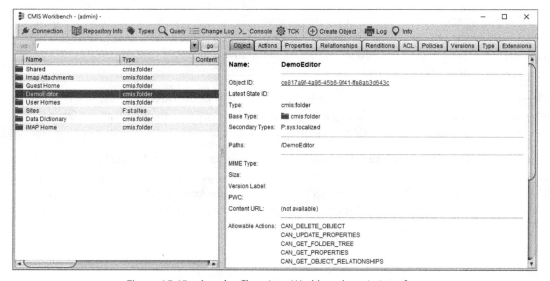

Figure 15.45 – Apache Chemistry Workbench main interface

We can of course check the document added, even if the interface is rudimentary (it is based on pure CMIS):

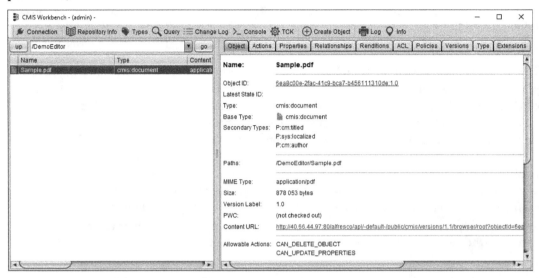

Figure 14.46 – The document created in CMIS seen in Workbench

Of course, the web interface is much nicer and will be the one used for all standard users (but the GUI is not part of the CMIS standard, which only deals with requests and responses):

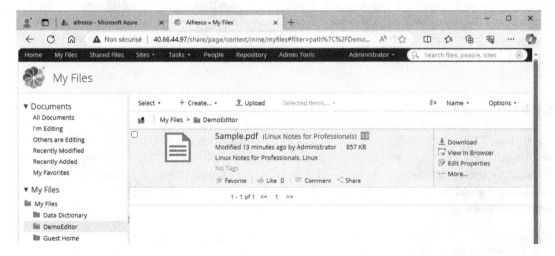

Figure 15.47 – The document created in CMIS seen in the Alfresco proprietary web dashboard

As this is not a book about CMIS and EDM, I will stop here with the explanations on this part of the information system and let you read some more specialized material on the subject (such as `http://docs.oasis-open.org/cmis/CMIS/v1.1/CMIS-v1.1.html` or the resources still available at `https://chemistry.apache.org/`) and get some knowledge about metadata management, schemas, secondary object types, and other advanced features.

One last thing: if you have to work in the field and appreciate the value of a good standard, you may be a bit desperate about the grim spectacle that is the EDM market and in particular its lack of care for CMIS. Even Apache Chemistry, the reference open-source implementation, is now an "attic" project. Is all lost? No, because being archived also means that the standard is complete and mature and does not need any newer versions because it supports all the functional needs. And if lots of EDM editors do not play the game of standards and orient their software into vendor lock-in methods, this simply means the CMIS standard is even more important if you know to stay compatible with many implementations (and there are many reasons to do so, because the ecosystem is quite large). Even if you have to create the connectors by yourself, the benefits of a single pivotal format are such in terms of responsibility separations, ease of evolution, and decoupling that it is still worth it. This is even more true when you have lots of applications/services that use the EDM function.

Externalizing queue management with RabbitMQ

Though we have extensively explained how bad the uneducated use of a **Message-Oriented Middleware** (**MOM**) can become, and shown that webhooks can already go a long way to solving choreography and asynchronous message management, we will still demonstrate in the sample information system a few exchanges of data using such a middleware. This will be done not only because they are an important part of the ecosystem but also because there remain some cases where their use is plainly justified. The problem is not in the MOM itself but rather in the fact that, when installed, they are used inadvertently and for all messages, even those for which they are clearly overkill and an unnecessary burden.

Starting up the MOM

In our case, I have decided to use RabbitMQ, as this is an excellent open-source MOM with lots of documentation and great ease of use for beginners. The fact that the .NET SDK is very good was also a reason for this choice. To start a Docker implementation of RabbitMQ, you can run the following command (or add the equivalent service into the Docker Compose YAML file; by now, you should know how to do so):

```
docker run -d --hostname my-rabbit -p 15672:15672 -p 5672:5672 --name
mom rabbitmq:3.13-management
```

The credentials by default are guest/guest.

Figure 15.48 – Connecting to the RabbitMQ console

This gives you access to a web interface where you can control your MOM, seeing the content of the queues as well as administrative features such as adding users or adjusting authorization and access:

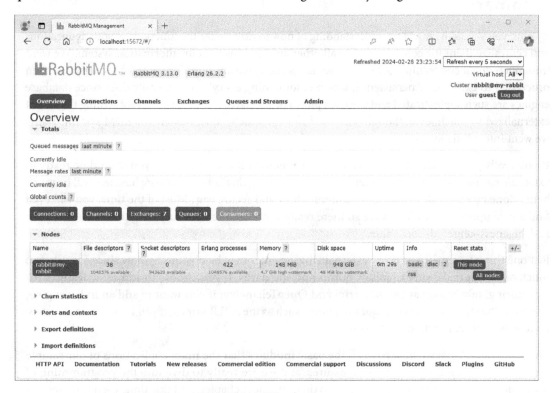

Figure 15.49 – RabbitMQ console overview page

Defining which streams should be handled, and operating a MOM

For now, there is no use in talking much more about RabbitMQ. Only when we show its use in practice will we come back to the subject, in particular, to do the following:

- Show how code can be used to send and receive messages

- See how the messages are treated

- Make a decision as to which streams of data should pass through the MOM

We will also quickly see some day-to-day operations with an MOM, such as operations with RabbitMQ, and how this type of middleware can be used to improve robustness and introduce some degrees of decoupling in the information system, which is our objective in this book. MOMs can be complex to manage, so we will try to stick to the simplest uses of them. If you need to use any of them in production, be aware that this is something that needs some customization and maintenance/management time. And, very importantly, never put any business functionality into an MOM or any other middleware: this type of logic should always be in the service.

Summary

By now, you should have a clear understanding of how to externalize the functions of your applications and centralize them for common use by all other applications in your information system. We went into much detail concerning IAM, because this is such an important function for security. We then proceeded to externalize persistence, which is something everybody generally does since database engines are such sophisticated and complex pieces of software. We went back to something that is not externalized so much, even though doing so brings so much value, namely the EDM system. There, we went into a little less detail.

If you really got this chapter and see how much benefit it can bring you, you should be ready to externalize some more. Do so! Remember that everything that is not your core business will be much better done by external specialists, in a more robust and secure way. Most of the time, you will even find a nice open-source alternative so there really is no need to code anything that is outside your real, business-centered, core value.

Just one thing to remember when externalizing: again, it may be obvious by now since I insisted so much on it, but make sure to follow norms and standards. If you choose to add some features for monitoring, take a look at OpenMetrics and OpenTelemetry; if you want to add an index engine, check out the standard ways of requesting them, such as the SOLR syntax; if you want to add statistics, review the SDMX standard; and so on.

This chapter is done: we now have all the basic modules that the main applications of our sample system will rely on for the common features, and we are ready to dive into the programming of those modules. This is what we are going to do in the next chapter, and this time, even if we rely on frameworks and existing dependencies, we will have to code the behavior, because this is where most of our processes and business rules will be.

16

Creating a Write-Only Data Referential Service

Now that the overall skeleton of the information system and, in particular, its common services are in place, we are going to dig a bit deeper into the technology and write some code for a service that we are not going to be able to get off the shelf, namely the **Master Data Management system**. In *Chapter 5*, we introduced the concept of a data referential service. In *Chapter 10*, we saw what master data management is about and what kinds of data referential services exist. In this chapter, we will transform the concepts into .NET code. It should be clear from what was presented before that we will use some kind of clean architecture, with the business domain being the most important part of the code. We will, of course, also separate the responsibilities of persisting, querying, and indexing the data, and use the right tool for each use case.

The chapter starts with an analysis of SQL's limitations and the reason why NoSQL can be the solution for many standard business IT systems. It also proposes a technical approach that makes it possible to have full traceability as well as the evolution of the data model over time. This is why a "write-only" data referential service is mentioned in the title. It does not mean that one is able only to send data and not read it: it would be a pointless application, in this case. It simply means that this data referential service will be a "know-it-all" application, never forgetting any information from the past. Deleting data will be possible in some kind of way, but I am not going to anticipate any more about the rest of the chapter here.

Along the way, we will progressively add some code to a standard web API project in .NET in order to implement a data referential service for books.

We'll cover the following topics in this chapter:

- Preparing a skeleton for an ASP.NET service
- Exposing an API
- Adding persistence that records every change
- Handling modifications

Technical warmup

Preparing the technical environment is necessary for those of you who want to reproduce the programming example. This section gives a few pointers on how to start, with a setup that is as simple as possible, and of course, using only free software in order for all of us to be able to follow along on our computers without any limitations due to costs or license restrictions.

The only two necessary tools are the following:

- .NET SDK 8.0 (released on November 14, 2023, which was right in time for the writing of this technical part of the book)

- Visual Studio Code (in the case of the author, version 1.84.2, but that does not really change anything) – the plugins for VS Code are optional; none are necessary to execute the following exercise

As the installation of these two applications does not need any particular competence, we will not provide any details here and will consider these prerequisites as installed and functional.

Getting connected to the Git repository

After launching Visual Studio Code, start by cloning the repository containing all the code associated with the book (https://github.com/PacktPublishing/.NET-Architecture-for-Enterprise-Applications) or create your own repository and clone it locally if you prefer to follow the instructions that follow and re-create the code yourself. The second option is, of course, recommended from a pedagogical point of view. The option to clone a Git repository is normally available on the welcome page of VS Code:

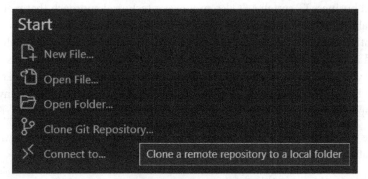

Figure 16.1 – Cloning option on the welcome page

If it is not, one can use the shortcut *Ctrl + Shift + P* to display the command menu, and then enter `git clone` to find the right command. *Enter* allows you to validate this choice of command and then provide the URL of the repository. The last thing you need to provide is the local directory in which the code will be cloned, and after a few moments, the following dialog box will be displayed, stating that the local Git clone is ready and you can edit it with VS Code:

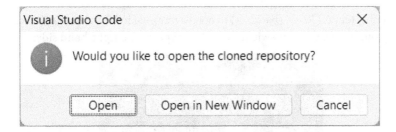

Figure 16.2 – Proposing opening the cloned repository

The following allows you to open the code in restricted mode or with full editing features, depending on your level of trust:

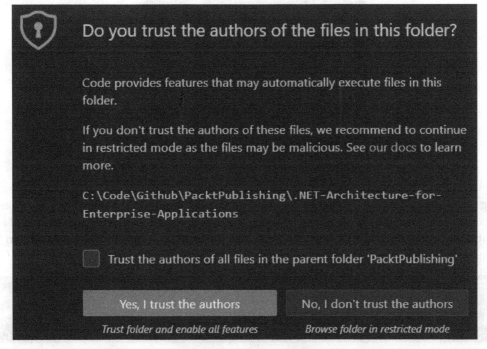

Figure 16.3 – Trusting the code

Creating the project skeleton

If you have decided to simply clone the existing code, the following commands should not be executed, as you already have the resulting code available. Otherwise, the following sections give you a step-by-step list of actions to reproduce the sample code provided online.

First of all, use the shortcut *CTRL+* ` (backtick) to open a terminal in Visual Studio Code. If the default shell does not fit your needs, you can switch to another type on the right-hand side:

Figure 16.4 – Choosing a different shell in Visual Studio Code

Typing `dotnet --version` will help you verify that .NET is correctly installed (you may have to disconnect and reconnect to a session in order for the command to appear in the PATH and be accessible without specifying the installation directory), and also to verify that the right version is being used in case you have several versions of the .NET SDK installed:

Figure 16.5 - Checking the .NET version

In this chapter, we will implement the `books` API. The command to start such a project is the following: `dotnet new webapi -o books`.

We will not go into much detail about the use of the `dotnet` command because this book assumes that you already have basic knowledge of .NET, but in order to help complete beginners without going into too much depth, here is the succession of commands one would run before entering the preceding command:

- `dotnet --help` to know which commands are available

- `dotnet new --help` to know which types of projects can be created with the new subcommand

- `dotnet new webapi --help` to know which options can be adjusted on a `webapi` project

Once the project is created, you will find the most interesting bit lies in `Program.cs`, namely the sample implementation of an API GET verb:

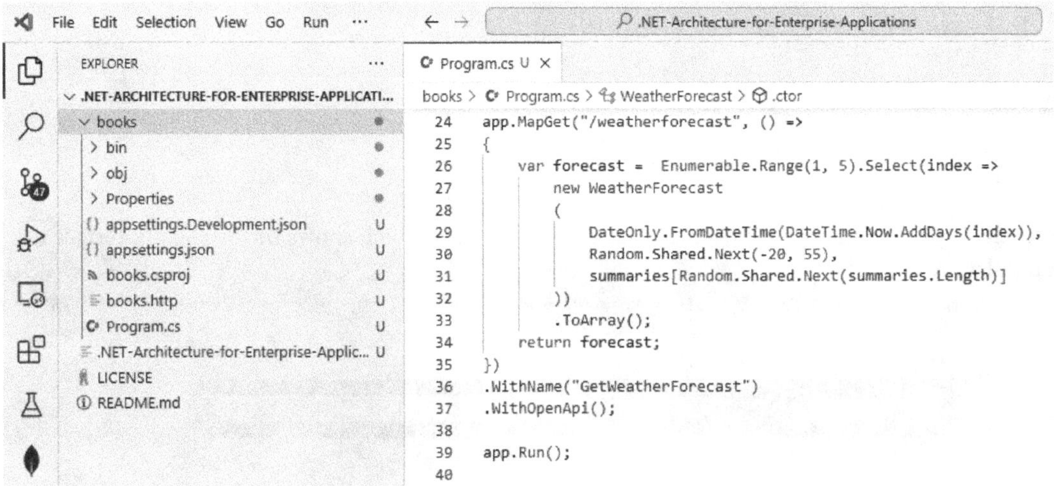

Figure 16.6 – Sample code

How to run the example

The last thing we will demonstrate as generic guidance on C# and Visual Studio Code is how to run the project. To do so, place the cursor in the terminal and enter the following command: `dotnet watch`. This will run the project (as the alternative `dotnet run` command would) and instruct Visual Studio Code to spy on project modifications to take them into account and refresh the project, which avoids unnecessary stops and reruns.

The following output should then appear:

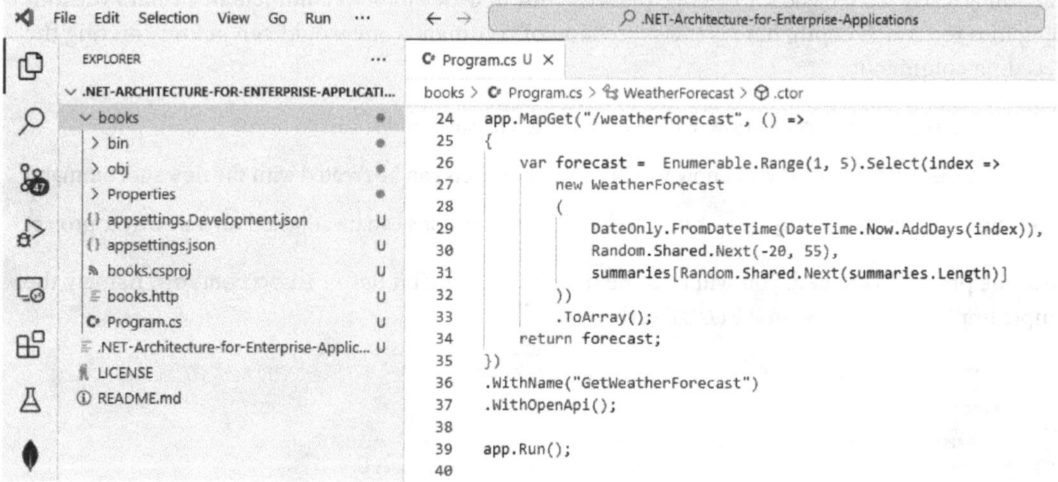

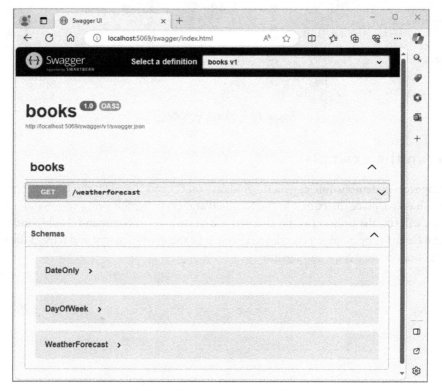

File Edit Selection View Go Run ···
.NET-Architecture-for-Enterprise-Applications
EXPLORER
.NET-ARCHITECTURE-FOR-ENTERPRISE-APPLICATI...
books
bin
obj
Properties
appsettings.Development.json
appsettings.json
books.csproj
books.http
Program.cs
.NET-Architecture-for-Enterprise-Applic...
LICENSE
README.md
Program.cs U
books > Program.cs > WeatherForecast > .ctor
app.MapGet("/weatherforecast", () =>
{
var forecast = Enumerable.Range(1, 5).Select(index =>
new WeatherForecast
(
DateOnly.FromDateTime(DateTime.Now.AddDays(index)),
Random.Shared.Next(-20, 55),
summaries[Random.Shared.Next(summaries.Length)]
))
.ToArray();
return forecast;
})
.WithName("GetWeatherForecast")
.WithOpenApi();
app.Run();
Figure 16.7 – Running in watch mode

Normally, a page should also appear in your default browser, which shows the Swagger interface to manipulate APIs:

Swagger UI
localhost:5069/swagger/index.html
Swagger
SMARTBEAR
Select a definition
books v1
books
1.0
OAS3
http://localhost:5069/swagger/v1/swagger.json
books
GET
/weatherforecast
Schemas
DateOnly
DayOfWeek
WeatherForecast
Figure 16.8 – Swagger interface

Expanding the command, click on **Try it out**, and then **Execute** runs a call to the API command that has been generated in the code skeleton, and the interface shows the results of this call. As can be seen, the call went OK (return code **200**) and some data was sent back – in our example, some weather-related information:

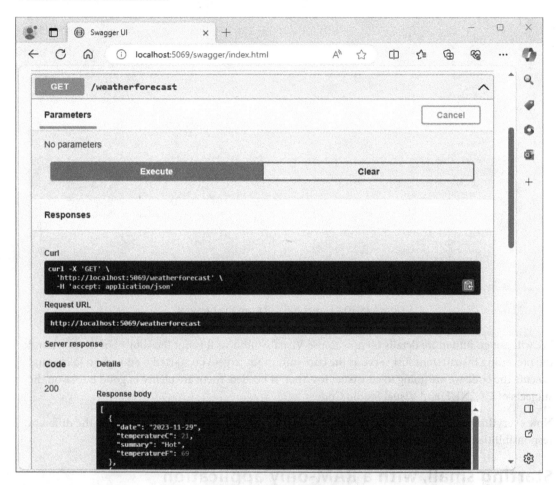

Figure 16.9 – Sample call

A direct call to the API can, of course, be run from the browser since it is a simple GET command:

Figure 16.10 – Direct API call from the browser

We will not go into more details on how to use Visual Studio, as it is not the subject of the book and the preceding instructions just serve as the bare minimum to get you started – enough to follow and execute the code we are going to construct together. If needed, there are plenty of good books on the subjects of C#, .NET, and Visual Studio Code.

Now everything is in place, we are going to add to the skeleton code provided by .NET the different responsibilities of a data referential service that we talked about in the previous chapter.

Starting small, with a RAM-only application

We will do this step by step, in the manner of an agile project, and also in order to try and make sure the project works after each addition. Though this is not a guarantee that all responsibilities will remain independent (coupling is very easy to add, and can be hard to remove), it is the minimum that should be done: programming two responsibilities at the same time would really be begging for trouble...

One of the consequences of this decision is that the very first version of the data referential service for books that we are going to program will not bear any persistence! All data will be hardcoded and only available in memory, which means we will not have to deal with a database for now. This is quite a radical decision, but completely adequate with an agile approach where taking the opinion of the customer (in our case, the `DemoEditor` director and representatives) on the way the data is designed and made available by the data referential service is our first concern, and does not need true, physically stored data to be used.

Representing the data

As I just said, the very first responsibility of our project will be to correctly represent the data. In *Chapter 9*, we spent quite some time using DDD to help create a functionally correct, technically independent entity definition. Here, we will translate this as simply as possible, by creating records, just as proposed in the sample code. The `WeatherForecast` record is thus replaced by the following:

```
record Book(string BusinessId, string? ISBN, int? NumberOfPages,
DateOnly? PublishDate, EditingPetal? EditingProps, SalesPetal
SalesProps)
{}

record EditingPetal(int? NumberOfChapters, string? Status)
{}

record SalesPetal(decimal? Price, decimal? Weight)
{}
```

For the sake of simplicity, we will not use all the attributes that have been mentioned previously. The same goes for the petals, with only two of them coded. As you might guess, `Price` and `Weight` could have been much more carefully designed, taking into account units. Even the title of the book is not included, you might notice. But all this is done on purpose as a real business model of a book could become quite complex, and we really do not want to make it difficult to follow the process of writing the code.

The same thing can be said for technical refinement and some of you might ask why a class is not used instead of a record for the books. This is indeed what is quickly going to happen when we need to add business rules and other behaviors. But for this very first step, we do not need it yet so we apply, as much as possible, the **YAGNI** (short for, **You Ain't Gonna Need It**) rule. If we need it in the future (and we most certainly will), there will not be any difficulty in making the switch to a full-fledged class definition. Similarly, we will not talk about business classes yet, plain old C# objects, data transfer objects, or any such accidental complexity.

Another remark one could make when studying the proposed record definitions is about the very liberal use of question marks, specifying the nullable character of most attributes. That may sound weird, particularly with all the efforts made in the latest versions of C#/.NET to get rid, as much as possible, of `null` values and all their associated quirks and bugs. But, there is a huge difference between an attribute of a class made for algorithmic use, where nullability should be removed as much as possible, and an attribute of a class made for business-aligned, functional data representation. As explained, technical considerations must be removed from business and never interfere with the correct representation of a functional model. In business terms, there are actually very few properties that are indeed absolutely mandatory. Sure, they can become mandatory due to business rules in certain contexts, statuses, or levels in the object life cycle, but being **always** mandatory is something that can only be said about IDs. Do you think `ISBN` should not be null? Incorrect – an ISBN is not allocated to a book until it is officially registered with international or local authorities. Do you think the number of pages is always known? When the status of a book is still `Idea`, there is no information on this. And we'd better separate the `0` value, in order to state that, when the status is `Writing`, there have been no pages produced yet.

In order to get a working example, the rest of the code will be changed as follows:

```
var statuses = new[]
{
    "Idea", "AuthorChosen", "ContentDefined", "Writing", "Editing",
"ReadyToPrint", "Available", "RetiredFromSales", "Archived"
};

app.MapGet("/books", () =>
{
    var books = Enumerable.Range(1, 3).Select(index =>
        new Book
        (
            index.ToString(),
            "978-2-409-03806-" + index.ToString(),
            Random.Shared.Next(400, 850),
            DateOnly.FromDateTime(DateTime.Now.AddDays(index)),
            new EditingPetal
            (
                Random.Shared.Next(5, 30),
                statuses[Random.Shared.Next(statuses.Length)]
            ),
            new SalesPetal
```

```
                (
                new decimal(Random.Shared.Next(1000, 10000) / 100),
                new decimal(Random.Shared.Next(200, 3500)))
            )
        ))
        .ToArray();
    return books;
})
.WithName("GetBooks")
.WithOpenApi();
```

You will notice that the values for `Status` are provided as strings. Translation could possibly be an issue, but it is in fact not, as these values are simply readable, fixed code, and the translation may be left for later manipulation by the client software. An argument may be that, since they are values, this should not be a `string` but rather an `enum`. But the real question to ask ourselves is how helpful this is in the end. A development-oriented person will think of the ease of manipulation (avoiding possible typing mistakes, or at least making them obvious at compile time). But if we think in business-first terms, changing these `string` values into an `enum` means that adding a new value to the list will force you to build and deploy a new version of the application. This is quite a high functional toll to simplify the life of one developer, isn't it? This is why status values will remain a `string`. And this time, this is not while waiting for a later, more advanced, version of the code; it is a **design choice** to operate with string values, in order to ease evolution. All negative aspects can be taken care of in the code, typically by unifying these values and forcing developers to manipulate as many values as from this array, as we did above. If the "magical string" remains in one place, there will be almost no problems. If one has to spread string values in different parts of the code, issues remain limited as long as we keep the contract-first approach and the contract that rules them all remains the OpenAPI JSON contract. By the way, JSON Schema has its own way of specifying an `enum`, so this might be a good way to go, depending – again – on the business orientation.

Running this code will show the following results, where the request has been correctly executed and the response, this time, is data that is related to our business domain of book editions, with some structured JSON representing books in an array:

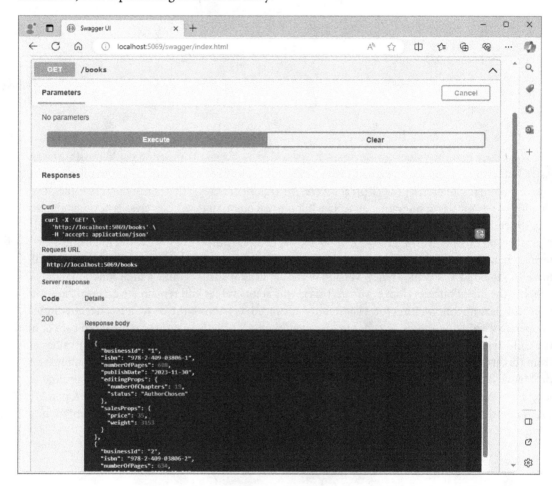

Figure 16.11 – Example result with generated books

This is the most basic working example, but it works and it takes on the most important responsibility of a data referential service, which is correctly representing the data. As an important note, we are using the .NET feature proposed in the code template to automatically generate the OpenAPI contract from the code. **This is of course not compatible with the contract-first principle** but we have kept it for now – again, to gradually show the improvement, and also because it allows for easy testing of the application. Also, even though we have not formally put it into an OpenAPI file, the contract was indeed designed before the code, because the record lines were written to represent the desired schema, and the different methods that we will support were precisely designed in the previous chapter. We repeat them here for reference:

- GET on /books
- GET on /books/{id}
- GET on /books/{id}?valuedate={date}
- GET on /books/{id}/history
- POST on /books
- PUT on /books/{id}
- DELETE on /books/{id}
- PATCH on /books/{id}
- PATCH on /books/{id}?valuedate={date}

Reading a single piece of data

Fulfilling the responsibility to correctly represent the data has forced us to partly implement another responsibility of a data referential service – namely, listing the entities, which is part of the global responsibility of data manipulation, sometimes called **CRUD** – short for **Create, Read, Update, and Delete**. We will carry on with this responsibility and add the capacity to read a single instance by its identifier. To do so, we start by externalizing the definition of the list of books (hardcoded for now) from the first method we implemented and put it in the general code. Then, we add a second API method – this time with a parameter. Here's the result (we of course only show relevant extracts of code in order to focus on the important bits):

```
var statuses = new[]
{
    "Idea", "AuthorChosen", "ContentDefined", "Writing", "Editing",
"ReadyToPrint", "Available", "RetiredFromSales", "Archived"
};
```

```
var books = Enumerable.Range(1, 3).Select(index =>
    new Book
    (
        index.ToString(),
        "978-2-409-03806-" + index.ToString(),
        Random.Shared.Next(400, 850),
        DateOnly.FromDateTime(DateTime.Now.AddDays(index)),
        new EditingPetal
        (
            Random.Shared.Next(5, 30),
            statuses[Random.Shared.Next(statuses.Length)]
        ),
        new SalesPetal
        (
            new decimal(Random.Shared.Next(1000, 10000) / 100),
            new decimal(Random.Shared.Next(200, 3500))
        )
    ))
    .ToArray();

app.MapGet("/books", () =>
{
    return books;
})
.WithName("GetBooks")
.WithOpenApi();

app.MapGet("/books/{id}", (string id) => {
    return books.FirstOrDefault(book => book!.BusinessId == id, null);
})
.WithName("GetBook")
.WithOpenApi();
```

When saving the file, since we used `dotnet watch` to run the project, Visual Studio Code may tell you that the modification is so important that it cannot hot-reload the application, and propose automatically rebuilding it in these cases. I recommend doing so in order to make this exercise easier. If all goes well, the content of the browser should update and the second functionality can be tested as shown above, by providing an existing index:

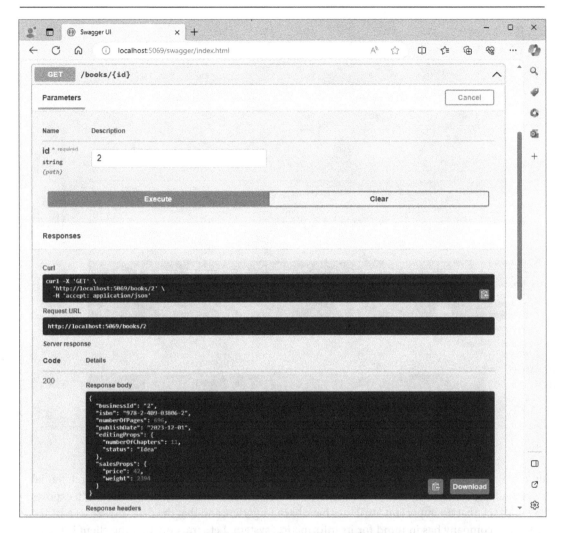

Figure 16.12 – Result – getting one book

Testing is always a good thing and we could try a non-existent index to verify that the resulting response will be `null` as stated in the code. For now, this is done manually, but testing is such an important function that there is a section dedicated to automating this later. But in the first steps of the exercise, we start by testing it ourselves. The result should be something like the following:

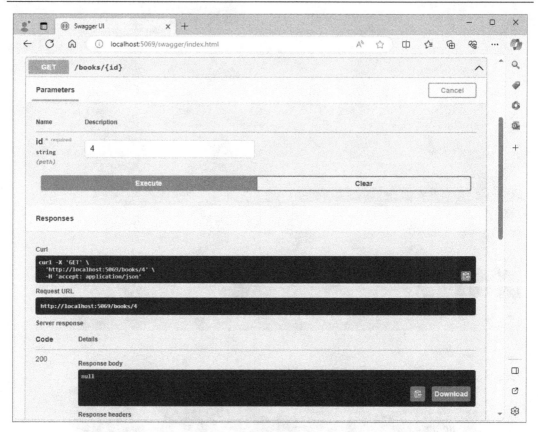

Figure 16.13 – Result with null

This very first – arguably small but usable for verification purposes – version of the books referential service could now be demonstrated to the DemoEditor board in order to check that the way it exposes data is correct, that no important attributes have been forgotten, and that the data "flower" is adapted to what the company has in mind for its information system. Let's imagine that the client is happy with this and requests a second sprint to develop a version of the data referential service that can store data this time. This is what we are going to do in the following section.

Adding persistence

The logical next step after reading sample in-memory data is to let the API user create its own data entities and store this data in a persistent manner, which means it will still be there when the application goes down and is restarted, which was admittedly a big limitation of our first version. With an agile approach, we could be even more conservative and start with a version that deals with in-memory data creation and modification, only then adding some persistence means. But it is a deliberate choice to carry out these two steps at the same time – because our data referential service will not work in the same way as lots of existing applications that store successive states of data: it will save modifications of data over time, and deduce states from this, still storing them in order to accelerate further querying of the data. The consequence is that the data schema will evolve, so it would be a waste of time to show the creation and modification of entities first and then change how it works.

We have already talked about what a good referential service is, in *Chapter 10*, and how it participates in the master data management "know-it-all" approach, in *Chapter 5*. We also explored the logical evolution of these into CQRS architectures. In the following section, we are going to implement a simple approach that's not as sophisticated as CQRS since the DemoEditor board's needs do not justify that level of complexity (remember, always start with business needs). In particular, a data stream middleware would be overkill for an information system like the one we are building. In fact, this will also be the case for lots of companies trying to use it. I hope the following example will show you that a careful design with eventual consistency is – most of the time – much less expensive and as good an approach as drawing out the "big guns" such as Apache Kafka.

Storing operations rather than states

Using a real MDM approach means the database will be organized in quite a different way from what you may be accustomed to. Generally speaking, a database schema is deduced from the business definition of an entity. If you needed to store a book with its ISBN identifier, you would create a `books` data table and add a data column named `isbn` with the right data type. As has been abundantly explained, this is limited since storing states as replacements of previous states results in forgetting the history of the data and makes it hard to deal with concurrency and so on.

The module in charge of persistence here will be designed in a different way, in order to be a write-only, forget-nothing, patch-based master data referential service. It will use three different data tables to keep modification actions, resulting states, and the best-so-far state for each of the instances of business data. This is quite a big change, as the architecture must evolve from storing a state to documenting an operation. In particular, the management of value dates will be essential, as we need this major criterion in this new approach.

A schema might be useful to explain what we are going to do a bit better. When using a traditional state-based approach, the content of the database evolves as follows when a modification command is issued:

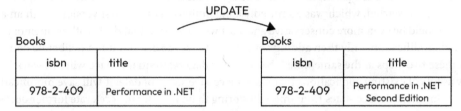

Books

isbn	title
978-2-409	Performance in .NET

UPDATE

Books

isbn	title
978-2-409	Performance in .NET Second Edition

Figure 16.14 – Updating data in a traditional database

As you can see, the data describing the book is replaced by another set of data. If, for example, the name of the book evolves, as above, from **Performance in .NET** to **Performance in .NET, Second Edition**, then the first data is simply replaced by the new data and there is no possibility of reverting to the previous state – the former data is lost.

With our MDM approach, this is what is going to happen in the database:

```
[{
    change author : { ....}
    valuedate: 2024-01-01
    patch: [{
            path= /
            op= add
            value= {
                isbn: 978-2-409
                title: Performance in .NET
            }
        }]
}]
```

UPDATE

```
[{ same as left },
{
    change author : { ....}
    valuedate: 2024-03-25
    patch: [{
            path= /title
            op= replace
            value= Performance in .NET, Second editon
        }]
}]
```

Figure 16.15 – Updating data in a referential-style database

Here, the information about changing the title is a piece of JSON that will be added to the existing list of changes stored in the data table – nothing is lost; there is only the addition of data (I have just written "`same as left`" in order to reduce the space needed for the figure).

This in itself would suffice to retain all of the data needed, but merging all the changes every time someone requests the API for a state would be highly inefficient, so the `books-changes` data table is generally complemented by a `books-states` data table, in order to keep the successive states generated by the changes:

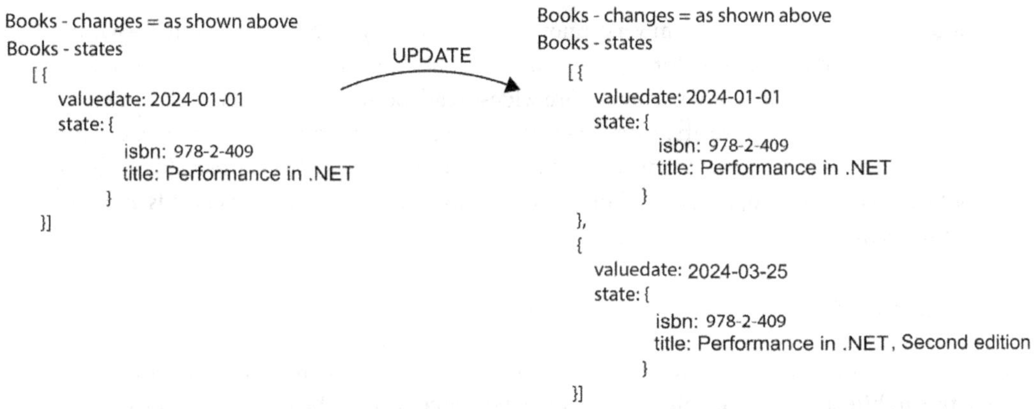

```
Books - changes = as shown above                    Books - changes = as shown above
Books - states                  UPDATE              Books - states
   [{                                                  [{
        valuedate: 2024-01-01                               valuedate: 2024-01-01
        state: {                                            state: {
              isbn: 978-2-409                                     isbn: 978-2-409
              title: Performance in .NET                          title: Performance in .NET
        }                                                   }
   }]                                               },
                                                    {
                                                         valuedate: 2024-03-25
                                                         state: {
                                                               isbn: 978-2-409
                                                               title: Performance in .NET, Second edition
                                                         }
                                                    }]
```

Figure 16.16 – Updating states of data

Finally, a second persistence optimization is generally done to take into account the fact that, very often, the API GET calls will request the latest state of an entity. Instead of finding the maximum date value of the entity and querying the corresponding state, thus extracting the content of the entity, it is easier to simply store the best-so-far states of each entity in a dedicated data table. In our case, this looks like the following:

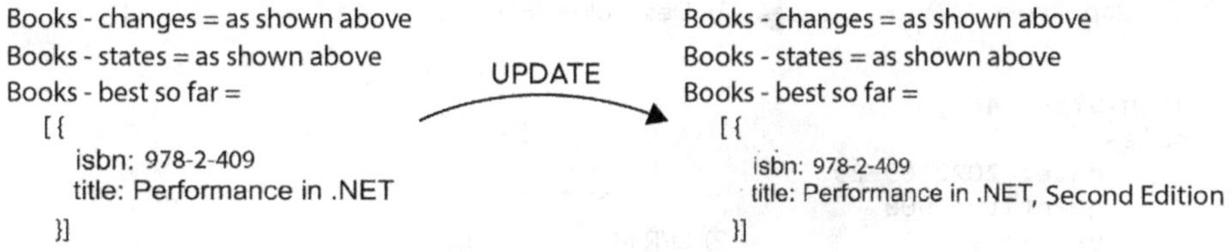

```
Books - changes = as shown above                    Books - changes = as shown above
Books - states = as shown above                     Books - states = as shown above
Books - best so far =            UPDATE             Books - best so far =
   [{                                                  [{
        isbn: 978-2-409                                     isbn: 978-2-409
        title: Performance in .NET                          title: Performance in .NET, Second Edition
   }]                                                  }]
```

Figure 16.17 – Updating best-so-far states

You will notice that the best so far actually reflects the change in states of the book and basically comes back to the "good old way" of storing states that are replaced by each other over time, except that in this case, no data is lost – the changes are stored, and the best so far is simply the last known state of the data.

Even though the BSF states look very much like the data stored in traditional databases, remember it is simply a local cache alongside the other two data tables. There is no doubt the whole lot looks more complex at first, but keep in mind that it only reflects the complexity of the business. This is not accidental technical complexity, but simply the right representation of what the business is about: entities change all the time and we need to respond to changes rather than just states, let alone being able to trace modifications for regulatory purposes. The fact that it avoids costly lock-based transactions is simply a nice consequence of well-designed content, removing the need for technically complex actions afterward (optimistic and pessimistic transactions, with lock management and compensation operations are definitely accidental complexity that should be avoided).

As a side note, this approach might seem very innovative to some of you, but it really is not – a minority of companies have worked this way for ages. It just turns out that it is not a very widespread way of doing it. I hope this book will help it become more widespread. Some applications, such as VestPocket (`https://khalidabuhakmeh.com/vestpocket-file-based-data-storage-for-aot-dotnet-applications`), should also help in making this happen. If, on the other hand, it looks too complicated, please read on – you will soon see all the benefits of this approach in line-of-business software.

Choosing NoSQL

As you might have guessed from the representation of data as JSON-serialized entities in the preceding diagrams, the architecture we are going to use for persistence will not be tabulated data storage and SQL, but a document-based, JSON-oriented, NoSQL database. This means, by the way, that we should have talked about collections rather than data tables, but this is just a detail.

Why this choice of database? Simply because, as was explained in *Chapter 10*, it is much more adaptable to line-of-business, MDM-capable data storage. It is also much more adaptable to the fact that we are going to receive the data from the API in JSON format. If we were to store it in a SQL database, the flow of data would be something like this when creating a book with a few lines of sales:

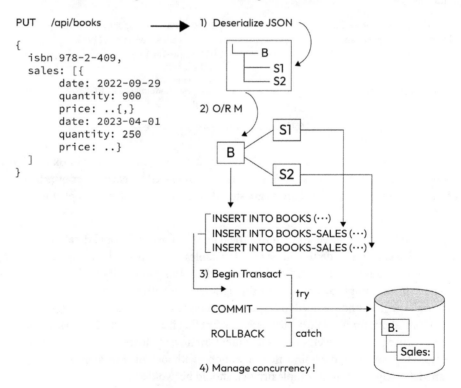

Figure 16.18 – Sending data in a traditional RDBMS

First, the server would need to transform the data sent in JSON (or XML, for that matter) into an object in memory. This is called deserialization and can be a complex task that takes some time and can cause difficulties. Then, the entity obtained is sent to an **object-relational mapper (ORM)**, which will transform this memory structure into a set of modifications that are to be sent in different data tables. Since we need to maintain consistency, these commands will be embedded in a transaction. We are not done yet because concurrency should be handled with specific code, which can itself become very complex to manage.

Of course, the same complicated path would have to be handled the other way around when reading the equivalent information:

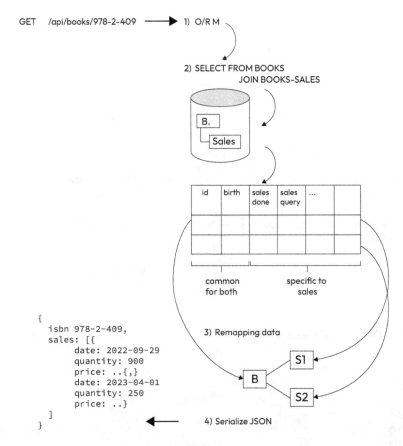

Figure 16.19 – Reading data in a traditional RDBMS

In this case, we would use the ORM to select the elements from the tables and SQL to automatically join them into a tabular view, then transform linear data into a structured memory model, and finally serialize this into JSON content that will be sent as the response of the API call.

Even though the storage organization schemas are admittedly more complex, the similar creation operation is already much easier in practice:

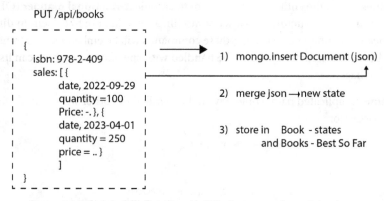

Figure 16.20 – Sending data in a NoSQL document-based database

Indeed, in this case, the JSON content is simply sent to the MongoDB insertion operation. Of course, we would recalculate the new state, put it in the list of states, and update the best so far, but this part is only because we're using a record-all referential.

If we wanted to be more precise, this is in fact what would really happen, but I anticipated a bit of the following content:

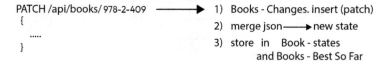

Figure 16.21 – Updating changes to data in a NoSQL database

Instead of sending the value of the entity, the change would be sent, and this is what would be inserted into the database.

The equivalent reading is even better:

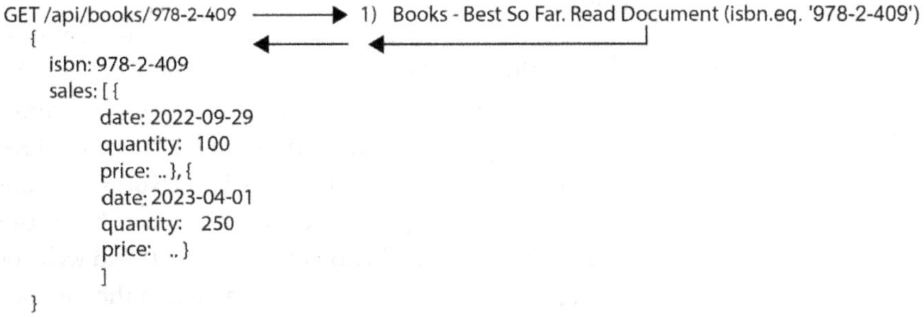

```
GET /api/books/978-2-409  ────────►  1)  Books - Best So Far. Read Document (isbn.eq. '978-2-409')
     {                     ◄──────── ◄─────────────────────────────┘
         isbn: 978-2-409
         sales: [ {
                 date: 2022-09-29
                 quantity:  100
                 price: .. }, {
                 date: 2023-04-01
                 quantity:   250
                 price:   .. }
                 ]
     }
```

Figure 16.22 – Reading data in a NoSQL database in the form of a single document

This time, since the reading of the MongoDB document directly sends some precomputed best-so-far states, there is simply nothing to do but read the data and send it. That's right – no transaction, no ORM, no complex serialization patterns. All of these are already great arguments for NoSQL storage. But wait! There is more... What was the hexagonal architecture used for again? Extracting the business model from all the dependencies, such as persistence and other interfaces. Why did we bother to do this? We were bothered because the business model was so entangled in lines of code about all the peripheral functions that it became difficult to make it evolve, or test it. But what happens when we simplify the persistence code of the model so radically and the tens of lines of .NET to handle the ORM, the transactions, the deserialization, and so on are finally reduced to a single `insertDocument`? Well, it radically changes the approach because this means that, if you have a limited scope, you are now able to modify the storage with just a few code changes. This is, of course, if your domain is reduced, but this is exactly what we are talking about when applying the principles of DDD to create (not so micro-)services, or if we obey the "two-pizza team" rule: the domain should be reduced in order to obtain a modular information system, and its interfaces should be standardized in order to make it easy to evolve and compose it.

What is the consequence of this? Since it is so easy to change the persistence implementation in this architecture, there is not even the need to go all the way into the clean/hexagonal architecture and extract any persistence calls from the business entities as is recommended: simply making these calls in the form of .NET interfaces and not concrete implementation like the Mongo client's is enough. Indeed, if one needs to switch to another document-based database implementation, there will only be a few calls to change in the class. Accepting unit tests will be as easy as creating mocks for these interfaces. Incidentally, it happens that Mongo calls have now become standard, so there is no real use for dependency inversion between the business entity model and persistence functions.

I know that this conclusion may sound heretical to lots of expert software engineers, but I will stand my ground on this one. After almost 40 years of coding, I have seen so much over-sophistication of code that I favor my mechanical engineering approach where only the module definitions and interfaces are important, and the implementation is not so important. After all, when you have a car and the oil filter does not work anymore, you do not try to clean it – layers of paper after another; you simply discard the old oil filter and replace it with a new one, owing to the fact that the interface to the engine is perfectly standardized. Why would we work in another way in the software industry and try by all means to fix the oil filter, or even make it so sophisticated that it can self-heal? This would simply be money not well invested. If the services are small enough, cleanly cut, and well-interfaced with one another, it is much cheaper to simply code the services again. In fact, in the case of the final version of the books referential service, you will see by the end of the chapter that a good software programmer could redo it in any language in one day.

A few more words on NoSQL and persistence

Before we go on, I need to add a few words on NoSQL because I have often been exposed to the objection that document-based databases allow you to mix anything in a collection and, because of this, can easily become a mess of heterogeneous data. This is theoretically true, but in practice only happens to people who have not understood the way NoSQL should be used. Let me explain this in more detail...

A huge advantage of NoSQL over SQL is that, as its name suggests, it does not force structured content. Of course, you'd better keep things clean and not push any kind of mixed data into a collection, but this relaxed constraint allows, for example, you to mix individuals and companies into an `actors` collection. You may ask why on earth one would do this, particularly if you come from a long tabular data background. But again, functional, business-aligned, data usage is all that matters. So, if you happen to have individuals and companies that act as customers of your information system, suppliers (or both), irrespective of their nature and common data that is used for functional processes (bank accounts, addresses, contacts, etc.), then there is an obvious advantage to putting them together. Owing to the Not Only Structured approach, this means you can still have specific data attributes for each type (family name for individuals, registered name for organizations, etc.).

Again, I understand that this may sound crazy for lots of readers who have been trained on structured, tabular data, where the schema exists before the data and constrains every aspect of it. But think about what would happen if there were no computers, and people still used paper cards to register a customer or supplier – are you sure everything would fit in nice cases? Of course not – there would always be side notes on the back of the cards and some of the lines would not be filled in with the right data. This is not only about data cleanliness – most of the time, the data being useful to users is most important. On paper, if someone was not happy with the label "accounting contact," they would happily strike through the text and write "small company: same as the boss" above it. Why not do the same with databases and be a bit more relaxed while – of course – still being strict about important business rules (customers must have a valid billing address, etc.)?

SQL and a strict tabular constraint may have their utility in some cases but, as far as master data is concerned, they are generally more of a problem due to oversized constraints. If you have worked with SQL long enough, you certainly will have seen some huge, complicated requests that occupy tens – sometimes hundreds – of lines. This alone should make one think about the adaptation of the technology to use cases. Again, SQL is great for some uses and has a long record of stable, robust persistence. But there should not be an obsession with it, and NoSQL is there for a good reason: it allows for better implementation of lots of persistence use cases. Most of the time, trying it results in adopting it and resistance is only due to difficulties in managing the change.

SQL has been the only persistence mode for a long time, so it is difficult to get one's mind free of it, and even more so from the constraints it silently carries with it, such as table-based design, transaction-oriented ACID consistency, and so on. But when you free yourself from this way of thinking, you will start thinking of even "weirder" persistence models than NoSQL. For example, what about using Git for a data referential service? Think about it: aren't commits very similar to the concept of changes presented above? The state is simply the complete code base at a given time, with the best so far being the state of the code base in the most current commit of the master. Also, think of the power of adding branches to an MDM referential service. For books, it would allow us to semantically show the different editions of a book, each one based on the preceding version. For budgets, for example, it would allow the users to test some hypotheses in parallel, then abandon all branches that are not good ideas and merge the ones that are chosen to be integrated into the official budget. Accountants could work in parallel on different parts of the budget without being obliged to wait for the others to edit the central Excel workbook or risk losing their work. In case they accidentally modified a given amount at the same time, the merge feature would let them merge their two solutions. Sounds great, doesn't it? All this would be brought to you for free since all these features are already integrated into Git. Again, this is an example of the power of modeling your business problem free of any technical hypothesis and, only after that, finding the right technology to fit the business needs.

Remark on multitenancy

One last small architecture consideration before we go back to C# code is the question of multitenancy. This is the kind of feature that should be thought of at the beginning of a project since – like security – it is so hard to implement afterward in an application (note that this is not the case for performance, and premature optimization is the root of all evil). This does not mean it should be implemented. Again, this simply depends on the business.

In our case, we will not implement multitenancy in the books and authors referential service because we are working on the hypothesis that such applications are not available off the shelf for DemoEditor. I do not mean they do not exist in the real world; there may be some excellent generic ERP systems for booksellers and editors. Since I showed in the previous chapter how we can select and add existing modules for our demo information system, I want to show in this chapter what happens when this is not the case. Generally, it is when we get to the core business of a company that we find existing software is not available. Some building blocks may be there, but the value of a business is generally where it differentiates itself from others, and this means specific, hand-made, company-customized software.

Of course, the reasoning would be completely different if we were working on the hypothesis that we are a software vendor trying to create a generic books referential service and sell it to many editors. In that case, we would invest in the software in the hope of selling and deploying it to as many editors as possible, and we would need to think of a way to handle that volume in an efficient way. That could, of course, be to simply do nothing and let the users install the data referential service application as many times as needed, even though that would not be very resource-efficient, as the need for memory and network ports would be multiplied by the number of tenants. In the most extreme case, where thousands more tenants are expected, it would be much better to think about multitenancy in terms of software isolation, coding the central process to be able to handle multiple tenants and ensure the complete isolation of data (which is admittedly a challenge – this is not a complex task, but the risk associated with a failure would be huge in terms of the impact on intellectual property/business secrecy). In most of the gray area between these two extremes, letting Docker do the isolation would be a good move. It would not be as resource-intensive as creating separate virtual machines, but at the same time would bring you a great level of isolation in a very safe, yet easy, way to deploy new tenants. In all cases, this should be decided at the beginning of the design because, as explained above, this is something that is complex to integrate into a working mechanism.

For the rest of this book, we will come back to our example information system and look at crafting our very own books referential service, which only needs to accommodate one tenant – DemoEditor.

Creating the data business model

In the first part of this chapter, we started modeling the books with a record, as this was the simplest way to go and what was proposed in the sample project template from Microsoft. We explained then that this would, of course, have to evolve into a real class in order to support business rules and other features, and this is what we are going to do right now.

In order to keep the code tidy, we will put a new Book.cs file that will contain the following class description in the Model folder:

```
public class Book
{
    public string BusinessId { get; set; }
    public string? ISBN { get; set; }
    public int? NumberOfPages { get; set; }
    public DateOnly? PublishDate { get; set; }
    public EditingPetal? Editing { get; set; }
    public SalesPetal Sales { get; set; }
}
```

To get it working, we will then add `EditingPetal.cs` and `SalesPetal.cs` files in the same folder, with the following contents, respectively:

```
public class EditingPetal
{
    public int? NumberOfChapters { get; set; }
    public Status? Status { get; set; }
}

public class Status
{
    public string Value { get; set; }
}

public class SalesPetal
{
    public MonetaryAmount? Price { get; set; }
    public decimal? WeightInGrams { get; set; }
}

public class MonetaryAmount
{
    public decimal Value { get; set; }
    public string MonetaryUnit { get; set; }
}
```

As you may have seen, `Status` is now put in a class that only contains a value. As explained previously in this chapter, it is a design choice to not use an `enum` for the status but a `string`. In addition, it is also an extensibility choice here to create a dedicated class for the status content. This guarantees that, if we need to add names, possibly in different languages, or other associated data, it will always remain possible while maintaining backward compatibility. It happens quite a lot that what we initially classify as simple attributes are in fact objects on their own. A good example of this is nationality: we may think that using a country code is enough, but countries may be addressed with two-letter (ISO 3166-1 alpha-2 norm) or three-letter (ISO 3166-1 alpha-3 norm) codes. They also have names that vary from one culture to another. A country is also associated with a flag, a top-level domain, and so on. So really, a country is a class of properties and not just a code. This actually goes for a lot of other characteristics we tend to think of as simple: monetary units, colors, and so on.

This is why, for example, `Price` is not defined as `decimal` but as `MonetaryAmount`, itself made of a `Value` and a code from ISO-4217:2015 norm in order to represent the currency (this standard states that US dollars are represented by USD, European Euro currency by EUR, etc.). Again, this is a design choice as we expect `DemoEditor` to sell books in an international context. On the other hand, the design choice for the weight of the book led to an opposite decision, with the `WeightInGrams` attribute clearly expressing that the internal standard we intend to always abide by is the International Units one.

Some may argue about simplicity concerns and the YAGNI rule for `Status` – since the class only contains one attribute, what is the use of this class, and isn't it too much to prepare for extensibility if we do not know that we will need it? Again, this is a design choice and the hypothesis here simply was that statuses were closer to price than to weight and that we had better take care to not obstruct evolution. By the way, the class is kept simple since there is only a `Value` attribute in the `Status` class for now, which means the only additional cost is the typing (which can be suppressed through a property) and the additional serialization level.

The sample list's content adapts to reflect the shift from records to classes in `Program.cs`, requiring no other changes for now.

```
var books =  Enumerable.Range(1, 3).Select(index =>
    new Book()
    {
        BusinessId = index.ToString(),
        ISBN = "978-2-409-03806-" + index.ToString(),
        NumberOfPages = Random.Shared.Next(400, 850),
        PublishDate = DateOnly.FromDateTime(DateTime.Now.
AddDays(index)),
        Editing = new EditingPetal()
        {
            NumberOfChapters = Random.Shared.Next(5, 30),
            Status = new Status() { Value = statuses[Random.Shared.
Next(statuses.Length)] }
        },
        Sales = new SalesPetal()
        {
            Price = new decimal(Random.Shared.Next(1000, 10000) /
100),
            Weight = new decimal(Random.Shared.Next(200, 3500))
        }
    })
    .ToArray();
```

Since the business model is now ready (in a reduced form for simplicity reasons) and the MongoDB database was installed in the previous chapter, we now have everything we need to add write operations to the API and implement them with true storage.

A switch in thinking – modify first, create afterward

Generally speaking, when an API is designed, the creation of the resource with POST is the first operation that is implemented. But, since we have defined the data referential service as a series of modifications, we are going to first implement the PATCH command. You read that right: not the PUT command, which replaces the complete state of the object – which is precisely what old storage implementations do – but the PATCH command (a lesser known but indeed legitimate HTTP verb), which allows us to send modifications of the content of a resource, which are the changes we need to store in the first collection, and that will eventually result in a change of state and will be the primary object to reflect the history of the modification of the resource representing the business entity. In fact, as you will see, we will simply not allow the PUT approach of replacing the whole resource.

When sending a PATCH operation to an API, the content of the request body should represent a set of changes and, guess what? There is a standard for this: RFC 6902, which is also called **JSONPatch**. In fact, we already introduced this in the earlier examples. The following content is an example of JSONPatch data:

```
[{
    "path": "/title",
    "op": "replace",
    "value": "Performance in .NET, Second Edition"
}]
```

The standard defines the three attributes shown and also what values op can take, namely add, replace, remove, and the lesser used copy, move, and test. The JSONPatch grammar also defines that the content of path should follow RFC 6901, and of course, all the details of behaviors/ expressions associated with the standard. Beware that JSONPatch is a wild animal and that resource patching is, sadly, not very often used for now (despite it being better aligned to business modifications), so it may be less simple to use than other operations, or you may find out the support in frameworks is not so advanced. For example, at the time of writing at least, Microsoft .NET does not completely support what is needed for JSONPatch management in the System.Text.Json library and you will need to resort to the "legacy" NewtonSoft.Json module.

Since the content will be a bit more complex, we will leave the project created with the Minimal API and create another one based on controllers. We thus type dotnet new webapi --use-controllers -o books-controller and navigate to the newly-created project. After copying the existing Models folder and renaming the controller from the weather forecast example provided to our books subject, we obtain the following code base:

```
using Microsoft.AspNetCore.Mvc;

namespace books_controller.Controllers;

[ApiController]
```

```
[Route("[controller]")]
public class BooksController : ControllerBase
{
    private readonly ILogger<BooksController> _logger;

    public BooksController(ILogger<BooksController> logger)
    {
        _logger = logger;
    }

    // The operations will be described step by step in this chapter
}
```

In order to implement the PATCH verb in the controller, we add the following method, which we will complete step by step in order to explain how it works:

```
[HttpPatch]
[Route("{entityId}")]
public async Task<IActionResult> Patch(
    string entityId,
    [FromBody] JsonPatchDocument patch,
    [FromQuery] DateTimeOffset? providedValueDate = null)
{
    // The code content of the function will be added below
}
```

In order for the preceding method to simply compile, we need to add the right package from the JsonPatchDocument class, with the following command:

```
dotnet add package Microsoft.AspNetCore.JsonPatch
```

We can now come back to the body of the Patch method, and the very first command we will include is to verify the patch argument is correct:

```
if (patch is null) return BadRequest();
```

After that, we will take care of adding a value date if it is not provided by the caller. Indeed, most of the time, the client calling the API will not really consider it important to specify the date associated with the modification action and will trust the server to take care of this. In this case, we will simply use the clock of the server itself:

```
DateTimeOffset valueDate = providedValueDate.
GetValueOrDefault(DateTimeOffset.UtcNow);
```

Our next step will be to create a class to store a change, which is more than we did for the JSONPatch content, as at least the value date should be there as well, and the identifier of the object is also necessary in order to be able to determine the changes applied to a given business entity:

```
public class ChangeUnit
{
    public string EntityId { get; set; }
    public DateTimeOffset ValueDate { get; set; }
    public BsonArray PatchContent { get; set; }
}
```

This allows us to continue writing the Patch function content with the following:

```
ChangeUnit patchAction = new ChangeUnit()
{
    EntityId = entityId,
    ValueDate = valueDate,
    PatchContent = new BsonArray(patch.Operations.
ConvertAll<BsonDocument>(item => item.ToBsonDocument()))
};
var collectionActions = Database.GetCollection<BsonDocument>("books-
changes");
await collectionActions.InsertOneAsync(patchAction.ToBsonDocument());
```

After creating the change unit, we insert it into the books-changes collection of our MongoDB database. Of course, as it is, this code would not compile. We need to make a few changes to do so, which are detailed below.

Linking the books-changes collection to the database

So far, we have mostly shown what kind of interactions we want to have with the database, but we have not explained how we are actually going to pilot the database. This is done by using a client (sometimes also called a driver) speaking in MongoDB language. First of all, we need to add the package for MongoDB:

```
dotnet add package MongoDB.Driver --version 2.18.0
```

> **Note**
>
> **Version 2.19.0** has a function that prevents the serialization of all objects that are not explicitly allowed to be serialized (see https://medium.com/it-dead-inside/net-mongodb-driver-2-19-breaking-serialization-errors-b456134a1a2d). Though this is great in production, we'll try to stay away from too many technical details here; understanding how to create a patch-based data referential service is already complicated enough.

Also, since the `Database` identifier is not known yet, we need to add a class member to define it, by modifying the beginning of the class like this:

```
[ApiController]
[Route("[controller]")]
public class BooksController : ControllerBase
{
    private readonly string ConnectionString;

    private readonly IMongoDatabase Database;

    private readonly ILogger<BooksController> _logger;

    public BooksController(IConfiguration config,
ILogger<BooksController> logger)
    {
        _logger = logger;
        ConnectionString = config.
GetValue<string>("BooksConnectionString") ?? "mongodb://
localhost:27017";
        Database = new MongoClient(ConnectionString).
GetDatabase("books");
    }

    // Rest of the class
}
```

In order for the `ChangeUnit` class to be correctly handled by MongoDB, we need to decorate it with the following:

```
using System.ComponentModel.DataAnnotations;
using System.Text.Json.Serialization;
using MongoDB.Bson;
using MongoDB.Bson.Serialization.Attributes;

public class ChangeUnit
{
    [BsonId()]
    [BsonRepresentation(BsonType.ObjectId)]
    [JsonIgnore]
    public ObjectId TechnicalId { get; set; }

    [BsonElement("entityId")]
    [Required]
    public string EntityId { get; set; }

    [BsonElement("valueDate")]
    [Required]
    public DateTimeOffset ValueDate { get; set; }

    [BsonElement("patchContent")]
    [Required]
    public BsonArray PatchContent { get; set; }
}
```

This change compared to the initial version of the class, without any dependency on storage-related attributes, is what I was talking about when explaining that the code here would not be "clean code" or follow the hexagonal architecture with a centralized business entity class without coupling with interfaces. Again, this choice is made as the size of the service makes it easy to evolve the whole source code without having to change the API interface. Such an effort in dependency inversion, with the loss of code readability, is interesting when the complexity of the code base justifies it. With a total code base of around 1,000 lines or less, it is overkill. However, that does not mean that we should dive head first into maximum coupling either, so the type of the `Database` field is an interface (`IMongoDatabase`) in order to make it easy to mock a connection or plug another driver into a different database system that implements the MongoDB standardized grammar:

```
IMongoDatabase
```

Note that `TechnicalId` has been added in order to accommodate the technical identifier MongoDB creates for a stored entity, but is decorated with `JsonIgnore` in order for it not to pollute our serializations. As for the other elements, the `BsonElement` attribute enables the adjustment of the name in the serialization for the storage. Since we are going to serialize our model classes, we need to also decorate them as was done for `ChangeUnit`; the `Book` class then resembles this (the `using` commands are omitted to minimize the code):

```
public class Book
{
    [BsonId()]
    [BsonRepresentation(BsonType.ObjectId)]
    [JsonIgnore]
    public ObjectId TechnicalId { get; set; }

    [BsonElement("entityId")]
    [Required(ErrorMessage = "Business identifier of a book is
mandatory")]
    public string EntityId { get; set; }

    [BsonElement("isbn")]
    public string? ISBN { get; set; }

    [BsonElement("title")]
    public string? Title { get; set; }

    [BsonElement("numberOfPages")]
    public int? NumberOfPages { get; set; }

    [BsonElement("publishDate")]
    public DateTime? PublishDate { get; set; }

    [BsonElement("editing")]
    public EditingPetal? Editing { get; set; }

    [BsonElement("sales")]
    public SalesPetal Sales { get; set; }
}
```

Also, note that `Title` has been added in order to support the testing scenario that was schematized at the beginning of this chapter and will be implemented and tested a bit later. `PublishDate` has changed its type from `DateOnly` to `DateTime` as there are known issues with serializing the former (see https://github.com/dotnet/runtime/issues/53539 for details). Of course, such breaking changes would not be done in production – we're simply showing here the thought process and adjustments that are needed during the design process.

The `EditingPetal` class is modified like this, so as to include the adapted decorator attributes:

```
using MongoDB.Bson.Serialization.Attributes;

public class EditingPetal
{
    [BsonElement("numberOfChapters")]
    public int? NumberOfChapters { get; set; }

    [BsonElement("status")]
    public Status? Status { get; set; }
}

public class Status
{
    [BsonElement("value")]
    public string Value { get; set; }
}
```

The `SalesPetal` class is adjusted to contain this code, with the same goal as above:

```
using MongoDB.Bson.Serialization.Attributes;

public class SalesPetal
{
    [BsonElement("price")]
    public MonetaryAmount? Price { get; set; }

    [BsonElement("weightInGrams")]
    public decimal? WeightInGrams { get; set; }
}

public class MonetaryAmount
{
    [BsonElement("value")]
    public decimal Value { get; set; }

    [BsonElement("monetaryUnit")]
    public string MonetaryUnit { get; set; }
}
```

Finishing the patch operation implementation

All this being set, we can now come back to the code of the `Patch` method and add the following, which retrieves the existing best-so-far state of the business entity from the identifier passed as an argument:

```
var result = Database.GetCollection<Book>("books-bestsofar").Find(item
=> item.EntityId == entityId);
var book = result.FirstOrDefault();
```

If the book does not exist in the database already, we need to add it and retrieve it right away in order to be in the same situation before the next round of code that will manipulate the `book` variable:

```
if (book == null)
{
    book = new Book() { EntityId = entityId };
    var bestknown = Database.GetCollection<BsonDocument>("books-
bestsofar");
    await bestknown.InsertOneAsync(book.ToBsonDocument());
    result = Database.GetCollection<Book>("books-bestsofar").Find(item
=> item.EntityId == entityId);
    book = result.First();
}
```

The next step in the function will be to apply the patch that has been provided as an argument to the current state of the book:

```
patch.ApplyTo(book);
ObjectState<Book> nouvelEtat = new ObjectState<Book>()
{
    EntityId = entityId,
    ValueDate = valueDate,
    State = book
};
```

This `ObjectState` generic class is constructed to handle the value of a generic business object referenced by its identifiers with a precise date value. The resulting code is like this:

```
using System.ComponentModel.DataAnnotations;
using System.Text.Json.Serialization;
using MongoDB.Bson;
using MongoDB.Bson.Serialization.Attributes;
```

```
public class ObjectState<T>
{
    [BsonId()]
    [BsonRepresentation(BsonType.ObjectId)]
    [JsonIgnore]
    public ObjectId TechnicalId { get; set; }

    [BsonElement("entityId")]
    [Required]
    public string EntityId { get; set; }

    [BsonElement("valueDate")]
    [Required]
    public DateTimeOffset ValueDate { get; set; }

    [BsonElement("state")]
    public T State { get; set; }
}
```

Once this is done, the new state is added to the list of states in the database's dedicated collection:

```
var collectionEtats = Database.GetCollection<BsonDocument>("books-
states");
await collectionEtats.InsertOneAsync(nouvelEtat.ToBsonDocument());
```

The best-so-far version of the business entity should then be updated to account for the application of the patch:

```
var collectionBestKnown = Database.GetCollection<Book>("books-
bestsofar");
await collectionBestKnown.FindOneAndReplaceAsync<Book>(
    p => p.EntityId == entityId,
    book
);
```

Finally, we return the object in order to show potential changes due to business rules (even if, for now, we have not defined any):

```
return new ObjectResult(book);
```

In order for the `Patch` function to work correctly, and in particular, for the deserialization of the `patch` argument to happen without any problems, we need to step away from the embedded `System.Text.Json` module and turn to the good old `NewtonSoft` library, since only this legacy serialization engine supports `JSONPatch` grammar. To do so, it is necessary to add the following function in `Program.cs`:

```
builder.Services.AddControllers().AddNewtonsoftJson();
```

And, of course, `dotnet add package Microsoft.AspNetCore.Mvc.NewtonsoftJson` will have to be run in order for everything to compile and, hopefully, work fine.

Creation as a special case of modification

In an MDM-enabled data referential service, receiving a POST creation order is in fact just a particular case of PATCH in which the initial state is simply empty. In order to account for this and the overall greater importance of changes related to states, we will base a second function to expose the POST operation on the code of the PATCH equivalent:

```
[HttpPost]
public async Task<IactionResult> Create(
    [FromBody] Book book,
    [FromQuery] DateTimeOffset? providedValueDate = null)
{
    // If no entity identifier is provided, a business rule states
that a GUID should be attributed
    if (string.IsNullOrEmpty(book.EntityId))
        book.EntityId = Guid.NewGuid().ToString("N");

    // Creating an entity is in fact nothing else than patching an
empty state
    JsonPatchDocument equivPatches = JSONPatchHelper.CreatePatch(
        new Book() { EntityId = book.EntityId },
        book,
        new DefaultContractResolver() { NamingStrategy = new
CamelCaseNamingStrategy() }
    );

    // After creating the equivalent patch, we thus pass this over to
the PATCH operation
    return await Patch(book.EntityId, equivPatches,
providedValueDate);
}
```

The JSONPatchHelper class that generates the patch from an original, modified version of the object is used to transform all attributes provided in the book variable into operations following RFC 6902. This class has been retrieved from https://stackoverflow.com/questions/43692053/how-can-i-create-a-jsonpatchdocument-from-comparing-two-c-sharp-objects, and its content is the following:

```
using Microsoft.AspNetCore.JsonPatch;
using Newtonsoft.Json.Linq;
using Newtonsoft.Json.Serialization;
using System;
using System.Collections.Generic;
using System.Linq;
using System.Text;

namespace books_controller.Tools
{
    public class JSONPatchHelper
    {
        public static JsonPatchDocument CreatePatch(object
originalObject, object modifiedObject, IContractResolver
contractResolver)
        {
            var original = JObject.FromObject(originalObject);
            var modified = JObject.FromObject(modifiedObject);

            var patch = new JsonPatchDocument() { ContractResolver =
contractResolver };
            FillPatchForObject(original, modified, patch, "/");

            return patch;
        }
```

The CreatePatch method is used to produce a patch from two versions of a given object. The resulting patch will be the list of commands that should be applied to go from the initial to the second version of the object. It calls a recursive method that will first take care of the nodes that have been removed:

```
        static void FillPatchForObject(JObject orig, JObject mod,
JsonPatchDocument patch, string path)
        {
            var origNames = orig.Properties().Select(x => x.Name).
ToArray();
            var modNames = mod.Properties().Select(x => x.Name).
ToArray();

            // Names removed in modified
```

```
foreach (var k in origNames.Except(modNames))
{
    var prop = orig.Property(k);
    patch.Remove(path + prop.Name);
}
```

Then it will go over the nodes that have been modified between the first and second versions:

```
// Names added in modified
foreach (var k in modNames.Except(origNames))
{
    var prop = mod.Property(k);
    patch.Add(path + prop.Name, prop.Value);
}
```

Finally, it will analyze the nodes that exist in both versions and drill into them to find what the differences are. This is where the recursion will happen, with the function calling itself when it finds content that is different (and not atomic, otherwise, it would record a change in the patch):

```
// Present in both
foreach (var k in origNames.Intersect(modNames))
{
    var origProp = orig.Property(k);
    var modProp = mod.Property(k);

    if (origProp.Value.Type != modProp.Value.Type)
    {
        patch.Replace(path + modProp.Name, modProp.Value);
    }
    else if (!string.Equals(
        origProp.Value.ToString(Newtonsoft.Json.
Formatting.None),
        modProp.Value.ToString(Newtonsoft.Json.Formatting.
None)))
    {
        if (origProp.Value.Type == JTokenType.Object)
        {
            // Recurse into objects
            FillPatchForObject(origProp.Value as JObject,
modProp.Value as JObject, patch, path + modProp.Name + "/");
        }
        else
        {
            // Replace values directly
```

```
                                    patch.Replace(path + modProp.Name, modProp.
    Value);
                            }
                        }
                    }
                }
            }
        }
```

The controller is now ready in its second version. The first one was a memory-based implementation and, this time, it uses a MongoDB database as its persistence backend. By the way, you will need to start the container if it has not been done (see *Chapter 15* if necessary). We could say that this was our second sprint of an agile method, and what remains is the demo of the features that have been created.

Summary

This chapter presented the first steps to implement a data referential service. It is only logical that this one was one of the most complex chapters because data referential services are at the heart of information systems. For a company such as DemoEditor, one can imagine how author and book management are important – they are simply the enterprise's bread and butter and should be dealt with accordingly.

Persistence was chosen carefully and it happens that NoSQL is often a good choice for functional, business-aligned entities, though SQL is still very much used for reporting and big data. Standards for querying such as GraphQL and Open Data Protocol were discussed and they are the open door to interoperate our data referential service with virtually any client.

Strong and standardized definitions of entity objects are of utmost importance, as you have seen according to the number of norms and RFCs used in the sample object definitions throughout this chapter, putting recommendations from *Chapter 2* to work.

Creating a data referential service for books was already a hard task in itself, so we'll close this chapter here and leave it to the next chapter to test the first results, add deletion and query features (because writing without eventually reading does not really make sense), and finally do a whole round of validation of the outcomes. Then, another chapter will deal with all the remaining features of this important part of the information system.

17

Adding Query to the Data Referential Service

In the previous chapter, we have presented the first and most important features of a data referential service, in particular the fact that a true MDM-compliant application will record every single change and allow full traceability and data consistency. But there are a few things missing from the data referential service.

After looking at what has been produced, this chapter will explain how to manage the deletion of data, which is a harder problem than it looks, and add reading capabilities to the data referential service. Indeed, the data is stored in a particular way to keep a full knowledge of the actions on the content, but this means there is some work to do before exposing this data for read operations. Also, since performance is so important, we will spend some time taking pagination into account, of course, in a standard way.

Finally, we will take some time to test the application that has been created.

Here is what we will cover in this chapter:

- Adding features to the first version of the data referential service
- Querying as an extension of the reading responsibility
- Testing

Adding features to the first version of the data referential service

The first version of the data referential service that was created in the previous chapter is for now only able to create data. In this section, we will first check that it works fine using a more industrial approach than the simple manual tests shown before, and then we'll discuss the delete operation.

Testing the data referential service

Having a more industrial approach to validation is really about automatic testing, and this is what we will do in this present section. The very first test is to run the application and check that the OpenAPI/Swagger interface works fine:

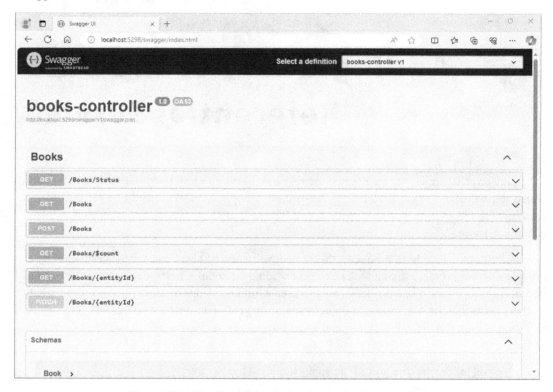

Figure 17.1 – The OpenAPI interface for the books controller

After that, we can test a creation operation on Postman by using the following setup, in which we choose the POST operation to send a creation command to the /Books URL of the server host you want to test (in a normal Postman setup, we would use a variable for the host to make it easier to change all occurrences at once, but it is clearer here to show the actual value), and provide a JSON serialization corresponding to what it expected to describe a book (if you select the **raw** format, the interface will normally auto-detect that it is **JSON** and adjust the format of the body):

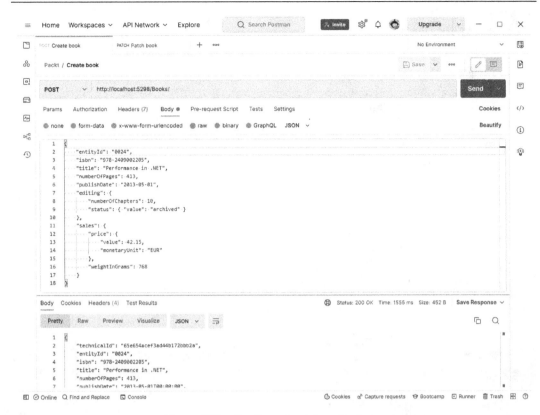

Figure 17.2 – Creating a book from Postman

You can specify any JSON serialization of a book since most of the attributes are not mandatory, but here is a sample text that can be used for your first test:

```
{
    "entityId": "0024",
    "isbn": "978-2409002205",
    "title": "Performance in .NET",
    "numberOfPages": 413,
    "publishDate": "2013-05-01",
    "editing": {
        "numberOfChapters": 10,
        "status": { "value": "archived" }
    },
    "sales": {
        "price": {
            "value": 42.15,
            "monetaryUnit": "EUR"
        },
```

```
            "weightInGrams": 768
        }
    }
}
```

Of course, the operation that interests us most in testing is the PATCH one, where we send the JSON content that respects the JSONPatch syntax and adjust the verb to **PATCH**:

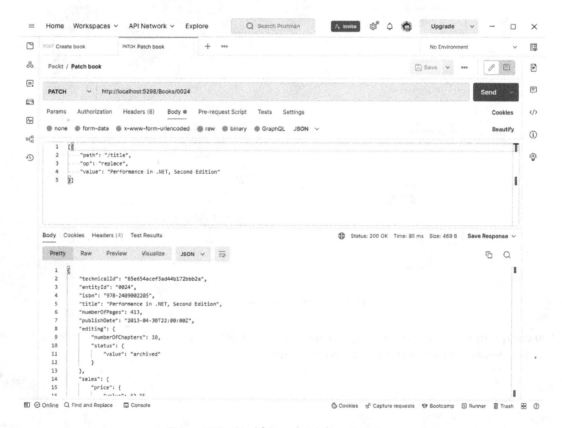

Figure 17.3 – Modifying a book from Postman

There again, you can operate your own test changes, but a simple patch could be realized with the following JSONPatch content, which will modify the title of the book designated in the URL indicated at the top of the interface (you should adjust that to include the identifier of the book previously created):

```
{
    "path": "/title",
    "op": "replace",
    "value": "Performance in .NET, Second Edition"
}
```

Compass or any other MongoDB console can be used in order to check the results are as expected. We start by verifying that there indeed have been two entries recorded in the books-changes collection, for the same entity, 0024, but with different timestamps and thus technical identifiers in MongoDB:

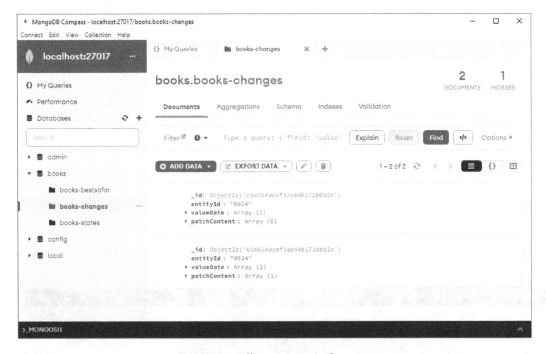

Figure 17.4 – Changes seen in Compass

A more detailed look at the first change shows that the creation of the entity has indeed been turned into JSONPatch content with six entries, one for each of the attributes that were defined in the entity upon creation:

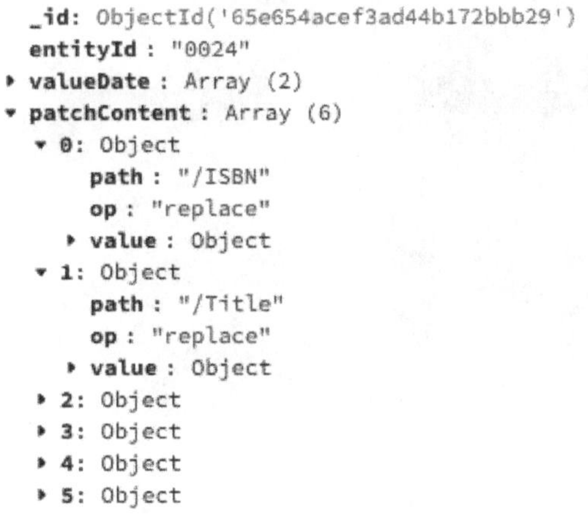

```
_id: ObjectId('65e654acef3ad44b172bbb29')
entityId : "0024"
▶ valueDate : Array (2)
▼ patchContent : Array (6)
    ▼ 0: Object
        path : "/ISBN"
        op : "replace"
      ▶ value : Object
    ▼ 1: Object
        path : "/Title"
        op : "replace"
      ▶ value : Object
    ▶ 2: Object
    ▶ 3: Object
    ▶ 4: Object
    ▶ 5: Object
```

Figure 17.5 – Details of the first change

We can then proceed to check that there are also two successive states of the entity, as shown here:

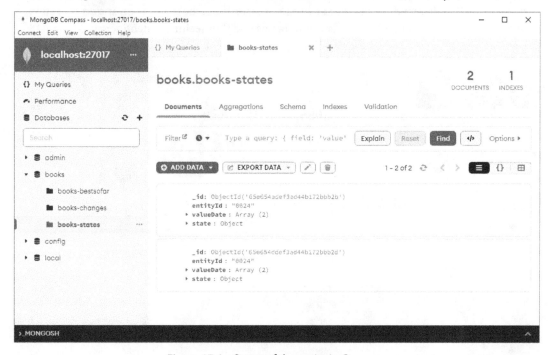

Figure 17.6 – States of the entity in Compass

And, as expected, there is only one entry in the books-bestsofar collection, since it is supposed to always be the latest. It is essential to verify that title has been changed as expected from our second Postman entry, in order to validate the correct functioning of the API:

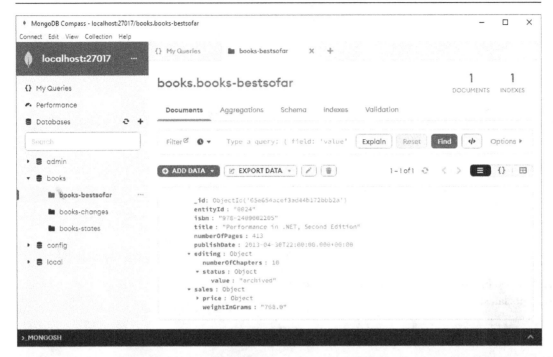

Figure 17.7 – Best-so-far state of the book entity seen in Compass

In order to show another way to test, we are going to implement a first and naive approach to reading data, with the following three operations: retrieving the whole list of entities, retrieving the number of entities, and retrieving a given instance (the first and third uses the Find operation of the Mongo driver, while the second one will use the dedicated CountDocuments function):

```cs
[HttpGet]
public IActionResult Get()
{
    return new JsonResult(Database.GetCollection<Book>("books-
bestsofar").Find(r => true));
}


[HttpGet]
[Route("$count")]
public long GetBooksCount()
{
    // The count is of course done on the best so far collection,
otherwise, it would add up historical states
    return Database.GetCollection<Book>("books-bestsofar").
CountDocuments(r => true);
}
```

```
[HttpGet]
[Route("{entityId}")]
public IActionResult GetUnique(string entityId)
{
    // As long as no parameter is added, the behavior is simply to
find and retrieve the best-known state
    var result = Database.GetCollection<Book>("books-bestsofar").
Find(item => item.EntityId == entityId);
    if (result.CountDocuments() == 0)
        return new NotFoundResult();
    else
        return new JsonResult(result.First());
}
```

A standard browser is enough to test the GET commands. A single GET operation on the entity will show its latest content:

```
1  {
2      "technicalId": "65e654acef3ad44b172bbb2a",
3      "entityId": "0024",
4      "isbn": "978-2409002205",
5      "title": "Performance in .NET, Second Edition",
6      "numberOfPages": 413,
7      "publishDate": "2013-04-30T22:00:00Z",
8      "editing": {
9          "numberOfChapters": 10,
10         "status": {
11             "value": "archived"
12         }
13     },
14     "sales": {
15         "price": {
16             "value": 42.15,
17             "monetaryUnit": "EUR"
18         },
19         "weightInGrams": 768
20     }
21 }
```

Figure 17.8 – Getting an entity

If we call the list of books, we will not see a big change, since there is only one in the collection for now, but note that the JSON is slightly different, as the [and] symbols around the entity serialization show the inclusion in an array:

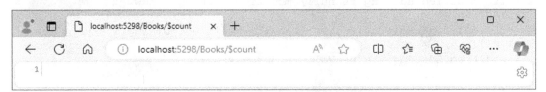

```
 1  [
 2      {
 3          "technicalId": "65e654acef3ad44b172bbb2a",
 4          "entityId": "0024",
 5          "isbn": "978-2409002205",
 6          "title": "Performance in .NET, Second Edition",
 7          "numberOfPages": 413,
 8          "publishDate": "2013-04-30T22:00:00Z",
 9          "editing": {
10              "numberOfChapters": 10,
11              "status": {
12                  "value": "archived"
13              }
14          },
15          "sales": {
16              "price": {
17                  "value": 42.15,
18                  "monetaryUnit": "EUR"
19              },
20              "weightInGrams": 768
21          }
22      }
23  ]
```

Figure 17.9 – Listing books

Finally, the counting of books is also tested (if you have noticed this is not the **Open Data Protocol** (**ODP**)-advised grammar, nice catch – we will correct this later in this chapter):

Figure 17.10 – Calling the count ODP operation on the API

Most of the operations are now ready. but we still need an important modification operation, namely the DELETE order, as well as a more complete way of reading data. Those are the subjects of the next sections.

The particular case of deletion

DELETE can mean a couple of things, and only the contract will specify exactly what is meant:

- Keep the data in the referential service, but make it not available to standard users anymore.

- Actually delete the entirety of the data (this is for when a mistake has been made or data should really be erased). This is a tricky one because GDPR can mean only obfuscating the data is possible. Also, a mistake might be interesting to keep, just in order to track possible consequences and solve them afterward.

In short, an MDM approach to data prevents you from simply implementing the DELETE operation as a simple, pure deletion of all the data. I propose here a simple approach, which would of course come from product owners in a real project, and that consists of archiving the entity:

```cs
[HttpDelete]
[Route("{entityId}")]
public async Task<IActionResult> Delete(
    string entityId,
    [FromQuery] DateTimeOffset? providedValueDate = null)
{
    // We retrieve the object in order to apply a modification,
setting the status to archived
    var result = Database.GetCollection<Book>("books-bestsofar").
Find(item => item.EntityId == entityId);
    if (result.CountDocuments() == 0)
        return new NotFoundResult();
    Book book = result.First();

    // A second version of the entity is created in order to apply the
modification to it
    Book modified = result.First();
    if (modified.Editing is null) modified.Editing = new
EditingPetal();
    if (modified.Editing.Status is null) modified.Editing.Status = new
Status();
    modified.Editing.Status.Value = "archived";

    // A JSONPatch is then generated to carry the status modification
    JsonPatchDocument equivPatches = JSONPatchHelper.CreatePatch(
        book,
        modified,
        new DefaultContractResolver() { NamingStrategy = new
CamelCaseNamingStrategy() }
    );

    // After creating the equivalent patch, we thus pass this over to
the PATCH operation
    DateTimeOffset valueDate = providedValueDate.
GetValueOrDefault(DateTimeOffset.UtcNow);
    return await Patch(book.EntityId, equivPatches, valueDate);
}
```

Of course, the `Get` and `GetBooksCount` operations should also evolve in order to take into account the fact that archived books should not be seen by default (and, incidentally, the Postman example used before should evolve in order for the book created to be visible):

```cs
[HttpGet]
public IActionResult Get(
    [FromQuery(Name = "$orderby")] string orderby = "",
    [FromQuery(Name = "$skip")] int skip = 0,
    [FromQuery(Name = "$top")] int top = 20)
{
    // The Find method of the MongoDB driver supports ordering and
    takes into account archived books that are not seen by default
    var query = Database.GetCollection<Book>("books-bestsofar").Find(r
=> r.Editing == null || r.Editing.Status == null || r.Editing.Status.
Value != "archived");

    // The rest of the function does not change
}


[HttpGet]
[Route("$count")]
public long GetBooksCount()
{
    // The count is of course done on the best so far collection,
    otherwise, it would add up historical states
    return Database.GetCollection<Book>("books-bestsofar").
CountDocuments(r => r.Editing == null || r.Editing.Status == null ||
r.Editing.Status.Value != "archived");
}
```

However, the `GetUnique` method is not modified: it is considered that, if someone is able to address a book by its precise identifier, then it does not make sense to hide it if it is archived. Again, this is a business hypothesis, just like the whole idea of an archiving implementation for the DELETE verb: some other people may decide to add an `active` attribute, set some other property, displace the entities into separate "trashcan" collections, and so on. Again, it only matters that what you do represent is the business' way of thinking. If product owners tell you that deleting an insurance file means only making it invisible to normal employees and having it be recoverable by somebody from the archive department, then this is what your code should mimic. This is the heart of business/IT alignment.

In some cases, it might be useful to propose a complete date removal. A possibility, in this case, is to still use the DELETE verb, as it is the semantically correct one, but adding an argument to the function that specifies the removal should be total, a bit like this:

```cs
[HttpDelete]
[Route("{entityId}")]
public async Task<IActionResult> Delete(
    string entityId,
    [FromQuery] DateTimeOffset? providedValueDate = null,
    [FromQuery] bool fullDeleteIncludingHistory = false)
{

    if (fullDeleteIncludingHistory)
    {
        // If this flag is activated (and it could be linked to a
special authorization), the object is indeed deleted
        await Database.GetCollection<ChangeUnit>("books-changes").
DeleteManyAsync(Builders<ChangeUnit>.Filter.Eq(item => item.EntityId,
entityId));
        await Database.GetCollection<ObjectState<Book>>("books-
states").DeleteManyAsync(Builders<ObjectState<Book>>.Filter.Eq(item =>
item.EntityId, entityId));
        await Database.GetCollection<Book>("books-bestsofar").
DeleteManyAsync(Builders<Book>.Filter.Eq(item => item.EntityId,
entityId));
        return new OkResult();
    }
    else
    {
        // Here would be the initial implementation of the function
already shown above, but not repeated in order to avoid overloading
the text with technical code that would not be useful
    }
}
```

A good practice when creating an API implementation is to add at least one Postman request for each operation, the minimum being to test the nominal case where everything works fine. If you do so, your Postman collection should look like the following by now:

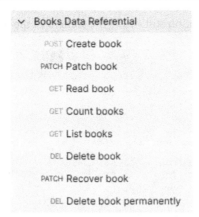

Figure 17.11 – Postman collection for the books data referential service

A Postman call shows that the call to DELETE changes the status of the book:

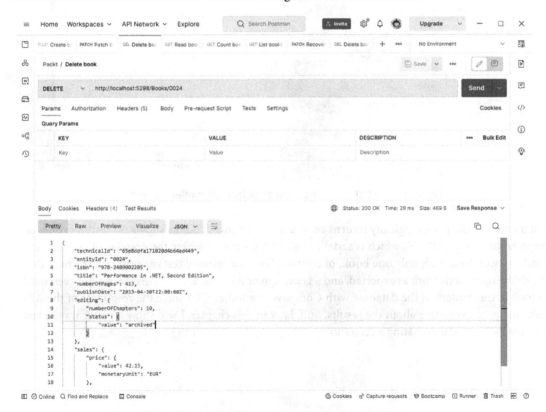

Figure 17.12 – Book turned into archived status by a DELETE request

As expected, the unique read operation still shows the content of the book:

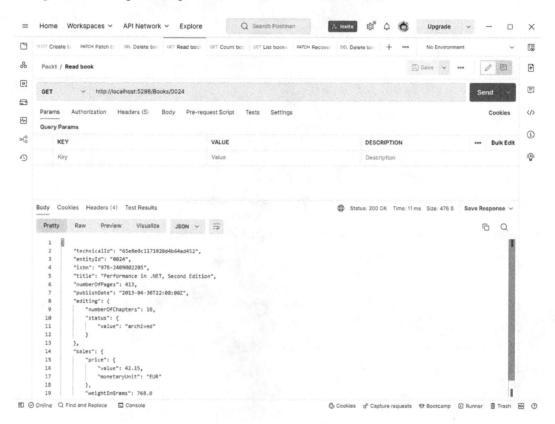

Figure 17.13 – Calling an archived book by its identifier shows it

But a count of the books logically returns zero, and displaying all books shows no result, expressed by an empty array in JSON, which is serialized as [] (in the case where the exercise of creation and patching was done with only one book, of course). For now, manual tests are enough to check that the delete operations work as expected, and a good sign of API readability is that we do not even need to look at the content of the database with Compass or another GUI since the responses of the API calls are quite expressive about the results. Still, later in this chapter, I will show a better way to test API operations, by automating a scenario.

As final words in this section, if we had to implement a "delete all books" function, we would discuss with product owners the use of this function and maybe even its purpose. In fact, there seem to be few cases where this is useful, so the best thing may be to just not implement anything. This is what we will do here, for the sake of abbreviating this long chapter, but also to illustrate that another consequence of a good business/IT alignment is that, when the business does not really need a feature, it is way better not to even think of it, let alone add it to the existing operations.

Now that we have covered everything about writing information in the data referential service, we will deal with reading it, otherwise it would not make much sense to have written anything at all. Generally speaking, read operations are explained before write operations because they are easier, but I thought it would be more logical in this case to go the other way around, as I needed to create some data before reading (which is more realistic, by the way, since it is closer to what happens in production, either manually or by migration). Also, in this case, where we record changes but read the deduced states, it is much easier to understand this way.

Querying as an extension of the reading responsibility

The product owners or customers can now create, patch, and delete some data while persisting it. Let's say that the request for the next programming sprint is for customers to be able to query books from the data referential service API. How to read a single book entity was shown earlier, but querying data in all its forms is so much more complex than reading one entity that it justifies a dedicated section, in which we will show how data should be filtered, paginated, exported, and so on.

Using ODP for a standard query API

One of the main themes of this book still holds true in this section: one should always use norms and standards in order to be compatible with as many uses as possible without having to take some time analyzing them. This is why, when it comes to supporting the querying of data through an API, we turn to the SQL of the API, which is ODP, and in particular the documentation of the associated protocol (`http://docs.oasis-open.org/odata/odata/v4.01/odata-v4.01-part1-protocol.html`).

We could also turn to GraphQL, which is even more powerful, but this would push us too far away from the main objective of this chapter, which is to create an MDM-enabled data referential service. We will therefore stick to ODP and quickly show how basic manipulations can be realized.

Pagination, filtering, and more

ODP is great because it standardizes a whole lot of commands for querying data through HTTP. It even enables **pagination**. The following code could be added to our controller in order to allow this type of request; the second and third arguments of the function will respectively handle the pagination offset and number of entities, and the MongoDB library methods allow us to simply pass them as numbers to the Skip and Limit functions, which apply to the query we have generated using the first argument, which will order the data (we could also add a filter):

```cs
[HttpGet]
public IActionResult Get(
    [FromQuery(Name = "$orderby")] string orderby = "",
    [FromQuery(Name = "$skip")] int skip = 0,
    [FromQuery(Name = "$top")] int top = 20)
{
    // The Find method of the MongoDB driver supports ordering
    var query = Database.GetCollection<Book>("books-bestsofar").Find(r
=> true);
    if (!string.IsNullOrEmpty(orderby))
    {
        string jsonSort = string.Empty;
        foreach (string item in orderby.Split(','))
        {
            if (item.ToLower().EndsWith(" desc"))
                jsonSort += "," + item.Split(' ')[0] + ":-1";
            else if (item.ToLower().EndsWith(" asc"))
                jsonSort += "," + item.Split(' ')[0] + ":1";
            else if (item.Length > 0)
                jsonSort += "," + item + ":1";
        }
        query.Sort("{" + jsonSort.Substring(1) + "}");
    }

    // It also supports active pagination
    var result = query.Skip(skip).Limit(top).ToList();
    return new JsonResult(result);
}
```

The parameters allow us to express this kind of request:

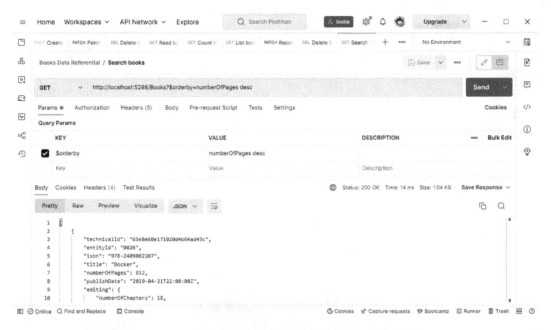

Figure 17.14 – Ordering books by descending number of pages

Active pagination can also be realized using combinations of $top and $skip, respectively, to retrieve only a given number of entities after an offset, thereby allowing us to progress through the list, which lightens the burden of the server exposing the data. A very simple example follows (the output has been partially collapsed in order to show the number of entities sent back by the server):

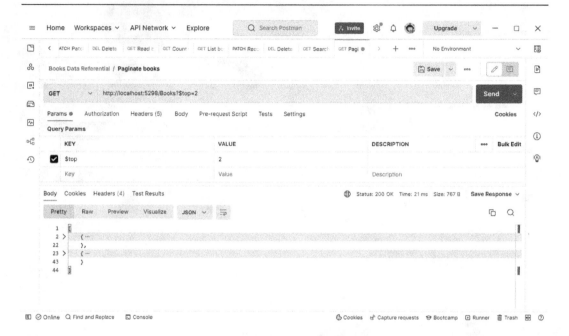

Figure 17.15 – Example of pagination

The ODP syntax also sports a `$filter` condition (the equivalent of a `WHERE` clause in SQL), with a grammar for the function that is so detailed that any implementation should use existing SDKs in order to be sure to implement it correctly. Again, this is not our subject here, but it's important to show one particular case of filtering data, and this is the one about value date. Indeed, the capacity to not forget any change and to go back in history is very important for a data referential service and makes filtering data stand out in comparison to usual data persistence means. Although the same call without the time argument is used much more often, there is real value in being able to query `http://localhost:5298/Books/0024?providedValueDate=2024-03-06T23:00:00`, for example.

A naive implementation of such a behavior would be the following evolution of the `GetUnique` function (this is one of many other functions that should evolve in the same manner):

```cs
[HttpGet]
[Route("{entityId}")]
public IActionResult GetUnique(string entityId, [FromQuery]
DateTimeOffset? providedValueDate = null)
{
    if (providedValueDate is null)
    {
        // As long as no parameter is added, the behavior is simply to
find and retrieve the best-known state
```

```
                var result = Database.GetCollection<Book>("books-bestsofar").
Find(item => item.EntityId == entityId);
            if (result.CountDocuments() == 0)
                return new NotFoundResult();
            else
                return new JsonResult(result.First());
        }
        else
        {
            // Naive implementation at first, to be evolved to a single
query only when actual performance is at stake
            var states = Database.GetCollection<ObjectState<Book>>("books-
states").Find(item => item.EntityId == entityId);
            if (states.CountDocuments() == 0)
                return new NotFoundResult();
            else
            {
                var results = states.SortBy(item => item.ValueDate).
ToList();
                if (providedValueDate < results[0].ValueDate)
                    return new NotFoundObjectResult("Provided value date
is prior to entity creation");
                int index = 0;
                do
                {
                    index++;
                }
                while (results[index].ValueDate < providedValueDate);
                return new JsonResult(results[index].State);
            }
        }
    }
}
```

As you can see, if there is no value date passed as an argument, we turn back to the initial code. But if there is one, then a very simple algorithm is applied. First, if there is no data in the list of states, then we return the corresponding NotFound message. But if there are some states, after ordering them, we descend down the list until we pass the value date we have fixed, and then we send the last state we found in the loop, which would be the state the entity was in at the value date or before.

Even if we start as unsophisticated as possible (remember the **KISS principle**), a minimal business rule should be to test that the value date makes sense in the history of the object. If the value date is before the object was created, for example, there should be some message like the following, allowing us to understand that the situation is different from when the identifier does not correspond to an actual object, though the result is quite similar:

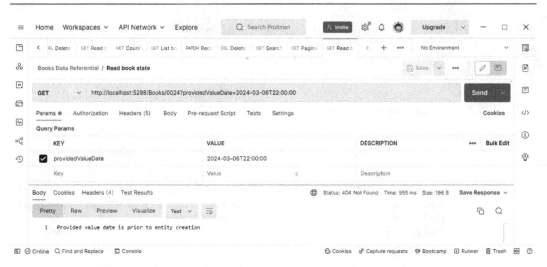

Figure 17.16 – Error when trying to access a state prior to book creation

We could elaborate on this implementation and work out the fact that the value date is not really a date in administrative form but rather an instant in the ISO-Time standard, or find out how we should write the response messages more precisely with a structured approach (there are some competitive standard proposals), but we will leave it here for the querying parts of the API and turn to the reporting aspects.

Links to business intelligence and big data systems

Single queries are used for machine-to-machine communication, but API read operations, particularly those exposing volumes of data, can also be used for reporting. A software tool should still be used to manipulate the data, but the interface would then be read by a human.

One of the most well-known and still heavily used approaches is the flat-file export approach. We could (and should, if there are some legacy uses in our information system) handle content negotiation in HTTP to respond to API requests with XML, or even CSV, instead of JSON. In order to respect the HTTP standard, this should be done by sending accepted MIME types in the `Accept` header of the request, for example, `application/xml` or `text/csv`. The response would thus include a `Content` header with the MIME type used to answer.

But as more and more business intelligence and reporting software tools are aware of API sources, it becomes logical to plug them directly into the API. For example, we could use the Power Query module of Excel to read books from our API. Let's see how we can turn our structured data into tables in Excel and manipulate data. First of all, in the **Data** tab of Excel, you need to choose the **From Web** data source provider:

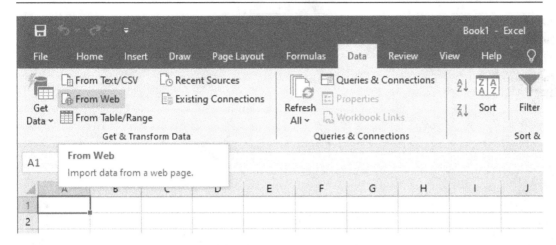

Figure 17.17 – Using the web data provider in Excel

Then, you provide the URL of the Books API:

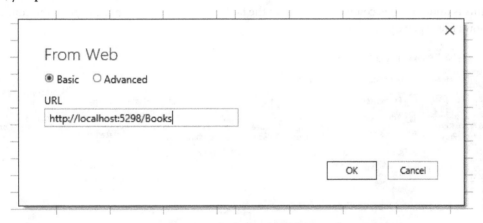

Figure 17.18 – Choosing the URL for the books data referential service

The transformation tool of Power Query appears and will let you extract the JSON structured format into a table. At first, the module sees a list of records, but it needs to turn them into a table to access the content. This is done with the **To Table** command:

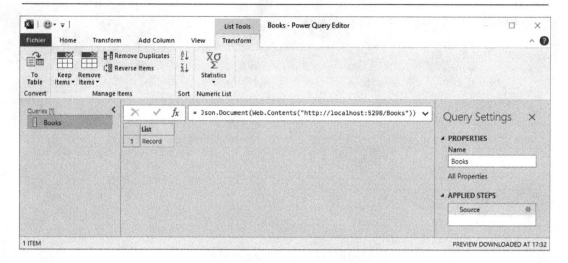

Figure 17.19 – Power Query interface

Once this is done, a single column is created in the table, but a small icon on the right of the column title allows the expansion of the record content:

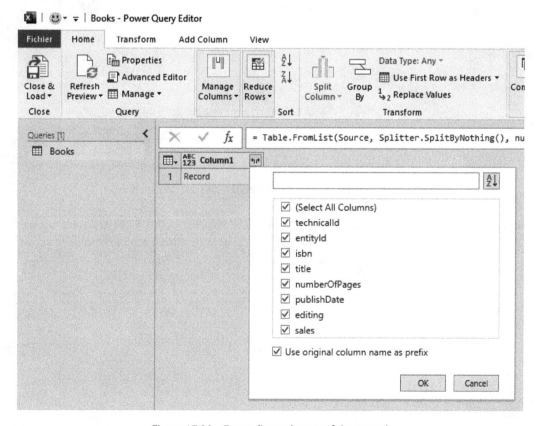

Figure 17.20 – Expanding columns of the record

When the records are expanded, this is done one level at a time, and at first, we see the root content. The `Editing` and `Sales` petals need to go through the same manipulation if we decide to get some data from there:

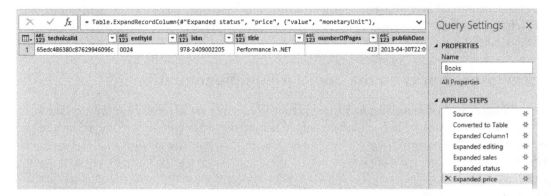

Figure 17.21 – Columns obtained

A few manipulations, such as the removal of useless columns or changing the type of data, allow us to obtain our data:

Figure 17.22 – Display of Power Query after column adjustments

When everything is ready, clicking on the **Close and load** button (which you can see in *Figure 17.20*, and that always sits in the top-left corner of the Power Query interface) starts the actual loading of the data from the API (the Power Query module uses a sample of data to prepare its transformations) and puts them into a table that we can edit, or use to add calculated column – an operation that could also be done in Power Query itself:

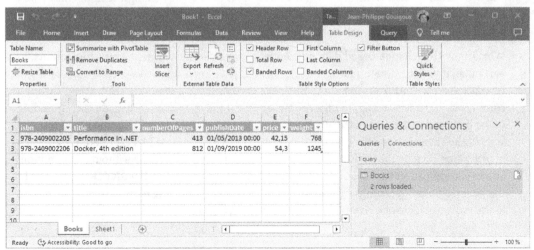

Figure 17.23 – Final imported results

As far as big data systems are concerned, API requests can also be used to feed data in, and this is precisely one of the cases where knowing the history of the data is important: not only could a data lake handle the evolution of the books information better, but in addition, knowing where changes have happened would help with getting only small amounts of incremental data instead of calling the data referential service for huge volumes of data when most of it has already been provided through previous calls.

A few last remarks on the books API implementation

Lots of things could be said about how to program the data referential service and how to improve the preceding code, which has been kept as simple as possible to not prevent the comprehension of the most important concepts; there could be a complete book on how to code such an MDM-enabled API. In this last section on the implementation of the API (after this, we will turn to how to test and deploy it), I will simply go over a few of these evolutions that are usually made in industrial-grade data referential services:

- First, the use of a complete SDK to implement ODP or GraphQL is highly recommended: it would take ages and create lots of bugs to reimplement the whole grammar analysis system, and it is already challenging enough to apply the SDL to the context of the data we have at hand.

- A PUT operation could be handled, and its implementation would consist of transforming it to a patch operation by comparing the object sent to the best definition of the same entity that we have so far. Of course, there will always be some ambiguity depending on what version of the entity the client based their new version on, and new changes can happen in between. One way or another, we will be faced with concurrency issues again, but the great thing about having a change-based system is that anyone can then go back in time and analyze all data changes, for example, to retrieve data that has been erased by a PUT instance that was based on an old state. Note that, in this kind of implementation, the PUT instance cannot be considered idempotent anymore without having a complete comparison of its content to the best so far data, which will avoid adding useless history when the entity does not actually change. Idempotency could then be implemented in the PATCH instance as well, considering that a modification with a zero list of atomic changes will not be recorded.

- Working with a change-based data referential service can have a strong effect on clients, in particular, if you use a state-management system in your applications, such as Redux or Fluxor. In this case, all modifications to the business entity on the client side are tracked through commands and a reducer applies the modifications. If the commands are made corresponding to the patching of the associated attributes, there can be an interesting alignment between the state of the client and the states of the server, possibly completing a PATCH operation for each modification done in the GUI. This makes it easy to implement interfaces without a **Save** button, as each modification command can be connected to a server call with a PATCH operation. In this case, some optimizations may appear logical, such as the erasing of two states if one reverts to the previous value of a given attribute. But, if you really want to avoid concurrency problems and obey traceability rules, the temporary state should appear in history, even if it existed for just a few seconds because you never know what business operation can be run by another client of the data referential service in the meantime. If this is too much of a burden in your context, stick to the **Save**-button approach or at least propose a way in the GUI to make several changes without saving them immediately (through a timer or a dedicated editing mode). This can be very helpful when small modifications can start complex computations and data refreshes on the server side and you do not want to pay too high a performance toll.

- Since an object will be created from a single `PATCH` command, and creating an object is simply a special kind of patch where the initial entity is empty, the frontier between creation and modification is blurred; this can make it tough to reason in terms of data duplicates. In fact, once we're used to the approach, it actually makes it simpler because duplicates are no longer objects that have been created twice or that have become identical to each other through modification. Duplicates, in this approach, are identified by a business rule that compares objects to each other. Business/IT alignment is also there to provide this kind of rule or, even better, provide a business-unique identifier (such as ISBNs for books, in our case) that makes finding duplicates a breeze.

- Our "best-so-far" value is meant to be the latest, but technically we should always apply a restriction based on the value date when reading data. This is important because there is nothing stopping us from entering value dates in the future when making changes, such as in banking. This means that when we call the "best-so-far" reading operation, we could be seeing the value of an entity that is in the future. If we want to ensure that the default reading is the closest state in the past at the time of the request, the read operation should always consider the date of the client.

- Authorizations have been largely ignored here but, in real production data referential services, they are often used in conjunction with the ability to change the history of an object, make changes to old versions, or add changes in the future, completely remove the history, and so on.

- The implementation with three collections is not transactional but could be made eventually consistent without too much effort by considering that the addition in `books-changes` alone is enough to retrieve the whole history of the data. One way to deal with this would be to have an asynchronous process that recalculates the states and best-so-far images in the background. In order to have fresh data all the time, this should be run every time a change appears. One often-used way to avoid performance issues when recalculating long histories is to use intermediate states, for example, every 50 changes. This way, if the client requests the state at version 162, the algorithm can start with a cached version 150 and only apply the patch from the next 12 changes.

All these remarks are the reason why the code shown in this book remains very simple and the architecture is not as decoupled and clean as should be in a large, production-ready service: if we had taken all this into account, the main subject of the book (in this chapter, showing how a good MDM should be implemented in order to solve the business/IT alignment) would have been lost in noise made by technical code.

The next part of the chapter will also reflect this, as it is about testing and, in actual production quality, it would have needed a complete chapter in itself. Since the goal here is not to be exhaustive but to show best practices, and also in the interests of maximal simplicity, this section on testing will not be too long. Still, it will allow us to close the loop and check that everything we have prepared in this data referential service is working smoothly.

Testing

Testing software is of utmost importance and it is impossible to say it enough. Forget the myth of the 10x developer: the real hero of the next era of industrialized software is the QA engineer! Again, this is a huge subject that would be worth another book talking about testing strategies, shift-left campaigns, security and performance, and the balance between unit/system/integration/end-to-end tests, but this is not the right place to do so, and I will therefore concentrate on the part of testing that is linked to business/IT alignment, and that means focusing on the contractual interfaces between the modules, namely the APIs.

Testing at the API level

Although other tests are also important and there should be a range of different types of tests to ensure the right coverage, testing at the API level is certainly the most important type to invest in when building an API-based information system. Indeed, services are related to one another, so testing them right at the API level allows us to verify that the whole chain of service is working and, at the same time, provides a view of the results from the functional perspective, if the alignment is correct: when telling the director that a test has shown a problem in a function, it is always better to say, "This is inside the books referential service; the function providing the total amount of sales is broken," than, "The connection string is not recognized because the password has changed and the push to the data lake is broken." Again, business/IT alignment allows a profound exchange between owners and technicians, and this is one of the most valuable advantages of having a well-designed API.

I should be clearer about API tests: what we really want to test is the adequacy of the implementation with the contract. The API contract does not really have to be tested per se: it should come from the product owners and, even if their business comprehension should be challenged, once the contract is signed, it is considered the lingua franca for the information system. This is why, by the way, it is so important to get this right: if you have a problem in your understanding of the business itself, there is absolutely no way that the IT system that you build as a symbolic representation of this business reality will be helpful!

One can of course create some tests for the compatibility of versions of the API. When lots of them are added, it may be worth automating the verification of the backward compatibility of the API, by automatically checking that all previous attributes and commands are there, that new attributes are not mandatory, and so on. But this kind of thing should happen scarcely, as it is not a good sign if your API contracts vary too often.

Most of the tests will be on the implementation of the API and whether it correctly implements the contract. That means of course that the endpoint that constitutes the system under test should obey the OpenAPI contract but also all the constraints of the contract that are not written in the OpenAPI file; these are generally referred to as secondary features, although these are actually as important as the technical response itself. The vocabulary used is, in my opinion, not right, because nobody in their right mind would consider that performance and security are *secondary*. There is also everything that could not be clearly expressed in the OpenAPI format, because lots of things remain to be standardized in this grammar – the latest version might require support for webhooks, for example. Anyway, the single OpenAPI file will never be 100% complete to define the full contract, the SLA, and so on. It is thus important to document the API use, and automated tests are a great way to do this. When working with integrators, if you provide the OpenAPI file, it is already a good sign. But if you also provide a comprehensive Postman collection helping the integrators understand the different calls that can be made with actual values, and they can use this to validate the system works as expected by adding their own calls, then you will get excellent cooperation.

I talk about Postman (but it could be any other similar platform) because, if you have made the effort of testing your implementation as you build it, as I have shown here, you end up with a great test harness in the form of your collection, and in particular the possibility of launching a *Runner*, which involves the automatic sending of all requests in sequence, with a unique log of results. This is the equivalent of a NUnit, but for API calls. How you do it is as follows: in the collection of requests, click on the ellipsis on the right and select the **Run collection** command in the contextual menu:

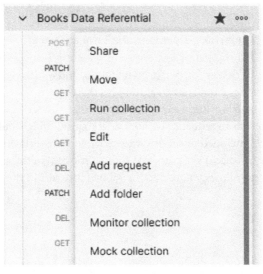

Figure 17.24 – Running a collection in Postman

The GUI displayed after running this command, shows how rich the command is:

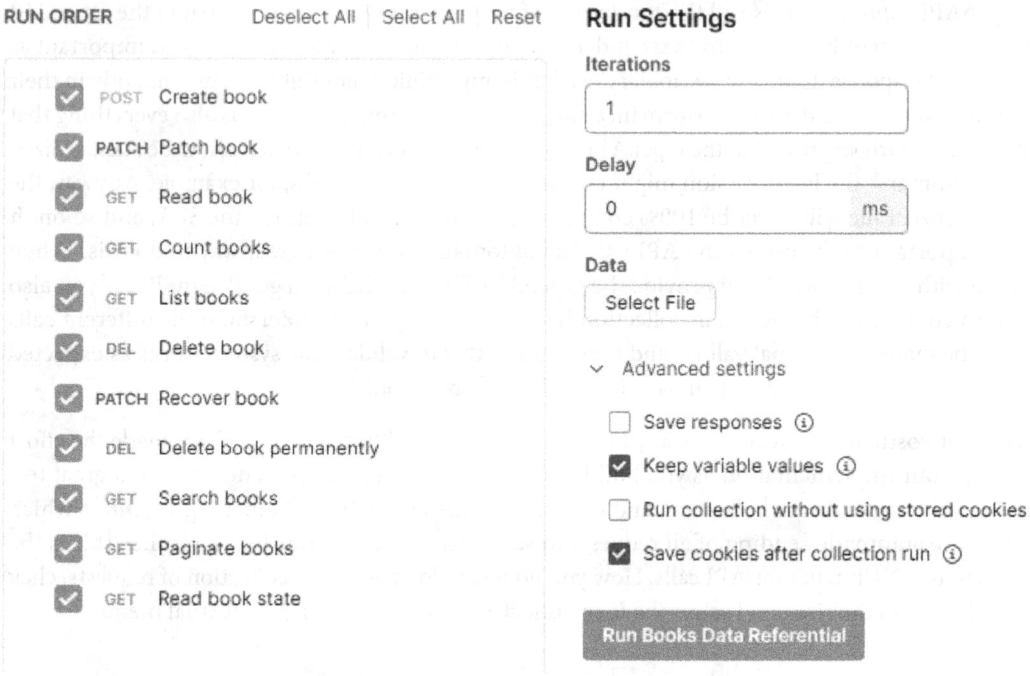

Figure 17.25 – Options for the Postman Runner

You can define iterations with delays for performance tests, add a file with multiple data lines in order to run the tests on batches of dedicated values, save the response in order to debug problems afterward, and more. For this to work, it only makes sense that you write asserts in your requests. Indeed, simply executing requests in sequence without verifying anything would not be a test harness. Luckily, you have a lot of help included in the software to perform asserts once the request has been run. For example, the `Status code: Code is 200` command in the snippet's proposal on the right-hand side of the **Tests** tab allows you to add the following assert in the **Create book** request:

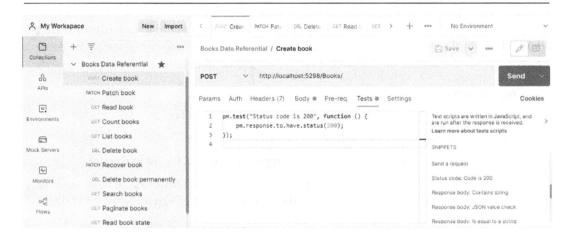

Figure 17.26 – Adding an assert to a test request

You can use the same method to add more assertions to the response content and even evaluate the call performance. You can then review the results in the dedicated tab of the response section, which is located at the bottom by default. Remember that your assertions should always be organized from the most important and general to the most detailed:

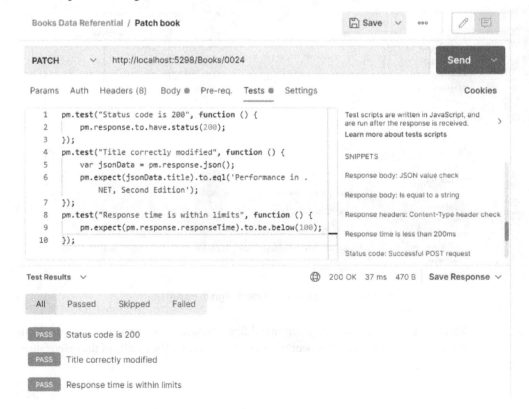

Figure 17.27 – Applying test asserts to a request

These features allow you to create more sophisticated test scenarios, for example, to verify the correct working of the soft/hard delete that has been put in place in the books referential service by doing the following:

1. Creating a book

2. Checking that it can be read and appears in the count and list of books

3. Soft deleting it (which means it will be put in archives)

4. Checking that it still can be read but does not appear anymore in the list and count

5. Recovering the book by patching its status to available (by the way, a better way to do so would be to re-create the state prior to archiving, since there would be no need to retrieve the target status)

6. Verifying that the read operations work just like after the initial creation

7. Performing a hard delete on the book and its whole history

8. Verifying that not only do the list and count not show it but also that even a direct read of the book identity does not work anymore and sends a **404 Not Found** message.

When executing Postman Runner, this kind of scenario appears as follows; the number of asserts that passed or failed enables us to quickly know whether the whole API's implementation works:

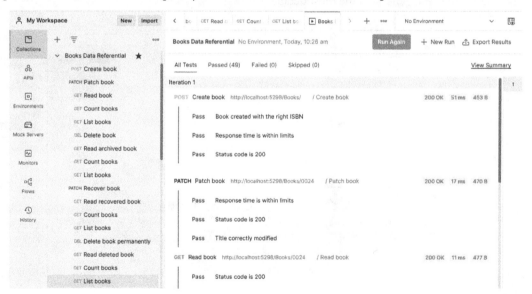

Figure 17.28 – Complete test campaign in Postman

Of course, the Runner can be launched from a command-line interface, which means all the campaigns can be run from your continuous integration workflow and thus increase the quality of the deliverables.

Managing dependencies

Since we have already talked about unit tests, behavior-driven development, and Gherkin grammar in *Chapter 13*, we are not going to dig further into the huge subject of testing and quality assurance here, but will simply look at a big change in the testing of APIs when using a modular architecture.

Generally, when one does not test at the unit level but a higher one, the question arises of how to take dependencies into account. Indeed, testing integrated software always means testing all of the modules working together and not necessarily being clear about which part is the culprit if the test fails. In this case, how could we say that we have tested the API, for example, if it is based on a database that also can fail? The solution is to strongly reduce the possibility of the dependency itself failing, and there have been several ways to do so proposed.

The first approach was to create a dedicated dependency for the test. In our case, this would mean a database specially prepared for the tests. Luckily, in our small test campaign, we started from an empty state, so it was not difficult to create such a database. But if you test correctly, you will use a database with production-like data, complete with errors and quirks; you should also use high volumes of data in order to verify performance. When you do so, there is another difficulty that comes up, namely that, once the database goes through the tests, its state is not the same as when entering them. Also, rollback is simply not available because there may have been many transactions during the test. So, generally, we resort to crafting a test database, backing it up, sending databases made from the backup to the tests, and then throwing them away. Storing, maintaining, and versioning these test databases was a job on its own...

This is why a second approach, with mocks and stubs, arose, where we would replace the dependency to a given module with one that simply mimics its behavior from code. Instead of using a real database connection in the API code (in our example case), we would carefully use dependency injection for all of those dependencies (which is well aligned with clean architecture) and inject a real database driver in production, but use a specialized bit of code for tests that would reproduce the expected behavior. This was a great step forward that helped a lot in the spreading of testing best practices, but creating the right mock comes with some design challenges and, in particular, mocks would tend to present "too perfect" situations. For example, they would have no latency compared to a real database; and they would also short-circuit the database driver, which is still a part of the system under test (though it's admittedly not supposed to fail).

When you think about it, Docker solves most of the problems of the first approach, and so it should certainly be reconsidered with Docker in mind. After all, versioning is easy for containers, a database inside a container is exactly the same as a database outside a container (in fact, production may be containerized, though I would not necessarily recommend it for databases), and even better, restoring the initial state of a database is as simple as removing its top volume after use and running again, with such a small addition as `-rm` in the `docker run` command, or ending your tests with a `docker rm -fv db` instruction to remove the data changes and start afresh on the next run of the container.

There is even a great tool called **TestContainers** (https://testcontainers.com/) to help you create containers on the fly from your unit tests in C#! It is the perfect way to get out of the unit/system dichotomy of tests while remaining as close as possible to real-world situations for your quality assurance.

Summary

After this second consecutive chapter on the data referential service, we now have something that works for most of the necessary features, or at least the ones that are the most important to demonstrate how to record changes. We finished this chapter with testing in order to verify that everything works as expected.

This chapter was also an opportunity to look at the use of standards; the implementation of read capacity has introduced the use of the ODP OASIS-defined standard. Respect for such norms allows much better interoperability and is a must-have feature for any application that seeks to implement MDM.

Creating one data referential service for books was a hard task in itself, hence it required two chapters. In the next chapter, we will deploy this service and then create a second service, which will be a bit more sophisticated (but we will go through the explanation more quickly since the first one was explained in so much detail in this chapter).

18

Deploying Data Referential Services

In the previous chapter, we saw how to create a data referential service. We left questions about how to plug this MDM service into authentication, authorization, and generally speaking, how to deploy it in the information system for this chapter.

We will see how to produce Docker images for data referential services (the books data referential service, but also the authors data referential service that we will deal with in the next chapter). These images will then be used in the following chapters as two of the main services of our sample information system. By the end of the next chapter, we will be able to deploy all the bricks needed for DemoEditor, and the following chapters will be about making interfaces between them, integrating them into complex business processes that necessitate functions brought by several of them, etc.

We'll cover these topics in this chapter:

- How to deploy the data referential service we have created in the previous chapter
- How to integrate a way to check that the service is working fine
- How to add authentication to the data referential service
- How to make it support authorization, first with roles and after that with more granular rights

Installing the data referential service in the information system

Let's imagine our DemoEditor customer is happy with the way the books data referential service works and wants to start production with it. This is generally when you will explain to them that some important steps remain, some of which are quite technical but still necessary, and some of which your customer should appreciate the importance of right away, such as plugging into the IAM and the authorization modules. Otherwise, everyone with access to the API can see the entire content, which is rarely adequate.

In this equivalent of a fourth agile sprint, we will adjust these features and create what is necessary to deploy our API code and database.

Docker deployment

Since we deployed most of the necessary modules of our sample information system on Docker in *Chapter 15*, it is only logical that we use the same approach for our new application, namely the books data referential service API. To do so, we of course need a `Dockerfile` file with all that is necessary to compile a Docker image from the ASP.NET Core source code files. Any Docker plugin will do the trick, but I would like to show you this new command from Docker, which is `docker init` (https://docs.docker.com/reference/cli/docker/init/). This command walks us through a wizard to create everything we need.

Here is the `docker init` command in the VS Code terminal:

```
PS C:\Code\Github\jp-gouigoux\DemoEditor> docker init

Welcome to the Docker Init CLI!

This utility will walk you through creating the following files with sensible defaults for your project:
  - .dockerignore
  - Dockerfile
  - compose.yaml
  - README.Docker.md

Let's get started!

WARNING: The following Docker files already exist in this directory:
  - docker-compose.yml

? Do you want to overwrite them? Yes
? What application platform does your project use?  [Use arrows to move, type to filter]
> ASP.NET Core - (detected) suitable for an ASP.NET Core application
  Go - suitable for a Go server application
  Python - suitable for a Python server application
  Node - suitable for a Node server application
  Rust - suitable for a Rust server application
  PHP with Apache - suitable for a PHP web application
  Java - suitable for a Java application that uses Maven and packages as an uber jar
  Other - general purpose starting point for containerizing your application
  Don't see something you need? Let us know!
  Quit
```

Figure 18.1 – The docker init command in the VS Code terminal

Here is the output; it shows the resulting files from the previous command:

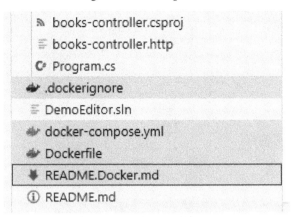

Figure 18.2 – Files created by the docker init command

After providing the simple parameters requested (type of project, version of the framework, etc.), the command takes care of creating several files, among which is a comprehensive `Dockerfile` that looks as follows (comments are not reproduced in order to make this code shorter, but they are provided on GitHub and will help you to understand how to adjust behavior):

```
# syntax=docker/dockerfile:1

# Comments are provided throughout this file to help you get started.
FROM --platform=$BUILDPLATFORM mcr.microsoft.com/dotnet/sdk:8.0-alpine
AS build
COPY . /source
WORKDIR /source
ARG TARGETARCH
RUN --mount=type=cache,id=nuget,target=/root/.nuget/packages \
    dotnet publish -a ${TARGETARCH/amd64/x64} --use-current-runtime
--self-contained false -o /app
FROM mcr.microsoft.com/dotnet/aspnet:8.0-alpine AS final
WORKDIR /app
COPY --from=build /app .
USER $APP_UID
ENTRYPOINT ["dotnet", "books-controller.dll"]
```

Creating the Docker image for our books data referential service is as easy as typing the following command and waiting for the result:

```
docker build -t demoeditor/books-api:0.1 .
```

The rest of the integration into the sample information system consists of adding the corresponding service into the `docker-compose.yml` file, which describes the services included in our application for now, together with the ports they expose and other customizations. This should resemble this:

```yaml
version: '3.0'
services:
  db:
    image: mongo:4.4
  dbgui:
    image: mongo-express:1.0.2
    ports:
      - 8081:8081
    environment:
      - ME_CONFIG_BASICAUTH_USERNAME=forster
      - ME_CONFIG_BASICAUTH_PASSWORD=gOU89tRF05
  iam:
    image: quay.io/keycloak/keycloak:23.0.7
    ports:
      - 8080:8080
    environment:
      - KEYCLOAK_ADMIN=forster
      - KEYCLOAK_ADMIN_PASSWORD=4AFXbm5vX7YFjN0rMYKK
    command: start-dev
  mom:
    image: rabbitmq:3.13-management
    ports:
      - 15672:15672
      - 5672:5672
  books:
    image: demoeditor/books-api:0.1
    ports:
      - 80:8080
```

You might have noticed that I have not included the environment variable value to point the books service at the db service. This is because it is generally recommended to keep the default value for production purposes (Docker creates a DNS entry to point at services by names, so we should use this facility) and set the environment variable value when it is needed for local tests. This is why the code in the BooksController.cs file should be modified like the following:

```csharp
public BooksController(IConfiguration config, ILogger<BooksController>
logger)
{
    _logger = logger;
```

```
        ConnectionString = config.
GetValue<string>("BooksConnectionString") ?? "mongodb://db:27017";
        Database = new MongoClient(ConnectionString).GetDatabase("books");
}
```

If you want to avoid the preceding manipulation, you can also set an alias called db that points to localhost in your hosts file. You have perhaps also noticed that I changed the service name from mongo to db, and accordingly, express to dbgui. The reason for this may sound like a trivial detail, but it is in fact extremely important because it shows that the service really is the database responding to the MongoDB standard API and that the implementation (which is Mongo 4.4 and above) could easily change to CosmosDB or any other database that follows the same standard. It's the same as for the IAM: most of the people using the information system do not have to know whether we use Apache Keycloak, LemonLDAP::NG, or any other implementation of the OpenID Connect standard.

Adding a status operation to the API

It is a good practice, when adding a service, to also provide a way for an external orchestrator such as Docker or Kubernetes, and also for monitoring systems, to know whether the service is up and ready. Indeed, checking that the port answers may not be precise enough. In our case of a books data referential service that is highly dependent on the database behind it, an implementation of the status function could be something like the following code, which checks whether the database connection pings correctly:

```
[HttpGet]
[Route("Status")]
public ActionResult Status()
{
    // If the ping to the database works, the controller is considered
as working fine
    if (Database.RunCommandAsync((Command<BsonDocument>)"{ping:1}").
Wait(1000))
        return Ok();
    else
        return StatusCode(500);
}
```

We can then use the health check functions of Docker by adding the following commands to the Docker Compose definition of the books service:

```
books:
    image: demoeditor/books-api:0.1
    build:
        context: ./books-controller
        dockerfile: Dockerfile
```

```
    ports:
      - 80:8080
    restart: unless-stopped
    healthcheck:
      test: curl --fail http://localhost:8080/Status || exit 1
      interval: 30s
      retries: 5
      start_period: 10s
      timeout: 1s
```

If you run docker compose up -d, after a bit of time, you should see the following (notice the healthy mark on the Books API service) when you run a docker ps diagnostic command:

```
CONTAINER
ID  IMAGE                          COMMAND                 CREATED
            STATUS                          PORTS
48fb6be17a69  demoeditor/books-api:0.1  "dotnet books-contro…"
26 seconds ago  Up 23 seconds (health: healthy)  0.0.0.0:80->8080/tcp
```

This is the sign that everything should be fine as far as deployment is concerned and that we can carry on to the next step of deployment, which is plugging the data referential service into its dependencies and, in particular, the IAM.

Plugging the data referential service into the IAM

In this section, we'll learn how to prepare the IAM to be used by the books referential service. We will also integrate the code with the IAM, add the RBAC authorization based on IAM attributes, and look at ABAC's additional capabilities.

Preparing the IAM to be used by the books referential service

The next very important step in building the books data referential service is to plug it into the IAM for the obvious security reasons that were mentioned at the beginning of this section. To do so, we will lean on the first step of parameters that was done on Keycloak in *Chapter 15* and create what is called a client in the demoeditor realm. Since many applications will call the IAM in the end, it is indeed important that Keycloak remains in control of who requests an identity token from it, and this is done through **clients**.

In the **Clients** menu, we can see that there are already some default clients configured:

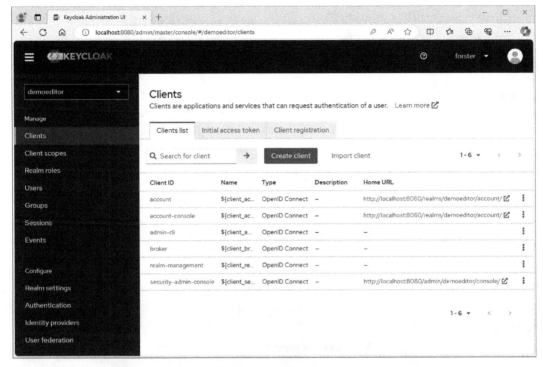

Figure 18.3 – Clients configured by default in a Keycloak realm

We will add our own client by clicking on **Create client** and filling in the first step of the wizard, like this:

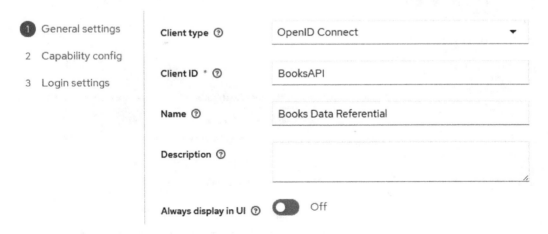

Figure 18.4 – Creating a client for the books referential service

As you can see, on the General settings page, these are the fields to fill in:

- **Client type**: **OpenID Connect**
- **Client ID**: BooksAPI
- **Name**: Books Data Referential

The second step of the wizard can be left as is for our purposes, except for **Client authentication**, which should be activated, but please read the information (by clicking on the question mark) to get a better understanding of these options if you do not know the OpenID Connect protocol:

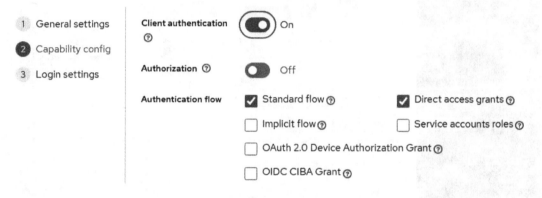

Figure 18.5 – OpenID Connect options for the client

Finally, fill in the following information to indicate the URLs that will be used by the ASP.NET authentication mechanism to communicate with the IAM service. Then, click on **Save**:

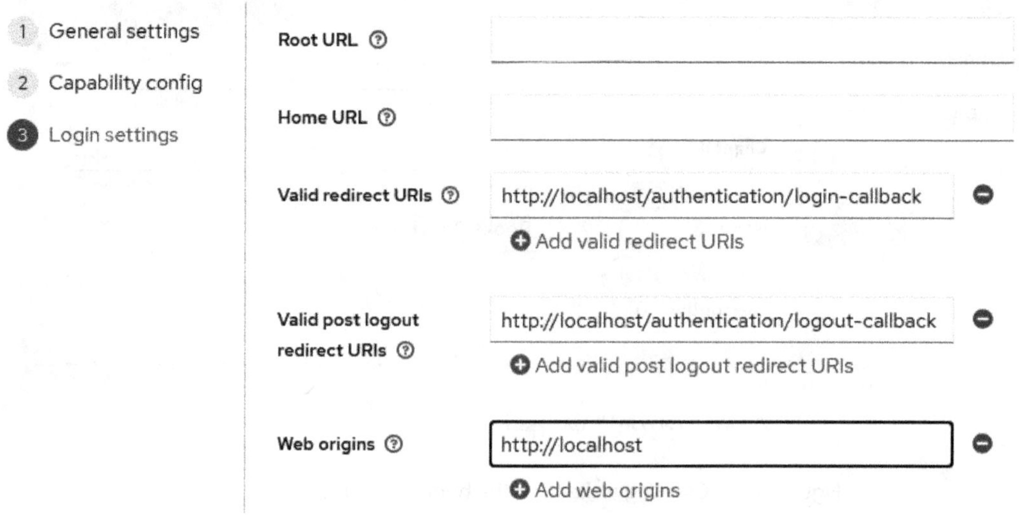

1 General settings	Root URL ⑦
2 Capability config	Home URL ⑦
3 Login settings	Valid redirect URIs ⑦ http://localhost/authentication/login-callback ⊖
	⊕ Add valid redirect URIs
	Valid post logout redirect URIs ⑦ http://localhost/authentication/logout-callback ⊖
	⊕ Add valid post logout redirect URIs
	Web origins ⑦ http://localhost ⊖
	⊕ Add web origins

Figure 18.6 – Connecting options for OIDC

As you can see, on the **Login settings** page, we fill in these fields:

- **Valid redirect URIs**: `http://localhost/authentication/login-callback`
- **Valid post logout redirect URIs**: `http://localhost/authentication/logout-callback`
- **Web origins**: `http://localhost`

Back in the interface on the page corresponding to the client that has just been created, we are going to create some associated roles in order to better control the authorizations. To do so, click on the **Roles** tab, then the **Create role** button, and fill in the form fields, with **Role name** as `editor` and **Description** as `Users with this role can realize most operations on the books`:

Create role

Role name *	editor
Description	Users with this role can realize most operations on the books

Save   Cancel

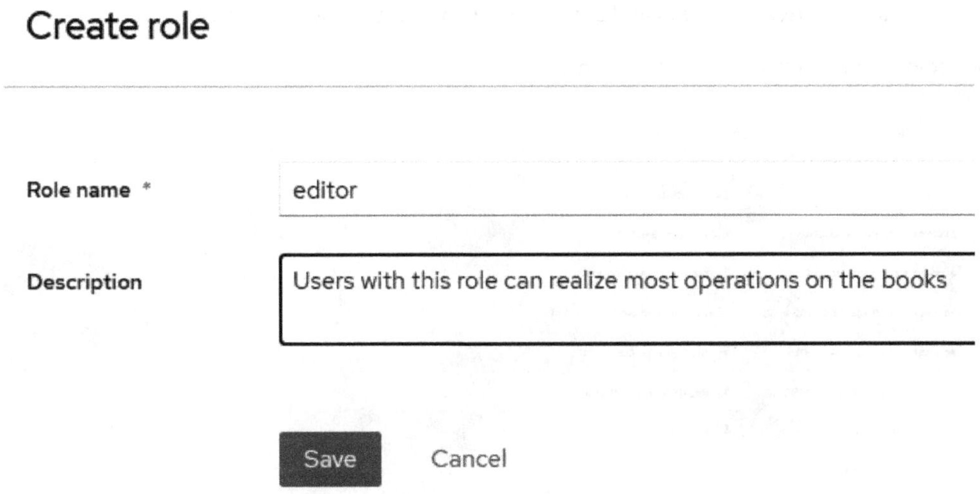

Figure 18.7 – Creating the editor role

Then create a second role, called `author`, with a description such as `Users with this role can only read the books for which they are one of the authors, and deposit some content`. In order for these roles to be used, we need to associate some of them with users. Go to the **Users** menu of the realm, select one of the external users that was created in *Chapter 15*, and go to the **Role mapping** tab:

demoeditor
Manage
Clients
Client scopes
Realm roles
Users
Groups
Users > User details
dr.s.johnson@edu.albertson.gov
Enabled
tails
Attributes
Credentials
Role mapping
Groups
Consents
Identity provi
Search by name
Hide inherited roles
Assign role
Unassign
Name
Inherited
Description
default-roles-demoeditor
False
${role_default-roles}

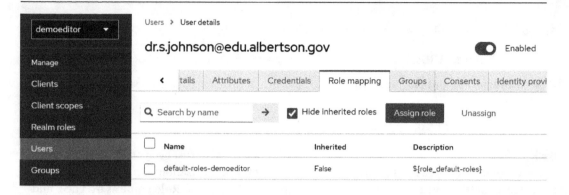

Figure 18.8 – Users form with the Role mapping tab active

Change the filter to **Filter by clients** and associate **author** with the selected user:

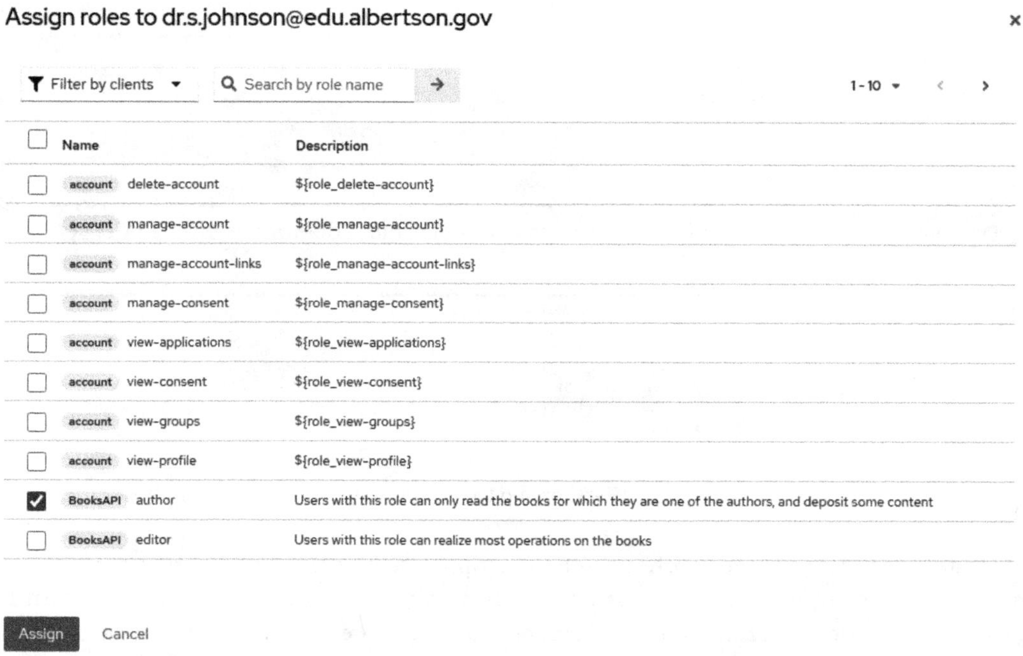

Assign roles to dr.s.johnson@edu.albertson.gov
Filter by clients
Search by role name
1 - 10
Name
Description
account delete-account ${role_delete-account}
account manage-account ${role_manage-account}
account manage-account-links ${role_manage-account-links}
account manage-consent ${role_manage-consent}
account view-applications ${role_view-applications}
account view-consent ${role_view-consent}
account view-groups ${role_view-groups}
account view-profile ${role_view-profile}
BooksAPI author Users with this role can only read the books for which they are one of the authors, and deposit some content
BooksAPI editor Users with this role can realize most operations on the books
Assign
Cancel

Figure 18.9 – Assigning the author role to an external user

Then do the same with an internal user (internal users are provisioned upon first login through the delegated identity provider, in our example Office 365, as explained in *Chapter 15*) and assign it the **editor** role.

Now that the IAM is ready, we are going to switch back to the code and add what is needed for it to use the authentication client that we have just created.

Integrating the code with the IAM

First, we are going to block the entire API if access is anonymous, which means there has been no authentication. To do so, we add the `Authorize` attribute at the top of the controller code:

```cs
using Microsoft.AspNetCore.Authorization;

namespace books_controller.Controllers;

[Authorize]
[ApiController]
[Route("[controller]")]
public class BooksController : ControllerBase
```

In order for this to have effect, and to let users in if they have authenticated with the client we customized in the previous section, we need to add the following to `Program.cs`:

```cs
app.UseAuthentication();
app.UseAuthorization();
```

In the same file, a bit above, the following lines will make the connection to the IAM realm:

```cs
builder.Services.AddAuthentication(JwtBearerDefaults.
AuthenticationScheme).AddJwtBearer(options => {
    options.Authority = "http://iam/realms/demoeditor/";
    options.Audience = "account";
    options.RequireHttpsMetadata = false;
}).AddCookie();
```

The third option (`RequireHttpsMetadata`) should never be used in production, but we can use it in our exercise because it is not useful to add HTTPS support. For this to compile, you will need to type `dotnet add package Microsoft.AspNetCore.Authentication.JwtBearer` in the terminal.

After restarting the API, a bit of customization with Postman will now be useful because if we run the command as before, we will be hit by an (expected) **401 Unauthorized** error, as follows:

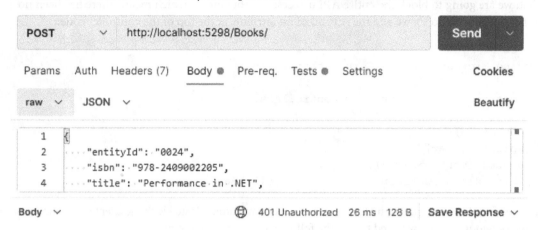

Figure 18.10 – Call in Postman without authorization

Since authorization is for now provided as soon as authentication is proved, we simply have to authenticate using Postman for this to work again. By default, all requests should be in the mode, where authentication is inherited from the parent, which is the collection. We can then activate the **OAuth 2.0** inside it:

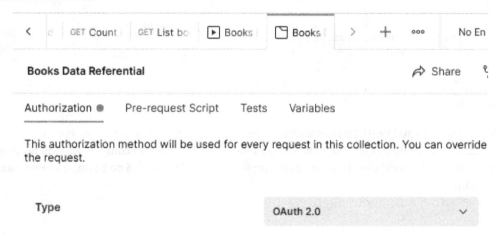

Figure 18.11 – Changing authentication to OAuth 2.0 in the Postman collection

Going down the interface, you will find that some parameters, such as the authentication and access token URLs, the client identifier, and the client secret, should be added:

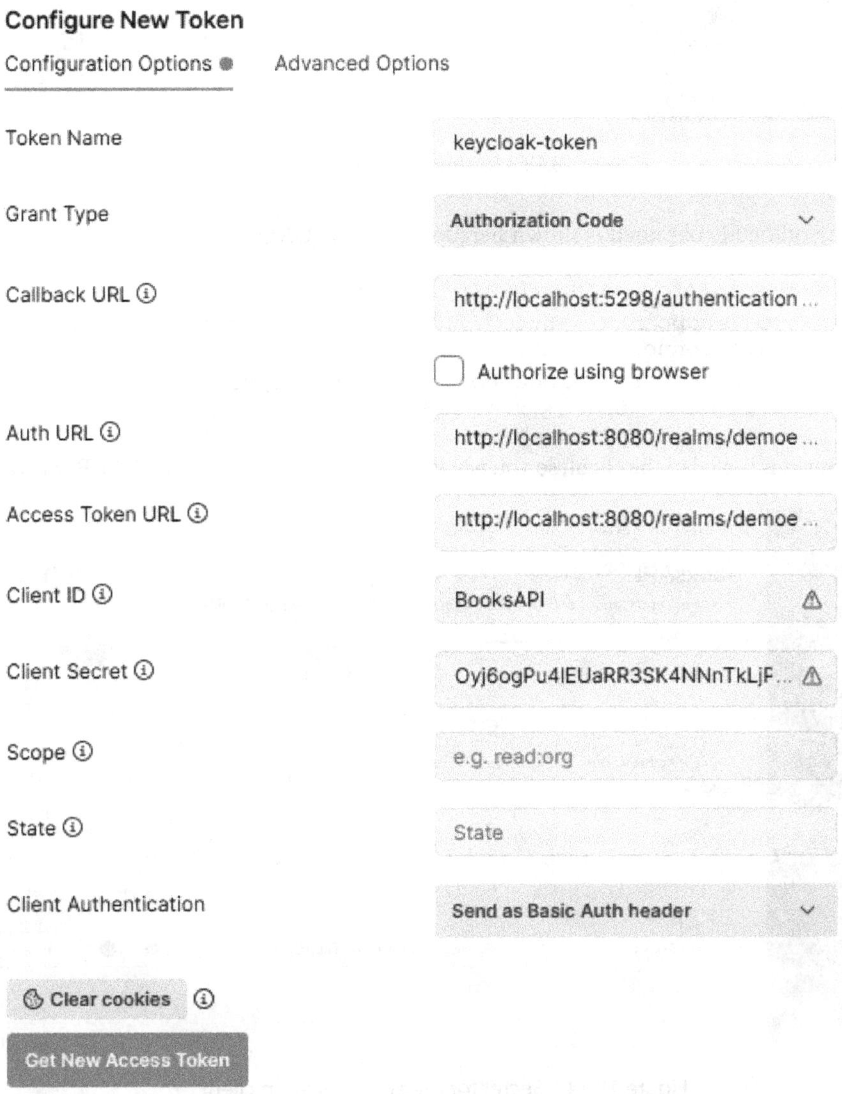

Configure New Token

Configuration Options ●   Advanced Options

Token Name	keycloak-token
Grant Type	Authorization Code ⌄
Callback URL ⓘ	http://localhost:5298/authentication ...
	☐ Authorize using browser
Auth URL ⓘ	http://localhost:8080/realms/demoe ...
Access Token URL ⓘ	http://localhost:8080/realms/demoe ...
Client ID ⓘ	BooksAPI ⚠
Client Secret ⓘ	Oyj6ogPu4lEUaRR3SK4NNnTkLjF... ⚠
Scope ⓘ	e.g. read:org
State ⓘ	State
Client Authentication	Send as Basic Auth header ⌄

🕲 Clear cookies ⓘ

Get New Access Token

Figure 18.12 – OAuth 2.0 parameters in Postman

Most of these parameter values will be found in the endpoint configuration that you can display in Keycloak in the **Realm settings**:

Figure 18.13 – Realm settings in Keycloak

The beginning of the file obtained is shown here to clarify the URL:

```
"issuer": "http://localhost:8080/realms/demoeditor",
"authorization_endpoint": "http://localhost:8080/realms/demoeditor/
protocol/openid-connect/auth",
"token_endpoint": "http://localhost:8080/realms/demoeditor/protocol/
openid-connect/token",
```

As for the secret, it is linked to the client, so you will find it in the **Credentials** tab of the **BooksAPI** client:

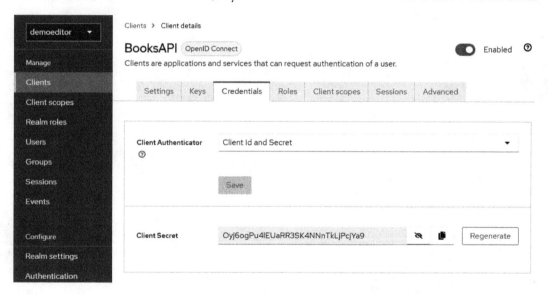

Figure 18.14 – Secret for the authentication client

The callback URL must be one of the authorized URLs, even if Postman will not use them directly to retrieve authentication codes. When everything is ready, you can click on **Get New Access Token**. This will open a popup navigator window in which you will have to fill in the credentials for a registered user:

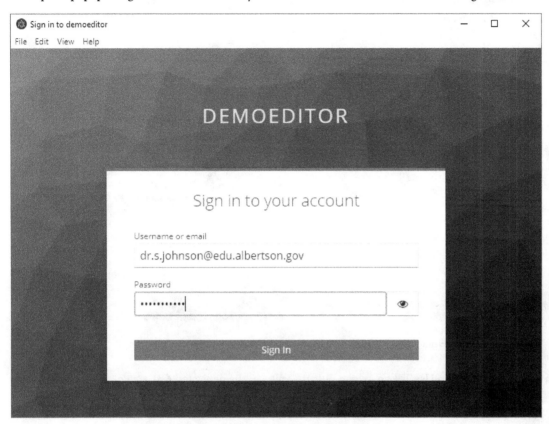

Figure 18.15 – Signing in to demoeditor for Postman

Postman will then display a success dialog, and then another one a few seconds after allowing us to store the access token retrieved in the Postman interface:

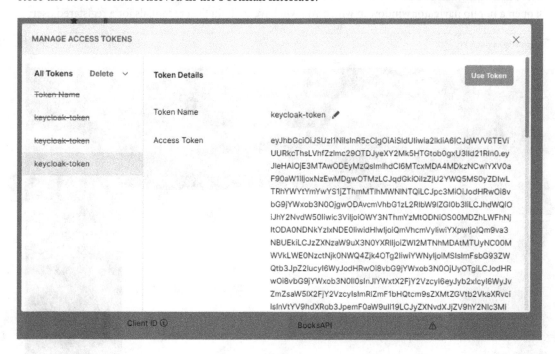

Figure 18.16 – Access token retrieved from Keycloak by Postman

We of course click on **Use Token**, which puts the value in the dedicated text area of the interface:

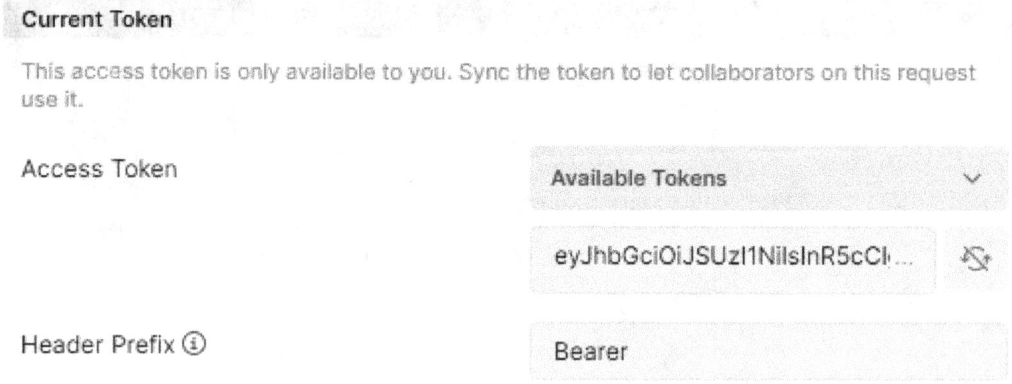

Current Token

This access token is only available to you. Sync the token to let collaborators on this request use it.

Access Token

Available Tokens ⌄

eyJhbGciOiJSUzI1NiIsInR5cCI...

Header Prefix ⓘ

Bearer

Figure 18.17 – Access token set for Postman use

If all goes well, you are then authorized again to the requests:

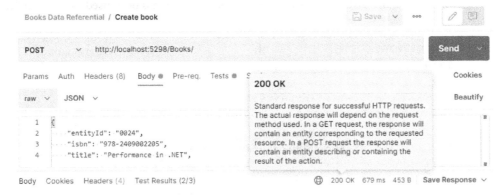

Figure 18.18 – Authorized call from Postman

We have done what is needed for basic authorization (purely based on authentication), but in order to have a well-aligned system where the books data referential service is as loosely coupled as possible with the IAM and authorization management, it would be great to go further with the separation of responsibilities, and this is what we will do with roles.

Adding RBAC authorization based on IAM attributes

If you are curious and paste the content of the access token you just retrieved into a decoder (such as https://jwt.io), you will see the encoded values:

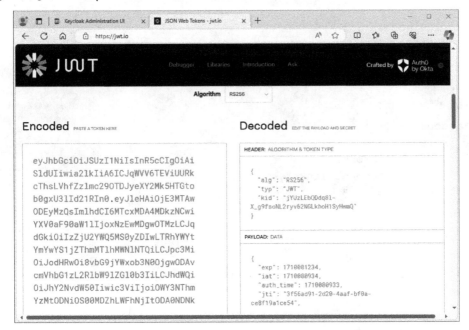

Figure 18.19 – JWT.io view

The JWT consists of the following text, and you will notice that it contains the roles that were defined in Keycloak, in particular, the author role for our external user that has been used as an example:

```json
{
  "exp": 1710081234,
  "iat": 1710080934,
  "auth_time": 1710080933,
  "jti": "3f56ad91-2d20-4aaf-bf0a-ce8f19a1ce54",
  "iss": "http://localhost:8080/realms/demoeditor",
  "aud": "account",
  "sub": "9f758fc3-83b9-406a-aa62-80443dc21414",
  "typ": "Bearer",
  "azp": "BooksAPI",
  "session_state": "eb613a00-1524-41ed-a477-6945d8f98986",
  "acr": "1",
  "allowed-origins": [
    "http://localhost:5298",
    "http://localhost"
  ],
  "realm_access": {
    "roles": [
      "offline_access",
      "default-roles-demoeditor",
      "uma_authorization"
    ]
  },
  "resource_access": {
    "BooksAPI": {
      "roles": [
        "author"
      ]
    },
    "account": {
      "roles": [
        "manage-account",
        "manage-account-links",
        "view-profile"
      ]
    }
  },
  "scope": "email profile",
  "sid": "eb613a00-1524-41ed-a477-6945d8f98986",
  "email_verified": true,
  "name": "Stacy Johnson",
```

```
"preferred_username": "dr.s.johnson@edu.albertson.gov",
"given_name": "Stacy",
"family_name": "Johnson",
"email": "dr.s.johnson@edu.albertson.gov"
}
```

In order to use roles in the authorization management of the books referential service, the following lines should be added to `Program.cs` first:

```cs
builder.Services.AddAuthorization(o =>
{
    o.AddPolicy("author", policy => policy.RequireClaim("user_roles",
"author"));
    o.AddPolicy("editor", policy => policy.RequireClaim("user_roles",
"editor"));
});
```

These definitions of what are called **policies** will then be applied to the operations inside the `BooksController` class. For example, all write operations that absolutely need an editor role to be accessed will be decorated like this:

```cs
[Authorize(Policy = "editor")]
[HttpPatch]
[Route("{entityId}")]
public async Task<IActionResult> Patch(
    string entityId,
    [FromBody] JsonPatchDocument patch,
    [FromQuery] DateTimeOffset? providedValueDate = null)
```

Other operations that are less restricted (counting books, reading the data of a particular book, even if we are going to restrict this further) are left without policy restriction. And we even make a case with the health check operation, which really should be completely public and accessible to anyone able to call the URL, using the `AllowAnonymous` attribute, which is the opposite of `Authorize`:

```cs
[AllowAnonymous]
[HttpGet]
[Route("Status")]
public ActionResult Status()
{
    // If the ping to the database works, the controller is considered
as working fine
```

```
        if (Database.RunCommandAsync((Command<BsonDocument>)"{ping:1}").
Wait(1000))
            return Ok();
        else
            return StatusCode(500);
    }
```

As you might have noticed, the roles we need are associated with the client in `resource_access`. It is thus important to transfer the roles onto the user claims by adding the following line in `Program.cs`:

```cs
builder.Services.AddTransient<IClaimsTransformation,
ClaimsTransformer>();
```

And, of course, we need to create the class used in this line of code, which should be implemented as follows:

```cs
using Microsoft.AspNetCore.Authentication;
using System.Security.Claims;
using System.Text.Json;

namespace books_controller
{
    /// <summary>
    /// Reference: https://docs.microsoft.com/fr-fr/aspnet/core/
security/authentication/claims
    /// </summary>
    public class ClaimsTransformer : IClaimsTransformation
    {
        private string PrefixeRoleClaim { get; init; }
        private string OIDCClientId { get; init; }
        private string SuffixeRoleClaim { get; init; }
        private string TargetUserRolesClaimName { get; init; }

        public ClaimsTransformer(IConfiguration config)
        {
            string ModelePourRoleClaim = "resource_access.BooksAPI.
roles";
            PrefixeRoleClaim = ModelePourRoleClaim.Substring(0,
ModelePourRoleClaim.IndexOf("."));
            SuffixeRoleClaim = ModelePourRoleClaim.
Substring(ModelePourRoleClaim.LastIndexOf(".") + 1);
            OIDCClientId = "BooksAPI";
            TargetUserRolesClaimName = "user_roles";
        }
```

```
        public Task<ClaimsPrincipal> TransformAsync(ClaimsPrincipal
principal)
        {
            ClaimsIdentity claimsIdentity = (ClaimsIdentity)principal.
Identity;
            if (claimsIdentity.IsAuthenticated)
            {
                foreach (var c in claimsIdentity.Clone().
FindAll((claim) => claim.Type == PrefixeRoleClaim))
                {
                    JsonDocument doc = JsonDocument.Parse(c.Value);
                    foreach (JsonElement elem in doc.RootElement.
GetProperty(OIDCClientId).GetProperty(SuffixeRoleClaim).
EnumerateArray())
                        claimsIdentity.AddClaim(new Claim(this.
TargetUserRolesClaimName, elem.GetString() ?? String.Empty));
                }
            }
            return Task.FromResult(principal);
        }
    }
}
```

When explaining RBAC with an example of Open Policy Agent in *Chapter 13*, it was hypothesized that roles could inherit from each other. This was quite a bold statement because it would need to be true for the entire information system. If the reality (which only business people can tell; the answer cannot be technical) is that this is really the books referential service – and maybe some other services but with different rules – where it should be applied, then the policy definition is typically where we would do this by applying an algorithm that's a bit more complicated than adding a policy to a role. This way, Keycloak would not have the responsibility to bear several roles, automatically assign them, or manage some kind of inheritance that would not even be its business. It would be done by the books data referential service, which decides how the role composition should work for its own business silo, which is much more aligned with the principle of responsibility separation.

Let's try testing access controls again with Postman in order to check if this works as expected. If we are connected as dr.s.johnson@edu.albertson.gov, who has the author role and should be prevented from carrying out operations restricted to the editor roles, we can count books (the URL is /Books/$count and the HTTP status code returned is **200 OK**):

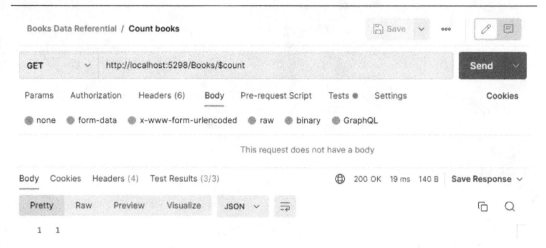

Figure 18.20 – Operation an author can access

But if we try to **PATCH** a book (which exists, by the way, as the count showed, and the fact that the return code is not 404), we now get a **403 Forbidden** response:

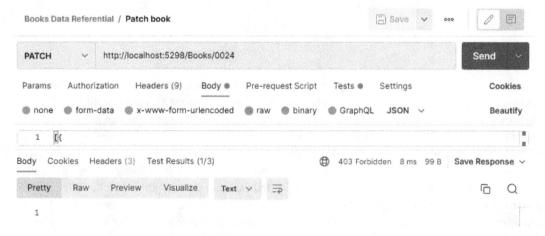

Figure 18.21 – Accessing an unauthorized command

Notice that this is not a 401 code anymore, as the authentication gate has been passed, but the authorization gate has prevented the successful execution of the request. Our authorization management has started to take shape, but we are still missing one important type of rule: those that are not simply based on a role.

Polishing with additional ABAC authorizations

As mentioned, the rule that needs to be addressed is the one stating that authors can only read their own books. For now, our authorization management suffers from a serious limitation, which is that authors can read any book. So, they cannot list books, but they can access everything, provided they guess the identity number, which is not good at all. Internal identifiers cannot be trusted to obfuscate data, particularly when they are so simple, such as a counter. GUIDs might be useful in this approach, but only actual authorization makes sense. Lack of visibility should never be taken as a guarantee of authorization enforcement.

So, we need to add another rule in order for the book to not be available to users who have the `author` role but are not one of the authors of the book. We could do this directly in the code:

```
[HttpGet]
[Route("{entityId}")]
public IActionResult GetUnique(string entityId, [FromQuery]
DateTimeOffset? providedValueDate = null)
{
    if (providedValueDate is null)
    {
        // As long as no parameter is added, the behavior is simply to
find and retrieve the best-known state
        var result = Database.GetCollection<Book>("books-bestsofar").
Find(item => item.EntityId == entityId);
        if (result.CountDocuments() == 0)
            return new NotFoundResult();
        else
        {
            Book book = result.First();
            if (HttpContext.User.IsInRole("author"))
            {
                if (book.Editing != null
                    || book.Editing.Author != null
                    || HttpContext.User!.Identity!.Name != book.
Editing.Author.Name)
                    return new ForbidResult();
            }
            return new JsonResult(book);
        }
    }
    else
    {
        //The rest of the code was omitted for simplicity
    }
}
```

Since there is an [Authorize] attribute on the controller, it is inherited here. This is why we can use the null-forgiving !. operator in User and Identity: we know we will reach this code only if authorized and thus authenticated. On the other hand, if we do not have any author information for the book, the request will logically be forbidden to anyone. Still, the preceding code remains extremely naive, as it considers the name of the user in the IAM and the name of the main author in the editing petal of the book referential service should be exactly the same for the user to gain the right to access the book. This approach becomes complex when using plain text instead of identifiers, as it can lead to case sensitivity issues. Additionally, this method fails to address the deeper problem of how authors and their associated users are related.

Let's imagine we, as the implementers of this rule, are having a discussion with the product owners, namely the editing director of DemoEditor. The business should drive us to make the right decision for the company. The first question would be about how to identify a user who has an author role. We could use a unique identifier from Keycloak, but that would mean coupling to this technology, so this is ruled out. Using an email address sounds like a good way to go, as there is quite a good guarantee of identifying the person behind the user. In the JWT, we not only receive this claim by default, but we even have a claim from the identity provider that the email address has been verified (which means a roundtrip of exchange has been made through the mail address by the user management system, ensuring the user indeed has access to this email address):

```
( ... )
"email_verified": true,
"name": "Stacy Johnson",
"preferred_username": "dr.s.johnson@edu.albertson.gov",
"given_name": "Stacy",
"family_name": "Johnson",
"email": "dr.s.johnson@edu.albertson.gov"
( ... )
```

As the business owners agree that an author should never have access to the information system without a valid email address and that this is enforceable for editors, this looks like the right solution. The case of email addresses is not important, and this is enforced by the RFCs, so we should take this into account in the rule implementation. As we will see later in the chapter, this analysis will have to go further (and, in particular, take care of the fact that we have not defined the Author property yet) but, for now, we can expand the authorization code like this:

```
Book book = result.First();
if (HttpContext.User.IsInRole("author"))
{
    if (book.Editing != null
        || book.Editing.Author != null
        || HttpContext.User.Claims(c => c.Type == "email") != book.
Editing.Author.Email
```

```
         || HttpContext.User.Claims(c => c.Type == "email_verified") !=
"true")
    return new ForbidResult();
}
return new JsonResult(book);
```

As usual, when we want to be as aligned as possible with the business, there is still something missing, and the product owners will remind us of the rule that a blocked author should not have access to the data, even for their own books. Things start to get a bit more difficult to do in the code because we might not have this information (such as the **blocked** status) in the Author property of the book (the email address is quite stable, so we store it in the Author property – and we will see how in the coming section – but this **blocked** status may change rapidly, and we should check for it). Also, since the directors of DemoEditor seem to have diverging and dynamic opinions about the management rules for the authorization, this is typically where we would externalize all this to a BRMS.

In our example, that would mean including OPA in the system. Adding a service to Docker Compose is easy and has been covered before, but the design of the rule is interesting. Since we have only to deal with a unique book, the volume handling issues we talked about at the end of *Chapter 13* do not apply here. In fact, even the ability to list books has been denied to authors, so the problem is out of the way.

Let's just take some time to reflect on the business alignment of this decision. If we consider a book to be equivalent to a contract with the editor, the fact that the author needs the editor to provide them with the right identifier is not really a problem (in fact, it would even allow better management of the editions of the books). Indeed, the situation when the author connects to their view of the information system would be a bit like when someone connects to the extranet of their electricity company and checks their bills: most of the time, people will only have one contract, but if they have a second house, they will need to register in the extranet with the identifier provided by the electricity company on the paper bill, for example. This is why an author should not have access to the listing GET operation on the API.

There is also another way of analyzing the situation (and again, it is the job of business experts to decide which is the right analysis): we could argue that authors should not be dependent on the editor to provide an identifier, and it is generally a bad practice that an identifier is used as some kind of secret. In this case, the best move we could make would be to use the ISBN as the exposed identifier in the URL pointing to a given book.

But let's go back to authorization with OPA. Supposing the data.json content is up to date, we should find in it the data about the book, which would be something like this:

```
{
    "entityId": "0024",
    "isbn": "978-2409002205",
    "title": "Performance in .NET",
    "links": [
        {
```

```
                    "rel": "self",
                    "href": "https://demoeditor.org/api/books/0024",
                    "title": "Performance in .NET",
                    "summary": "First edition of a book about performance
management in .NET"
                }
        ],
        "numberOfPages": 413,
        "publishDate": "2013-05-01",
        "editing": {
            "numberOfChapters": 10,
            "status": { "value": "available" },
            "authors": [
                {
                    "rel": "dc:creator",
                    "href": "https://demoeditor.org/api/authors/0012",
                    "title": "JP Gouigoux"
                }
            ]
        },
        "sales": {
            "price": {
                "value": 42.15,
                "monetaryUnit": "EUR"
            },
            "weightInGrams": 768
        }
    }
}
```

We should also find the full definition of the author of this book (even if I only show an extract of the final full definition we should have to accommodate all the business features):

```
{
    "entityId": "0012",
    "links": [
        {
            "rel": "self",
            "href": "https://demoeditor.org/api/authors/0012",
            «title»: «JP Gouigoux»
        }
    ],
    «firstName»: «Jean-Philippe»,
    "lastName": "Gouigoux",
    "userEmailAddress": "jp.gouigoux@frenchy.fr",
```

```
    "restriction": "none"
}
```

The input.json content sent to OPA to check for authorization would be something like the following:

```json
{
    "input": {
        "user": {
            "exp": 1710081234,
            "iat": 1710080934,
            "auth_time": 1710080933,
            "jti": "3f56ad91-2d20-4aaf-bf0a-ce8f19a1ce54",
            "iss": "http://localhost:8080/realms/demoeditor",
            "aud": "account",
            "sub": "9f758fc3-83b9-406a-aa62-80443dc21414",
            "typ": "Bearer",
            "azp": "BooksAPI",
            "session_state": "eb613a00-1524-41ed-a477-6945d8f98986",
            "acr": "1",
            "allowed-origins": [
                "http://localhost:5298",
                "http://localhost"
            ],
            "realm_access": {
                "roles": [
                "offline_access",
                "default-roles-demoeditor",
                "uma_authorization"
                ]
            },
            "resource_access": {
                "BooksAPI": {
                "roles": [
                    "author"
                ]
                },
                "account": {
                "roles": [
                    "manage-account",
                    "manage-account-links",
                    "view-profile"
                ]
                }
            },
```

```
                    "scope": "email profile",
                    "sid": "eb613a00-1524-41ed-a477-6945d8f98986",
                    "email_verified": true,
                    "name": "Stacy Johnson",
                    "preferred_username": "dr.s.johnson@edu.albertson.gov",
                    "given_name": "Stacy",
                    "family_name": "Johnson",
                    "email": "dr.s.johnson@edu.albertson.gov"
                },
                "operation": "read",
                "resource": "books",
                "book": "978-2409002205"
        }
}
```

Instead of just passing the identifier of the user, as we did in the first OPA example, we would pass all of the JWT content. Indeed, not only is this the most standard-compliant approach, but in addition, it opens up additional rules that the business might be interested in. For example, we could argue that, once the director of the company has been informed that the `iss` attribute corresponds to the issuer of the identity (in this case, the Keycloak server), they might add a rule enforcing that only the `DemoEditor` identity provider should be trusted when allowing an external author to read the content of the book (but not the editing attributes, for example, where an email address verified by an external issuer of token would be enough).

As for the `policy.rego` content, we will keep some of the rules we wrote in *Chapter 13*, such as the ones that check whether the book is associated with the author or that the author is not blocked:

```
author_books contains book if {
    some author in user_author
    some b in data.books
    b.editing.author == author.id
}

authors_on_books_they_write if {
    some role in user_roles
    role == "book-writer"
    some author in user_author
    some b in data.books
    b.editing.author == author.id
    b.id == input.book
}

no_author_blocking if {
    some role in user_roles
```

```
        role == "book-writer"
        some user in user_author
        user.restriction == "none"
}
```

We will modify some Rego syntax to check the attachment of the author and user with their email addresses:

```
matching_author if {
    some book-author in b.editing.authors
    book-author.rel == "dc.creator"
    some author in data.authors
    some link in author.links
    link.rel == "self"
    link.href == book-author.href
}

matching_user if {
    some author in data.authors
    lower(author.userEmailAddress) == lower(input.user.email)
}
```

Finally, the code will evolve into something like the following (the injection of the `httpClient` variable and the design of `OPAResult` are omitted, as well as the `opa` service in the Docker Compose definition):

```
Book book = result.First();
if (HttpContext.User.IsInRole("author"))
{
    var response = await httpClient.PostAsJsonAsync("http://opa/v1/
data/app/abac", input);
    OPAResult result = await response.Content.
ReadAsAsync<OPAResult>();
    if (!result.allow)
        return new ForbidResult();
}
return new JsonResult(book);
```

I am also not going to go into the details of graceful degradation because it is not related to our main subject of business/IT alignment, but we should of course remember that authorization is not granted if the call to OPA times out for one reason or another. If the SLA needs to be improved, we could also imagine having a simple fail-over authorization system that would work in certain cases (for example, read access to only public information on a book)

Why are we doing all this?

After spending so many pages implementing authentication, identification, and authorization correctly, you may be wondering "Why don't we simply put a user and hashed password in some database table and check them?" Well, what used to be possible a few years (alright, decades) ago is not possible anymore for security reasons. If you store the password in your application, everything sounds easy at first, but then you have problems when customers and regulators demand that you enforce password robustness and expiration dates. How about forcing the user to change their password at first login? This will only lead to more trouble because users will want to have a dedicated interface to manage their profiles. Guess what: such an interface is already available in Keycloak, at `http://localhost:8080/realms/demoeditor/account/#/` (please adjust depending on the realm name and the exposition port if you have not chosen the same ones as proposed in this exercise):

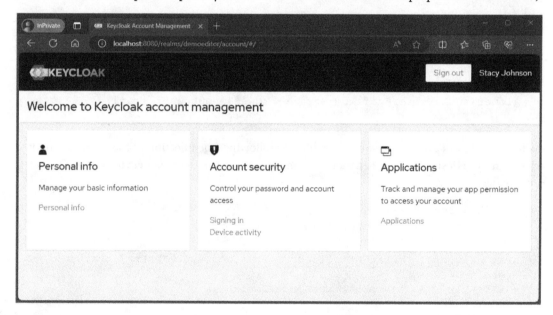

Figure 18.22 – User profile management interface

In Keycloak, the user logged in (**Stacy Johnson**, in our case) will not only be able to change her password but also set up multi-factor authentication:

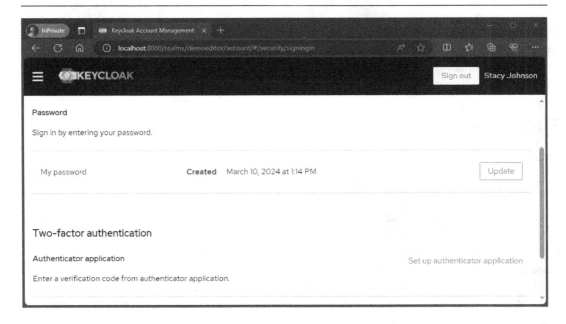

Figure 18.23 – A user changing their credentials in a dedicated console

How long would it take to set this up in your code? More practically, how long would it take to do it right and reduce the number of vulnerabilities, like Apache has done on Keycloak for years? And we are not done yet; you can also follow up on devices where you have logged in and sign out if needed, manage your application permissions, and more:

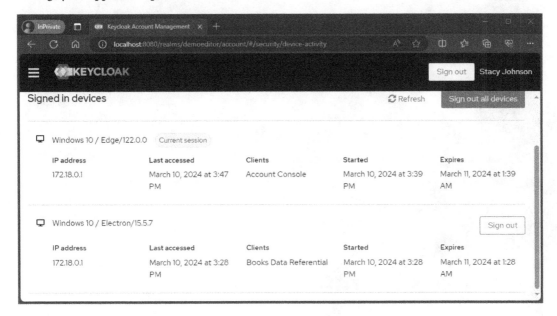

Figure 18.24 – Signed-in devices in the Keycloak user console

In short, IAM is not something you should try and do by yourself or even in a large group or company. Even if you think you need very limited features, the lightweight aspect of Keycloak and its ease of use, after the initial configuration, will make you better off than with an old-fashioned homemade user management system.

Summary

In this chapter, we have addressed the installation part of data referential service management, deploying (with Docker containers) the first data referential service we created in the previous chapter for book management, and adding authentication features and authorization rules to it.

We have shown the potential pitfalls in the deployment, but also some tricks about industrializing the API presence in the information system. The part about IAM was a big one, and this shows how important (and complex) this feature of a good referential service can be. Luckily, we can externalize these features, which means that all the hard work of MFA, securing passwords, providing GUI to modify personal profiles, adding identity federation and delegation, and so on, is now done by a dedicated server. But simply integrating this IAM is not easy and should be carried out with great care.

Now that you are equipped with the knowledge to deploy a data referential service, we will show in the next chapter a second data referential service, this time about authors. You may wonder why we will create a second server that looks like the first one, but bear with me, and you will see that lots of things remain to be discovered on data referential service, and the second one has some peculiarities that are worth knowing.

19

Designing a Second Data Referential Service

In the previous two chapters, we have shown how to create a data referential service, using the books data management as a target of analysis and then development, and how to deploy such an API server. In this chapter, we will extend our new knowledge about creating data referential services by creating a second one – this time, for authors. You may wonder why we do this, since the example organization obviously needs to have them but the concepts have already been explained in the previous chapter. So, what can be interesting to add here? In fact, this second data referential service that we are going to design has some peculiarities due to its link with the books data referential service and will need some additional features that we did not cover in the first one. In addition, these features are quite particular and require some specific explanation about their design in order to achieve our long-term goal of clean separation of responsibilities.

Once you have finished reading this chapter and used the same method as shown in the previous one to deploy the second data referential service, the two servers will be used in the following chapters as two of the main services of our sample information system. Since all the bricks needed for `DemoEditor` will be ready by then, the following chapters will be about making interfaces between them, integrating them into complex business processes that necessitate functions brought by several of them, and so on.

We'll cover the following topics in this chapter:

- Creating a second data referential service
- Linking it to the first one through standardized HTTP links
- Adding a local cache to avoid systematic calls
- Using events to refresh the cache only when needed
- Dealing with issues that may arise in the Webhooks-based architecture

Building the authors data referential service

If I have done my job correctly, you should by now be aware of all the different components of a real data referential service, which is to a database as a car is to an old bicycle. It would be pointless to go again on all the concepts to create the authors data referential service. After all, only the JSON Schema is needed in theory to completely build such a piece of software and we already defined it in the previous chapter, so I am just showing again an example of an author instance as follows:

```
{
    "entityId": "0012",
    "links": [
        {
            "rel": "self",
            "href": "https://demoeditor.org/api/authors/0012",
            «title»: «JP Gouigoux»
        }
    ],
    «firstName»: «Jean-Philippe»,
    "lastName": "Gouigoux",
    "userEmailAddress": "jp.gouigoux@frenchy.fr",
    "restriction": "none"
}
```

Links between data referential services

By the way, why did we show the example again, since we were building the books (and not authors) data referential service? Simply because there are some links between them, as shown in an even clearer manner by the presence of links to the authors in the content of a book:

```
"editing": {
    "numberOfChapters": 10,
    "status": { "value": "available" },
    "authors": [
        {
            "rel": "dc:creator",
            "href": "https://demoeditor.org/api/authors/0012",
            "title": "JP Gouigoux"
        }
    ]
},
```

I have not insisted on this particular grammar, but this is a very important way of thinking about data in an information system, as it has been explained already several times in this book when talking about eventual consistency. Links can be used to ensure a higher degree of availability when they carry business data that is highly bound to the linked object and not supposed to change. This is what we do in the preceding JSON content, using the `title` attribute to show some almost fixed data on an author. After all, the most frequent case where one changes name is when marrying, so this is not supposed to happen every night; in addition, it is generally accepted that addressing someone by their new name can take some time and that using the previous last name can be accepted for a short period.

Now, do you remember this line of code?

```
HttpContext.User.Claims(c => c.Type == "email") != book.Editing.
Author.Email
```

I was anticipating that we could also add the email address inside the `authors` links, since it is also quite a stable attribute for a person, as we even use it for identification purposes. So, in addition to the standard `rel`, `href`, and `title` attributes of a standard link, we could inherit this into `AuthorLink` in the OpenAPI schemas and add an `email` attribute. This way, the link would resemble this:

```
"editing": {
    "numberOfChapters": 10,
    "status": { "value": "available" },
    "authors": [
        {
            "rel": "dc:creator",
            "href": "https://demoeditor.org/api/authors/0012",
            "title": "JP Gouigoux",
            "email": "jp.gouigoux@frenchy.fr"
        }
    ]
},
```

In most cases, we would not even need to call the authors referential service to get this data. In fact, we could do this for some other attributes, but we need to take care of keeping a balance: if the data is too volatile, the local copy will often be outdated. In some cases, it does not matter, but in others, that may be a source of problems: imagine, for example, that the `blocked` status of an author is not updated soon enough and we continue to authorize them to some content!

Also, if we store too much data about the author on the link in the book, then why bother keeping a data referential service since – in the extreme case – all the data is stored somewhere else? In multiple instances, by the way, this could be a small reason to keep a unique data referential service. I say a small reason because we are talking of very limited storage quantities, at least when we do not add binary documents. This is where duplication of data hits the most but, as far as the data is concerned, one could even argue that this is better to keep local author data on books since there are not that many cases where authors write many books, so duplication is low. In addition, maybe keeping the state of the author at the time they wrote the book would be a business requirement in the future, so why keep these states in a centralized system? The question really makes some sense and, in a real information system, should be discussed at length. But, for our example, we have made the hypothesis of a second referential service typically to talk about these issues.

Another way to handle this need for local author data would be to use a cache. The same questions of consistency will arise, but they can be treated in a different way.

Adding a local cache

If we decide to add a local cache of authors inside the books referential, the very first question that arises is how much of the data should we cache, horizontally and vertically?

When talking about how many authors, the reasonable answer sounds like all the ones associated with an existing book. But again, the devil lies in the details. How good is it to keep an author in the cache if the only book they are associated with has an `archived` status? Not much...

When talking about how much data of an author we should keep, the question is harder. We need to know whether the cache should be total (all the changes, the states, and the best so far) or whether only the best-so-far state is enough. Maybe it would be only the states used in the books links. These links would also help us find out what is really useful in the cache for the books operations. Perhaps the books referential service never uses the addresses of the authors?

Perhaps also there should be some authorization restriction as to what the books referential service should keep in its cache. For example, whatever happens, the owners of the authors data referential service may insist that bank coordinates never leave the referential service and that accounting employees are the only ones to manipulate this sensitive data on authors. When accessing the authors data referential service synchronously because there has been a request to the books referential, things are easy for authorization management: the JWT access token is simply passed and authorizations will apply just as usual. But what if a cache is kept and some out-of-band process updates it at regular hours? Then, we would use a service account, which would have its own rights and privileges.

All of these questions should be answered before deciding how to cache the data. In some cases, a single in-memory cache of only the data requested may be enough. In some other cases, it might be useful to have an almost complete copy of the data and its history at hand. All cases are different and should be treated according to business needs, as always.

Using Webhooks for cache quick refresh

Before we proceed any further, let's imagine we are in this second case where a local cache should be kept in its almost complete perimeter, except for confidential attributes. And let's make the hypothesis that only the best-so-far states are useful. In this case, we could add a local collection in MongoDB (if the cache needs to be persistent) and call it `authors-bestsofar-cache`.

Let's also extend this situation by hypothesizing that data should be kept as fresh as possible. Instead of a polling system that reads data at frequent intervals, possibly optimizing by scrutinizing the history of data and only querying the best so far when there indeed have been new changes, it is much more efficient to register data changes through a Webhooks mechanism.

There is no point in showing a complete cache and Webhooks mechanism here: I will only show a few code skeletons in order to pinpoint the challenges in this kind of architecture. First, we would need to implement the callback in the books data referential service in order to handle the changes on the authors. A good practice would be to create a dedicated controller for this (and this in itself is a response to "*Is there always only one controller in a data referential service since we are supposed to handle only one business entity?*"):

```
using Microsoft.AspNetCore.Mvc;
using MongoDB.Driver;

namespace books_controller.Controllers;

[ApiController]
[Route("books/[controller]")]
public class AuthorsCacheController : ControllerBase
{
    private readonly string ConnectionString;

    private readonly IMongoDatabase Database;

    private readonly ILogger<AuthorsCacheController> _logger;

    public AuthorsCacheController(IConfiguration config,
ILogger<AuthorsCacheController> logger)
    {
        _logger = logger;
        ConnectionString = config.
GetValue<string>("BooksConnectionString") ?? "mongodb://
localhost:27017";
        Database = new MongoClient(ConnectionString).
GetDatabase("books");
    }


    [HttpPut]
    public async Task<IActionResult> Receive([FromBody] AuthorCache
author)
    {
        if (author is null) return BadRequest();
```

```
        await Database.GetCollection<AuthorCache>("authors-bestsofar-
cache").FindOneAndReplaceAsync<AuthorCache>(item => item.EntityId ==
author.EntityId, author);
        return new OkResult();
    }
}
```

This introduces the `AuthorCache` class, which is a reduced version of `Author`, since the books referential service may not be interested in storing all data coming from the authors referential. We will show further that the authors referential service will not send everything anyway for confidentiality reasons. However, each of the participants should keep its own choice on what they use in the attributes. For now, `AuthorCache` contains the following attributes, as well as functions to convert to and from `Author`, since the contract on the authors API defines the full `Author` schema, even if the content may not always be complete, depending on authorization rules (this distinction is very important to remember – the contract does not define *what* is going to be sent, but *how* this is going to be sent if it is):

```
using System.ComponentModel.DataAnnotations;
using System.Text.Json.Serialization;
using MongoDB.Bson;
using MongoDB.Bson.Serialization.Attributes;


// This class corresponds to an author as cached by the book
referential
// It may not contain all the data in the complete author model
public class AuthorCache
{
    [BsonId()]
    [BsonRepresentation(BsonType.ObjectId)]
    [JsonIgnore]
    public ObjectId TechnicalId { get; set; }


    [BsonElement("entityId")]
    [Required(ErrorMessage = "Business identifier of an author is
mandatory")]
    public string EntityId { get; set; }


    [BsonElement("firstName")]
    public string? FirstName { get; set; }


    [BsonElement("lastName")]
    public string? LastName { get; set; }


    [BsonElement("userEmailAddress")]
```

```
    public string? UserEmailAddress { get; set; }

    [BsonElement("restriction")]
    public string? Restriction { get; set; }

    public AuthorCache(Author author)
    {
        EntityId = author.EntityId;
        FirstName = author.FirstName;
        LastName = author.LastName;
        UserEmailAddress = author.UserEmailAddress;
        Restriction = author.Restriction;
    }

    public Author ConvertToAuthor()
    {
        return new Author() {
            EntityId = this.EntityId,
            FirstName = this.FirstName,
            LastName = this.LastName,
            UserEmailAddress = this.UserEmailAddress,
            Restriction = this.Restriction
        };
    }
}
```

Now that the cache mechanism is in place, we are going to attach its content to the events when authors change.

Implementing callbacks in the authors referential

Now the books referential service is ready to receive the data from the authors referential service when an author changes, we are going to create the corresponding call in the authors referential. The added code will be in the Patch function for the authors:

```
[HttpPatch]
[Route("{entityId}")]
public async Task<IActionResult> Patch(
    string entityId,
    [FromBody] JsonPatchDocument patch,
    [FromQuery] DateTimeOffset? providedValueDate = null)
{
```

```
    // All this part is the same as what has been done for the books
referential, but adapted to the authors, so we do not copy it here in
order to clean up display and concentrate on the rest of the function,
which is the most important in the context

    // When sending author information to services that have
registered on the webhook, a business rule states that no address
should be communicated
    Author strippedAuthor = (Author)author.Clone();
    strippedAuthor.Contacts = null;

    // If all went well, we send the registered callbacks associated
with the webhook for change events
    // No sophistication here, we simply send as a sequence, without
error management, because eventual consistency is ensured some other
way
    HttpClient? client = null;
    foreach (Uri callback in OnChangeCallbacks)
    {
        if (client is null)
            client = _clientFactory.CreateClient("Callbacks");
        await client.PutAsJsonAsync<Author>(callback, strippedAuthor);
    }

    // Finally, the object is returned in order to show potential
changes due to business rules
    return new ObjectResult(author);
}
```

The Author class is defined as follows, with simple properties for each of the business entity attributes and a Clone method:

```
using System.ComponentModel.DataAnnotations;
using System.Text.Json;
using System.Text.Json.Serialization;
using MongoDB.Bson;
using MongoDB.Bson.IO;
using MongoDB.Bson.Serialization.Attributes;

public class Author : ICloneable
{
    [BsonId()]
    [BsonRepresentation(BsonType.ObjectId)]
    [JsonIgnore]
    public ObjectId TechnicalId { get; set; }
```

```
    [BsonElement("entityId")]
    [Required(ErrorMessage = "Business identifier of an author is
mandatory")]
    public string EntityId { get; set; }

    [BsonElement("firstName")]
    public string? FirstName { get; set; }

    [BsonElement("lastName")]
    public string? LastName { get; set; }

    [BsonElement("userEmailAddress")]
    public string? UserEmailAddress { get; set; }

    [BsonElement("restriction")]
    public string? Restriction { get; set; }

    [BsonElement("contacts")]
    public ContactsPetal? Contacts { get; set; }

    public object Clone()
    {
        string serialized = JsonSerializer.Serialize(this);
        return JsonSerializer.Deserialize<Author>(serialized);
    }
}
```

In addition to the same simple attributes that can be found in `AuthorCache` in the books controller, we can see that we have additional data in the `Contacts` petal. This is defined as shown below:

```
using MongoDB.Bson.Serialization.Attributes;

public class ContactsPetal
{
    [BsonElement("adresses")]
    public List<Address>? Addresses { get; set; }

    [BsonElement("status")]
    public List<Phone>? Phones { get; set; }
}

public class Address
```

```
{
    [BsonElement("streetNumber")]
    public string? StreetNumber { get; set; }

    [BsonElement("streetName")]
    public string? StreetName { get; set; }

    [BsonElement("cityName")]
    public string? CityName { get; set; }

    [BsonElement("zipCode")]
    public string? ZipCode { get; set; }

    [BsonElement("country")]
    public CountryLink? Country { get; set; }
}

public class Phone
{
    // Type of phone, following IANA type defined in RFC 2426
    [BsonElement("type")]
    public string IANAType { get; set; }

    // Phone number, following E.123 standard
    [BsonElement("number")]
    public string Number { get; set; }
}
```

As usual, we try to use as many standards as possible (country code from ISO 3166, RFC 2426 for the phone types, E.123 for telephone numbers, etc.). By the way, addresses can also be standardized, but there are competing formats and international incompatibilities, so I did not favor one over another here and tried to stick to a neutral pivotal format. The `CountryLink` class inherits `Link` and specializes it with the three-letter country code:

```
using MongoDB.Bson.Serialization.Attributes;

public class CountryLink : Link
{
    // 3-letter country code, as per ISO3166
    [BsonElement("code")]
    public string ISOCode { get; set; }
}
```

The `Link` definition blindly follows the RFC:

```
using MongoDB.Bson.Serialization.Attributes;

public class Link
{
    [BsonElement("rel")]
    public string Rel { get; set; }

    [BsonElement("href")]
    public string Href { get; set; }

    [BsonElement("title")]
    public string Title { get; set; }
}
```

Let's not come back to the "meat" of the previous code and explain it in a bit more detail:

```
// If all went well, we send the registered callbacks associated with
the webhook for change events
// No sophistication here, we simply send as a sequence, without error
management, because eventual consistency is ensured some other way
HttpClient? client = null;
foreach (Uri callback in OnChangeCallbacks)
{
    if (client is null)
        client = _clientFactory.CreateClient("Callbacks");
    await client.PutAsJsonAsync<Author>(callback, strippedAuthor);
}
```

Again, my purpose is not to try and define a fully functional Webhook management system (it is way too complicated for this book), but just to show what should be taken care of if we follow the rule of a strict separation of responsibilities. First, it is good practice to define a dedicated client for the callbacks (in `Program.cs`):

```
builder.Services.AddHttpClient("Callbacks").
AddPolicyHandler(GetRetryPolicy());
```

No base address has been specified because the callbacks can go anywhere, but we apply a retry policy that is defined like this:

```
static IAsyncPolicy<HttpResponseMessage> GetRetryPolicy()
{
```

```
    return HttpPolicyExtensions
        .HandleTransientHttpError()
        .OrResult(msg => msg.StatusCode == HttpStatusCode.NotFound)
        .WaitAndRetryAsync(5, retryAttempt => TimeSpan.FromSeconds(0.1
* retryAttempt));
}
```

This way, when calling the callbacks, if there is a slight failure in the network or on the receiving end, the request will be sent again, up to five times and with an increasing delay of 100 ms, then 200 ms, and so on. The possible repeating of a call is normally not a problem, since we use a PUT verb to state that the call should be idempotent. For this to work, we will need to add the package called `Microsoft.Extensions.Http.Polly` (which implements all kinds of retry policies and makes them available as middleware units) and the following statements at the top of the `Program` class:

```
using System.Net;
using Polly;
using Polly.Extensions.Http;
```

There will, of course, be times when this retry policy will not be enough, but we will deal with eventual consistency in a later section. For now, we turn back to our code that sends callbacks and notice that they are simply sent in sequence, and without any error management. Some logs could be useful, but even that remains to be weighted against business needs, as eventual consistency will get the data clean at regular intervals.

One last thing we need to put in place in the authors data referential service is the way to add callbacks to the list, and this will be done with a `Subscribe` function:

```
private List<Uri> OnChangeCallbacks = new List<Uri>();

[HttpPut]
[Route("Subscribe")]
public IActionResult Subscribe([FromQuery] string callbackURL)
{
    // Very naive implementation, without persistence
    OnChangeCallbacks.Add(new Uri(callbackURL));
    return Ok();
}
```

Again, this is a very naive implementation here because, if the service stops, the callbacks – in memory only – are lost. In a real situation, we would organize persistence for them, but this would not add any value to the content of the present book, as this is mostly technical complexity. There are many ways to do this, among which I can recommend `https://blog.stackademic.com/`

```
implementing-an-effective-asp-net-core-webhook-system-for-event-
notifications-6d5b93761640.
```

Everything is prepared in the authors referential service that will emit the callback when the books referential service (or any other service) has registered to the webhook. To do so, we will thus turn back again to this books data service and add the necessary calls.

Registering to webhook in the books referential

We could register to the webhook for any change in the authors by creating a `Subscribe` call upon the start of the ASP.NET service, using `IHostedService`, but I figured it would be more interesting to show a more specialized use of the webhook, where parsimony is preferred and callbacks are registered only for the authors actually associated with a book. To implement this, we will create the `Subscribe` call, but only upon the creation of a book (the particular case of a change of author during the writing of a book is left aside in our example), and filter on the main author of the book. This will be even cheaper in callback bandwidth than registering on changes for authors with at least one book active. The code is added like this:

```
[Authorize(Policy = "editor")]
[HttpPost]
public async Task<IActionResult> Create(
    [FromBody] Book book,
    [FromQuery] DateTimeOffset? providedValueDate = null)
{
    // If no entity identifier is provided, a business rule states
that a GUID should be attributed
    if (string.IsNullOrEmpty(book.EntityId))
        book.EntityId = Guid.NewGuid().ToString("N");

    // Creating an entity is in fact nothing else than patching an
empty state
    JsonPatchDocument equivPatches = JSONPatchHelper.CreatePatch(
        new Book() { EntityId = book.EntityId },
        book,
        new DefaultContractResolver() { NamingStrategy = new
CamelCaseNamingStrategy() }
    );

    // Registering to author changes
    HttpClient? client = null;
    if (book.Editing != null && book.Editing.mainAuthor != null)
    {
        if (client is null)
            client = _clientFactory.CreateClient("AuthorsWebhook");
```

```
        await client.PutAsync("?callbackURL=http://demoeditor.org/
books/authorscache&$filter=href eq '" + book.Editing.mainAuthor.Href +
"'", null);
    }


    // After creating the equivalent patch, we thus pass this over to
the PATCH operation
    return await Patch(book.EntityId, equivPatches,
providedValueDate);
}
```

Again, we create a dedicated client for the webhook-related calls. This appears in a clearer way in the books referential service than in the authors one because we will need a second type of client (as will be explained shortly):

```
builder.Services.AddHttpClient("AuthorsWebhook", client => client.
BaseAddress = new Uri("http://demoeditor.org/authors/Subscribe"));
builder.Services.AddHttpClient("Authors", client => client.
BaseAddress = new Uri("http://demoeditor.org/authors")).
AddPolicyHandler(GetRetryPolicy());
```

URLs are hardcoded for the sake of readability. In the real code, they will be in the configuration, with default values corresponding to the convention used by Docker Compose services. Showing them here explains better that the `AuthorsWebhook` HTTP client instances will be used to call the `Subscribe` function. In the preceding code, it can be noticed that two arguments are passed in the URL:

1. `callbackURL` allows the subscriber to specify the URL on which it wants to be notified when the event happens, which, in our case, is the PUT endpoint that has been exposed by the `AuthorsCacheController` class.

2. `$filter` (to use the ODP standard syntax) allows the subscriber to specify a condition on which it should be called; in our case, this states that the author should be the one for the book (`$filter` has not been implemented on the authors data referential service in order to simplify code, since it does not bear any functional difficulty or business alignment analysis need).

All is now set. This code is normally enough to get the information every time there is a change in an author from a book created in the referential service and store this in the cache. We now need to show one last thing – which is not directly related to the webhook system but is the reason why we went to all this trouble – which is how to reduce time-based coupling between the two referential services.

Using the cache in the books referential

Now that the authors are partially cached (not all the authors and also not all the data in their instances), we can use this cache to avoid a failure in the books referential service if the authors referential service does not answer. Imagine a scenario where we implement the $expand feature on the books referential, allowing us to retrieve not only the complete book attributes on a GET call but also the details on the book's main author. A simple way to accommodate the additional data is to add the following last attribute in the AuthorLink class:

```
using MongoDB.Bson.Serialization.Attributes;

public class AuthorLink : Link
{
    [BsonElement("userEmailAddress")]
    public string UserEmailAddress { get; set; }

    [BsonElement("fullEntity")]
    public Author? FullEntity { get; set; }
}
```

This way, when a GET verb is called on https://demoeditor.org/books/ 0024?$expand=mainAuthor, the result would be something like the following:

```
{
    "entityId": "0024",
    "isbn": "978-2409002205",
    "title": "Performance in .NET",
    "links": [
        {
            "rel": "self",
            "href": "https://demoeditor.org/api/books/0024",
            "title": "Performance in .NET",
            "summary": "First edition of a book about performance
management in .NET"
        }
    ],
    "numberOfPages": 413,
    "publishDate": "2013-05-01",
    "editing": {
        "numberOfChapters": 10,
        "status": { "value": "available" },
        "mainAuthor": {
            "rel": "dc:creator",
```

```
                    "href": "https://demoeditor.org/api/authors/0012",
                    "title": "JP Gouigoux",
                    "userEmailAddress": "jp.gouigoux@frenchy.fr",
                    "fullEntity": {
                        "entityId": "0012",
                        "firstName": "Jean-Philippe",
                        "lastName": "Gouigoux",
                        "userEmailAddress": "jp.gouigoux@frenchy.fr",
                        "restriction": "blocked"
                    }
                }
            },
            "sales": {
                "price": {
                    "value": 42.15,
                    "monetaryUnit": "EUR"
                },
                "weightInGrams": 768
            }
        }
    }
```

This may sound like there is not much difference from the data we already had since only `firstName` and `lastName` are made additionally available, but keep in mind that this is because we have chosen to consider that the `Contacts` petal should be kept confidential. If we had completed the design of the `Author` class, we might have ended with much more data, like the following examples:

```
{
    (...)
    "editing": {
        "numberOfChapters": 10,
        "status": { "value": "available" },
        "mainAuthor": {
            "rel": "dc:creator",
            "href": "https://demoeditor.org/api/authors/0012",
            "title": "JP Gouigoux",
            "userEmailAddress": "jp.gouigoux@frenchy.fr",
            "fullEntity": {
                "entityId": "0012",
                "firstName": "Jean-Philippe",
                "lastName": "Gouigoux",
                "userEmailAddress": "jp.gouigoux@frenchy.fr",
                "restriction": "blocked",
                "skills": {
```

```
                    "domainsOfExpertise": [".NET", "Tests", "Docker",
"Kubernetes", "OpenData", "PowerBI"],
                    "degrees": [
                        {
                            "title": "Engineering degree in Mechanical
Systems",
                            «year»: 1997,
                            «school»: {
                                «name»: «Université de Technologie de
Compiègne»,
                                "country": {
                                    "rel": "country",
                                    "href": "http://demoeditor.org/
common/countries/FR",
                                    "title": "France",
                                    "code": "FRA"
                                }
                            }
                        },
                        {
                            "title": "Master of Science in Advanced
Automation and Design",
                            "year": 1997,
                            "school": {
                                "name": "Cranfield University",
                                "country": {
                                    "rel": "country",
                                    "href": "http://demoeditor.org/
common/countries/GB",
                                    "title": "United Kingdom of Great
Britain and the Northern Ireland",
                                    "code": "GBR"
                                }
                            }
                        }
                    ]
                }
            }
        }
    },
    (...)
}
```

In the preceding example, the full entity attribute indeed justifies its name, because it reproduces the complete definition of the targeted author, with their basic identity, along with their skills, possible degrees, and so on.

The following is the code that has been added (between the first and last lines, which have been copied for reference on where to insert it) to support this feature in the `GetUnique` method of the `BooksController` class:

```
Book book = result.First();
string? expand = HttpContext.Request.Query["$expand"];
if (book.Editing != null && book.Editing.mainAuthor != null
    && expand != null && expand.Contains("mainAuthor"))
{
    HttpClient client = _clientFactory.CreateClient("Authors");
    try
    {
        Author? mainAuthor = await client.
GetFromJsonAsync<Author>(book.Editing.mainAuthor.Href);
        if (mainAuthor != null)
        {
            // As a second level of security, the books referential
service immediately deletes confidential data in case the authors
referential service has a security problem
            mainAuthor.Contacts = null;

            // Then, the author is used in the expand but also updated
in cache
            book.Editing.mainAuthor.FullEntity = mainAuthor;
            await Database.GetCollection<AuthorCache>("authors-
bestsofar-cache").FindOneAndReplaceAsync<AuthorCache>(item => item.
EntityId == mainAuthor.EntityId, new AuthorCache(mainAuthor));
        }
    }
    catch
    {
        // Naive retrieval of the entityId from the href link
        int indexLastSlash = book.Editing.mainAuthor.Href.
LastIndexOf("/");
        string authorEntityId = book.Editing.mainAuthor.Href.
Substring(indexLastSlash + 1);

        // If, for whatever reason, fresh author data cannot be
retrieved, we failover on the cache
        var res = Database.GetCollection<AuthorCache>("autho
rs-bestsofar-cache").Find<AuthorCache>(item => item.EntityId ==
authorEntityId);
```

```
        AuthorCache authorCache = res.FirstOrDefault();
        if (authorCache != null)
        {
            book.Editing.mainAuthor.FullEntity = authorCache.
ConvertToAuthor();
        }
        else
        {
            // The choice of implementation in case of failing to
    expand the author is to raise an error
            return new NotFoundObjectResult("Impossible to expand
    author information");
        }
    }
}
return new JsonResult(book);
```

The comments in the code should be enough to understand how it works, and again, the implementation is very simple in order to talk about the business alignment problems and not the technical parts. What is important is that we first make a call to the authors data referential service but, this time, with a retry policy that is a bit reduced (3 calls, at 100-ms intervals), although still existing. However, it may not be necessary to make the call as the cache provides security.

```
static IAsyncPolicy<HttpResponseMessage> GetRetryPolicy()
{
    return HttpPolicyExtensions
        .HandleTransientHttpError()
        .OrResult(msg => msg.StatusCode == HttpStatusCode.NotFound)
        .WaitAndRetryAsync(3, retryAttempt => TimeSpan.
FromSeconds(0.1));
}
```

If this call works, mainAuthor.fullEntity is filled with the content retrieved, as is the cache for future use when connectivity is lost. Note that we double-check the absence of Contacts, even if this is not the responsibility of the books referential; this does not mean that there is a sharing of responsibility: it is still up to the authors data referential service to never let this data out to external services. However, when it comes to security, it's one of the rare cases where too much attention may not be enough.

If, on the other hand, the call does not function well after the three tries, then we revert to the cache mechanism, which should normally be quite fresh since it is updated at every modification of an author. The data should normally be present in the cache since the book has been created prior to reading it, but in the particular case that it is not, we would raise a 404 Not Found error. We could make this even more sophisticated by working on the value dates to check whether the data is fresh enough in the cache, or pile up states in the cache; again, all this should be decided according to business needs. If the product owner decides the last data in the cache is fine, so be it. If they decide the cache should not present data that is more than a day old, this is their choice again. Also, be careful to correctly explain the difference between cache freshness and the data's last change: an author may be dated back a few months because it has not changed at all for this update, and this can be the freshest data we have in the cache, taken from a few seconds ago from the authors referential.

Additional issues on webhooks management

We have now completed our simplistic implementation of webhooks to synchronize partial data between two referential services in separate services while keeping a complete separation of responsibility and lowering coupling to the minimum. However, lots of considerations still remain to be taken into account, including the following:

- Filtering on the types of events on which we want to listen. In the previous implementation, the hypothesis was that all we could register to was a change of author, only being able to filter on the content. In reality, we might want to filter on the creation/modification/deletion of an author. We might want to restrict filters on attributes after a change or before a change; we might even want to mix both conditions.

- Projecting data right from the subscription would also be useful; indeed, why receive an almost full definition of an author in a callback if all we are interested in is its unique identifier because we are implementing a counter of authors? This could be done by implementing $select along with $filter in the Subscribe function.

- Unsubscribing should also be taken into account in a complete implementation.

- If a subscription is dumped by mistake in the authors data referential service, the books referential service should be able to add it again. In fact, it might be useful to recall the authors referential service from time to time so that the subscription is still active. We might even be forced to deal with this because of the JWT preemption date. Indeed, we talked about authorization on Subscribe but how about authorization on the callbacks? This has been set aside, with an AuthorsCacheController class not secured at all, but this would, of course, not be the case in production. How do we handle this? With which user? And what do we do when the token is out of date? A long-duration token can only postpone the problem. Registering again at regular intervals could be the solution, but this again means that there should be a clear separation of responsibility between the two parties, with a contract in the middle stating how long the tokens are accepted.

This would usually bring the analysis to an embedded orchestration system, such as **Quartz.NET**, which is a new subject in itself.

Even when all this is taken into account, some other questions would arise on the implementation itself. When showing the OPA-based authorization management, we saw that it should be aware of the last state of the data. Would it be a good idea for the authorization server to subscribe to any change of data it needs to make authorization decisions?

Finally, you have certainly noticed that the classes were duplicated in the `Models` folder of the two projects. Since they correspond to the .NET implementation of the schema in the OpenAPI contract, it would make sense to share them between all the projects needing them for serialization, for example, in a `DemoEditor.SharedModels` library. But we should also take care of its versioning alongside the versioning of the API contracts.

In short, event-driven architecture is much more than what has been shown here in technical terms but, as usual, the most important aspect is to tackle the subject by clearly separating the responsibilities, and the best way to do so is to also think in functional terms.

Dealing with eventual consistency

Several times in this chapter, we have talked about eventual consistency as the last resort solution to loss of connectivity, missing webhooks, bad callbacks, and so on. This notion of eventual consistency is particularly important and deserves to be treated in a dedicated section of this chapter because this is a great illustration of the difference between functional complexity and accidental complexity in lots of information systems.

Indeed, one of the reflexes of many architects, in this case, is to use a **Message-Oriented Middleware** (**MOM**), such as ActiveMQ or RabbitMQ. Since we installed one in Docker Compose in *Chapter 15*, you most certainly are wondering why we did not use it here, by the way... Well, the reason is that these messages are not that important. Sending a contract to be signed by an author is important, retrieving some content they deposit on the system and putting this into the books referential service and the **Electronic Data Management** (**EDM**) is important. But synchronizing authors on books is definitely not important. How do we know? Again, ask the business owners. If you simply tell them that data might be out of date, sure they will tell you it should not be. But tell them the truth about the maintenance cost of a MOM and they will quickly ask the real questions: how frequent will this be? How long will the data be outdated? By how much?

In our case, I can imagine having a discussion with the `DemoEditor` director and explaining that, since they are working in an internal network, there will be less than one in a million lost packets, and that the most common case of losing a callback will be when one of the referential services is under maintenance. I would also explain to the director that we have a process running every night that sends all up-to-date data to the relevant referential. And I would finally add that in the statistics they can check on the referential service (remember the Power Query interface?), numbers show that around 20% of the authors change in a given year, with an average stable state of 6 months. Can you imagine how quick the decision will be? Any chance that a complex server with a high maintenance cost such as a MOM would be chosen in this situation? No way!

You could argue that, since we need RabbitMQ to secure contract signing and book manuscript deposits, we might as well use it also for the authors' cache refresh. But the cost of the MOM is not only in its installation: it is also in manipulating messages, finding old ones in dead-letter queues, connecting to an additional console, re-authenticating messages for which the security token is too old, and so on. Compare this to just letting the orchestrator send a nightly refresh (or even send it manually if you really need to), and you will see that the choice should really be dictated by business concerns and not technical architecture views.

I might sound very critical against MOM and architects but what I warn you against is simply overengineering and not using the right technology for the right feature. There is a good reason why even banks recalculate their balances only once a night.

A data referential service is not just about CRUD

At this point of the chapters about data referential services, it seemed important (thanks to Guillaume for pointing this out) to stress that, although we have mostly talked about the four main operations (**Create/Read/Update/Delete**, or **CRUD**), data referential services are, of course, much more than this. Explaining how these fundamental operations work was in line with the "record all" approach and allowed us to talk about the inner structure of the service, as well as the data schema, the validation criteria, and a lot of other characteristics. However, it left aside another important part, which is the inclusion of business processes and business rules.

In the chapters dedicated to BPM and BRMS, I already explained that externalization was costly and that, most of the time, these functions would be embedded in the code. After all, setting up and maintaining a full BPM engine or a dedicated BRMS is only worth it when processes and rules change frequently, or when there are some particular requirements of traceability or versioning of their content. In the vast majority of information systems, these features are located inside the data referential services, right in the code.

If you want to stay as close as possible to the REST principles, all things should be resources and attributes, so business operations would not be sent as functional verbs but rather with POST, GET, PUT, PATCH, or DELETE (and I know that this sounds strange when you are used to web services that correspond to operations, such as SOAP ones that carry a soap:action attribute, but this is the REST way to go and it has many advantages, such as statelessness, and thus better scaling):

- Closing a contract would use PATCH on the status attribute of the contract. Typically, this change of status would start some other operations in the data referential service (for example, making the contract read-only for most profiles) and also in other services (through the use of webhooks).

- Validating a proposal could use POST of a decision (secondary resource) that is attached to the proposal resource. Again, this could change some exposed characteristics of the proposal when someone reads it (for example, removing the validity end date of the proposal).

- Ordering phone numbers in a given business order (home, then cell, for example) could be triggered by PATCH on an order attribute that would only be exposed to modification and not reading. If phone numbers are then reordered manually, this inner attribute would be modified by the code to contain a value such as `none` or `without`.

There are also a lot of cases where the data referential service does much more than simply store the data sent to it using its CRUD features, and this is also where lots of business processes and business rules are applied. Here are a few examples of such operations, from the simplest to the most complicated:

- Validating some inputs depending on business rules. Most of the time, checking the type is correct is enough, but there are also lots of cases where business rules are in place, such as checking that the date of birth of someone is not in the future, or that an end date is indeed after the start date, and so on.

- Simple business processes such as the life cycle of an object can also be implemented in simple code inside a data referential service, for example, by checking that a modified status is after the actual one. There is indeed no way that a book, for example, can revert from a "ready to print" to a "draft" status.

- Refreshing some attributes based on others is also a common feature that is based on business rules and not about CRUD at all. A simple example is when a tax rate has been changed by `PUT` and `PATCH` and the total amount including tax should be changed. There are multiple ways to apply this, but one that is quite common is to apply the calculation in the code and store the updated value. Another one is to wait for the next `GET` command to calculate it, but that does not change much due to the fact that it is a business rule executed in the code of a data referential service.

- Things then get a little more complicated when the computation includes data coming from multiple services. If we talk about a few services that have to exchange data, it is not that hard to orchestrate their exchanges in an efficient way. This is where we are going to use direct calls or local caches, such as the one shown previously in this chapter.

- If there are more services that are included in the computation, there might be a need for something a little more complex, such as a service dedicated to assembling the data from the others. This would typically be a service specializing in reporting, which would gather the data from other services, prepare it, aggregate it if necessary, and sometimes update some results by registering to the modification of the data sources (through webhooks or other event-based mechanisms).

- One more complication is when lots of services provide some data that should be aggregated into complex calculations with many inputs, possibly changing frequently. Since computing everything every time there is a change is out of the question, there should be a "smart" engine capable of waiting for enough data changes to run a batch of calculations, but also being aware of the dependencies between the jobs and capable of executing them in the most efficient order. This is typically where solutions such as Apache Kafka would be used.

Though there is no actual need to show this kind of operation in detail in the book (because it would not change the comprehension of what a data referential service is and how it acts inside a modern information system), the companion source code on GitHub shows some of the features that have been explained in the present section, which will hopefully make it more concrete if needed. It will also stress again the important point that, even if we have mostly talked about CRUD operations, data referential services in reality handle lots of business processes and rules (and in almost all information systems nowadays, much more than externalized BPM or BRMS engines).

Summary

In this chapter, we used what we already learned about building a data referential service (the one about books) and extended the approach to generate a second data referential service, which is about authors this time. Books and authors are the two main business entities to be managed by an editor, so by the end of the present chapter, we are quite complete in terms of **Master Data Management** (**MDM**), which means we are going to be able to proceed, in the next chapter, to add a GUI to our information system since the raw data part is taken care of. This is logical when the API is ready to give users a "real-world" way to access and modify data and also realize their business processes.

You also have certainly noticed (and this would be even more obvious if you follow the complete code in the GitHub repository associated with this book) that the two referential services have a lot in common. In fact, a huge majority of the code is the same, except that we use the `Book` data structure on one side and the `Author` schema on the other. This is not the result of some random situation, but the materialization of the fact that a good alignment on the business tends to erase a lot of the accidental, technical complexity. In fact, in the utopic information system, MDM is generic for every entity. This is still far-fetched but, as we saw in this chapter, creating a correct data referential service is not that hard.

A possible evolution in the design of data referential services, in general, could be simply to generate this data referential service from an OpenAPI contract. Though the result would not be dynamic, a good code generator such as T4 or Jinja could already go a long way by outputting all the code necessary to realize what has been done previously without any developer typing a single line of code. Some projects have already dug into this way of doing and I encourage you to try them in order to speed up the process of creating data referential services. But please do not create yet another framework: what is missing today in information systems is not technical tools but actual, usable, business-aligned referential services. And I can tell you from experience that a good referential service with clean, widely usable, and interoperated data is a treasure for most companies who have made the effort to make them appear, however old-fashioned its technical implementation is.

20
Creating a
Graphical User Interface

Do you remember this diagram from *Chapter 8*? It uses the hexagonal architecture to show all the interfaces around the business domain of books:

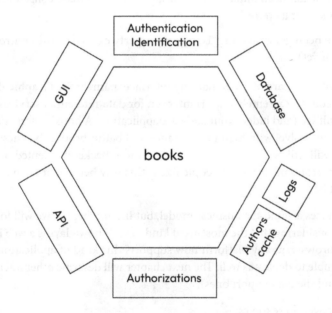

Figure 20.1 – Hexagonal architecture schema for the books domain

In the previous chapter, when working on the data referential services, we covered, of course, the API interface (which is so central to this service-base architecture that it should almost sit in the center of the hexagon), but also many of the other interfaces:

- Persistence has been plugged into a NoSQL database, with a coupling that some would consider too high but which is really not a blocking problem, as long as the referential service remains focused on only one business entity and is quick to re-implement if ever necessary.

- Authentication and identification have been delegated to the IAM implemented in Apache Keycloak, and the plugging of this interface is done through OpenID Connect/JSON Web Token standards, which virtually eliminates coupling.

- Authorization has been implemented with **Role Based Access Control** (**RBAC**) and a bit of **Attribute Based Access Control** (**ABAC**), using, respectively, the roles provided by the IAM and the additional authorization management from OPA (arguably more coupled, since this is still a proprietary interop, but this is already way better than having the business rules hardcoded in the referential service).

- The authors cache has been implemented through a local database collection and a webhook-based mechanism for its refresh.

- Even logs have been generated (a `_logger` property existed in the controllers, even if we have not used it yet).

In this chapter, we will take care of the remaining interface, namely the **Graphical User Interface** (**GUI**). A GUI is, of course, extremely important, even for data referential services, because this is how the final users will see (and judge) your software application. A book's referential service can be well-designed, ready to evolve, with high performance and battle-proof robustness, but if the GUI that calls it is ugly, it will still be considered a bad application. Backend-oriented people such as me think this is a shame; frontend-oriented people like it this way because it favors their job – that's simply just the way it is!

There are other interfaces around the business model, but in this chapter, we will focus on the main graphical interface, considering that the most-used kind of GUI nowadays is a web interface created to be displayed in a browser, as any platform now supports this kind of application, and this is the most versatile and simple to deploy as well. The next chapter will describe other interfaces (including both a second GUI and the data import ones).

In this chapter, we'll cover these topics:

- A standard web single page application
- Connecting the GUI to the backend

A standard web single page application

The client expressed satisfaction with the last sprint demo and has made necessary corrections to the old legacy Excel file. The next step is the migration process, which is not far away. In this step, the new data referential service will become the master, and the old legacy Excel file will continue to be used to produce existing reports as a clone. But first, the customer wants to get out of the technical process and start enrolling the teams in the project. To do so, a GUI is needed so that anyone can access data in a new manner and start getting their head around the new functions of an API.

When advising the customer on this sixth sprint, it was proposed to start right away on a mobile-first approach. The director of `DemoEditor` said they would normally not use any mobile device to access books and authors, and that the integrators should stick to a web interface. In the exchange, it was proposed to indeed avoid choosing the complexity of a native application, but they should still try to make the interface as responsive as possible and test the GUI from the beginning on mobile devices, although in their web browsers. The director agreed this was not a bad idea, since the additional cost is very limited, and this would also go in the direction of light and ergonomic interfaces, even when the large screens allow for more data to be displayed.

Defining the GUI

For this first attempt at providing a visual frontend to our sample information system, it was decided to use the Blazor technology, in its WebAssembly-based single page application variant. The choice of technology does not really matter for our client, just like for the services, but since we have used the Microsoft .NET stack, for example, it is a logical choice to use a technology that allows us to develop a browser-embedded application, while still using C# and all its tooling.

In visual terms, the GUI will be extremely simple, with one page allowing you to manage author information, another one that is similar but for books, and a third one where the user will be able to search for either of these entities. There's nothing else to do for now since the data referential services are not complete. It's not necessary to display complete interfaces here, as our objective is to demonstrate how to design the client in a way that respects business/IT alignment and ensures a clear separation of duties. It's worth noting that a considerable amount of work has already been done to make the API available for data retrieval upon which the clients will be built.

The basic GUI navigation in the form of a storyboard would be like this:

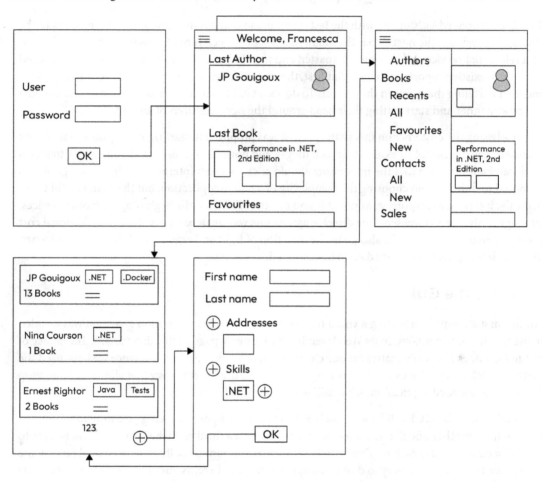

Figure 20.2 – Storyboard for a simple application for editors

The user, once logged in, will be welcomed with a page showing their last viewed author and book, and then some favorites. The menu would offer access to authors, books, contracts, and sales (depending on the authorizations), with submenus to rapidly access some specific operations on these business entities. When choosing the menu, a list would be proposed, together with an icon enabling the creation of a new entity. The creation of entities for a particular business can be modified and viewed from a list by using a screen that shows the attributes of a given entity. Of course, in a real application, filters on a list would be present and navigation would be particularly elaborate. In the following example, we will stick to the simplest application possible, illustrating only the choices that are necessary to maintain a good business/IT alignment.

Preparing IAM for the client

As explained, the very first thing users will see when they start the application will be a login page. In fact, they will see the login page provided by the IAM. When dealing with applications, specifically Single Page Applications that run on a browser or mobile applications on a user's device, it's not safe to assume that the client platform can keep a secret. Thus, we need to use a dedicated client to customize the IAM system. This client will be referred to as Portal. This name represents the GUI that will combine the visual management of different domains into one interface.

We will return to the Keycloak management console and define a second **OpenID Connect** client, in addition to `BooksAPI`:

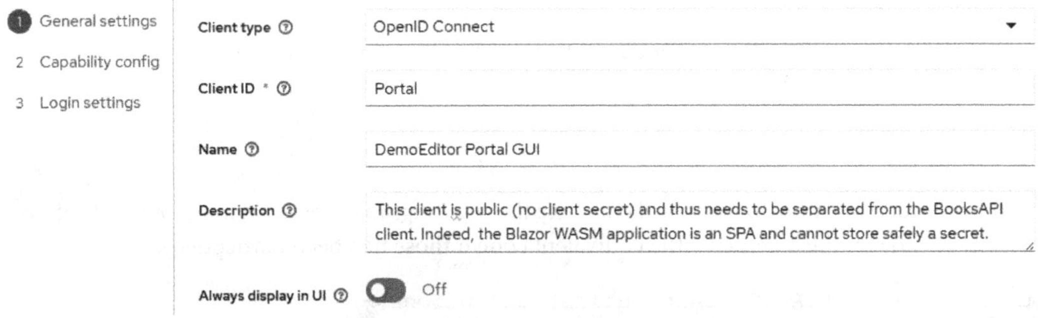

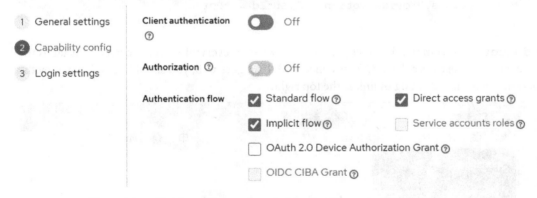

Figure 20.3 – Creating a new client in the Keycloak console

The second page of the wizard is where we choose to not have any client authentication:

Figure 20.4 – The Capability config panel for the client wizard in Keycloak

We will also have to add the callback login and logout URLs, as we did previously for the client created in *Chapter 18*. To do this, we will first need to know the URL of the application, which will appear when we follow the steps in the upcoming section.

Generating the client skeleton

To create this single page application, we will use a slightly more elaborated creation command line than the one previously used, as follows:

```
dotnet new blazorwasm --auth Individual --authority http://
localhost:8080/realms/demoeditor/ --client-id BooksAPI --use-program-
main -o portal-gui
```

The generated application skeleton will embed some `appSettings[.Development].json` files, in which you will find the provided settings for authentication:

```
{
  "Local": {
    "Authority": "http://localhost:8080/realms/demoeditor/",
    "ClientId": "Portal"
  }
}
```

The following code in `Program.cs` will consume these settings in order to establish authentication at runtime (I show in the comments the equivalent code if those had been hardcoded):

```
builder.Services.AddOidcAuthentication(options =>
{
    builder.Configuration.Bind("Local", options.ProviderOptions);
    // options.ProviderOptions.Authority = "http://localhost:8080/
realms/demoeditor/";
    // options.ProviderOptions.ClientId = "Portal";
});
```

A `dotnet run` command will let you discover whether everything works correctly as far as the connection to the externalized IAM is concerned. If all goes well, you should be welcomed by the following screen, with a **Log in** link at the top right:

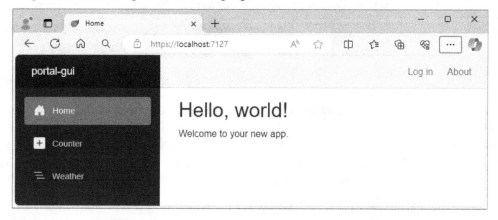

Figure 20.5 – A basic template of a Blazor application with a login link

Clicking on this link should redirect you to the form exposed by Keycloak (note the change in host), where you can log in as a user – for example, the director of `DemoEditor`:

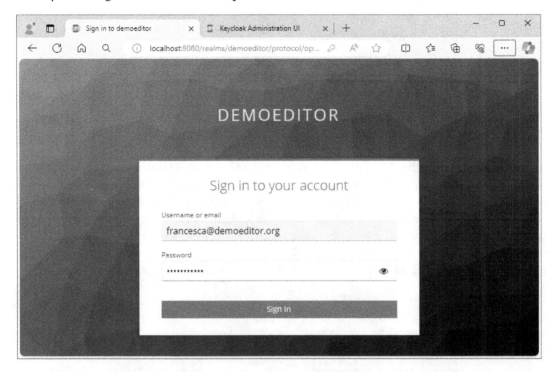

Figure 20.6 – Logging into the Blazor app

The greeting to Francesca Presidio, director of `DemoEditor`, shows that the authentication process was successful:

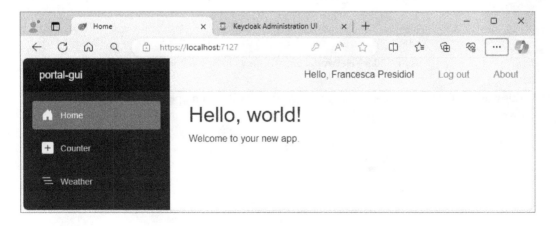

Figure 20.7 – The Blazor app with DemoEditor's director connected

As usual, a `docker init` command will prepare everything for containerization, and `docker-compose.yml` is updated to reflect the additional service:

```
portal:
  image: demoeditor/portal-gui:0.1
  build:
    context: ./portal-gui
    dockerfile: Dockerfile
  ports:
    - 80:8080
  restart: unless-stopped
```

However, this time, we will make another change in the Docker Compose setup in order to show that expanding the system needs to be accompanied by a growing separation of concerns to maintain our business/IT alignment. In this case, it happens to also be a security good practice – we will separate services with network-based frontiers. Indeed, the GUI should never have direct access to the database but always pass through the API gateway (for now, a call to the API, but we will show how to improve this approach in the next chapter). By default, the Docker Compose startup creates a dedicated overlay network for the application. Now, we will define two networks, with the following services inside each:

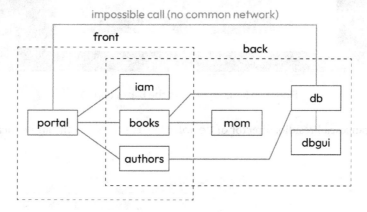

Figure 20.8 – Network compartments for the services

How this works is that the `docker-compose.yml` file will be equipped with a (simple) definition of networks:

```
networks:
  front:
  back:
```

Each service will participate on one or two networks, depending on its position in the system (I only show the network definition of some services, and not their other attributes, to make this shorter):

```
version: '3.0'
services:
  db:
    image: mongo:4.4
    networks:
      - back
  iam:
    networks:
      - front
      - back
  books:
    networks:
      - front
      - back
  portal:
    networks:
      - front
```

The skeleton of the application is now ready and can plug into the authentication and retrieve partial identification, but we need to retrieve the roles in order to complete identification and deal with a bit of authorization on the client side, although this is not straightforward, as we will explain in the next section.

Taking into account authorization concerns

We cannot really discuss authorization management for a client, since rights enforcement is the responsibility of the server side of an application. Nonetheless, a client surely can participate in making authorization easier for users – the situation is the same as validation, where the server is ultimately responsible for checking data correctness, but clients certainly can help pre-check data in order to save a network round trip and speed up the GUI. In the case of authorization, a client who knows what is granted and what is denied can, among other ergonomic benefits, reduce the number of visible menus and avoid proposing an option that would not be authorized when activated.

This is why we are going to use the roles associated with the identification in our Blazor application in order to adjust the GUI appearance. In fact, we will go even further than that, in the sense that buttons to access unauthorized features will be hidden, but we will add another layer of security so that even if the user directly accesses the right URL in the browser, they will be refused access to the page. Of course, that does not mean we should trust this wholeheartedly and abandon API-level security, since an attacker could simply call the API directly, thereby finding a shortcut to all the security layers in our dedicated client. The idea here is to pile up as many security layers as possible to make it difficult for the attackers (and also guide legitimate users).

This section will also be the opportunity to adjust some IAM designs. Indeed, when the only client plugged into Keycloak was the books repository, it was logical to call the `BooksAPI` client, but this proved inappropriate later on when the authors referential service appeared. Does this mean we should create a `demoeditor` OIDC client? In fact, no, because we still need to assuage the authentication client's concerns with more granularity than at an organizational level. We can indeed consider that "internal" services will all have the same security characteristics, but this is definitely not the case for the GUI applications. This is not only because they are going to be more exposed (and it would be practical to be able to remove them from IAM without unplugging the internal services) but also because they do not have the same capacity to keep a secret. This is why the client for the APIs was customized with **Client authentication** activated, but we toggled it off for the `Portal` client:

Figure 20.9 – The Client authentication setting in Keycloak

We could consider the use of a single authentication client for all GUIs, just like we grouped all APIs on one client, but again, thinking in terms of a business and aligning it with the technical aspects clearly shows that there should be a client for each GUI application. Indeed, a web portal, an extranet, a mobile application, or yet another type of client will have different ways to handle IAM, and some devices will be more secure than others, which is a strong argument for separate authentication clients. Thus, we will leave the `Portal` client only for our Blazor application and create other dedicated clients – for example, one called `Mobile` when we talk about a mobile device application later in the chapter (how to create other dedicated clients will not be shown in detail, as it was covered in previous chapters and it should be quite clear how to do it by now but remember that it should be done).

This decision has a consequence in terms of role definitions. You may remember that, when we created the initial authentication client called `BooksAPI`, we defined two client roles, named `author` and `editor`, which allowed us to have a clean, well-separated authorization management shared between the referential services and the IAM. Now that we are adding a second authentication client, does this mean we should duplicate these roles (and take advantage to append them with a `director` role that we recently introduced in the functional discussion)? Again, this technical decision should not be thought of in technical terms (performance, duplication, etc.) but in functional terms. Let's simply ask the users and directors of `DemoEditor` what they think of it, and chances are that they will tell us the roles may be *used* differently from one application to another, but that they are *defined* for a given user in the organization. In other terms, if a sample person named Manfred has their rights set to `editor`, this should be true everywhere, whether they use a mobile application or the web intranet.

Further discussion with the stakeholders might raise the question of whether roles should be associated with groups. If so, the simple fact that Francesca (our sample user, who is the director of `DemoEditor`) belongs to the `board` group would give her `director` access to `BooksAPI`, the `Portal` clients, and anything else. In our sample `DemoEditor` context, this would be refused for several reasons (again, this is purely an arbitrary example, and you should make your own decision based on the business context of your organization; I am just showing a particular case):

- The team is adamant that roles are associated with the person (or rather, the user, even if they do not see precisely the difference), and they should not differ from one application to another. Otherwise, some behaviors might be difficult to understand – in particular, the idea that someone may have more rights using one application instead of another might seem absurd to them.

- When the technical consultant explains to the company's team that, for security reasons, they may have to reduce rights when someone connects on the extranet instead of from the internal, safer network, they reply that most of them work from home and that they are considering giving up on going into the office, so this change will become irrelevant. And anyway, they could also use different user accounts, which some of them already do for highly secure features such as running payments. So, again, this is not a relevant point of discussion for them.

- Another person in the team points out the fact that, if roles are associated with clients, then it will be necessary to create them and assign them to users in the Apache Keycloak interface when a new application appears, and they are really not keen on this, as IAM management is not in their comfort zone. Again, this applies to general roles that apply everywhere.

- The last straw comes from the director, Francesca, who explains that her highly trusted assistant Stephanie needs to have `director` access, even if she does not belong to the board. So, really, roles should be used for her and not groups.

All this discussion concludes that roles should really be at the realm level, and luckily, this is possible in Keycloak, which has a dedicated menu. After a few easy changes (so easy they will not be detailed here), we obtained such a configuration of the realm roles, where we can clearly see the **author**, **editor**, and **director** roles in addition to the previously existing ones that are embedded in Keycloak and there by default:

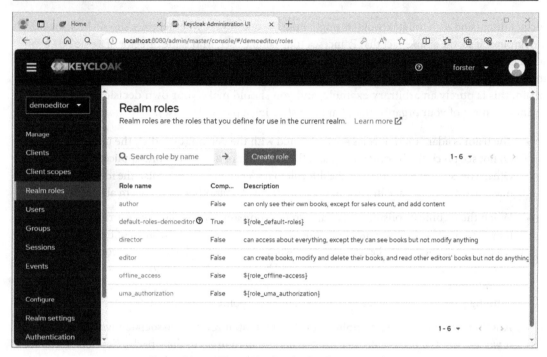

Figure 20.10 – The realm-level roles for DemoEditor

Of course, this change means that the mapping of roles should be adjusted accordingly in the existing service. Yes, this is a breaking change, but this was done on purpose because it was interesting to show that, even with a good analysis when creating a referential service, the lack of a global overview can lead to a bad design decision. This is the difficult art of information system design, where we try to always balance a DRY/YAGNI approach with enough vision to reduce churn. But no one gets it right all the time, even with decades of varied experience and advanced skills, as well as a knowledge of the available technologies and listening to and deeply understanding business/functional needs. In our example, the `BooksAPI` client should be adjusted to map realm roles instead of client roles. While we are at it, it would be better to rename the client as well, since it is used for the books referential service but also for the authors. Finally, the code that was added to retrieve the client roles and make them available in a .NET authorization system should be modified as well.

Note that, if the `DemoEditor` team were to think of a given role that only makes sense in some clients (let's say `archivist`, which could be imagined as a role for book management only), nothing would prevent you from adding this role only to a given client and not at the realm level. Keycloak is able to show both in a given token, in two separate sections. Yet, it is necessary to map roles to the client scope, whether they are defined in the client only or for the whole realm. This is done in the **Client scopes** tab of the client definition and, more precisely, in the scope called **Portal-dedicated** in our case:

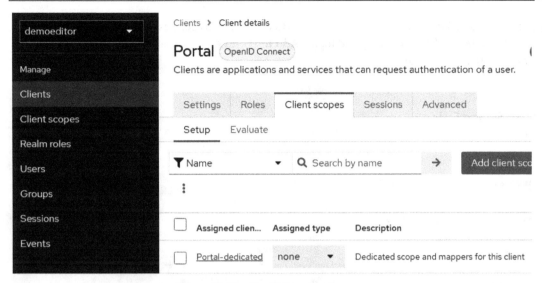

Figure 20.11 – Modifying client scopes

Once the corresponding link has been clicked, we will reach an interface that allows us to add a predefined mapper, and by navigating through the different pages of the list (displaying the pages in sets of 10), we can find the mapper for **realm roles**:

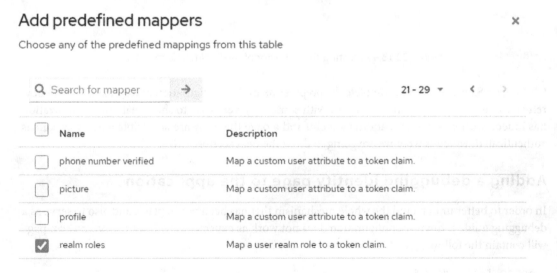

Figure 20.12 – Choosing a predefined mapper

Once this predefined mapper is selected and an Add operation has been executed, we have to click on the newly added **realm roles** entry and toggle **Add to ID token** on because, by default, the mapper only affects the access tokens:

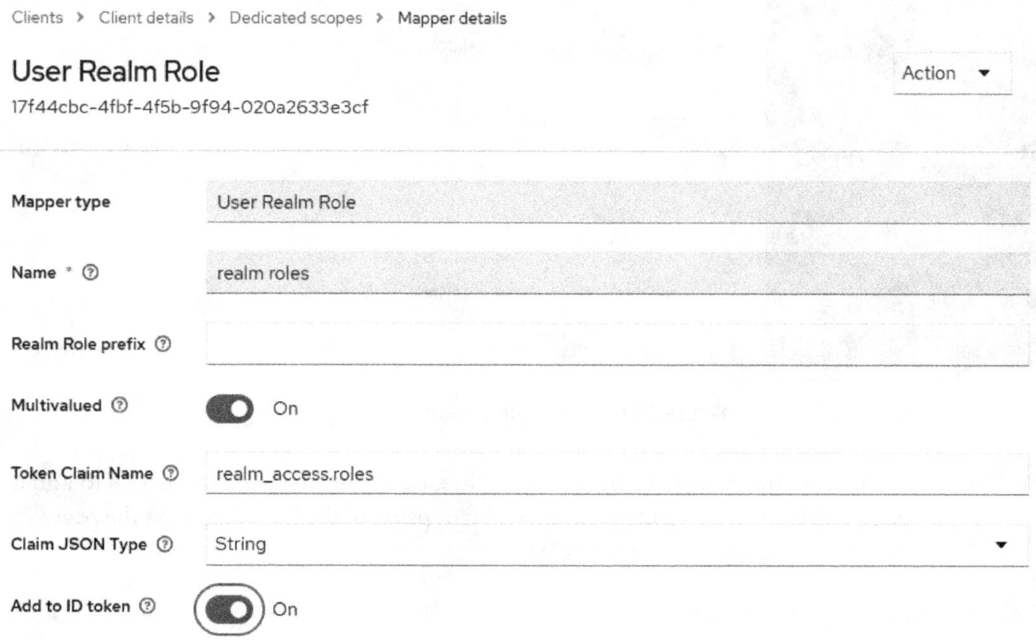

Clients > Client details > Dedicated scopes > **Mapper details**

User Realm Role

17f44cbc-4fbf-4f5b-9f94-020a2633e3cf

Action ▼

Mapper type	User Realm Role
Name * ⑦	realm roles
Realm Role prefix ⑦	
Multivalued ⑦	🔵 On
Token Claim Name ⑦	realm_access.roles
Claim JSON Type ⑦	String ▼
Add to ID token ⑦	🔵 On

Figure 20.13 – Adjusting the behavior of the Realm Role mapper

Clicking on **Save** allows us to complete the preparation of the configuration in Keycloak. We will now return to the Blazor application and start with something very easy to show that our customization has indeed worked – namely, adding a menu and a page that only are accessible when someone is authenticated, which is what we are going to do in the next section.

Adding a debugging identity page to the application

In order to better understand the whole mechanism that has been put in place, and also to serve as a debugging guide if the role configuration does not work as expected, the Identity.razor page will contain the following code:

```
@page "/identity"
@attribute [Authorize]
@using Microsoft.AspNetCore.Authorization

<PageTitle>Identity</PageTitle>
```

```
@* Reference: https://code-maze.com/blazor-webassembly-role-based-
security-with-identityserver4/ *@
<h3>Identity of the connected user</h3>

<AuthorizeView>
    <h2>
        Hello @context?.User?.Identity?.Name,
        here's the list of your claims:
    </h2>
    <ul>
        @if (context?.User?.Claims != null)
        foreach (var claim in context.User.Claims)
        {
            <li><b>@claim.Type</b>: @claim.Value</li>
        }
    </ul>
</AuthorizeView>
```

Note the second line (@attribute [Authorize]) that creates a level of authorization even if, as we explained before, a client should never be trusted for actual authorization management, as its code is executed on a machine that is not under the control of the application owner, unlike a server. We are also going to add this kind of behavior to the new menu that will be added in NavMenu. razor, making it only appear that the user is authenticated (note that this is not a real security block and only an interesting behavior that hides the link, even if it does not prevent users from manually changing the URL if they want to circumvent this):

```
<AuthorizeView>
    <Authorized>
        <div class="nav-item px-3">
            <NavLink class="nav-link" href="identity">
                <span class="bi bi-person-nav-menu" aria-
hidden="true"></span> Identity
            </NavLink>
        </div>
    </Authorized>
</AuthorizeView>
```

The following content should also be added to NavMenu.razor.css for the icon to appear correctly:

```
.bi-person-nav-menu {
    background-image: url("data:image/svg+xml,%3Csvg xmlns='http://
www.w3.org/2000/svg' width='16' height='16' fill='white' class='bi bi-
person' viewBox='0 0 16 16'%3E%3Cpath fill-rule='evenodd' d='M8 8a3
3 0 1 0 0-6 3 3 0 0 0 0 6m2-3a2 2 0 1 1-4 0 2 2 0 0 1 4 0m4 8c0 1-1
1-1 1H3s-1 0-1-1 1-4 6-4 6 3 6 4m-1-.004c-.001-.246-.154-.986-.832-
1.664C11.516 10.68 10.289 10 8 10s-3.516.68-4.168 1.332c-.678.678-.83
```

```
1.418-.832 1.664z'/%3E%3C/svg%3E");
}
```

The preceding content can be retrieved from `https://icons.getbootstrap.com/icons/person/`, should you want to create new entries in the application with a dedicated icon. Once this is done, running the application again shows an introductory menu (which only appears after logging in), and that leads to the following interface that lists all the claims of the **ID** token:

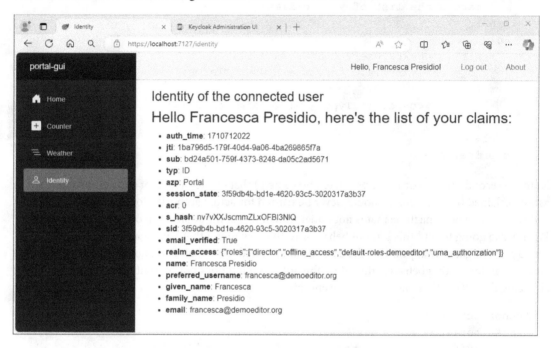

Figure 20.14 – Debugging the identification claims of the connected user

Note the following claim, which corresponds to the roles expected, as it contains not only the default roles sent by Keycloak but also the `director` role that we added to the `francesca` user:

```
realm_access: {"roles":["director","offline_access","default-roles-
demoeditor","uma_authorization"]}
```

To show that the interface can also adapt to the presence (or absence) of roles, which is what we want to achieve to obtain a strong IT/business alignment, we will add another menu, but this time, that should only be visible when a given role is present. In our example, only the users with the `director` role should have access to the sales, so we modify the content of the `<Authorized>` tag in `NavMenu.razor` like this:

```
<AuthorizeView>
    <Authorized>
```

```
        <div class="nav-item px-3">
            <NavLink class="nav-link" href="identity">
                <span class="bi bi-person-nav-menu" aria-
hidden="true"></span> Identity
            </NavLink>
        </div>
        @if (context.User.IsInRole("director"))
        {
            <div class="nav-item px-3">
                <NavLink class="nav-link" href="sales">
                    <span class="bi bi-currency-dollar-nav-menu" aria-
hidden="true"></span> Sales
                </NavLink>
            </div>
        }
    </Authorized>
</AuthorizeView>
```

A simple `Sales.razor` page is created, only to show how access to this page is regulated through the `Authorize` attribute:

```
@page "/sales"
@attribute [Authorize(Roles = "director")]
@using Microsoft.AspNetCore.Authorization

<PageTitle>Sales</PageTitle>

<h3>Confidential information about sales</h3>
```

In order for this to work, we need to add the following code to `Program.cs`:

```
builder.Services.AddOidcAuthentication(options =>
{
    builder.Configuration.Bind("Local", options.ProviderOptions);
    options.UserOptions.RoleClaim = "realm_access.roles";
});

builder.Services.AddApiAuthorization().
AddAccountClaimsPrincipalFactory<RolesClaimsPrincipalFactory>();
```

The class that is referred to in the preceding code should be added to the project with code resembling the following (this is a very crude implementation, without error management, and is used only to show the principle that we adhere to in order to achieve a correct – that is, guided by functional considerations – separation of responsibility between the IAM and the client application):

```
using Microsoft.AspNetCore.Components.WebAssembly.Authentication;
using Microsoft.AspNetCore.Components.WebAssembly.Authentication.
```

```
Internal;
using System.Security.Claims;
using System.Text.Json;

namespace portal_gui
{
    public class RolesClaimsPrincipalFactory :
AccountClaimsPrincipalFactory<RemoteUserAccount>
    {
        public
RolesClaimsPrincipalFactory(IAccessTokenProviderAccessor accessor) :
base(accessor)
        {
        }


        public override async ValueTask<ClaimsPrincipal>
CreateUserAsync(RemoteUserAccount account,
RemoteAuthenticationUserOptions options)
        {
            var user = await base.CreateUserAsync(account, options);
            if (user?.Identity != null && user.Identity.
IsAuthenticated)
            {
                var identity = (ClaimsIdentity)user.Identity;
                var resourceaccess = identity.FindAll("realm_access");
                string Contenu = resourceaccess.First().Value;
                JsonElement elem = JsonDocument.Parse(Contenu).
RootElement;
                foreach (JsonElement role in elem.
GetProperty("roles").EnumerateArray())
                    identity.AddClaim(new Claim(options.RoleClaim,
role.GetString() ?? String.Empty));
            }
            return user;
        }
    }
}
```

Once we test this, we can see that Francesca is authorized to access this page since she is the director of DemoEditor (there's no need to reconnect; a simple refresh of the application should be enough to display the menu):

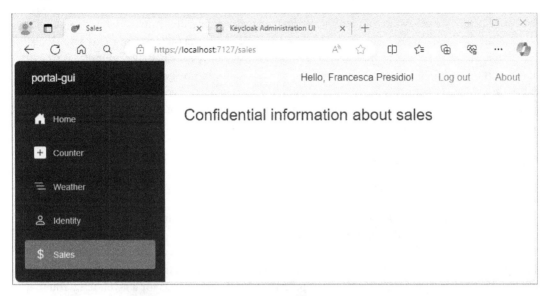

Figure 20.15 – Showing the page where access is authorized by roles

Let's try and connect with the same page once logged out and connected again, this time with a user called **Emilio**, as editor. Not only does the menu not display but also, if we try to force the application to go to the page by manually changing the URL, Blazor reacts as expected:

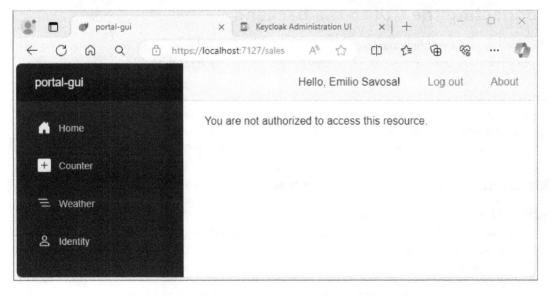

Figure 20.16 – Trying to access an unauthorized Blazor route

A last word before we get away from authorization management and dive into the data manipulation feature of the GUI – there are not any standards at the time of writing (at least not to my knowledge) on how to describe permissions in an application. In fact, even roles are actually bound to Keycloak and not part of the OIDC standard. Nonetheless, we still can find some general idioms and recurring ways to describe the operations associated with permission management. For example, lots of GUI frameworks use some kind of `Can[Operation]` function to determine whether an element of GUI should be visible or not. Generally, these are plugged into a `/HasPermission` API that outputs (with a `{ result: true }` response, for example) whether a given access is allowed or not when a request contains something like this:

```
{user:null, permissions:["Open","Save"]}
```

When `user` is `null`, this API understands that it should use the identity of the user connected. Otherwise, it would provide the result for a given request on a given user, but only if the authenticated user has the right to read permissions of other users. In some cases, in order to avoid frequent unit calls (this is the same complexity that Open Policy Agent has, as shown in *Chapter 12*), we could use the `Permissions` API operations to list simultaneously all permissions for a given user. You may also find `Capacity` in some clients instead of `Permissions`, which again shows that there is no standardization in this part of the authorization management domain, which is a big lack. If you want to dig deeper into this subject, `https://devblogs.microsoft.com/odata/odata-and-oauth-protecting-an-odata-service-using-oauth-2-0/` is a great read for starters.

Connecting the GUI to the backend

We now have the first implementation of the GUI, with forms and even some logic associated with the authorization levels, but we do not yet have any data coming from the backend. This is what we are going to realize in the following sections. After adding authorization management to the backend, we will detail how the GUI will call the API to get some information and also focus on an important feature, pagination. This will be also the opportunity to think about the GUI in terms of autonomous components, which is as important as the separation of concerns in the API (which is the domain where we have mostly applied this principle, for now, even if it is a general principle that applies everywhere).

Modifying authorization in the backend

So far, we have connected the backend API to a client called `BooksAPI`. Not only does this name not make sense anymore, since we should also authorize the authors data referential service, but it is also more logical to align authorization rules based on roles between the GUI and the APIs, now that we have turned to realm-level roles. So, we will start changing the way the two API servers retrieve the role claims, like this:

```
using Microsoft.AspNetCore.Authentication;
using System.Security.Claims;
using System.Text.Json;
```

```
namespace books_controller
{
    public class ClaimsTransformer : IClaimsTransformation
    {
        public Task<ClaimsPrincipal> TransformAsync(ClaimsPrincipal
principal)
        {
            ClaimsIdentity claimsIdentity = (ClaimsIdentity)principal.
Identity;
            if (claimsIdentity.IsAuthenticated)
            {
                foreach (var c in claimsIdentity.Clone().
FindAll((claim) => claim.Type == "realm_access"))
                {
                    JsonElement elem = JsonDocument.Parse(c.Value).
RootElement;
                    foreach (JsonElement role in elem.
GetProperty("roles").EnumerateArray())
                        claimsIdentity.AddClaim(new Claim("user_
roles", role.GetString() ?? String.Empty));
                }
            }
            return Task.FromResult(principal);
        }
    }
}
```

This code transforms the roles coming from the realm into roles at the user level since the methods that are used in the code only understand this level. This would be enough to switch to realm roles, but since we have introduced a new role called director, we will show how it should be handled the right way, which is – as you guessed – determined by a strong alignment to business rules. What are the characteristics of the director role? Mainly, it has access to sales and statistics, but as far as the books and authors data referential services are concerned, director mostly gets authorizations from the editor role. If you recall from *Chapter 12*, we expressed this in the following structure, where book-edition has book-direction as its parent, which means all access is moved up in the hierarchy:

```
"roles": {
    "book-direction": { "access": []},
    "book-sales": { "parent": "book-direction", "access": [{
"operation": "all", "type": "books.sales" }]},
    "book-edition": { "parent": "book-direction", "access": [{
"operation": "all", "type": "books.editing" }]},
```

```
      "book-writer": { "parent": "book-edition", "access": [{
"operation": "read", "type": "books.editing" }, { "operation": "all",
"type": "books.content" }]}
}
```

What is the right way to implement this upward transmission of access? If we want to be as aligned as much as possible with the business functional rule, there should be some code somewhere that gives a director automatically the editor role. Aligning with business requirements, the inheritance of roles should only occur for books and author data management, rather than being universal in IAM. Adding conditions each time we have [Authorize] would make a mess of our code. The closest to the functional requirement expression is to use the policies and make it so that, if a user bearing the director role comes into play, they automatically gain the author policy. In order to apply this new requirement, we first create the following classes that implement this behavior:

```
using Microsoft.AspNetCore.Authorization;

public class EditorRequirement : IAuthorizationRequirement
{
    public EditorRequirement()
    {
    }
}

public class EditorAuthorizationHandler :
AuthorizationHandler<EditorRequirement>
{
    protected override Task HandleRequirementAsync(
        AuthorizationHandlerContext context,
        EditorRequirement requirement)
    {
        foreach (var role in context.User.Claims.Where(claim => claim.
Type == "user_roles"))
        {
            if (role.Value == "editor") context.Succeed(requirement);
            if (role.Value == "director") context.
Succeed(requirement);
        }
        return Task.CompletedTask;
    }
}
```

This way, the requirement succeeds not only if a `user_roles` claim with the `editor` value is obtained, but also if another claim with the same key and `director` as a value is found. Taking this customized requirement into account, we need to modify `Program.cs` accordingly:

```
builder.Services.AddSingleton<IAuthorizationHandler,
EditorAuthorizationHandler>();

builder.Services.AddAuthorization(o =>
{
    o.AddPolicy("author", policy => policy.RequireClaim("user_roles",
"author"));
    o.AddPolicy("editor", policy => policy.Requirements.Add(new
EditorRequirement()));
    o.AddPolicy("director", policy => policy.RequireClaim("user_
roles", "director"));
});
```

By the way, we could have established the same mechanism between `author` and `editor`, since the hypothesis was that editors automatically obtain access from the `book-writer` entity shown previously. The principle would be exactly the same, so I'll let this be an exercise for you. Another small change is that we switched to policy-based authorization for the `ImportBooksCatalog` function and defined the corresponding policy in the preceding code. Also, note that we could have used an API gateway between the GUI portal client and the API endpoints, but we decided to go the way of direct access in order to quickly illustrate CORS issues that may arise and, in addition, not complexify the architecture too much yet.

Listing books from the portal GUI

In order to slightly clean the basic skeleton of the GUI, we first get rid of the `Counter.razor` page and remove the corresponding menu from `NavMenu.razor`, and in order to display a list of books, we are going to modify the page and menu used for the weather report. Thus, the default `Weather.razor` file is renamed `Books.razor`, and the top of the page is modified according to the new destination of the component:

```
@page "/books"
@attribute [Authorize(Roles = "editor")]
@using Microsoft.AspNetCore.Authorization
@inject HttpClient Http

<PageTitle>Books</PageTitle>

<h1>Books</h1>

<p>This component shows the list of books retrieved from the API</p>
```

The `NavMenu.razor` evolves to the following behavior, where the books list can be accessed by editors and the sales remain a reserved domain for the directors:

```
<div class="@NavMenuCssClass nav-scrollable" @onclick="ToggleNavMenu">
    <nav class="flex-column">
        <div class="nav-item px-3">
            <NavLink class="nav-link" href="" Match="NavLinkMatch.
All">
                <span class="bi bi-house-door-fill-nav-menu" aria-
hidden="true"></span> Home
            </NavLink>
        </div>
        <AuthorizeView>
            <Authorized>
                <div class="nav-item px-3">
                    <NavLink class="nav-link" href="identity">
                        <span class="bi bi-person-nav-menu" aria-
hidden="true"></span> Identity
                    </NavLink>
                </div>
                @if (context.User.IsInRole("editor"))
                {
                    <div class="nav-item px-3">
                        <NavLink class="nav-link" href="books">
                            <span class="bi bi-list-nested-nav-menu"
aria-hidden="true"></span> Books
                        </NavLink>
                    </div>
                }
                @if (context.User.IsInRole("director"))
                {
                    <div class="nav-item px-3">
                        <NavLink class="nav-link" href="sales">
                            <span class="bi bi-currency-dollar-nav-
menu" aria-hidden="true"></span> Sales
                        </NavLink>
                    </div>
                }
            </Authorized>
        </AuthorizeView>
    </nav>
</div>
```

The `<Authorized>` element delimits the HTML content that shall be visible only when a used is correctly logged in, while the `@if (context.User.IsInRole())` part of the code introduces an additional condition based on the role (or even roles, for that matter) that the connected user presents. In order to implement the same rule in the preceding code and ensure that the `director` role inherits the access of the `editor` role, we modify the `RolesClaimsPrincipalFactory` class by adding an `editor` claim if a `director` role is present:

```
foreach (JsonElement role in elem.GetProperty("roles").
EnumerateArray())
{
    identity.AddClaim(new Claim(options.RoleClaim, role.GetString() ??
String.Empty));
    if (role.GetString() == "director")
        identity.AddClaim(new Claim(options.RoleClaim, "editor"));
}
```

Now, let's return to our `Books.razor` component. Changing the class model from the Microsoft-proposed example, `WeatherForecast`, to something that better suits our exercise will be done at the bottom of the file, by adding the following code:

```
public class Book
{
    public string EntityId { get; set; }
    public string? ISBN { get; set; }
    public string? Title { get; set; }
    public int? NumberOfPages { get; set; }
    public DateTime? PublishDate { get; set; }
    public EditingPetal? Editing { get; set; }
    public SalesPetal Sales { get; set; }
}

public class EditingPetal
{
    public int? NumberOfChapters { get; set; }
    public Status? Status { get; set; }
    public AuthorLink? mainAuthor { get; set; }
}

public class AuthorLink : Link
{
    public string UserEmailAddress { get; set; }
    //public Author? FullEntity { get; set; }
}

public class Link
{
```

```
    public string Rel { get; set; }
    public string Href { get; set; }
    public string Title { get; set; }
}

public class Status
{
    public string Value { get; set; }
}

public class SalesPetal
{
    public MonetaryAmount? Price { get; set; }
    public decimal? WeightInGrams { get; set; }
}

public class MonetaryAmount
{
    public decimal Value { get; set; }
    public string MonetaryUnit { get; set; }
}
```

The DRY principle would be applied by many architects, and a common class could be shared between the client and the server, which Blazor templates propose by the way. This helps us to stay synchronized between the API and the portal GUI. The classes in which we have put the serialization attributes could then inherit from these model classes, or we could use a utility such as AutoMapper (https://automapper.org/) to automatically create the right correspondence. There are many ways to achieve this, but if you really want to stick to the fundamentals of the contract-first approach, the logical way is to create these classes automatically from the OpenAPI contract definition. Again, there are many great tools (https://quicktype.io/, for example) to do so without having to type anything.

For simplicity reasons, we stopped the deserialization at the content of the full author entity, which would be reactivated only if there was a query with the $expand=mainAuthor argument that has not been implemented. There is no particular difficulty in creating this code, neither technically nor in terms of the right business alignment, so we leave it as is. Also, it will be more adapted to the spirit of the web to present a link to the component describing the author.

Previously, the Books.razor component downloaded weather forecasts from a file exposed on the same server that allowed the download of the Blazor single page application. Now, it will retrieve a list of books from another server:

```
private Book[]? books;

protected override async Task OnInitializedAsync()
```

```
{
    books = await Http.GetFromJsonAsync<Book[]>("/Books");
}
```

For this to work, the `Program` class must be modified with the following code:

```
builder.Services.AddScoped(sp => new HttpClient { BaseAddress = new
Uri("http://books") });
```

This is really where the Docker Compose approach helps greatly. In a normal environment, it would have been necessary to adjust the base URL depending on environment variables or a setting, making this adjustable depending on the situation. When you use Docker Compose, the container can be anywhere in a cluster and even in multiple instances, since the included DNS will always point at the right position by simply using the name of the service, as defined in the `docker-compose.yml` file, of which we recall the corresponding portion here:

```
version: '3.0'
services:
  books:
    image: demoeditor/books-api:0.1
    build:
      context: ./books-controller
      dockerfile: Dockerfile
    ports:
      - 81:8080
```

Of course, it may be necessary to adjust the URL for debugging or some other situation where we need to be local. But even this has become not so important, as it is now possible to debug directly from Visual Studio Code into a container.

Finally, the last part we need to implement is the display of the books in an HTML list, just as it was done for the weather forecasts in the Microsoft-provided example which we only have to adapt:

```
<table class="table">
    <thead>
        <tr>
            <th>Id</th>
            <th>ISBN</th>
            <th>Title</th>
            <th>Pages</th>
            <th>Published</th>
            <th>Chapters</th>
            <th>Status</th>
            <th>Author</th>
            <th>Price</th>
```

```
                <th>Weight</th>
            </tr>
        </thead>
        <tbody>
            @foreach (var book in books)
            {
                <tr>
                    <td>@book.EntityId</td>
                    <td>@book.ISBN</td>
                    <td>@book.Title</td>
                    <td>@book.NumberOfPages</td>
                    <td>@book.PublishDate</td>
                    <td>@book.Editing.NumberOfChapters</td>
                    <td>@book.Editing.Status.Value</td>
                    <td><a href='/books/@book.Editing.mainAuthor.Id'>@
book.Editing.mainAuthor.Title</a></td>
                    <td>@book.Sales.Price.Value @book.Sales.Price.
MonetaryUnit </td>
                    <td>@book.Sales.WeightInGrams gr</td>
                </tr>
            }
        </tbody>
    </table>
```

Ideally, all the code necessary to get the books data from the server and display it in the form created in the GUI should now be in place, and we can test the result.

Solving authentication issues

If you run the sample application now, there will be an error like the following when loading the books page (you can view it in the browser developer tools, in the **Console** tab):

```
Access to fetch at 'http://localhost:5298/Books' from origin 'https://
localhost:7127' has been blocked by CORS policy: No 'Access-Control-
Allow-Origin' header is present on the requested resource. If an
opaque response serves your needs, set the request's mode to 'no-cors'
to fetch the resource with CORS disabled.
```

This is because CORS security is not set up. To correct the problem, we have to add the two following lines of code in `Program.cs` of the books (and authors) API projects, the first one placed before the `Build()` function and the second one after:

```
builder.Services.AddCors();

var app = builder.Build();
```

```
app.UseCors(options => options.WithOrigins("http://localhost:7127").
AllowAnyMethod());
```

Of course, the origin URL will be adapted to your context. In this example, I use a debug local address, but when using the Docker Compose application, the URL will be something like `http://portal`, since this is the name we used for the service. Using `if (app.Environment.IsDevelopment())` allows us to clearly separate the authorized addresses.

Another issue that must be dealt with is the fact that, for now, the HTTP client does not use any authentication, since we have created it like this:

```
builder.Services.AddScoped(sp => new HttpClient { BaseAddress = new
Uri("http://books") });
```

In order for this client to send the right token when calling the API, we need to change this line to something like this (this code is in the `Program` class also, but be careful – this time, it is the code from the portal GUI client):

```
builder.Services.AddTransient<CustomAuthorizationMessageHandler>();
builder.Services
    .AddHttpClient("BooksAPI", client => client.BaseAddress = new
Uri("http://books/"))
    .AddHttpMessageHandler<CustomAuthorizationMessageHandler>();
builder.Services.AddScoped(sp =>
sp.GetRequiredService<IHttpClientFactory>().CreateClient("BooksAPI"));
```

The `CustomAuthorizationMessageHandler` class should contain the following code, and a `dotnet add package Microsoft.Extensions.Http` command will be needed to compile:

```
using Microsoft.AspNetCore.Components;
using Microsoft.AspNetCore.Components.WebAssembly.Authentication;

public class CustomAuthorizationMessageHandler :
AuthorizationMessageHandler
{
    public CustomAuthorizationMessageHandler(IAccessTokenProvider
provider,
        NavigationManager navigation)
        : base(provider, navigation)
    {
        ConfigureHandler(
            authorizedUrls: new[] { "http://books/", "http://books/
Books" },
            scopes: new[] { "Portal-dedicated", "roles", "email",
"profile" });
```

```
    }
}
```

For the weird reason that authorized clients with tokens cannot be used for anonymous calls (see https://chrissainty.com/avoiding-accesstokennotavailableexception-when-using-blazor-webassembly-hosted-template-with-individual-user-accounts/ for more details), it is necessary to also add the following code to the Program class:

```
builder.Services.AddHttpClient("NeedsNoAccessToken", client => client.
BaseAddress = new Uri(builder.HostEnvironment.BaseAddress));
builder.Services.AddScoped(sp =>
sp.GetRequiredService<IHttpClientFactory>().
CreateClient("NeedsNoAccessToken"));
```

You can visit the Microsoft website (https://learn.microsoft.com/en-us/aspnet/core/blazor/security/webassembly/additional-scenarios?view=aspnetcore-8.0#configure-authorizationmessagehandler) resource to debug the strange behavior you might encounter when setting up authentication for communication between the Blazor client and the API server. In particular, it will recommend the following code adjustment (with a redirection when an exception is caught) to the way the API is called in the Books.razor component:

```
protected override async Task OnInitializedAsync()
{
    try
    {
        var client = HttpFactory.CreateClient("BooksAPI");
        books = await client.GetFromJsonAsync<Book[]>(@"http://
localhost:5298/Books");
    }
    catch (AccessTokenNotAvailableException exception)
    {
        exception.Redirect();
    }
}
```

The following additions are necessary at the top of the file for the compilation to work fine:

```
@using Microsoft.AspNetCore.Components.WebAssembly.Authentication
@inject IHttpClientFactory HttpFactory
```

With all these adjustments, ideally, you should be able to run the two projects again in local debugging mode, or inside Docker with docker compose up -d, logging into the Blazor application, and clicking on the **Books** menu to get the following display:

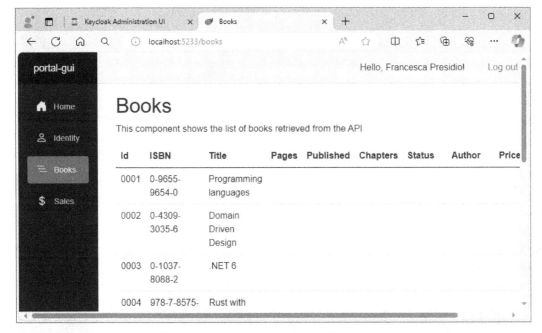

Figure 20.17 – Showing the books list in Blazor

The books that have been imported are missing quite a lot of data, but you can create a sample with more information by POST-ing the following to the API:

```
{
    "entityId": "0024",
    "isbn": "978-2409002206",
    "title": "Docker, 4th edition",
    "numberOfPages": 812,
    "publishDate": "2019-09-01",
    "editing": {
        "numberOfChapters": 15,
        "status": { "value": "available" },
        "mainAuthor": {
            "rel": "dc:creator",
            "href": "http://authors/Authors/0012",
            "title": "JP Gouigoux",
            "userEmailAddress": "jp.gouigoux@free.fr"
        }
    },
    "sales": {
        "price": {
            "value": 54.30,
```

```
        "monetaryUnit": "EUR"
    },
    "weightInGrams": 1245
  }
}
```

If you now refresh the list, a much better sample is displayed with, in particular, a link to the main author:

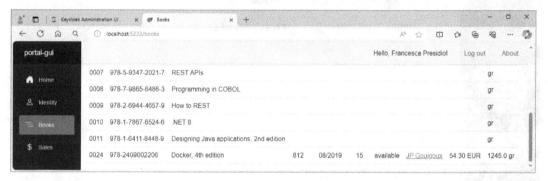

Figure 20.18 – More complete books on the list

The link will bring you to `http://localhost:5233/books/0012`, which has not been implemented yet in the application, but you should have all the knowledge and tips to add it without further help. For now, our list of books is not very intelligent – some styles could be added, the `Weight` column should not display simply **gr** if the value is not indicated, and there are many other things that can be improved. One worth discussing in this chapter, as it is highly related to a good IT/business alignment and separation of responsibility, is pagination.

Separation of concerns for the pagination feature

Pagination is a great exercise in the separation of responsibilities. In fact, with pagination, there actually are two different best practices, depending on the intelligence level that the client is trusted for. The problem is with the management of performance – at the moment, we only have a few books in the referential service, but what would happen if there were thousands of them? Performance would degrade as we load more and more data and, in addition, users would not find what they wanted if we presented them with a huge list without any filtering or ordering. This is why we should only send a few lines of data as a first page (the number depends on the client we have at hand – a PC browser with a big screen will support more lines than a mobile device, for example) and help users to navigate on the following pages of data or refine their query.

Ordering and filtering will be done using the `$orderBy` and `$filter` Open Data Protocol-compliant arguments, and you will easily discover how to plug these into a grid of data, but what interests us here is the separation of responsibility relating to the pagination feature. Is the client responsible for asking for fewer lines of data? Should the server be in charge? As we said previously, the client should be able to indicate what is best for their display. However, the server is ultimately responsible for its own performance and should not agree to requests from many clients, only to receive thousands of lines in one request; otherwise, it will slow down and harm the performance of other clients that "play by the rules." To complicate things even further, there are two possibilities for implementing pagination. Active pagination is when the client takes care of calculating the offset for the second page, while passive pagination is when the server is in charge of knowing which page is next. This may sound difficult, as the server is supposed to remember where each client is in the list of data pages, but we are going to show that it is possible to arrange this properly without too much code complexity.

Again, it helps to create a metaphor – imagine the client and the server as two persons in a physical office without a computer, and imagine how their functional, solely business-related conversation could be made as efficient as possible. The following steps would happen, showing clearly the separation of responsibility:

1. Andy needs to make some statistics about the content of books published in the past year. He goes to Stephanie, the person in charge of the books directory for the company (an old-fashioned DemoEditor), and requests the folders for the books published in 1973, asking first how many there are to get a rough idea of how much work he is going to have to deal with. In the same conversation, he asks Stephanie if he could get the first 20 books to work on them.

2. Stephanie indicates that there are 40 books for 1973 and that her policy is generally to hand over 5 book folders at a time, but she can provide more if needed. Nonetheless, she is allowed to go up to 10 book folders at once, but no more.

3. Andy gets the first 10 book folders and gets to work on them, calculating his statistics. He has already deduced that he will have to make four roundtrips in total, since he can take up to 10 books each time, and there are 40 books to analyze for the given year, 1973.

4. When Stephanie gave Andy the first folders, she placed a small slip of paper marked "Books of 1973 for Andy" in her physical block of cards and folders. She knows there is a good chance that Andy will return to get the next 10 books, and the slip of paper will make it easier for her to retrieve them.

5. When Andy comes back to Stephanie to get the second set of 10 books, she quickly finds her previous place in the book folders' cabinet and retrieves the book folders, providing copies to Andy. She, of course, does the same thing again with the slip of paper, moving it to the next position.

6. At the same time, Stephanie tells Andy that there are now 43 books in the 1973 archive folder. Since his original request, somebody from the archives dug out some old books that should have been filed this year, which is the reason why the total number of books has gone up.

7. Andy goes back to his desk with books 11 to 20 and works on them, and he tells his boss that he will have to do five roundtrips instead of the initial four that he had planned. His boss asks about why there is additional statistics work, and Andy explains this comes from something he has no control over, which is the number of books; even though it is not supposed to change

all the time, the best he can do is adjust his workload. The state of freshness of the statistics will be dated at the time of the last roundtrip he will make.

8. The work goes on and Andy makes a third and then fourth roundtrip. At some point, his boss asks if it is possible to join the statistics from 1973 and 1974, but Andy explains that Stephanie has prepared the batch of book folders for 1973; although it would be possible, it would mean abandoning the statistics and doing them again. It is decided that he should carry on with what he is doing.

9. Andy makes the final roundtrip to Stephanie, who confirms that there have not been any additional books from the archive, so the total is still 43 books. She passes the remaining books to Andy and throws out the slip of paper, as Andy is not going to come back for more books since he has had all the ones relevant to his initial request.

The translation of this purely functional approach into a technical exchange is as follows:

1. The client calls the books API with `http://books/Books?$filter=editionYear eq '1973'&$count=true&$top=20`.

2. The server sends only the 10 first books corresponding to the filter and provides a total count of 40 as part of the answer (in a metadata attribute, or something equivalent, but distinct from the book data itself). If the client had not specified `$top` of 10, it would have received 5 books only.

3. The client displays the 10 books and also links at the bottom, allowing the user to directly go to the pages required, finishing at number 4, since there are 4 pages in each of the 10 books.

4. The server also provides in the metadata a link, with `rel` equal to `next`, and an attribute, `href`, pointing to a special URL it crafted to deliver.

5. The client comes back to the server using this URL, and the server sends back the next 10 books.

6. The total count indicated in the metadata is now 43.

7. The client then refreshes the list of books with the next 10 pieces of data, but it also adds a `page` 5 link at the bottom to show that there is now more data than initially displayed.

8. The user skips to the next pages of data; at one point, they wonder about filtering on 1973 and 1974, but they know that this would break the list, so they don't change the filter.

9. Clicking on the last page displays only three books in the grid. The server does not provide any `rel=next` link, since this is the end of the request batch.

In "technical terms", 5 is the default page size, while 10 is the maximum server page size. The slips of paper are implemented through links, with a URL pointing at the next page, described using the `rel` attribute. The number of pages is deduced by the client from the total number of books made available, which can vary, along with the requests for additional pages. Although this is a simple example of a feature, I think it illustrates quite clearly how the technical implementation should always be driven by the functional decomposition of the requested feature. We are now well-equipped to implement a pagination system in the next section.

Note that we could try and improve performance by keeping a match between the client and the server's internal request for data, allowing the latter to keep a pointer to the current request resultset and not replay the query at each call. But doing so causes lots of other problems:

- The server would not be stateless anymore; therefore, for this to work, we would need to activate server affinity, which in turn harms scalability

- Clients abandoning the current query would leave the server with an open resultset, and the resource would stay active until timeout allows for recycling

- The database connection pool would not be used efficiently, as connections would remain open much longer, slowing down circulation

In our metaphor, imagine if Stephanie had to serve many people and used a slip of paper for each of them. At some point, her cabinet of folders could be overwhelmed with little slips of paper, making it difficult for her to skip to the appropriate section in the cabinet. In addition, if some people finished their requests abruptly, Stephanie would not know that she could remove the slip of paper. After some time, she knows she can remove them, but this is an additional burden for her, as she has to note the times on the slips of paper and review them periodically to know what should be gotten rid of.

Implementing pagination

The main goal of this book is to help clarify the architectural thinking of software applications inside an information system, but it would not be fair to leave it here and not provide an implementation of the pagination feature that we have decomposed. Actually, I hope this will be considered as a quality of this book, going all the way from high-level information system evolution and design strategy down to the lowest details on how to concretely realize API implementations. My consulting clients have told me that they appreciate my ability to communicate with both the board of directors and technical staff about the same topic during digital transformation projects. This is due to my emphasis on using precise language and the importance of semantics.

You may remember that we actually prepared our API implementation in the books referential service for active pagination:

```
[Authorize(Policy = "editor")]
[HttpGet]
public IActionResult Get(
    [FromQuery(Name = "$orderby")] string orderby = "",
    [FromQuery(Name = "$skip")] int skip = 0,
    [FromQuery(Name = "$top")] int top = 20)
```

In order to keep it simple here, we will not consider passive pagination, but since we have already shown how to use links, you should be able to do it by yourself. It is more interesting to discuss the client implementation of the pagination because, again, we need to think in terms of responsibilities and share them correctly. The pagination bar is a common component that can be used for books, authors, and any other business entities. Ideally, it should not be coupled to any functional entity. The following code shows no coupling:

```
<nav aria-label="Page navigation example">
    <ul class="pagination justify-content-center">
        @foreach (var link in _links)
        {
            <li @onclick="() => OnSelectedPage(link)" style="cursor:
pointer;" class="page-item @(link.Enabled ? null : "disabled") @(link.
Active ? "active" : null)">
                <span class="page-link" href="#">@link.Text</span>
            </li>
        }
    </ul>
</nav>
```

The code associated with this component (to be put inside the @code region of the PaginationBar. razor file) is not coupled to any entity either:

```
[Parameter]
public int CurrentPage { get; set; }
[Parameter]
public int TotalPages { get; set; }
[Parameter]
public int PageSize { get; set; }
[Parameter]
public int TotalCount { get; set; }

public bool HasPrevious => CurrentPage > 1;
public bool HasNext => CurrentPage < TotalPages;

[Parameter]
public int Spread { get; set; }
[Parameter]
public EventCallback<int> SelectedPage { get; set; }

private List<PagingLink> _links;

protected override void OnParametersSet()
{
    _links = new List<PagingLink>();
    _links.Add(new PagingLink(1, true, "<<"));
    _links.Add(new PagingLink(CurrentPage - 1, HasPrevious, "<"));
    for (int i = 1; i <= TotalPages; i++)
```

```
        if (i >= CurrentPage - Spread && i <= CurrentPage + Spread)
            _links.Add(new PagingLink(i, true, i.ToString()) { Active
= CurrentPage == i });
    _links.Add(new PagingLink(CurrentPage + 1, HasNext, ">"));
    _links.Add(new PagingLink(TotalPages, true, ">>"));
    base.OnParametersSet();
}

private async Task OnSelectedPage(PagingLink link)
{
    if (link.Page == CurrentPage || !link.Enabled) return;
    CurrentPage = link.Page;
    await SelectedPage.InvokeAsync(link.Page);
}
```

The preceding code is quite simple to understand, and the following code should normally lift any doubt about its functions; thus, I will not discuss it in detail. I'll just briefly mention the Spread value, which allows us to limit the number of links displayed if there are many pages. In order for this component to work, we need a definition for the PageLink class, which simply contains information corresponding to each of the links displayed in the pagination bar:

```
public class PagingLink
{
    public string Text { get; set; }
    public int Page { get; set; }
    public bool Enabled { get; set; }
    public bool Active { get; set; }
    public PagingLink(int page, bool enabled, string text)
    {
        Page = page;
        Enabled = enabled;
        Text = text;
    }
}
```

Finally, even if this should be easy to find, I prefer to add the using clauses to the top of the file, in order to accelerate the exercise:

```
@using Microsoft.AspNetCore.Components;
@using System;
@using System.Collections.Generic;
@using System.Linq;
@using System.Threading.Tasks;
```

Integrating this component into the page listing books is as easy as adding the following under the `<table>` object:

```
<div class="row">
    <div class="col">
        <PaginationBar
            CurrentPage="CurrentPage"
            TotalPages="TotalPages"
            TotalCount="(int)TotalNumberOfBooks"
            PageSize="PageSize"
            Spread="2"
            SelectedPage="SelectedPage" />
    </div>
</div>
```

We need to add the following class fields to the `@code` section to get the binding working:

```
private int? TotalNumberOfBooks = null;
private int CurrentPage = 1;
private int TotalPages = 1;
private int PageSize = 5;
```

The initialization method evolves in order to gather the count of books:

```
protected override async Task OnInitializedAsync()
{
    TotalNumberOfBooks = await Http.GetFromJsonAsync<int>(@"http://
localhost:5298/Books/$count");
    TotalPages = (int)Math.Ceiling((decimal)TotalNumberOfBooks /
PageSize);

    books = await Http.GetFromJsonAsync<Book[]>(@"http://
localhost:5298/Books?$top=" + PageSize);
}
```

Note that we could have included this inside the second HTTP call to make only one roundtrip to the server, as recommended in **Open Data Protocol**. I will leave this as an exercise for you, as it does not add anything to IT/business alignment and is a purely technical decision. Now, we can make our first test of the application, which would bring up the following display:

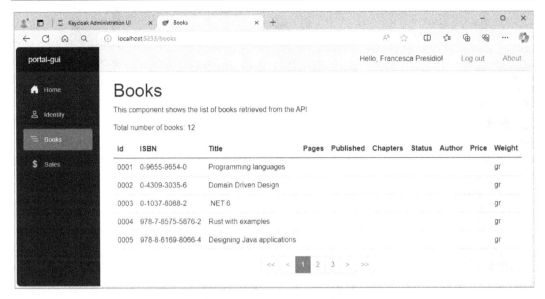

Figure 20.19 – The first page of books

Of course, we need to provide some additional code to complete the pagination, since we have not yet defined the following HTTP calls. We need to add the method to implement the change of page selection:

```
protected async Task SelectedPage(int page)
{
    TotalNumberOfBooks = await Http.GetFromJsonAsync<int>(@"http://
localhost:5298/Books/$count");
    TotalPages = (int)Math.Ceiling((decimal)TotalNumberOfBooks /
PageSize);

    CurrentPage = page;
    int top = PageSize;
    int skip = PageSize * (page - 1);
    books = await Http.GetFromJsonAsync<Book[]>(@"http://
localhost:5298/Books?$top=" + top + "&$skip=" + skip);
}
```

Adjusting the value of CurrentPage on the number of pages requested in the parameter is necessary in order for the selected link to appear as the active page. Other than that, the code is quite simple, as it recalculates the number of books (if this has increased, in order to maintain consistency with the number of pages) and calls the next batch of books, using the $top and $skip values, calculated by a very simple algorithm. The result when clicking page **3** in the interface will be the following:

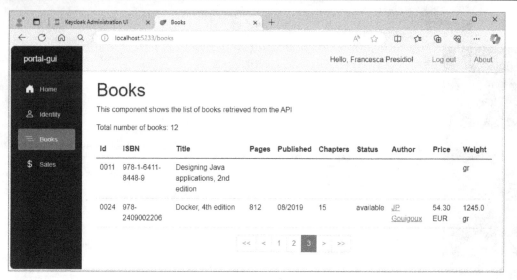

Figure 20.20 – The last page of books

The core of the web interface for the edition-related data management of `DemoEditor` is now ready, at least in its skeletal form. It would be a waste of pages to show a complete application, as all the important bits in terms of architecture and integration in a business-aligned information system were discussed when we implemented the features in this chapter. Of course, in reality, the application would also allow us to list authors, create business entities, read sales and other details of the books, and so on. This would be a lot of features, and in order to keep quality at a high level, tests are of uttermost importance!

Summary

In this chapter, we added a web-based GUI to handle the business entities for `DemoEditor` and hopefully showed that the method of separating technical design by functional responsibilities helps to keep the resulting system easy to implement and, most importantly, evolving. The same principles apply to both frontend and backend applications, which is quite reassuring.

Again, the choice is not between being coupled or integrated but, simply, between being aware or not of the impact of your design decisions and trying to reduce as much as possible the total complexity of the final information system, by making it strongly aligned with its intrinsic complexity. At the beginning of a project, this only takes a bit of brainstorming with the people who know the business. If this stage is delayed, this can result in information systems that are completely blocked by the business and disposed of, along with the huge amounts of financial investment they have gathered.

In the next chapter, we are going to show a second GUI and, in particular, how it can be developed while utilizing some of the skills learned in this chapter. We will also return to the subject of importing data, which we did not discuss in this chapter, as the data was shown in the application. Our purpose was to focus on the GUI and postpone discussing what was going on under the hood. This is what is coming up in the next chapter.

21
Extending the Interfaces

In the previous chapter, we described how to create a **graphical user interface** (**GUI**) and what responsibilities should be taken care of. In this chapter, we will address all the other interfaces in the hexagonal architecture of our sample business domain and, in particular, look at data imports and the mobile GUI.

In the previous chapter, we took data for granted and showed it in the interface captures, but where does it come from, since we have not yet created any forms for users to enter data with? If you remember, there was an import function that was talked about, and this is what has been used to generate the entries in the new data referential service. We will start this chapter by explaining in detail how this interface works and why it has been designed this way.

The ability to create dedicated interfaces is not only important because GUIs should be customized for the use of specific groups of users (a web application in a browser for people sitting in offices, a mobile application for traveling users, etc.), but also as a working proof that separation of responsibilities has been respected. This is the reason that this chapter builds on the previous one by describing the creation of an additional mobile application.

We will create a mobile application and demonstrate how content can be shared with the web app. We will also talk about how to create a very limited ultra-light GUI by simply using content negotiation. But before that, we will come back to an interface that has not been explained yet but would still be very useful for DemoEditor and has been used without any explanation in the previous chapter: the data import interface.

In this chapter, we'll cover these topics:

- Sending actual data in the data referential service, instead of arbitrary sample information
- Adding automated tests using the GUI
- Creating a mobile application with a technology that allows the reuse of already-created components
- Some additional considerations on the challenges and peculiarities of the design of interfaces

Importing some real data

The data import interface is the interface that allows us to import data into the data referential service. This kind of interface, though we left it aside until now, is important because it helps with the batch creation of data; it also enables data migration by making it possible to start working with a new referential service before the old one is completely abandoned.

The first step of migration

This interface is the one for legacy data imports. `DemoEditor` currently uses an Excel worksheet to manage the books, and although it is very limited (hence the new and modern information system we are creating), it still contains all the important data from the company. Since it is strongly advised not to use a Big Bang approach, the very first small step would be to still enter data in the legacy shared Excel file but have the new data referential service available to start some modern operations and interop.

To do so, it has been decided that, in this first migration period, the file will be imported every night using the new books data referential service. Logs will allow functional users to see what data causes problems and proactively repair them. At the same time, data will be available in the new data referential service in read-only mode (since the Excel worksheet will still remain the master data reference in this first step). This will enable our directors to start getting the reports and statistics they are eager to realize on Power Query by getting data from several sources. GUIs such as those we are going to create in this chapter will thus be able to get some feedback (the read-only mode is already doing 90% of the job).

The directors insisted that they should be the only ones authorized to perform this operation during the day because it would replace everything in the new NoSQL database with what comes from the Excel file, which can have some unexpected effects. At night, a planned task would execute the same migration operation under the service account of the editing director. These requirements are considered as the backlog of our fifth sprint, and this is what is going to be implemented in this first section.

Implementing data import

The data made available is in the following form:

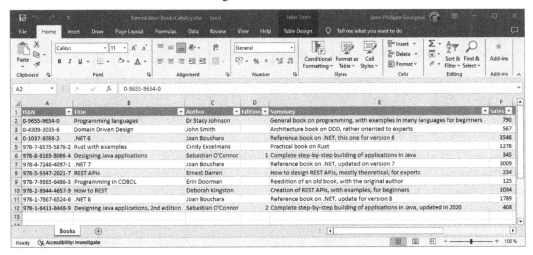

Figure 21.1: Books catalog in the form of the legacy Excel workbook

The directors have agreed that data should be retrieved step by step, and they know that lots of data are not clean, so they will have to get their hands dirty to correct some things. For example, authors are linked to books by their full names, which has always been a pain for them because of duplicates.

One of the first questions that arises when trying to implement this is where this code should be. One could argue that data import is going to be used only for a limited time and that it should not be inside the referential service but in a dedicated tool. This makes some sense, but provided we can limit the feature to a dedicated method inside the controller, the coupling should remain low. Another way to avoid being stuck with this code is to avoid anyone using it except for a few limited and agreed-upon users. Luckily, this is the case in our feature requirements since only directors will be able to use this function. This means it will be easy to delete when not used anymore, and this is the reason that we will keep it inside the data referential service, in addition to it being much easier to plug into existing commands.

The skeleton of the function will be as shown in the following code, with a first line to allow access only for the `director` role, a second one setting the PUT verb to listen, the third one defining the listening URL, and the final part creating the associated .NET method, together with the parameters that are going to be exposed in the query URL, and default to a provided value:

```
[Authorize(Roles = "director")]
[HttpPut]
[Route("Import")]
public async Task<IActionResult> ImportBooksCatalog([FromQuery]
string? localFileAddress = "/import/DemoEditor-BooksCatalog.xlsx")
{
}
```

The default name of the path to the file has been designed to make running the application in a Docker container easier, using the volume option to link to the local file:

```
docker run -d -p 80:8080 --name books -v /import:/mnt/c/Users/
jpgou/.../Resources:ro demoeditor/books-api:0.1
```

The first step in the function will be to stop and exit with a 404 Not Found error code if the file is not available:

```
if (!System.IO.File.Exists(localFileAddress))
    return new NotFoundObjectResult(string.Format("Could not find
books catalog at {0}", localFileAddress));
```

After that, and this may sound a bit extreme, we will simply clean every collection in the database. After all, we are still in the first phase of the migration where the new data referential service is not actually the referential service since the old legacy Excel workbook remains the master reference for the data and the ASP.NET server API is only a read-only clone of it:

```
await Database.GetCollection<ChangeUnit>("books-changes").
DeleteManyAsync(item => true);
await Database.GetCollection<ObjectState<Book>>("books-states").
DeleteManyAsync(item => true);
await Database.GetCollection<Book>("books-bestsofar").
DeleteManyAsync(item => true);
```

After adding a package called FastExcel, the following code can be appended to the function:

```
using (FastExcel.FastExcel excelReader = new FastExcel.FastExcel(new
FileInfo(localFileAddress), true))
{
    Worksheet worksheet = excelReader.Read("Books");
    var rows = worksheet.Rows.ToArray();
    for (int i = 1; i < rows.Length; i++)
    {
        var cells = rows[i].Cells.ToArray();
        Book b = new Book() {
            EntityId = i.ToString().PadLeft(4, '0'),
            ISBN = (string)cells[0].Value,
            Title = (string)cells[1].Value
        };
```

It reads the content of the file and, for each row in the Excel worksheet called Books, it will generate an instance of the Book class with some attributes extracted from the cells (this remains limited for now, but remember that we have agreed on a baby-steps approach, which is particularly suited to migration projects). To do so, a library that can read Excel files directly, without any dependency on Microsoft Office, is indicated. In a production environment, I would tend to use a well-known, supported, component such as SpreadsheetGear, but for such a small exercise, this open-source package will do the job, after a quick study of the code provided on GitHub. It's crucial to be vigilant with the recent public repository attacks. Please ensure everything we use is monitored carefully. Remember that being cautious is a valuable trait when designing software or administering servers.

Now that the basic instance of the book is ready, we are going to try and do something a little more complex, namely retrieving the author's information. Of course, that means that the import of the authors should be run before the import of the books. But there is a much bigger problem than this, which is that authors are indicated only by their full name in the legacy Excel books repository. Again, there should be a balance between the sophistication of the code to get more matches with existing authors and the cost of the code, particularly since it is perhaps only going to serve for migration, the volumes are quite low, and the data needs to be cleaned anyway. In short, there is not much incentive to use some kind of indexed or double-metaphone-based algorithm to hit a few tens of additional author matches. It is better to simply log the missing authors' links and provide the data stewards with the logs for them to correct the entries that cause errors. The following code does exactly this, by trying to get the authors from the data referential service, and if it does not match exactly on the full name (in the `Title` attribute for the modern persistence system), then a simple log is created:

```
HttpClient client = _clientFactory.CreateClient("Authors");
try
{
    Author? author = await client.
GetFromJsonAsync<Author>("?$filter=Title eq '" + (string)cells[2].
Value + "'");
    if (author != null)
    {
        // URLs are hardcoded in order to better explain what is done
by the code
        b.Editing = new EditingPetal() {
            mainAuthor = new AuthorLink() {
                Rel = "dc.creator",
                Href = "http://demoeditor.org/authors/" + author.
EntityId,
                Title = author.FirstName + " " + author.LastName,
            }
        };
    }
}
catch
{
    _logger.LogInformation("Could not find author {title} using the
full name", (string)cells[2].Value);
}
```

Finally, the function relies on the `Create` method to realize all the work of actual insertion, together with all the business rules (`Create` delegates itself the actual job of inserting documents in the collections to the `Patch` function, promoting further the separation of concerns). It has been decided by the business owners that the data migrated from the old books catalog should appear in the new referential service as if it had been created at the beginning of 2024, so the `valueDate` argument will be used in the call:

```
DateTimeOffset valueDateForMigratedData = new DateTimeOffset(2024, 1,
1, 0, 0, 0, new TimeSpan(1, 0, 0));
await this.Create(b, valueDateForMigratedData);
```

The function code results in a message along the `200 OK` status that informs the client of the number of lines that have been treated:

```
return new OkObjectResult(string.Format("{0} lines have been
imported", rows.Length));
```

Testing data import and next steps

Running the operation that we have created is quite easy, the only prerequisite being to log in with a user on which the `director` role has been affected (and created, for that matter, since we had not yet decided on it earlier). Please note that the `[Authorize]` attribute directly refers to the role rather than using a policy, just in order to show this is possible, and that this is acceptable for simple cases. The result is shown in the following Postman request:

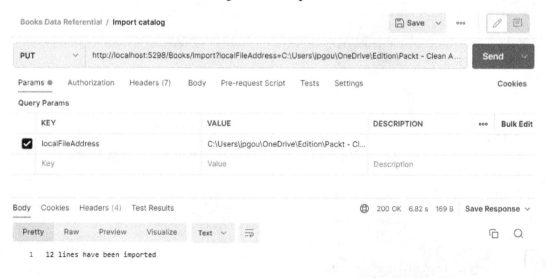

Figure 21.2: Postman request running the import on the Excel books repository

A few seconds after, the import is done and we can test the results by calling **GET** on /Books:

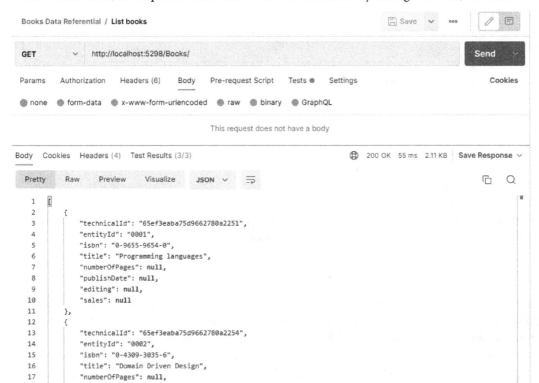

Figure 21.3: Listing the books after the import operation

Of course, several attributes are missing, which is quite normal since they were not indicated in the legacy books referential service, or have been left for later. Another question we might ask about this process is about performance. 12 books imported in a bit more than 6 seconds means that the process is not very quick. Again, remember that premature optimization is the root of all evil. We all know technical developers that will rush into the asynchronous execution of import, or even start some background service or orchestration of batch jobs in order to deal with the volume in a sophisticated way.

Let's just quickly do the math: if the timeout is set to 120 seconds (which is the default in IIS for query execution), that means we can import around 230 books. A discussion with the editing director may indicate that this falls a little short. How about using the authors cache instead of calling GET on the authors referential service? Doing so gets the time down to almost a fourth of the initial delay, which allows us to import almost 1,000 books. If this is not enough, adjusting the timeouts to 5 minutes will see us reach the 2,500 books limit, which is higher than the editing director guessed as the maximum number of books. There – the problem was solved without any complicated technical measures that would have been overkill.

I know I may sound a bit careless to lots of you who have (as I do) a technical background, always trying to find some quick and dirty solutions. But trust me on this one: I have been on the technical expertise side for at least 20 years and I know it is very tempting to pull out the big guns and over-engineer solutions. In fact, over-engineering is the real problem, because under-engineering is easily spotted and can be corrected quickly. Overly sophisticated solutions, on the other hand, cannot be corrected, simply because the time wasted is impossible to recover. The KISS and YAGNI principles should always be remembered in this kind of project, particularly when you know the software created is going to be thrown away after use.

Testing the GUI

This section will talk about the automated testing of a GUI. This is not strictly related to business/IT alignment but automated testing is so important for quality that it would have been a shame to not at least talk about it and show some examples in a book that strives to increase the overall quality of software architecture and make it reach industrial standards.

Automatic end-to-end tests of the GUI

One of the most well-known ways to test a GUI is to pilot its web interface with tools such as **Selenium**. Though performance is not the strong suit of such tests, they have the great advantage of being as close as possible to what a human would do in reality, as the "driver" really clicks on buttons and links, types text in the designated areas, and generally manipulates the application as would be the case in regular human use (but it's done much quicker, as there is no typing delay).

In order to unify what is going to be used in all the tests, it can be useful to create a base class like this:

```
using System;
using OpenQA.Selenium;
using OpenQA.Selenium.Chrome;

namespace EndToEndSeleniumTests
{
    public class TestBase
    {
        protected IWebDriver driver;
        protected string appURL;

        public TestBase()
        {
            var options = new ChromeOptions();
            options.AddArgument("--headless");
            driver = new ChromeDriver(options);
            appURL = Environment.GetEnvironmentVariable("TestUrl");
```

```
                if (string.IsNullOrEmpty(appURL)) appURL = "http://
localhost:5233";
                Console.WriteLine($"appURL is: {appURL}");
            }
        }
    }
```

This will run a `Chrome` driver, which is basically a memory-only browser used for tests. If you need to, you can make it visible in order to make debugging easier, but tests should generally do without this in order to attain maximal speed. After writing this first `TestBase.cs` file, we can create some test classes inheriting this base class and navigating to the Blazor page we are interested in testing the behavior of. Since pagination was the feature we just created, it is the logical target of our sample test:

```csharp
using System;
using Xunit;
using OpenQA.Selenium;
using OpenQA.Selenium.Chrome;
using OpenQA.Selenium.Support.UI;


namespace EndToEndSeleniumTests
{
    public class TestsOT : TestBase
    {
        [Fact]
        public void TestAffichagePaginationOT()
        {
            // We let a bit of time for the table HTML element to
appear in the
            driver.Navigate().GoToUrl(appURL + "/books");
            var wait = new WebDriverWait(driver, TimeSpan.
FromSeconds(10));
            wait.Until(d => d.FindElement(By.CssSelector("table")));

(...) // Rest of the test will be explained below

            // At the end of the test, we release the resource, which
is quite heavy
            driver.Quit();
        }
    }
}
```

The core of the test will be a scenario whose steps are going to validate the behavior of the pagination system. We start by verifying that there indeed are only five lines of data in the table:

```
Assert.Equal(5, driver.FindElements(By.CssSelector("table tbody tr")).
Count);
```

Next, we try and find the < link and check that it is disabled (which is logical, since we have arrived by default on the first page of data, so it should not be possible to click on the *previous* link):

```
Assert.Contains("disabled", driver
    .FindElement(By.ClassName("pagination"))
    .FindElement(By.XPath(".//li/span[text()='<']/.."))
    .GetAttribute("class"));
```

It makes sense that the link with 1 should be the active one:

```
Assert.Contains("active", driver
    .FindElement(By.ClassName("pagination"))
    .FindElement(By.XPath(".//li/span[text()='1']/.."))
    .GetAttribute("class"));
```

After that, we simulate an interaction by finding the 2 link and sending a click action on it, which is supposed to bring us to the second page of data. We verify this by waiting for the 2 link to become active (this is a CSS characteristic called active):

```
driver.FindElement(By.ClassName("pagination"))
    .FindElement(By.XPath(".//li/span[text()='2']"))
    .Click();
wait.Until(d => d.FindElement(By.ClassName("pagination"))
    .FindElement(By.XPath(".//li/span[text()='2']/.."))
    .GetAttribute("class").Contains("active"));
```

When this is done, we are still supposed to have five lines in the list of results (since there was a total number of books of 12), but this time, the < link is supposed to not be disabled anymore, since it now makes sense to be able to go back one page:

```
Assert.Equal(5, driver.FindElements(By.CssSelector("table tbody
tr")).Count);
Assert.DoesNotContain("disabled", driver
    .FindElement(By.ClassName("pagination"))
    .FindElement(By.XPath(".//li/span[text()='<']/.."))
    .GetAttribute("class"));
```

As far as the 1 link is concerned, it should not be active anymore but should not be disabled either:

```
string classLink1 = driver
    .FindElement(By.ClassName("pagination"))
    .FindElement(By.XPath(".//li/span[text()='1']/.."))
    .GetAttribute("class");
Assert.DoesNotContain("active", classLink1);
Assert.DoesNotContain("disabled", classLink1);
```

As mentioned previously, this kind of test emulates actual use well, but at the cost of very low performance. Fully testing the scenarios of a complex application may take some hours, and sometimes whole nights, which can be a problem in continuous integration and deployment environments.

Tests at the component level

Luckily, there is an alternative with component-level testing on Blazor. Though very close to actual manipulation, it is much quicker because the Razor components are loaded in memory without the burden of a whole browser (even though the browser can be made transparent, it is still there in Selenium, which has a great cost). This is possible thanks to a component called **BUnit** (see https://bunit.dev/ for more information).

An example of a BUnit test would be the following:

```
using Bunit;
using Xunit;
// Other using clauses omitted for simplicity

namespace Portal_Tests
{
    public class TestBooksList : TestContext
    {
        [Fact]
        public void DisplayOfTotalNumberOfBooks()
        {
            Services.AddSingleton<IBookService>(new
MockBookService());
            var cut = RenderComponent<Books>();
            cut.Find("p").MarkupMatches("<p>Nombre total d'OT : 12</
p>");
        }
    }
}
```

The **Component Under Test** (**CUT**) would be our Blazor page called Books, and we will assume that an indirection level would have been created for our HTTP client (which is for now hard-coded in the component) using IBookService. This way, we will be able to inject a mock of the service so that the test remains completely independent of the books API implementation (and also much quicker, because this way, no HTTP calls will be done). The mock is going to provide hard-coded data that respects the contract, making the test more efficient:

```
using ODataHelpers;

namespace MockImplementations
{
    public class MockBooksService : IBookService
    {
        public async Task<ODPList<Book>> GetAll(int skip, int top,
bool totalcount)
        {
            var data = new List<Book>();
            for (int i = skip; i < skip + top; i++)
                data.Add(new Book() {
                    EntityId = "00" + i.ToString("00"),
                    ISBN = "978-00000-" + i.ToString("00"),
                    Title = "Mock book number " + i,
                    // Rest of the fake book creation was omitted for
conciseness
                });
            var results = new ODPList<Book>()
            {
                Results = data,
                TotalCount = totalcount ? data.Count : null
            };
            return await Task.FromResult<ODPList<Book>>(results);
        }
    }
}
```

The interface definition is not very important, but here it is anyway:

```
public interface IBookService
{
    Task<ODPList<Book>> GetAll(int skip, int top, bool totalcount);
}
```

The definition of the generic `ODPList<T>` class is much more interesting to develop, as it contains the correct structure of the response expected by the component; indeed, while we are mocking the result from the API, we might as well make it obey the ODP standard (which we had left to the user in the previous implementation):

```
using System;
using System.Collections.Generic;
using System.Text.Json.Serialization;

namespace ODataHelpers
{
    public class ODPList<T>
    {
        [JsonPropertyName("@count")]
        public long? TotalCount { get; set; }

        [JsonPropertyName("value")]
        public IEnumerable<T> Results { get; set; }
    }
}
```

All this eventually allows us to write a second test, which this time is for validating the behavior of the pagination in the list of books:

```
[Fact]
public void PaginatingThroughBooks()
{
    Services.AddSingleton<IBookService>(new MockBookService());
    var cut = RenderComponent<Books>();
    cut.Find("table tbody tr td").MarkupMatches("<td>0000</td>");
    // The fourth li element corresponds to the page 2 link
    var element = cut.Find("nav ul li:nth-child(4) span");
    element.Click();
    cut.Find("table tbody tr td").MarkupMatches("<td>0001</td>");
}
```

As you can see, this is much more succinct than the Selenium equivalent, and we could go even further by directly addressing the properties of the CUT if we had to make asserts on them in addition to asserts on the HTML result of the component. We will stop this section about testing here, hoping that it will be enough to pique your curiosity and convince you – if you are not already – that tests are not as huge an investment as they are sometimes seen as being, as long as you use the right tool for the right use case and you get the beginning of the project right.

As test management concludes our approach to a web interface, it is time to imagine other possible interfaces for our API and referential services; mobile applications immediately come to mind. This is the subject of the next section.

Applying a MAUI mobile application

We are now at the seventh sprint of our development for DemoEditor. The directors are so happy with our GUI interface that they now want to have it everywhere with them, and ask for a mobile version of the application. Luckily for us, we designed right from the beginning in a "mobile first" way. Wait a minute… Did we?

The benefits of a responsive interface

In fact, everything was created in a mobile-first way except for the grid I created with simple HTML dumped in text. But let's just replace this `<table>` content in `Books.razor` with the following, in order for the table lines to be changed into data cards:

```
<div class="container">
@foreach (var book in books)
{
    <div class="row">
        <div class="col-xl">
            <div class="card">
                <div class="card-body">
                    <h5 class="card-title">@book.Title</h5>
                    <h6 class="card-subtitle mb-2 text-muted">@book.
ISBN (@book.EntityId)</h6>
                    <p class="card-text">Published @book.PublishDate?.
ToString("MM/yyyy")</p>
                    <p class="card-text">@book.NumberOfPages pages / @
book.Editing?.NumberOfChapters chapters</p>
                    <a href="/books/@book.Editing?.mainAuthor?.Id'"
class="card-link">@book.Editing?.mainAuthor?.Title</a>
                </div>
            </div>
        </div>
    </div>
}
</div>
```

This time, each row of data generates a card with well-adjusted information on each book, and all of a sudden, the web interface can be condensed, meaning you don't have to scroll around so much:

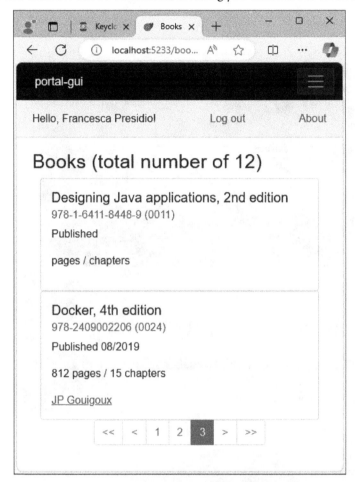

Figure 21.4: List of books, in reduced resolution

Actually, Blazor has already paved the way for a responsive application, since even the side menu appears as a burger menu when the width is reduced:

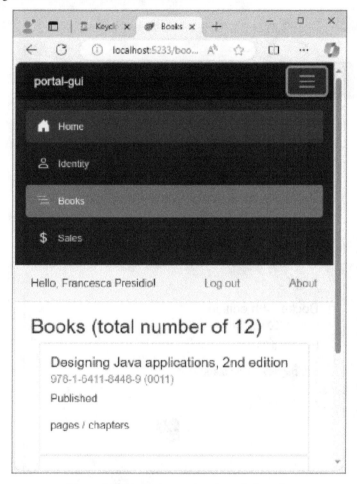

Figure 21.5: Burger menu in expanded mode

Even the Keycloak console GUI works fine on a mobile resolution, putting all the menus at the top and letting the user access the piled-up information sections below!

Figure 21.6: Keycloak responsive interface

We now know our Blazor components can adapt to low resolution and implement some responsive quality, but how good is this for our mobile requirement? It may look like these technologies don't have anything in common, but they actually do, as I am going to show you in the next section.

MAUI Blazor as a mobile platform

How convenient would it be to stick to the nice and cozy C# environment while creating a mobile application working on Android and Apple iOS? Meet the MAUI Blazor application! This platform enables you to publish a Blazor application inside a web browser that is invisibly embedded inside a native mobile application developed using the .NET MAUI framework. This way, everything we have done in the previous section can be reused directly on a modern smartphone. Such applications can even be executed as standalone Windows applications.

The prerequisites are quite simple:

- Add the .NET MAUI extension to your Visual Studio Code installation (go to the **Extensions** icon on the left menu, enter MAUI in the search box, and run the install command on the right extension in the list. Please be careful to choose the exact extension that was named previously and validated by Microsoft.)

- Follow the instructions upon completion

This is enough to execute on Windows, but if you want to deploy to Android, you will also need to run the dotnet workload install maui command. When this is finished, you can create an MAUI Blazor application and open it using the following commands:

```
dotnet new maui-blazor -o mobileapp
code -n mobileapp
```

You should then be greeted with an interface:

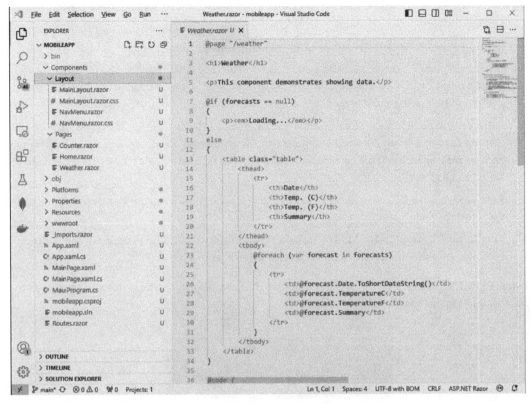

Figure 21.7: Visual Studio Code MAUI Blazor project

Does this look familiar? Apart from a few changes in the project structure to accommodate mobile-related specificities, it is the same content as our Blazor application. A simple press of *F5* will actually show that the result is pretty similar as well, by running the Windows application generated by MAUI:

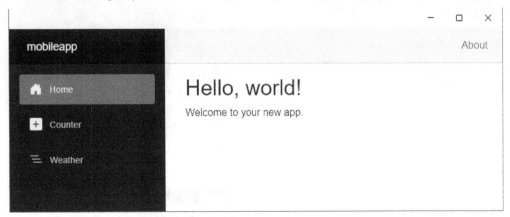

Figure 21.8: MAUI Blazor application running on a Windows target

If you want to run the application on Android, you will have to install the Android SDK and the Java SDK, then configure the emulator and change the target of execution. We'll execute the following steps to install the Android SDK and Java JDK:

1. Run the `dotnet build -t:InstallAndroidDependencies -f:net8.0-android -p:AndroidSdkDirectory="<AndroidSdkPath>" -p:JavaSdkDirectory="<JavaSdkPath>" -p:AcceptAndroidSDKLicenses=True` command (adjusting the folders depending on where you want to install the SDKs).

2. In the command palette of Visual Studio Code, use the `.NET MAUI: Configure Android` command to follow up on setup and follow the instructions (if necessary, you will find them on `https://learn.microsoft.com/en-us/dotnet/maui/get-started/installation`).

3. Restart Visual Studio Code and select the target (see the small brackets at the bottom right of the interface).

4. Finally, hit *F5*, select **.NET MAUI**, and after a bit of time for the first startup of the Android emulator, you should see the main menu and a .NET icon.

When the .NET application is run, you will find the expected display corresponding to the template application, but inside the Android emulator:

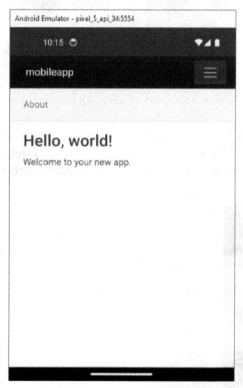

Figure 21.9: Blazor MAUI application inside the Android emulator

Note that setting all this up is still not a bed of roses and you may have to run the `sdkmanager` `"emulator"` `"system-images;android-34;google_apis;x86_64"` operation from `AndroidSDK\cmdline-tools\11.0\bin`, or run `emulator.exe -avd pixel_5_api_34 -partition-size 512` from `AndroidSDK\emulator`. All in all, though, the integrated help is well done and should be effective, even if you have limited knowledge of the Android SDK and emulator. Remember that this needs quite a lot of disk space, and you may receive strange errors if the emulator cannot run properly. You may have some crashes depending on the video board you are using, but a message explaining how to disable hardware acceleration that is in the case will help you get rid of this problem. Restarting Visual Studio Code might help in some cases.

Fusing MAUI and Blazor WebAssembly

Now that the skeleton of the application is ready, we can reuse the Blazor components that we have created for the Blazor WebAssembly-based application in the first part of this chapter. And what could be better than merging the components and the two application projects into one solution? This way, with a dedicated project containing the common pages, components, and class-injection commands, we could have two very small projects containing only the bootstrap modules that are specific to Blazor WebAssembly on one side and MAUI Blazor on the other side.

Creating a common library is as easy as adding a Razor component project and moving to it all the `.razor` files, the associated `.razor.cs` code files, the `wwwroot` folder with all its content, and even the `_Imports.razor` file at the root of the project. Since almost all the class injections can be put in the common library, we can put them in a dedicated class called, for example, `CommonProgram`:

```
public class CommonProgram
{
    public static void InitServices(IServiceCollection services)
    {
        services.AddSingleton<CustomAuthenticationService>();
        services.AddSingleton<IAuthenticationService,
CustomAuthenticationService>(sp =>
sp.GetRequiredService<CustomAuthenticationService>());
        services.AddSingleton<AuthenticationStateProvider>(sp =>
sp.GetRequiredService<CustomAuthenticationService>());
        services.AddTransient<BearerTokenHttpMessageHandler>();

        services.AddScoped<IAuthorizationService,
AuthorizationService>();

        services.AddAuthorizationCore();
    }
}
```

This way, the project corresponding to the Blazor WebAssembly application only contains a very limited `Program.cs` where we simply add `CommonProgram.InitServices(builder.Services);` and an `App.razor` file where we modify the `Router` attribute to take into account the pages contained in the common library:

```
<Router AppAssembly="@typeof(App).Assembly"
AdditionalAssemblies="new[] { typeof(MainLayout).Assembly }">
```

Since `MainLayout` has also been migrated to the common library, the assembly is the one containing all the pages with routing instructions that Blazor should take into account. In theory, this is enough to have everything running as before, but a bug still makes it necessary to keep a local copy of `wwwroot` for now.

As for the MAUI Blazor project is concerned, we can apply the same treatment and remove all common resources, which are usually grouped in the `Components` folder (containing the `Layout` and `Pages` subfolders) and the `wwwroot` resources folder. The `Platforms` and `Resources` folders stay in this project because, respectively, contain the classes that remain specific to the different platforms and the binary resources that are used for the mobile application (for example, the splash screen), as we can see in the following project structure:

Figure 21.10: Structure of the MAUI Blazor project, before common file extraction

As for the web application, it is necessary to adjust the `Routes.razor` file in order to take into account the fact that some components are not inside the present assembly but in the library that is added as a dependency and is now common to both projects:

```
<Router AppAssembly="@typeof(MauiProgram).Assembly"
AdditionalAssemblies="new[] { typeof(MainLayout).Assembly }">
    <Found Context="routeData">
        <RouteView RouteData="@routeData" DefaultLayout="@
typeof(Components.Layout.MainLayout)" />
        <FocusOnNavigate RouteData="@routeData" Selector="h1" />
    </Found>
</Router>
```

Since everything is now in the common library, we can run the Windows application and it indeed shows the expected results:

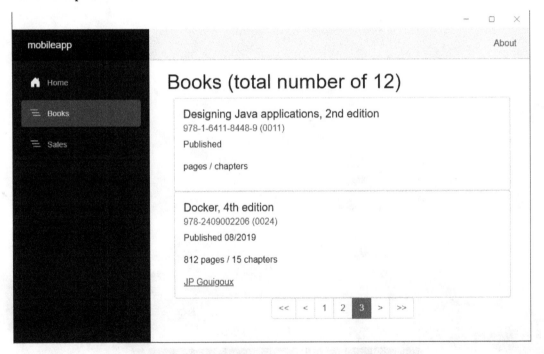

Figure 21.11: Books GUI inside the Windows Blazor MAUI application

And of course, the same application can be deployed to an Android device (here in the emulator for a better display of the result):

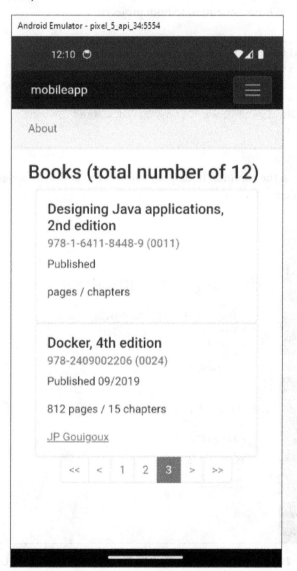

Figure 21.12: Books GUI inside the Android emulator

Beware that, when using the emulator, you cannot point to the localhost address. Depending on your configuration, you might be able to use your network IP address, or 10.0.2.2, or you may have to set up a proxy to reach it. In my personal case, I ended up deploying my Docker Compose stack on a virtual machine on Azure and pointing to its public IP from the mobile application. If you do so and do not expose the API endpoint using HTTPS, do not forget to modify the AndroidManifest. xml file in order to authorize the HTTP endpoint by adding the usesClearTextTraffic option as indicated here:

```xml
<?xml version="1.0" encoding="utf-8"?>
<manifest xmlns:android="http://schemas.android.com/apk/res/android">
    <application android:usesClearTextTraffic="true"
android:allowBackup="true" android:icon="@mipmap/
appicon" android:roundIcon="@mipmap/appicon_round"
android:supportsRtl="true"></application>
    <uses-permission android:name="android.permission.ACCESS_NETWORK_
STATE" />
    <uses-permission android:name="android.permission.INTERNET" />
</manifest>
```

This marks the end of this section about mobile application programming. It is not the purpose of this book to show how to design web applications or explain their code, but we needed to get hands-on in order to see how business/IT alignment and the separation of concerns apply all the way until the last level of concrete, applied details.

Your technology should always follow functional use and requirements, and it happens that Blazor, owing to its capacity to very clearly separate the visual components, their behavior, and how they are embedded in a supporting application, is particularly suited to our desire for the correct separation of concerns and evolution of information system. This is apparent in many ways that have been explained but are worth reminding you of:

- The ability to include RBAC and policy-based authorization in a purely declarative way

- Support for the OpenID Connect standard, which completely isolates the application from the IAM

- The capacity to plug two very different application skeletons (one web, one mobile) into the same graphical components, ensuring pixel-perfect visual compatibility and ease of development

- Tests that can be applied directly to the components library

There certainly are other technologies and platforms that allow us to do the same with different modules, but I have yet to see such a high level of integration as Blazor offers, and I have been looking since I started working almost three decades ago. What I also appreciate is the fact that the stack is based on standard and open source code, which is some reassurance, both for a future that is not going to resemble Silverlight's and for the hope that the stack's evolution will not be hindered by proprietary limitations. In fact, the rapidity of the combination of high-level features, such as hybridization between WebAssembly and Blazor Server, for example, is a very good sign of the health status of the platform. Only time will tell, though, whether it becomes a widespread approach (I know from experience that, in the software domain, thinking more than a few years in advance is like reading in a crystal ball). But for now, it perfectly suits the hypothetical needs of `DemoEditor`, and this is the platform on which I have personally seen the quickest rate of learning for junior teams.

Additional considerations for graphical interfaces

Before we close this chapter, a few considerations are in order about graphical interfaces when talking about business alignment and the correct separation of responsibilities. Since these considerations bear no relation to each other, they will be discussed in the following separate sections.

They will not be accompanied by code samples because it would bring us too far on the technical side. If you want to see them in action, please keep an eye on the accompanying source code and its evolution, which may encompass these subjects in the future. But for now, we will only cover how they should be implemented to respect the design principles we have developed in this book.

Decomposition of the frontend

One of the first reflections architects have when they start working with services (I prefer to simply talk about services, rather than SOA or microservices, which have become too oriented toward particular technologies and techniques) is that the modular decomposition of services that they implement into several APIs and their dedicated, separated implementations (included persistence) could be applied to the frontend part of applications.

Indeed, it would be practical to have separate life cycles for different bits of the GUI and be able to recompose them. This is particularly useful for large frontends, with many teams working on the same portal. This way, work is more fluid and separation of responsibilities is almost always a good thing.

The main point here is that, after modularizing the different parts just like we would do for a service (and we can certainly align them to the business domains as well), the time comes for recomposing the module and making the assembly work well together. This is where things become harder on the frontend. As far as the backend is concerned, we have lots of well-established ways to recompose: the orchestration or choreography of services, middlewares, interop connectors, etc. And the standards are numerous. On the frontend side, things are definitely more complicated; some of these issues are listed here:

- Iframe composition can cause security problems that must be addressed

- Composition also means having a common style, which is harder than just sharing a CSS file, since elements should use the same classes (this is strong coupling)

- Layout may be complicated, as visual designers will generally not be able to propose a screen where all the domains are nicely separated in rectangular blocks

- Visual interaction is also a concern, as the visual blocks should be able to talk to each other, and there is no standardized way to achieve this

Some standard solutions have been devised in the industry, such as WebComponents, but they are not widely spread and as irrefutable solutions as HTML or even some smaller-level standards such as OIDC. Still, standardizing the GUI is not only a matter of homogeneous appearance and ergonomics for the user but is also a way to save time and reduce errors in the GUI coding by keeping predictable behavior. Even in line-of-business applications where the look-and-feel is not as big a competitive advantage as in websites, trusting dedicated experts for the hard bits of HTML/CSS and providing their work as a skeleton that C# developers can embed in the Blazor application is the right way to go. Again, this is all about separation of concerns. Some tools, such as `https://storybook.js.org/`, can be a great help in creating web-based components.

Nonetheless, this approach is not standardized for now and remains the job of experts. As a consequence, investing in frontend modularization is not a controlled activity and bears risks and potentially high costs. You should do it only if you have a strong reason for it.

And sadly, if you follow the proverb, "*Throwable GUI, durable API*," there may not be a real justification for doing so. In lots of projects I have seen, such approaches were not justified by business reasons but only supported by technical teams, which is almost always a bad sign in a software project. In lots of cases, the GUI only acts as a glue that visually assembles graphical components: this does not mean that care should not be taken into separating components' life cycles but simply that the glue is highly coupled to the way the whole frontend behaves, and that this is not necessarily a problem as long as the skeleton is quite small in extent and can quickly be developed again using another technology or framework. This is typically what has been done in this chapter when switching to a mobile application where only the "glue" has been coded in a different way.

Content negotiation for a simple interface

Talking about reduced-footprint frontends naturally leads to *dumb clients*, where the browser basically takes absolutely no responsibility for how the application works and blindly obeys orders coming from the server. How is this possible in a web environment, without any connection from the server to the clients and while trying to be stateless, you might ask? It turns out this is possible by sending possible operations as links in each page returned by the server, alongside the data in HTML. This is precisely what we did when sending `rel=next` links for pagination, but it's extended so that it's used for every possible manipulation the designer wants to expose to the users. For example, links can be used to send to the modification page for the displayed resource, or to the creation link when on the list of existing objects. They can also be specified to navigate to other resources (this is what we prepared for by using a `rel=dc:creator` link between books and their authors).

In fact, every possible interaction can be specified as specialized links and the dumb client can simply display them with the `title` value provided by the server alongside the `href` value. This way, the UserAgent is not the application but the browser itself. Taken to the extreme, this is what is called the **Hypermedia As The Engine Of Application State** (**HATEOAS**) model and is considered the absolute best approach for those looking for dumb clients.

If we mix this approach with business-domain-based decomposition, we can go as far as trusting the API implementation for the return of a piece of frontend in HTML by using content negotiation (making the API support a value of `text/html` in the `Accept` request header). This remains quite futuristic at the time of writing and the general orientation of the software industry is toward more complexity rather than this radical-simplification approach, but this may prove highly valuable in some specific uses.

State management

As a complete opposite of the previous model of a dumb client that does not even deal with its state (it is continuously kept in the exchange with the server in the HATEOAS model), the modern single-page application tends to keep state in the browser. Since the HTTP standard has been created as stateless, this makes for a bit of extra manipulation, but there are solutions to deal with it in a nicely decoupled way.

One such approach is the **Redux pattern** (you will find an implementation for Blazor in the package called `Fluxor`), where a complete separation of responsibilities is obtained by using different objects for interactions in the frontend:

- `State` classes contain the definition of a given state (there can be several in an application and, as you have surely guessed, the way they are decomposed should follow the business domains in order to reduce coupling as much as the functional complexity allows)
- `Action` classes serialize the operations that are applied to the client
- `Reducer` classes are able to apply the action and define a new state coming from the previous one and the application of the command, thus enabling the transmission of a state

- `Effect` classes are used to define the behavior caused in the application by the execution of an action

If you need to use state in your application, consider exploring the state management options provided by the Redux pattern. You will find that, eventually, it leads to interactions in the frontend that are exactly aligned to the atomic modifications we have talked about in the *forget nothing* persistence model we studied in *Chapter 16*. If taken to its logical end, the client then sends `PATCH` commands to the server every time it detects a modification of one of the fields in the resource displayed. If wanted by the user, the interface might provide the capability of waiting for enough modifications to send a bigger atomic patch to the API. Choosing whether this is based on time elapsed, the size of the patch, or a manual intervention of the user is solely the responsibility of the client. When using this approach, one of the consequences is that everything is asynchronous in the client and there is no **OK/Save/Submit** button anymore, as each interaction leads to sending an order for the modification to the server.

Again, this might seem a bit premature to consider when facing old-fashioned business applications, but this is a pattern that tends to be frequent in newly designed line-of-business software frontends. And, again, it pushes strongly in favor of responsibility decomposition in terms of business functions. For example, such a model allows some kinds of users to apply complete transactions on objects while others may be traced in whatever interactions they are performing, even if they cancel an interaction right after the change (which can be important in highly regulated environments where traceability is of utmost importance). The server might also apply some buffering in the orders it transmits in order to improve overall performance, or even repackage orders coming from the same user; again, this is a discussion of the responsibilities that have to be carried out before implementing technical functions. However, the model allows any kind of modification afterward, which is a sign that the modular decomposition of responsibilities has been done correctly.

Separation of responsibilities for application logging

Another example of separation of responsibilities (and this will be the last example, but the list could go on) is the way logs are written in the application.

When considering it from a purely functional perspective, it seems that logging serves a multitude of use cases, all of which must be taken into account when designing the technical solution:

- Tracing code execution in the client for debugging purposes or helping an educated user understand what is going on within the user agent; this is typically done with `console.log` instructions.

- Tracing code execution in the server to follow the behavior of the backend application; this is typically implemented with `log4net`-like traces, and many sinks can be used to retrieve the traces, from a simple local file to an elaborated Prometheus system.

- Following a given problematic interaction is where it becomes more complex and coupled because the interaction starts on the client but soon reaches the server and the potentially numerous different API implementations that are needed. To be able to correctly realize this shared responsibility, the client must send a unique code for a given human interaction and every module needs to ensure that this unique code is passed along with everything they do, making it possible in a centralized tracking system to follow the unique interaction (and even, in some cases, to only trace this one interaction).

- Monitoring of resources is another responsibility, as it is lower level. But again, there is a need to cross-analyze these values with other traces in order to find out the origin of resource depletion, for example. This is why a centralized trace and monitoring system is often used.

- There are of course additional responsibilities, such as observing the system, adding limits and alerts, and sending notifications, and it is equally important that all those are lowly coupled (for example, beware of a monitoring system that has a proprietary notification system; high-priority alerts may soon increase in price if you use them for long).

- **Business Activity Monitoring** (**BAM**) (which basically involves doing some statistics about which software functions are used and when precisely the different interaction steps happen) is also an associated, but separate, responsibility. It is thus important that you can use it without activating verbose traces. Indeed, BAM is supposed to be a continuous activity, while traces may vary in verbosity depending on your degree of trust in the platform; the cost of storage you are prepared to assume; your ability to send important, regulatory-level information to cold storage; and many other criteria.

As you can see, there are many responsibilities behind the monitoring and tracing activities, and it is easy to create some coupling and make it hard to evolve if you do not pay attention at design time to the standardization of interfaces between different modules. Of course, it makes some sense to reassemble some traces in a unified console – in particular, to study the multi-service interactions – but remember this creates some technical coupling. However, centralizing the monitoring system helps reduce the complexity of managing storage, information retention, and so on (provided, of course, that the logs have the correct semantics and metadata, which is a strong incentive to use semantic logging).

This activity is such a complex subject that it warrants a complete study of responsibilities, just as we have done for the `DemoEditor` sample information system. There is a long list of subjects to which we could apply the same method for business/IT alignment, but we will close it here since the examples provided should have made it clear how the method works.

Summary

In this chapter, we have added a second GUI to handle the business entities for `DemoEditor` in order to illustrate the fact that a dedicated interface is the right move to go for the given users and that GUIs should be adapted as much as possible. In fact, they can even be thought of as disposable, since all the business rules and data management/persistence are handled by a unique server address by any number of clients.

We also showed how a technology like Blazor can help keep good modularization, although the number one criterion for such design qualities remains, of course, the functional decomposition. Now that `DemoEditor` has a backend, two frontends, an IAM, and a robust authorization system, we only need to put it all together, and this is what we will do in the next chapter, which will show how to integrate modules around business processes.

22
Integrating Business Processes

This chapter will close the loop around the sample application by applying the **Business Process Modeling (BPM)** approach to it. It will show how a software system can be piloted by BPM, with approaches such as orchestration or choreography. Note that I said BPM instead of **Business Process Model and Notation (BPMN)** because the process itself is more important than its technical representation. Indeed, a GUI is responsible for the screen display, but also – except in the extreme case of HATEOAS – for transitions between pages and thus running processes. This chapter will also show a sample application of this with different techniques and, in particular, demonstrate how webhooks can be applied to achieve good decoupling.

Since we have a small and simple sample information system, we will not resort to a complex BPMN engine setup but rather use a small, web-based solution that is more adapted to what we are trying to achieve (again, adjusting the tooling to the usage). A few examples of actions supported by this tool will be shown in order to explain the different possible situations this can be adapted to.

Sometimes, the steps will be the same between processes (they may even be generic between many organizations), which makes these processes great candidates for a dedicated service. We will discuss the importance of this particular service in detail, as its alignment with the business is crucial. Indeed, even the smallest amount of coupling will harm not only one process but potentially all the processes of the organization, due to their genericity. This is the reason why special care should be taken to clearly decompose their responsibilities.

Though this chapter is quite technical, you will find that there is much less .NET code, and this is deliberate. I wanted to show precisely that this is the most expensive and complex way to solve a problem, so it should be implemented only when there is no other way to do it. And it happens that we have lots of other ways to perform some process execution. Let's explore them!

In this chapter, we'll cover these topics:

- What kind of processes should be added to our set of services
- How we can use the GUI for most simple business processes
- When it is necessary to resort to external tools for process management
- How to use them for choreography or orchestration
- When the setup of a dedicated service for some special process steps is needed, and how to do so with the example of a middle office validation service

The need for processes

Processes are everywhere: every act we are doing, every task a computer is realizing, is generally part of a larger group of actions that contribute to reaching a given objective. As such, even the simplest information system, such as the one we created for our sample `DemoEditor` organization, contains processes. Of course, some are very small and not worth describing (an editor changing the address of an author, for example). However some of them bear lots of value for the organization, and special care needs to be taken when using them. Those are the ones we are going to deal with in this chapter because they are generally the ones that are implemented using software, in order to make them quicker or to ensure a perfectly repeatable way of running them.

Orchestration and choreography

As a quick reminder before we dive into specific processes and show how they can be integrated into the information system, there are two main ways to implement a process:

- Orchestration is when a unique actor in the system pilots the work of other actors in the system, telling them what to do, sending these orders in a given sequence, and eventually reading some information from the actors or the context to adjust the sequence
- Choreography is when all actors are considered equal in the system and each reacts to events coming from the others, which themselves are emitting events that other services may take into account

In both cases, there will of course always be some kind of trigger to start the process (the choreography chain has got to start from somewhere, and someone has to run the orchestration), and this is generally when we go back to the human part of the system, which is the user.

As we talked in detail about processes and BPMN in *Chapter 11*, we will leave it here for this introductory section and start analyzing some real cases (as real as our sample scenario can be).

Application to DemoEditor

I already provided a few processes describing the activity of `DemoEditor` in *Chapter 11*. If you remember, there was this BPMN diagram explaining the enrollment process for authors:

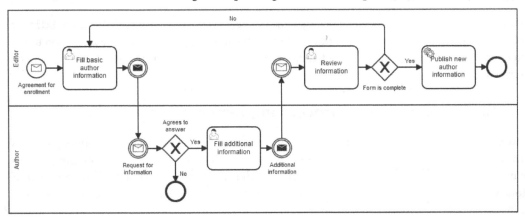

Figure 22.1 – Example of BPMN for author enrolling

During our conversation, we discussed the implementation of choreography events to improve the process of handling incoming paper communication and potential signing. This process could be easily described in the first layer of our four-layer map:

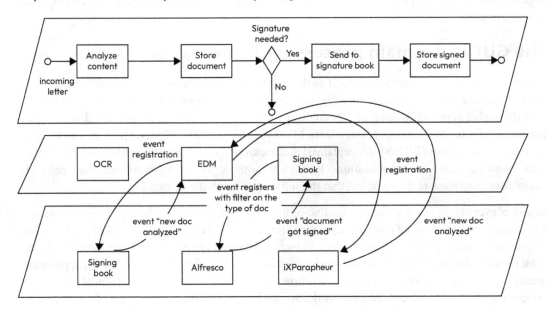

Figure 22.2 – Low coupling choreography

The first of these two processes will be implemented in the next sections. As for the second one, we would rather implement a contract-signing process depicting the relationship between an editor and an author, which will be more interesting in our case because it will show how an external service can participate in the orchestration.

We will also show clearly how a small process engine can help in two other processes that `DemoEditor` needs, namely the one to realize the nightly import of the books catalog during the migration to the data referential service replacement, and another one that we have not talked about but that will be described later in this book.

How to choose the right process implementation

As we said, we will show several (and admittedly very different) ways to include business processes in the information system. The question you may have in mind, then, is how to know which one is the best. It depends on quite a lot of factors, such as the complexity of the process, the number of actors, whether there are external actors, and whether you need to follow up on what is going on in the instances of the process.

In fact, there are so many criteria that it would be a real endeavor to design a decision table outputting the required method depending on all the possibilities. The best way I have found is to show several ways to do this, explaining for each of them why the chosen implementation is the right one for the process. Again, you will notice that this almost always boils down to business alignment, and the urge to stick as much as possible to what is functionally true when designing a technical response to requirements.

The GUI as the main process implementation

It may sound weird to talk about the GUI as the main tool for implementing processes, particularly since I have talked so much (in an entire chapter) about business process management and normalized, fully decoupled approaches using BPMN engines. But remember this was in the context of a utopic, perfect system made only with **Master Data Management** (**MDM**), BPMN, and a **Business Rule Management System** (**BRMS**). As explained at the end of *Chapter 11*, the actual use of a BPMN suite – which is quite a heavy application – is reserved for those processes that are varied or need such complex management of their context that the software is worth the investment.

In a lot of everyday business processes that line-of-business applications heavily implement throughout all software, the succession of tasks is simply supported by the sequence of screens linked to each other by buttons or other graphical components. In fact, in addition to visually showing data, this is the most expected feature of a GUI and one from which users expect a lot. The usability of an application greatly depends on how well it guides the user through a process and keeps them informed about their progress. It should be easy to follow the workflow and be aware of the current stage of the process.

Human processes versus automated processes

GUIs have another great quality when it comes to business process execution: since they are designed for human interaction, they allow smooth and efficient transitions between human and automated actions, depending on which one is the best. By the way, this is also why it is important to talk about BPMN even if we do not use execution engines: they enable a clear representation of the workflow and, though it can be explicitly stated which tasks are human-executed and which are fully automatic, it can also be left up to the actor executing the workflow to decide. In *Figure 22.1*, the icon showing envelopes and designated messages can represent anything, such as a call from one role to another, an email automatically sent, or a phone call.

Implementation through the user interface

Staying with the scenario represented in *Figure 22.1*, let's see how it could be implemented simply through a GUI, by using the user interface to unfold the tasks necessary to reach the objective.

The first task is done by the editor: once they have the agreement for enrolling a new author, they will create the entity in the data referential service, filling in some basic information they have about them from the enrollment process. Maybe they met somebody at a seminar, collected their business card, and talked only by phone before reaching an agreement on a subject for a new book. Since this first part is done, the editor will enter a new phase of relating with the newly appointed author and needs some more data about them.

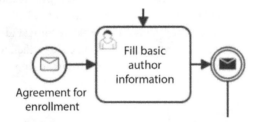

Figure 22.3 – First step of the process: filling in basic author information

The most logical way to execute this task is to provide the editor with a form to provide data about the author (first name, last name, or any other kind of data we can store in the data referential service, and that editors should generally know about their authors). The authors referential and its GUI are the right place to do so, and a graphical interface similar to the following could be provided:

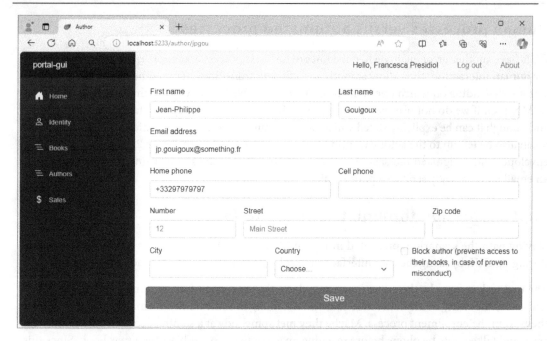

Figure 22.4 – Example of a form to record author information

This first step is quite easy to implement. It is important to separate the creation of the instance in the data referential service from the attribution of a functional identifier. Let's explain this: as you will certainly remember, the author business entity has an internal, technical identifier in the form of a GUID (in our case, provided by MongoDB) and a business-related identifier, which is a counter-based integer padded for a length of four digits, but could actually be any string (the object is purposely relaxed on this since strongly enforcing an integer actually does not have much business value):

```
public class Author : ICloneable
{
    [BsonId()]
    [BsonRepresentation(BsonType.ObjectId)]
    [JsonIgnore]
    public ObjectId TechnicalId { get; set; }

    [BsonElement("entityId")]
    [Required(ErrorMessage = "Business identifier of an author is
mandatory")]
    public string EntityId { get; set; }

    (...)
}
```

Providing the right value for `EntityId` is in fact harder than it looks. If you calculate the next value as soon as the user hits the **Create new author** button, then you may have a few holes in the integer sequence when the user abandons creation, whether voluntarily or forced by bugs or another external circumstance. On the other hand, if you decide to let the database do the job of selecting the next counter, you are putting a high toll on it because this means distribution is out of the question; there needs to be a centralized, unique view of the next value of the counter. It also adds difficulty when you are trying to persist a graph of data, and identifiers are necessary to define the links between nodes.

In this case, the (almost) best choice you can make is to let an asynchronous job handle the attribution of the business identifier after the creation of the entity. That may sound strange at first, but this is actually the most performant approach, because the process can be easily distributed, even at an international level, and the time an entity spends without a business identifier is in fact extremely reduced (a few seconds at most, which is the time necessary to refresh the form anyway). Pre-affecting ranges to the different actors in the distributed system will further improve performance, at the acceptable cost of temporary holes in the sequence (which is generally not a problem; permanent holes are).

At the beginning of the previous paragraph, I said "almost" because the actual best move one can make in dealing with this situation is to convince the users that they should get rid of their old counter-based business identifiers altogether. Of course, that does not mean forcing them to use an ugly GUID; but an easily readable, 8- or 10-character-long unique and arbitrary identifier, such as `AR8HGP57`, can be acceptable. In fact, most of the time, people love their old-fashioned, counter-based identifiers because they remind them of the global number of entities in the database and their progress in filling it (you can imagine `DemoEditor` would throw a big party when author numbered `0100` is created). Again, that may sound strange for people not used to this and would simply think of using `/api/authors/$count` as the basis for a celebration, but this kind of preference does exist and you could face a strong negative reaction if you do not take it into account. One potential approach is to explain to the director the technical cost of maintaining the sequence of numbers without gaps, assuming they have some understanding of the technical challenges involved.

I will not go into the low-level details of implementing an asynchronous business identifier attribution system; just know it is a good way to avoid contention problems in a database (and not just internationally distributed databases; a database shared by a few application servers may already expose the problem), and you can implement it in different ways:

- Have a single process close to the database, in charge of filling in the counters as soon as new entities appear.
- Trust the database itself to provide this counter (most will be able to do it well).
- Asynchronously send `increment`/`upsert` to your database and base the returned value on that returned by the previous entity.
- If you need a longer delay, make it so that the API can handle `/api/authors/{id}` with an `id` attribute that can be either the business identifier (in most cases) or the technical identifier (while waiting for the business identifier to be affected by the entity).

- Providing each of the actors with ranges of identifiers, thus dividing the number of calls to the generator and improving performance, at the cost of wider holes (though temporary).

- Use counters based on the modulo of the actors. In an example with three application servers, the first one would create the values 0001, 0004, 0007, and so on. The second one would create the values 0002, 0005, and so on. You get the idea for the third server. This way, there is no contention whatsoever, or any kind of dependency on the database to provide a unique counter. This is great for performance, but even round-robin load balancing may cause drift when servers are stopped for maintenance, and long-duration holes may appear in the sequence. While these holes wouldn't be permanent, because they would eventually be filled, they would bother people who are used to identifying entities based on their index number.

Passing the message to another role in the process

The first task of the process has now been completed and we arrive at the message sent by the editor/received by the newly enrolled author in the process:

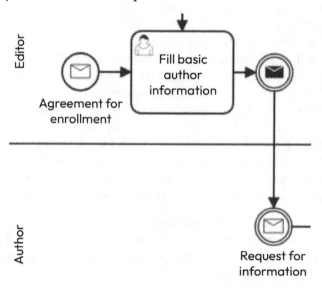

Figure 22.5 – Second step in the process: passing the token

This kind of transition is often called **passing the token**, as this is a good metaphor for the way software business process management systems work, passing a container of data from one task to another, each of them possibly reading the data coming from before and adding some data of their own, gradually filling in all the data the process instance will present by the end of their execution.

As I said before, BPMN does not state anything about the actual implementation of these messages. In a first, low-ceremony version, we could simply display a text message to the editor that they have to tell the author to log in and fill in their part of the form. The only question (which has nothing to do with the message-passing, but rather with authentication) would be how to be sure the right author has access to the rest of the process.

This question was briefly addressed in *Chapter 14*, explaining that the IAM can guarantee that a JWT has a claim containing the address, but also that this email address has been verified by the identity provider. This is one way to proceed: if the editor enters the valid email address for the author, this information is enough, along with a trusted identity provider that provides a verified email address in the JWT claims to authorize the account as being "attached" to the author business entity in the data referential service, and then gives the external user access to add information about this author (and this author only, of course).

There are, of course, other ways to proceed, for example, by asking authors to create their user accounts first and then letting the editors manually associate them with authors in the data referential service. The way this is done can be adjusted depending on the security context, the available tools, the capacity to customize the technical implementation of the process, and so on. The important part is that this "sub-process" should not lead to unwanted decisions about the main process of author enrollment, to keep coupling low and ease evolution in the future.

But back to our message-passing problem: we could rely on simple human-carried messages from the editor to the author, but inevitably, if the company grows, we would need to evolve to other ways to notify authors that they should add information to the forms. One possible way to receive notifications is through email. However, this method may develop into other channels such as SMS or multi-channel notifications. It may even evolve to allow recipients to choose their preferred channel based on the day of the week, group messages, and more. This is typically where we would use a dedicated external service/API for notifications.

Again, I am not saying this is what you should do; it really depends on the channels. If your company sends only a few emails per month, the logical implementation is to simply have a common library that calls your SMTP server. The only important thing to note is not to "insult the future" and to have at least a central point for such a function. This way, if you ever evolve to multi-channel or have a volume of emails that necessitates branching out to dedicated external online services that cover delivery management, anti-spam, and other advanced features, it will be easy to do so because all notifications will go through a single endpoint that can evolve without having any impact on the senders. As always, this indirection by a standardized interface or a pivotal format is only going to cost you a few minutes or hours of reflection and save you loads of trouble in the future. And if you think it is not worth it because your system will not change, remember, *"Change is the only constant"*!

Letting the author fill in the information in the same GUI

Now we are in the author's "swimlane" of the process. This stage of the process involves adding some information to the author data entity (which is, really, just information about themselves that they agree to provide to DemoEditor):

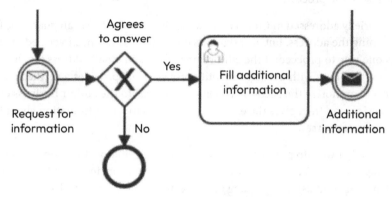

Figure 22.6 – Third step in the process: adding personal data about the author

This step brings two important architectural decisions:

- Since the function is the same (filling in data about authors), should the software feature be the same for editors and authors? How about possible differences in the fields that should be filled in by editors or authors or be open to modification by both roles? Should the GUI be specialized or should it stay generic with business-led behavior included in its general implementation?

- Also (and this is related to the previous point), should the infrastructure support two different applications to account for the fact that authors and editors will not reach the GUI from the same network? Security seems to command that a DMZ should be used in the case of external access, while the internal network could reach the API directly. And how does this still make sense in a "home office" situation where editors access the DemoEditor information system from their home internet connection, which is as unsecure as the authors' connections?

As usual, let's try to deal with these questions by keeping a non-technical, purely business-aligned point of view:

- The technical considerations about network segmentation do not make any sense from a functional perspective. Editors may plug into the information system from a low-security connection as well. This is the "zero-trust" approach of treating them exactly as any others; the fact that they are on the local network does not permit them any shortcuts that could compromise security.

- There can only be one data referential service since it is the single version of truth for author-related data. Clients will communicate with it through a contract-based interface. This is the API we designed. An API is not only a technical artifact; it is also a business-defined contract on how to deal with the data, with its precise definition.

- If there were several GUIs to point to the authors API, they should have their "author" form based on the version of the contract, which means the same graphical component would be used by all the concerned client applications.

- The real question from a purely business-oriented point of view is: should we let data from someone we do not trust as much as our employees enter directly into the data referential service? In this particular case, we may answer yes, as long as authorization ensures that only the user corresponding to the author (by an email address verified through a trusted identity provider such as Google, Facebook, or Microsoft) can modify their data. In a later section, we will study another hypothesis, where this incoming data would have been put on hold before entering the data referential service, waiting for approval to come from someone with the appropriate rights.

The technical choices that translate from the preceding functional decisions are the following:

- The GUI will be hosted in a DMZ network exposed on the internet. Only authenticated access will be allowed.

- Authorization rules will be set to allow editors to access all authors (read or write, depending on their pool of authors) and allow authors to only access (read as well as write) the data corresponding to their email address.

- The authors referential will be hosted on a separate network. Depending on the security requirements, access between the two networks can be blocked by IP filtering, firewalls, message queues offering inversion of stream, or any combination of these and other security features.

- Since the authors data referential service is a true MDM, the residual risk of a user tampering with data that they have authorized access to is not considered a harmful problem. Editors will check data from time to time and, if necessary, can block the author through the `restriction` attribute.

Validating and publishing information

Now that the third part of the process has been dealt with, we arrive back in the editor's "swimlane":

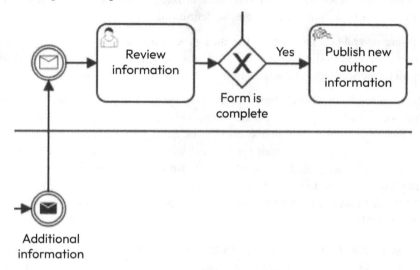

Figure 22.7 – Fourth step in the process: validating and publishing author data

As far as message passing is concerned, the same reasoning applies here as in the first role transition: the author might call the editor to tell them that they have finished filling in the form, they could send an email, an automated notification could be triggered once they declare their work finished on the interface, and so on. Everything is possible, and this is how the functional process is decoupled from the technical implementation: the BPMN standard does not say anything about how the task is realized.

Reviewing the information will, of course, be done again in the exact same interface. The only thing that could differ is how this interface is displayed, depending not on the swimlane per se, but rather on the authorization level of the person concerned. Editors, for example, could indicate domains on which they consider the author has the right expertise for writing a book, but maybe should not be allowed to view authors' bank details. On the other hand, authors should certainly be able to change their personal information, such as address, but should not be able to change – or even view – their internal identifier.

If the editor is not satisfied with the form filled in by the author, the process loops back to the beginning, which can be something purely manual, with the editor notifying the author of the need to fill in some additional data. This is where the notification system could be utilized with a dedicated message. In a very sophisticated form, the form could contain notes on each of the attributes to be filled in, in order to indicate what the editor expects. In this case, this would depend on the kind of form to fill in, a good amount of thinking about how to present the notification, the author graphical component, and the indications metadata would be needed to allow for further evolution of each module.

Finally, the last task, called **Publish new author information**, is marked as a service task, which means it is entirely automatic. All previous tasks were user tasks, so there was a human making decisions on how to proceed. As the BPMN syntax indicates this task should not necessitate any interaction, the associated implementation should be entirely automated. But in addition to not saying anything about how the implementation should work, this process does not give many clues about what "publishing" can represent in functional terms. Again, we will have to ask the owners of the process.

Maybe the chief editor will indicate that, once enrolled, the author's information must be presented to all editors, in order for them to know that this person, enrolled for a given book, in fact, has other expertise that may interest them for a future manuscript. "Publishing," in this case, mostly means the author's information should be readable (at least partly) by all editors and that they should be made aware of this newcomer in the pool of authors. The first part can be dealt with using authorizations; the second needs, again, some kind of notification. It could be through internal Teams messaging, adding a "new" icon to the author's fact sheet for each editor until they have read it, or maybe even putting all "new" authors at the top of a list that the editors are presented with when they connect to the intranet, and so on.

If we choose this last way of doing things, here is what a nice separation of responsibilities would be:

- Instead of "polluting" either the authors referential or the one listing the editors (we have not talked about it, since the editors have been, for now, only seen as users of the information system), we would create a dedicated service to record "novelties" (not only new authors but any kind of business entity that is new to the system).

- Enrolling a new author would send an API call to this new service with the identifier of the author. The service would record all the editors at that time and create a list of "viewed" tags for each of the editors in the list of users, except for the one who enrolled the author, because the author is obviously known to this user.

- This list would then be reduced (or the flag set to false) each time the author has been observed by an editor. If a new editor is recorded during this time, they would not have any "new" signal because all entries would logically be new to them. This way, we do not have to make the service observe this kind of event.

- The "novelty" service would also include a time orchestration feature that would erase all editors in the list associated with an author once a week has passed since the recording of this "novelty," which by then is not one anymore.

- Any feature that needs to know about the "new" character of an author (for example, the list of authors, or the form displaying one author in particular) would, when an editor logs in, look up the novelty service to know whether the "new" icon should be displayed. This call would be done asynchronously in order to never slow down the main information stream. Also, if the service does not respond, it will not generate any errors or changes in the interface, as this "new" icon is simply helpful information but does not change the business processes. Only a log would be recorded so that technical administrators know that something has not gone as desired.

- Access to the author's individual form (their "fact sheet") would send a signal to the "novelty" service that the editor has seen this author and should not display a "new" icon for them anymore. This would erase the entry in the list rather than setting the flag to false, in order to reduce memory use, even before the "novelty" feature is removed because it has been a week. Of course, this second call would also be "fire and forget," as an error would not be of high importance.

We have shown how simple GUIs often bear some kind of business process information, even when we do not think of them in terms of BPMN.

Advanced uses of the GUI for process implementation

In some cases, GUIs can handle some more complicated business processes, typically with many steps, many actors, or even both, and a long-term approach, where the execution of the whole process spans a period of days, weeks, or even years, even if it is not very frequent.

Implementation through a wizard

GUIs are a great solution when we need to implement many steps in order by the same actor. This is what is generally called a "wizard." This particular feature of a frontend application will assist the user in entering information and/or making decisions with multiple steps where each corresponds to a dedicated page of the GUI.

Buttons or links displaying **Next** or **Previous** will allow the user to navigate through the steps of the wizard, logically going forward when a step allows it (typically when some client-side validation has taken place) and possibly backward if needed. "Breadcrumbs," which are a way to indicate the position inside the different steps, may be used to send some feedback to the user, or even a percentage of the wizard that has been filled already. In the end, wizards may display a summary page with everything that has been added in order to enable the user to spot some mistakes and come back to the right steps where they can correct them. This last form is also where the user can submit the whole set of information.

A good wizard would allow execution in many sessions, without forcing the user to enter all the information at once or, even worse, retyping information they already provided when previously using the wizard. The smallest implementation of such a feature is to serialize the wizard content in the local storage of the browser. For a non-web application, if the need is to allow the filling in of the wizard from different user agents by the same person, or when multiple concurrent instances of the wizard should be possible, more sophisticated approaches with dedicated persistence would be put in place.

An example of such a wizard in our `DemoEditor` sample context would be the registration of a book proposal, which could convey the following four steps (the breadcrumb line shows that we are in the third step in the following screenshot):

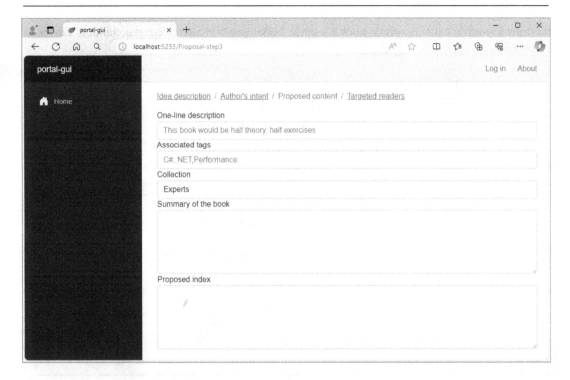

Figure 22.8 – Wizard-based implementation of a single-user process

Though generally limited to single-actor processes, this approach with a GUI remains quite powerful and much more adapted than an external BPMN engine.

Dealing with long-duration processes

People tend to expect BPMN engines to be worth the trouble of installing and maintaining when they have long-duration processes. When such process instances take days or months, they cannot stay in memory and have to be "frozen" or "dehydrated" when not used, then "thawed" or "rehydrated" (depending on the editor's jargon) when they should be added back to memory to receive modifications, such as the recording of values in forms or the fact that they can progress to a next step.

Let's draw a workflow for the whole book life cycle. Following `DemoEditor` stakeholders' advice, we may end up with something like this:

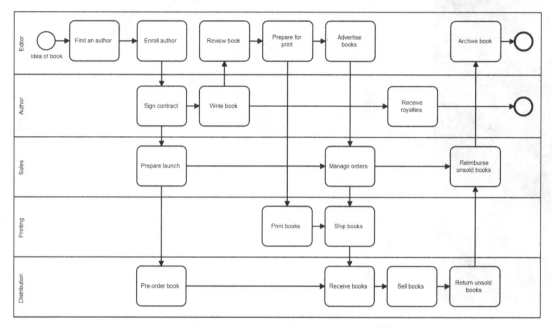

Figure 22.9 – General process of book publication

But again, when you think about this from a purely functional point of view, do we really expect that the different roles have some kind of token that is passed between each other along the life cycle of the book? No, because people work in parallel. And if we use the good-old method of imagining what would happen in an office without computers, people would most certainly coordinate with each other by asking questions such as *What is the status of this case?*

This word is interesting: *status*. A status qualifies the place of a thing inside its life cycle. In the case of a book, for example, it would state whether it is still an idea (only known by an editor and, maybe, a prospective author), whether it has been contracted as a future manuscript, whether it has been reviewed and is ready to print, and whether it is currently selling or has reached the end of its life cycle and is not available anymore. Does this sound familiar? In fact, this is something we analyzed in *Chapter 9* when talking about domain-driven design and how it should cover the entire life cycle of a business entity. The corresponding diagram looked like this:

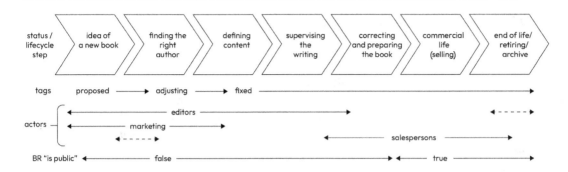

Figure 22.10 – Life cycle of a book, with its status

From top to bottom, we see the steps in the macro-process representing the life cycle of a book, from the idea to the archive, then the tags that are potentially associated with the book along this life cycle, the actors that work at the different stages, and finally, the value of business rules stating whether the book is considered public, depending again on its advancement in the life cycle.

Notice how the status changes during the life cycle, which is represented as a macro-process? Tags are used to provide some text associated with the status. In the book life cycle, we can identify the actors associated with each step, similar to how we can identify actors involved in a specific action in the BPM process.

All this means that a single piece of status information on a book entity and some authorization management to decide which role can read and/or write which book attributes depending on the status go a long way and can avoid – again – using such heavy and complex machinery as a BPMN engine.

In fact, the status attribute of the book data entity should be one of the most carefully implemented fields, as it will depend on lots of business rules and imply many changes as a consequence of its changing value. For example, reaching the ready-to-print status may need a dedicated sub-process where validation should first be given by the main editor of the book and completed by a sign-off from one of the directors. In turn, reaching this value may indirectly start some other processes, such as printing the book or allowing salespeople to place direct orders. In particular, I insist on the "indirectly," even if I hope that anyone who has read this book so far is totally convinced that the code implementing the business rules to reach the status should be completely decoupled from the bits of code that would react to this change of status. In this particular case, a good technical implementation would be through events and webhooks. Those are a very nice and easy-to-explain way of creating such transitions in a global process.

To give a bit of technical architecture, what would be put in place could be something like this:

- The data referential service book would allow any application to register for a change in the status of a book (or any books, for that matter). This could be part of a wider-purpose webhook that acts on any change in observed books.

- A $filter attribute upon registration may be added in order for the subscribing clients to specify that they only want to be notified when a particular step is reached, in our example, ready-to-print.

- When the status is reached, the callbacks for each registered client will be called. For those for whom this is simple information, a retry policy would be enough. Actors for which this piece of information is critical would use some kind of delivery-robustness feature, such as message-oriented middleware (we will show how a message queue could be used for this later in this chapter).

- Each of the clients would then do its own job on the book. Some of them would keep the copy of the book data that was sent to them together with the callback because they need data stability (we could think of the printing company relying on a given, signed-off version of the book entity to retrieve content). Some others would rather call the data referential service every time because they need to ensure the data is fresh.

- When the book reaches archived status, surrounding applications might be interested in this information to stop listening to future changes and maybe do some cleanup in their own copies of the data.

As we have seen in the initial sections of this chapter, there are many processes that can be dealt with simply through the GUI and a good data referential service. Still, business process externalization might be a good choice in some cases. It is just important to consider the fact that such engines are an additional server to take into the equation, adding some complexity to the system. This means the return in ease of evolution should be worth it and, generally, such solutions are used when processes are complex, change often, or need some specific features to continue running.

Externalizing processes in a dedicated tool

This section will show a few cases where an externalized piece of software may be indicated to implement the execution of a business process. BPMN engines readily come to mind, but they are quite complex servers and some lighter alternatives exist, as we are going to explain.

Using a BPMN approach, but not a BPMN engine

As explained in *Chapter 11*, the actual use of a BPMN engine is rarely justified, and process modeling is mostly used to reach a complete understanding of the functional use cases, which is necessary to reach good business/IT alignment. In our case, or rather in the case of DemoEditor, the processes to be executed are so simple that installing a full-blown BPMN engine such as **Bonitasoft** or **Bizagi** would be unreasonable. As shown previously, most of the execution of the processes will simply be realized by relying upon the different GUIs.

In fact, we are in the same situation as what was discussed when presenting business rules management systems. Despite Open Policy Agent being a great and sophisticated approach, care should be taken to only use it when actually justified. Complex authorization rules, with the necessity to justify allowing or refusing access, are definitely one of those cases, but we saw in *Chapter 18* that simple role-based access control was enough for `DemoEditor`'s requirements. Business processes at `DemoEditor`, likewise, do not really need a BPMN engine, and it would go against the whole simplicity of thinking presented in this book to try and squeeze one in just for the sake of demonstration.

A light alternative to BPMN orchestration

What is more reasonable is to show how a lighter alternative to BPMN 2.0-compatible, freeze-and-thawing-capable, full-featured engines can do this. And it happens that there is a very nice candidate for this position, known as **n8n** (`https://n8n.io`). I usually tend to use only free services in my book examples, but I am making an exception here due to the fact that the standalone execution of n8n in Community Edition will be free of charge for you if you run it in a Docker container. As far as the cloud service is concerned, fees apply after the trial period.

Other services, such as **Zapier** or **Microsoft Power Automate**, could have been used here, but I had to make a choice, and the simple, effective interface of n8n together with its versatility convinced me it was the right tool to demonstrate some light orchestration and choreography. Running n8n is as simple as entering the following command:

```
docker run -d --name n8n -p 5678:5678 docker.n8n.io/n8nio/n8n
```

Connecting to `http://localhost:5678/` then displays the welcome screen after a few seconds, once the application has warmed up:

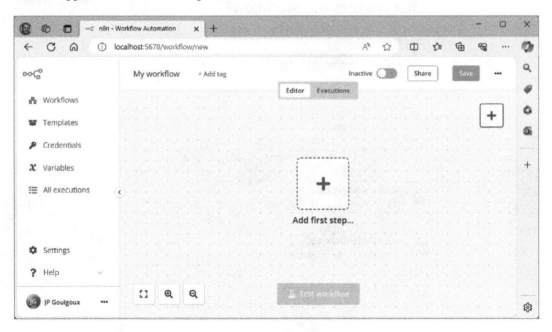

Figure 22.11 – n8n welcome screen

The interface of n8n is mainly organized around an editing panel where workflows can be created by adding chained tasks. Since these tasks can be about listening to events or executing operations, it is possible to use n8n to implement orchestration processes or choreographic ones indistinctly.

Choreographing the regular import of the book's catalog

The first workflow that will be demonstrated in n8n in relation to `DemoEditor` will be in the second category, which is about choreography services. As we will be sticking to simple workflows and prefer to show something related to our sample company, we are going to automate the import of the book's catalog.

If you remember from *Chapter 15*, we talked about the phase during the data migration to the data referential service when it would be useful for users to start using the new data referential service for read operations, but keep writing modifications in the old Excel-based books catalog. While importing the books every night, logs of errors would be presented by the data referential service, which would help users progressively correct the legacy data until everything can finally enter the new data referential service, passing the (higher) gates of data cleanliness. Instead of having the director run the Postman command every night, it would be much better to automate this, and this is what we are going to implement in this section using n8n.

Clicking on **Add first step...** on the welcome screen will bring up the following dialog:

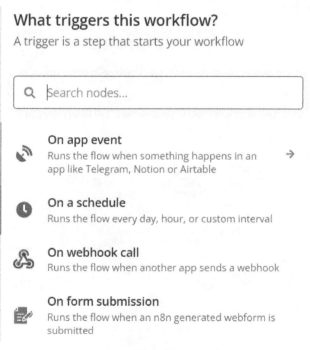

Figure 22.12 – Possible triggers

Choose the **On a schedule** trigger and accept the default parameters, which will trigger the workflow every day at **Midnight**:

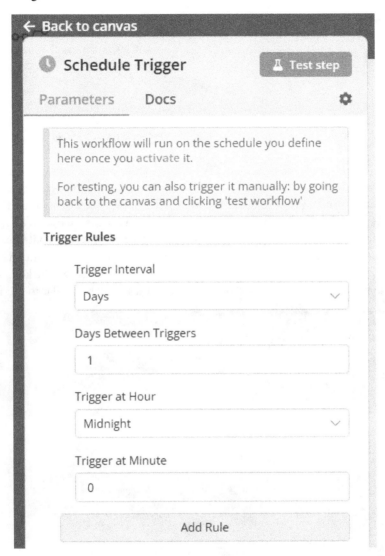

Figure 22.13 – Default timer trigger

Click on **Back to canvas** and add a second node to the workflow by clicking on the + sign that is displayed on the right of the starting node we just added, and select a **Files** node this time:

Figure 22.14 – Starting node and link to extend the workflow

Select **Read/Write Files from Disk**, then **Read File(s) From Disk**, and indicate a filename, such as /tmp/books.xlsx. If you remember in *Chapter 18*, we decided to import the existing books catalog on Excel during the migration phase to the new API-based data referential service, and this file was available somewhere on the network of DemoEditor. We could use the -v option on the docker run command, or any other way to make this file available from the n8n container, but it really does not matter here. Since we are practicing, you can even simply send the file to the container with docker cp DemoEditor-BooksCatalog.xlsx n8n:/tmp/books.xlsx. Click on **Test step** to check that the file-reading operation works correctly; the result should be similar to this screenshot:

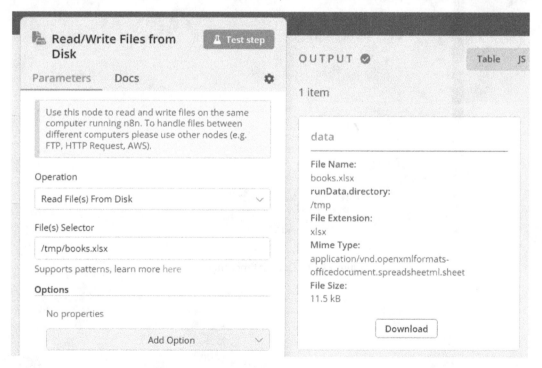

Figure 22.15 – Reading a file copied locally in n8n

Back to the canvas, the next operation is to add an **Action in an app** node, with the **HTTP Request** subtype. We will start the setup by indicating the URL for the Import operation and the verb (which is **PUT**, in this case), then choose a way to authenticate to the API:

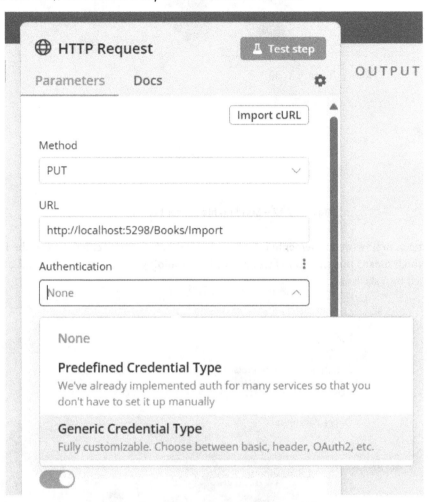

Figure 22.16 – Customizing an HTTP request

If the API expects to receive binary content, we would set up the rest of the attributes like this, since `data` is the name that has been proposed for the output of the previous node:

Figure 22.17 – Sending file content to HTTP

But if we go back to how we implemented the `Import` operation, we decided to read the file from its address, which means this should in fact be adjusted by adding a query parameter and using two fields separated by a slash as its value:

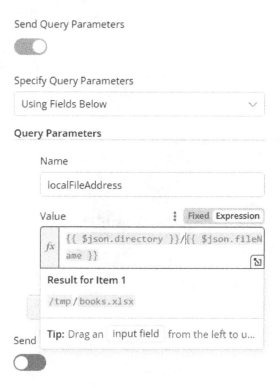

Figure 22.18 – Sending filename

The interface helps us to retrieve these parameter values by letting them be dragged and dropped from the **INPUT** section on the left-hand side:

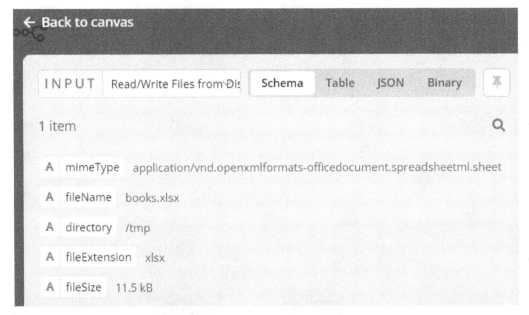

Figure 22.19 – Inputs from the previous node

The selection between reading a file content or a filename requires a business-aligned reflection. If we know from the start that there will be different versions of the file, with different conventional names, and that taking one or the other will depend on some parameters, such as the days the trigger happens and the user is authenticated, then we will have a strong reason to change the filenames. If not, then a simple upload of the fixed-address file content will be more suitable. This may sound like an unnecessary detail, but this is the piling of all such technical details that, in the end, may paralyze an information system or at least make it hard to adjust to users' requests.

When executing this step, we receive a positive response from n8n:

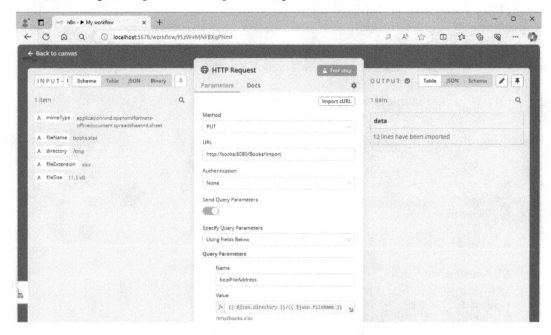

Figure 22.20 – Calling an import function from n8n

Notice that I changed the URL from the original one indicated previously. Indeed, since n8n and the books API are working in Docker containers, they have to be run on the same network to be able to see each other. This is why I included a new section called bpm in the docker-compose.yml file, with a reference to the needed image and the exposed port, as shown here, and ran it again:

```
bpm:
    image: docker.n8n.io/n8nio/n8n
    ports:
      - 5678:5678
    networks:
      - back
```

You will also see that I have removed authorization altogether to carry out this test. The reason is that within a Docker Compose context, there is no strict equivalence between localhost and the local machine that you are using. Since the OAuth 2 authorization process brings up a pop-up browser window to log in, IAM has to be exposed on the local machine. But since it is also used in the n8n application, it should be designated under the iam host alias provided by Docker Compose. The best way to get rid of all these complications is to deploy everything under a demoeditor.org domain name, but this (and the associated certificate creation and network parameters) would take us too far from our subject.

One last bit of work on the configuration of this example: for simplicity reasons, I have injected the Excel document inside the containers with `docker cp` but, as explained, we would normally do this with volumes to be clean. Simple links to a local path are not recommended because they do not scale when passing to a Docker cluster such as Swarm or Kubernetes. The best way is to use named volumes. For example, we could modify the `docker-compose.yml` file so that it creates a local volume that we could call `demoeditor`, and connect this volume to `n8n` (to analyze the content of the file) and to `books` (since this is the service that actually loads the workbook for import):

```
bpm:
  image: docker.n8n.io/n8nio/n8n
  ports:
    - 5678:5678
  networks:
    - back
  volumes:
    - demoeditor:/tmp/data
volumes:
  demoeditor:
    driver: local
```

After the next `docker compose up` command, the volume will be created, as we can check with `docker volume ls`, which returns something like the following output:

```
DRIVER     VOLUME NAME
local      cb04003616c6e2161d2550f349f6cd5872e18d9e5d0
ca8dc4af3ae4436b08b4e
local      eb2eacf217a730a64163a6c16d2a6a891db330ebec5
149cedc416fc98c60c366
local      net-architecture-for-enterprise-applications
_demoeditor
```

Since it is not possible to directly copy files into a volume, one way to do it is to use a container that is plugged into both a local directory through an anonymous volume and into the named volume we have created (in this case by connecting it to `/tmp/data`, as other containers will do) and run a Linux copy command on it:

```
docker run --rm -v ./DemoEditor-BooksCatalog.xlsx:/tmp/books.xlsx -v
net-architecture-for-enterprise-applications_demoeditor:/tmp/data
ubuntu cp /tmp/books.xlsx /tmp/data
```

The Ubuntu container is run with the `--rm` command because it is to be deleted as soon as the command has finished and the container has stopped. The bpm service supervising the n8n-based Docker container, on the other hand, keeps on working as long as the Docker Compose application is active, which means that the import of books data will now happen automatically every night.

Orchestrating the creation of a new project

Now that our first choreographic-style workflow is active, we are going to create a second one. To do so, you simply need to come back to the list of workflows using the first icon on the left menu and let the interface guide you (it is generally a good idea to give precise names to existing workflows):

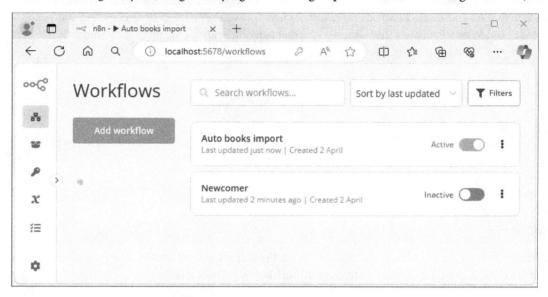

Figure 22.21 – Workflows interface in n8n

This second workflow will be an orchestrated one, which means that an initial operation by a human will start a process that will then execute several tasks, controlling how and when they start and are chained to each other. This is a major difference as this is generally associated with the fact that each instance of a process corresponds to an individual business case (hence the importance for the engine to be able to persist in these occurrences). The new workflow will start when somebody enters information in a form. Forms are handled internally by n8n, although it is, of course, possible to connect to external forms to gather data. In our case, we will use an internal form trigger as the starting point of our second workflow:

On form submission
Runs the flow when an n8n generated webform is submitted

Figure 22.22 – Form submission trigger in n8n

The form editor is so easy to use that it would be useless to describe how to create such a simple form. Enter the required details in this form and click **Submit form**:

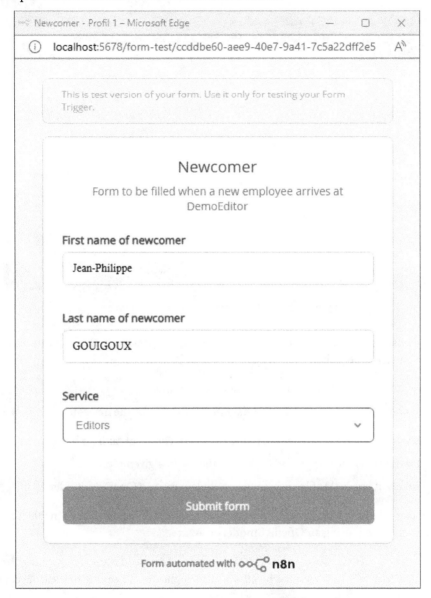

Figure 22.23 – Form to enter a newcomer information

Once submitted, the form sends data to the task in n8n, and the output panel shows this in **JSON** format:

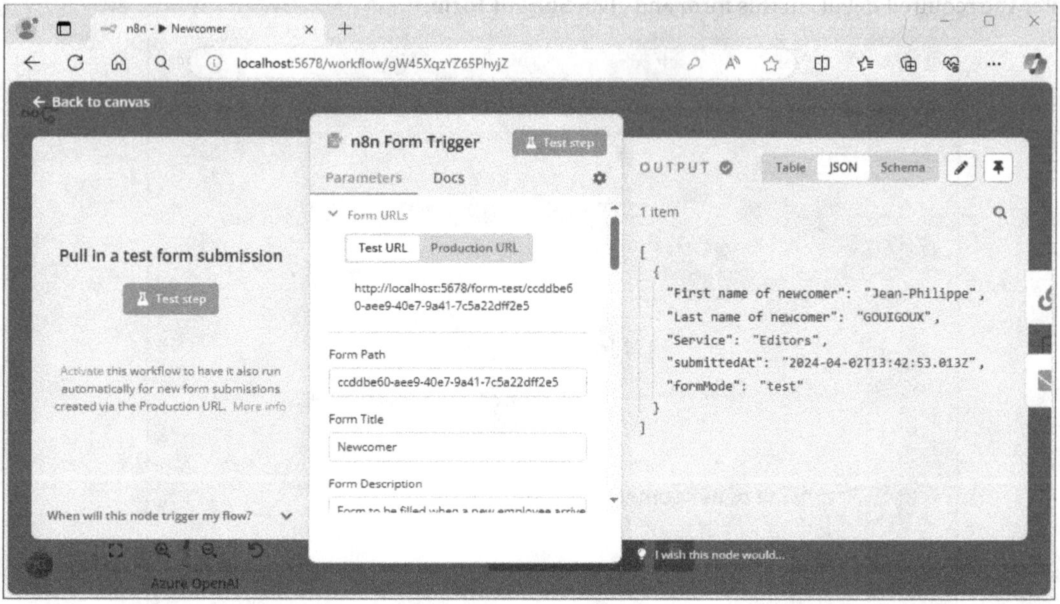

Figure 22.24 – JSON output of the form in n8n

The rest of the business processes should not be too difficult to design in the n8n interface. I have depicted the sequence of steps here:

1. Once the form is completed, two sets of tasks are run in parallel.

2. The first one is about collecting additional information about the newcomer.

    I.   It starts with sending an email to the person in charge of taking photographs.

    II. Then it waits for this person to deposit the file in a given place.

    III. Finally, it sends this file with an updated filename to a OneDrive folder.

3. The second set of tasks starts by summarizing the resume of the newcomer in 100 words for a quick presentation, using an OpenAI model on Azure.

    I.   It then sends a message to an **AMQP** (short for **Advanced Message Queuing Protocol**) topic that will be consumed by all software applications needing to create accounts for this user.

4. The two "lanes" finally merge (for now, they were operating in parallel).

5. As a consequence, the final task of sending general information to a Teams channel is postponed until both sets of tasks are completely finished.

This roughly translates into the following n8n workflow, where you can see red frames around tasks that are still missing some parameters to work (generally because authentication to the service has not yet been done):

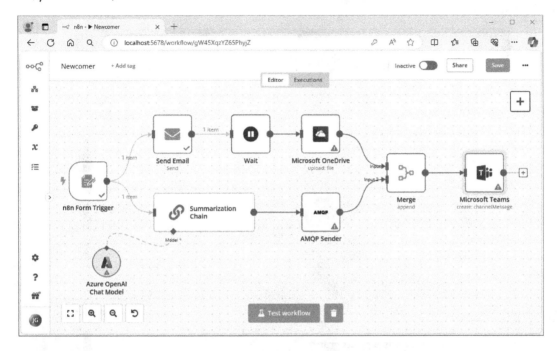

Figure 22.25 – Orchestration workflow in n8n

Though there is no need to go through all the customizable actions, I will discuss a few points that deserve attention. For example, the **Send Email** task needs, of course, an SMTP server to work. When doing some exercises like this, not only is it difficult to address a production-ready server, but in addition, we do not necessarily want the emails to actually be sent to real addresses, and modifying them for test addresses is not always easy to do and can be difficult to reverse. The solution is a great piece of software called **MailHog** that will serve as a local SMTP server. To add such a service to our application, we just need to add the following section and run `docker compose up -d` again:

```
mail:
    image: mailhog/mailhog
    restart: always
    ports:
        - 1025:1025
        - 8025:8025
    networks:
        - back
```

Once this is ready, the **Send Email** task can be customized with credentials pointing to the MailHog port. We also set from and to addresses, as well as a title and body for the email to be sent (notice that the content can use parameters coming from previous tasks, in particular, in our case, the identity of the newcomer):

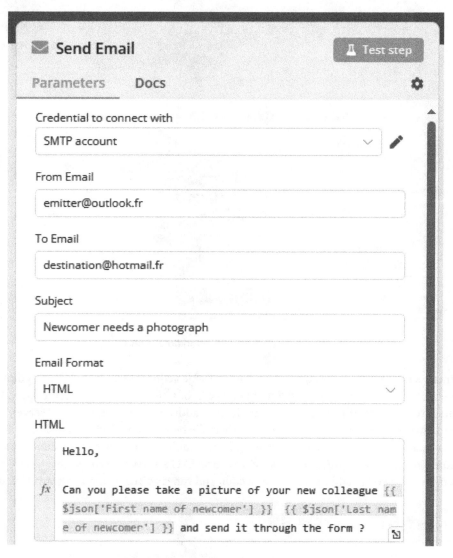

Figure 22.26 – Send Email customization dialog

The credentials dialog box is simply filled with this information:

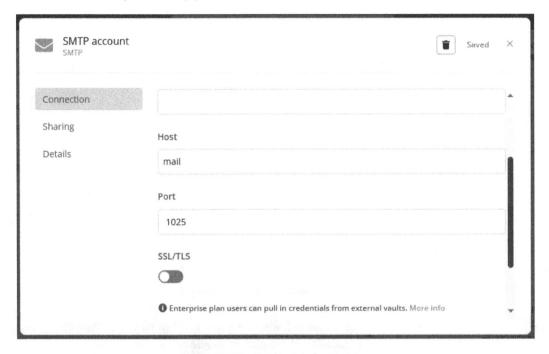

Figure 22.27 – Credentials for MailHog

As you can see, no username or password is used since MailHog is not supposed to be used for anything other than tests. The `mail` host should be used since this is the name of the service we have declared in the `docker-compose.yml` file, and `1025` is the default SMTP port number for MailHog that we have used. Once this step is tested, you can take a look at the `8025` HTTP port of MailHog to check out this result:

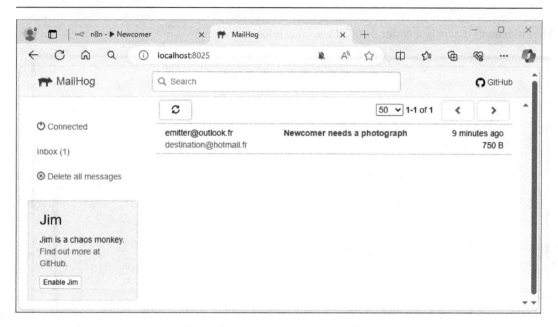

Figure 22.28 – Mail arrived on MailHog

A click on the email shows the expected content, where the first and last names of the newcomer created in the form have been used to indicate whose picture should be taken:

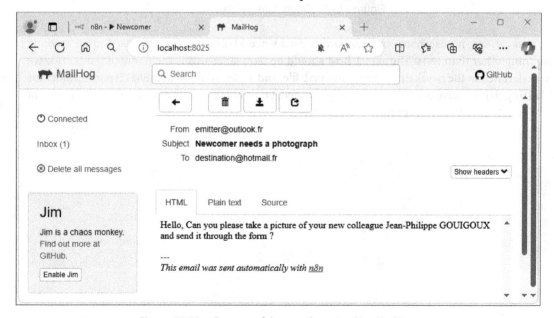

Figure 22.29 – Content of the email received by MailHog

One last thing that is interesting is that orchestrations do not necessarily prevent interactions. Having a pilot that decides when all tasks start does not mean it can also control how much time they take and when they will be able to start the following ones (even if the orchestration engine ultimately keeps control with timeouts). This happens all the time with tasks associated with human interaction. An example has been created to illustrate this in the workflow, with the **Wait** task after the email has been sent to the photograph asking for a picture of the newcomer. Since the person should manually take the picture, extract the file, and provide it to the workflow, the logical way to configure this waiting task was with another form that enables the user to provide the path to the picture file:

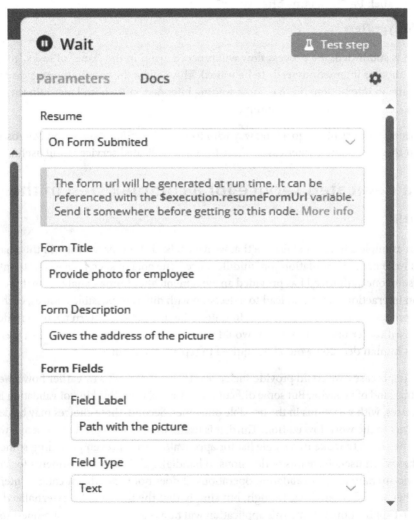

Figure 22.30 – Wait task in n8n

As you can see, here is the input that needs to be added to these fields:

- **Resume: On Form Submitted**
- **Form Title**: `Provide photo for employee`
- **Form Description**: `Gives the address of the picture`
- **Form Fields**:
  - **Field Label**: `Path with the picture`
  - **Field Type: Text**

Once the form is submitted, the process flow will proceed again in the "lane" of tasks, going to the next one until another interaction needs to be waited. This will be the case in the `Merge` task: even if there is no human interaction, there is some waiting intended, since both lanes join there and the last one will decide on when the flow continues.

As you have seen, it is possible to go a long way with tools such as n8n, Zapier, and Microsoft Power Automate. But there are some corner cases where having a dedicated service is still useful.

Adding a dedicated service for a complex step in the processes

Let's take as an example a functional feature that we have talked about previously (in this chapter and *Chapter 13* as well) named "validation" or "middle office." The goal of such a service is to interrupt a process until someone authorized has provided an agreement, after human analysis, that the flow can go on. Such an interaction could also lead to a decision with different possible choices, each of them switching the process to a given lane of tasks. In Enterprise Integration Pattern terms, a middle office is a content-based router brick. However, it would not be possible to provide business rules to route the message. A human decision would be required to replace those rules.

Of course, in simple cases, we could provide such a feature with n8n. We saw earlier how a `Wait` task could handle this kind of behavior. But some difficulties may arise. First, this kind of validation is needed in lots of processes, with variations in the possible outcomes. Second, these choices may be defined by someone external to the workflow edition. Third, it is useful to be able to present someone with all of the decisions they have to make in a single list for approval/rejection (even providing some ways to do so in batches when users have lots of decisions to handle). All these are arguments for creating a dedicated service to handle such validation operations. It does not mean that it cannot interact with n8n or any other workflow engines, though, but simply that the flow will be externalized at some point in a special application. Calling this application will be as easy as sending it a request to its API, and coming back to the workflow engine could simply be implemented by listening to a callback that would have been provided in the middle office webhook configuration. If this is not clear yet, do not worry: we are going to explain this in much more detail in the following section.

Business-aligned definition of a middle office

At first, as usual, we need to focus on the business-aligned features of a middle office. Setting aside everything that is technical, how can we describe what goes on in such a service? The metaphor of a computerless office could help us again here. Let's see how a middle office would work if there were no software:

- The legal department at `DemoEditor` is in charge of validating all kinds of contracts. These can be authoring contracts, signed between the editor and the author, or sales contracts, where salespeople deal with bookshops and distributors.

- On the first kind of contract, legal advisors are only requested to accept or reject the proposition as it is presented to them, with a message in case of rejection. The editor of the contract would, in this case, work on a new version and submit it again afterward. Otherwise, the next step is the validation by the director.

- As far as the second kind of contract is concerned, the opinion of the legal advisors is not binary, but should rather evaluate a level of financial risk between 1 (no foreseeable problem) and 5 (many high risks, which equates to a strong recommendation of not signing the deal). In all cases, the answer is routed to the sales department.

- Several requests can be made at the same time for each of the possible types, waiting for someone from the legal department to make a decision on them.

- Generally speaking, legal advisors specialize in either authoring law or commercial law, but some have both specialties and are allowed to deal with the two types of contracts we talked about.

- When requests come in, a physical folder with everything necessary (copy of the contract draft, contacts, and so on) is placed in a dedicated tray. People sending requests for review know exactly where to put them and are perfectly aware that sending a request in the wrong tray would mean much longer delays in getting an answer.

- In the case of authoring contracts, there is a convention that if the legal has not explicitly rejected it within a week, the contract is assumed to be valid. Legal advisors actually rely on this convention and only send a decision to the editor when a contract is rejected, together with some text explaining why in a note attached to the folder. All documents are returned to the editor, the name of whom is indicated on the first page, to track the case.

- In the case of a sales contract, the legal advisors must send a risk assessment evaluation in all cases. If they do not within months, their manager or the requesting salesperson notifies them when this is becoming urgent.

This first round of analysis clearly shows that there are three main business notions that interact to provide a middle office service:

- Entities we could call **requests** have a limited lifetime: they are created when a user asks for validation on a given business case, are addressed to people who have the ability to emit an opinion on them, and are finally sent back with an answer. Once this is done, they cannot be altered and there is not even a need to read them again. They can be considered archived or even deleted if the regulations allow so.

- Entities named **decisions** are part of the requests, which can contain zero or one of them and no more. Decisions thus follow the life cycle of the request they are associated with. Once a request is archived, the decision is too.

- Another entity has a longer life cycle and is the **template**: it defines the different types of requests and, for each, which possible decisions are possible. Templates also describe what data can/ must be carried by requests and decisions.

All other business entities necessary to implement the functional scenario mentioned have already been defined:

- Users are defined as belonging to groups, and the legal department is one of the groups

- Subgroups for authoring law and commercial law are in place, and some advisors may belong to both groups

- Authorization rules are in place to associate specific templates to specific groups or users, detailing the possible actions (create, read, and delete) on requests and decisions

Defining the middle office service API

Now equipped with this business-aligned analysis, we can start translating this into more technical terms, not talking about implementation yet but about the OpenAPI contract. The following is the graphical representation of a possible model of the business entities we talked about:

Figure 22.31 – OpenAPI contract for a middle office API

Notice that, once created, **templates** and **requests** cannot be modified, but only read or removed. This is to avoid a request for changes while being examined, which would possibly cause a wrong decision. As for **decisions**, they can only be created on a request and neither modified nor deleted once made. In fact, they cannot even be read because the action following the decision is not supposed to be taken by another service other than the middle office itself. Otherwise, the decisions could lead to behaviors that are not defined in the template.

Let's see the content of the template schema since we are talking about it:

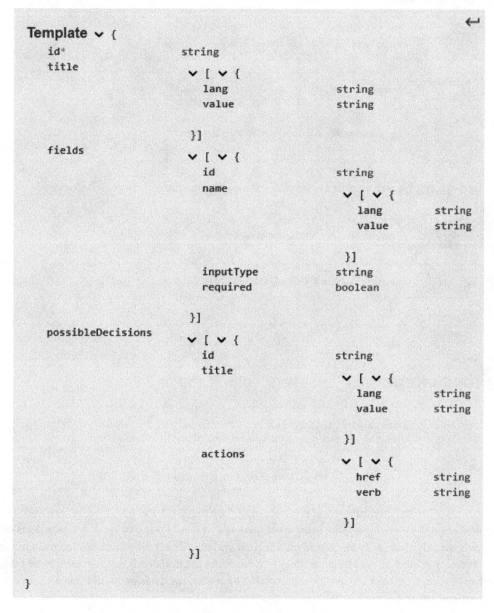

Figure 22.32 – Schema for a template

To give a better understanding of the preceding template schema, it is easier to look at the definitions of the two templates we have talked about for `DemoEditor`, in the format that validates the schema. The first one would be represented as this:

```
{
    "id": "urn:org:demoeditor:mo:authoring",
    "title": [
        { "lang": "en", "value": "Validation of authoring contracts" }
        { «lang»: «fr», «value»: «Validation des contrats d'auteurs» }
    ],
    "fields": [
        {
            "id": "urn:org:demoeditor:mo:authoring:contract",
            "name": [
                { "lang": "en", "value": "Reference code for the
contract to be reviewed" }
                { «lang»: «en», «value»: «Référence du contrat à
valider» }
            ],
            "inputType": "string",
            "required": true
        },
        {
            "id": "urn:org:demoeditor:mo:authoring:country",
            "name": [
                { "lang": "en", "value": "Code for country where
contracting laws apply" }
                { «lang»: «fr-FR», «value»: «Code pays de rattachement
légal du contrat» }
            ],
            "inputType": "string",
            "required": false
        },
        {
            "id": "urn:org:demoeditor:mo:authoring:author",
            "name": [
                { "lang": "en", "value": "Reference code for the
author" }
                { «lang»: «fr-FR», «value»: «Référence de l'auteur» }
            ],
            "inputType": "string",
            "required": true
        }
    ],
```

```
    "possiblesDecisions": [
        {
            "id": "urn:org:demoeditor:mo:authoring:decisions:reject",
            "title": [
                { "lang": "en", "value": "Contract rejected" }
                { "lang": "fr-FR", "value": "Contrat rejeté" }
            ],
            "actions": [
                { "href": "http://localhost:5678/form-waiting/9?contra
ctRef=${urn:org:demoeditor:mo:authoring:contract}&decision=rejected",
"verb": "POST" }
            ]
        },
        {
            "id":
"urn:org:demoeditor:mo:authoring:decisions:validate",
            "title": [
                { "lang": "en", "value": "Contract validated" }
                { "lang": "fr-FR", "value": "Contrat validé" }
            ],
            "actions": [
                { "href": "http://localhost:5678/form-waiting/9?contra
ctRef=${urn:org:demoeditor:mo:authoring:contract}&decision=validated",
"verb": "POST" }
            ]
        }
    ]
}
```

A template defines the input values that must be passed when sending a request of this type. In this case of an authoring contract request for review, it is not only the contract and author identifiers that must be sent but also the country under which the law should be verified. The template also expresses what the possible decisions are and what actions will be taken when a decision is sent. As explained before, it is essential that the service itself takes care of calling the receiver of the decision in order to reduce tampering. In addition, creating a template would, of course, be restricted to high-privilege users.

For simplicity reasons, the second contract definition will be shown without internationalization strings (the section with the lang attributes and the associated translations):

```
{
    "id": "urn:org:demoeditor:mo:sales",
    "title": "Validation of sales contracts",
    "fields": [
        {
            "id": "urn:org:demoeditor:mo:sales:contract",
```

```
            "name": "Reference code for the contract to be reviewed",
            "inputType": "string",
            "required": true
        },
        {

            "id": "urn:org:demoeditor:mo:sales:salesperson",
            "name": "Reference code for the salesperson",
            "inputType": "string",
            "required": true

        }
    ],
    "possiblesDecisions": [
        {
            "id": "urn:org:demoeditor:mo:sales:decisions:1",
            "title": "No foreseeable risk",
            "actions": [
                { "href": "http://localhost:5678/form-waiting/12?contr
actRef=${urn:org:demoeditor:mo:sales:contract}&decision=1&salesPersonT
oNotify=${urn:org:demoeditor:mo:sales:salesperson}", "verb": "POST" }
            ]
        },
```

```
Same decisions definitions with:
2: "Limited risks, without impact on image"
3: "Average risks, mostly financial but could impact image"
4: "Potential risks to the image of the company, on top of financial
impacts"
        {
            "id": "urn:org:demoeditor:mo:sales:decisions:5",
            "title": "Critical financial and image impact risk,
recommend to not proceed",
            "actions": [
                { "href": "http://localhost:5678/form-waiting/12?contr
actRef=${urn:org:demoeditor:mo:sales:contract}&decision=5&salesPersonT
oNotify=${urn:org:demoeditor:mo:sales:salesperson}", "verb": "POST" }
            ]
        }
    ]
}
```

In order to ease schema manipulation, the Decision object is defined separately from both Template and Request definitions, which use it:

```
Decision:
  type: object
```

```
        properties:
          id:
            type: string
          title:
            type: array
            items:
              type: object
              properties:
                lang:
                  type: string
                value:
                  type: string
          actions:
            type: array
            items:
              type: object
              properties:
                href:
                  type: string
                verb:
                  type: string
```

This way, `Template` only refers to it in its fields:

```
        possibleDecisions:
          type: array
          items:
            $ref: '#/components/schemas/Decision'
```

And `Request` adds it to its `decision` field, with the other attributes represented here:

```
    Request:
      required:
        - template
        - id
        - title
      type: object
      properties:
        template:
          type: string
        id:
          type: string
        title:
          type: string
```

```
    inputs:
      type: array
      items:
        type: object
        properties:
          id:
            type: string
          value:
            type: string
    decision:
      $ref: '#/components/schemas/Decision'
```

This way, a request for reviewing a contract can be sent to the middle office via a POST request to `http://demoeditor.org/api/mo/Requests/` with such a JSON body:

```
{
    "template": "urn:org:demoeditor:mo:authoring",
    "id": "73f71aee-5bf1-4856-9123-3f56ce2a59bb",
    "title": "Request for validation of 2nd edition of the book on
.NET performance by JP Gouigoux",
    "inputs": [
        {
            "id": "urn:org:demoeditor:mo:authoring:contract",
            "value": "http://demoeditor:org/contracts/
Contracts/2024-0046"
        },
        {
            "id": "urn:org:demoeditor:mo:authoring:author",
            "value": "http://demoeditor.org/authors/Authors/jpgou"
        },
        {
            "id": "urn:org:demoeditor:mo:authoring:country",
            "value": "FRA"
        }
    ]
}
```

You might wonder where the documents associated with this request are. After all, we talked about documents that needed to be analyzed by legal advisors and the request contains nothing but pointers. If you remember when we talked about the electronic document management system and the use of the **CMIS 1.1** standard, it was explained that dynamically querying documents from their metadata was ideal because it requires almost no coupling. This is why, if you really want to pass a URL with the documents to help the receiver (even if they should not have this convention of the information system), what you should really send is a CMIS query with something such as SELECT * FROM `demoeditor:CONTRACTS WHERE demoeditor:contractRef='2024-0046'`. This way, not only do you have something purely dynamic (while still being able to block on a given version of the contract, by pointing at a CMIS version, which would be a good idea from a functional point of view) but you can also use document-level authorizations to enable some particular advisors to review a special contract if this is your requirement. Combining middle office granular authorizations with document-level authorizations is a great example of defense in depth (also called multi-layer software security).

Once an advisor who has the rights to access the request (possibly putting a lock on it or affecting its property to oneself, to prevent concurrency) studies the metadata and associated documents, they make a decision and inform the middle office service of it by sending a POST operation on the `/api/mo/Requests/{id}/Decision` route, providing at least the ID of the decision, and possibly some additional information, such as a comment (this has not been modeled in the preceding schemas). The modification could stop at modifying the request as is:

```
{
    "template": "urn:org:demoeditor:mo:authoring",
    "id": "73f71aee-5bf1-4856-9123-3f56ce2a59bb",
    "title": "Request for validation of 2nd edition of the book on
.NET performance by JP Gouigoux",
    "inputs": [ NOT REPEATED HERE FOR SIMPLIFICATION ],
    "decision": {
        "id": "urn:org:demoeditor:mo:authoring:decisions:reject",
        "details": "The address of the author was missing"
    }
}
```

This would semantically be enough because the template identifier allows for a lookup of the correspondence between the decision code and the complete description of the decision, but it is generally recommended that the server reproduces the whole content of the related entity, for the sake of clarity:

```
{
    "template": "urn:org:demoeditor:mo:authoring",
    "id": "73f71aee-5bf1-4856-9123-3f56ce2a59bb",
    "title": "Request for validation of 2nd edition of the book on
.NET performance by JP Gouigoux",
    "inputs": [ NOT REPEATED HERE FOR SIMPLIFICATION ],
    "decision": {
        "id": "urn:org:demoeditor:mo:authoring:decisions:reject",
        "title": [
            { "lang": "en", "value": "Contract rejected" }
            { "lang": "fr-FR", "value": "Contrat rejeté" }
        ],
```

```
      "actions": [
            { "href": "http://localhost:5678/form-waiting/9?contract
Ref=${urn:org:demoeditor:mo:authoring:contract}&decision=rejected",
"verb": "POST" }
      ],
      "details": "The address of the author was missing"
   }
}
```

In some cases, it may be useful to modify a template after a request has been created. However, to ensure consistency and avoid confusion, we have made it impossible to modify templates once they have been created. It is still possible to delete and recreate templates, but this should only be done with caution, as it may affect any running instances of that process. If a functional administrator decides to make changes to a process while instances are still running, they should be aware of the consequences.

Implementing the middle office service

I considered not including this section because now that the whole definition of the API and the schemas of the entities have been designed, there is honestly not much to talk about. Apart from plugging into the IAM and persisting the entities in a NoSQL database, both operations that have already been shown for the data referential services, there is not much to be done to implement such a small API. Even the GUI would be quite simple, with the hardest bit being to be dynamically created, depending on the number of possible inputs and decisions (which is extremely easy to do with Blazor, by integrating a `foreach` loop in the Razor syntax).

Removing all outdated requests may sound a bit tricky since we need a signal, but we have seen with n8n how to create a workflow based on a time of the day or the week, and this feature could be handled in the exact same way as we did for the automated import of the legacy books list contained in an Excel worksheet.

Calling documents from the EDM would really be a matter of sending a CMIS HTTP request to retrieve the documents associated with the validation request and then another HTTP call to download the files to be displayed for analysis. If the EDM supports it, we could even call the preview feature to provide the user with a view of the document without having to download it and needing the associated application on their device. Again, there is not much work here as a simple URL composition would suffice. Could the stability of the document be a problem? Not really, since instead of letting the user call a fully dynamic EDM request, the validation request would simply be completed with a URL pointing to the version of the document, in a static way. We would just need to add a `contractURL` attribute to the request and change the implementation so that this supersedes the dynamic CMIS call.

How about managing concurrency? Sure, this is a difficult feature to put in place when you need to deal with transactions because your data is spread into several databases and consistency needs to be maintained. However, since the request and the associated decision are one and only one document in a NoSQL database, consistency is guaranteed. In a scenario where two legal advisors attempt to claim ownership of a request for validation simultaneously, only the first person to finish the (atomic) operation will be able to affect the request to their user. The second person will receive a cancellation response indicating that the request has already been claimed. Implementing this is only a matter of adding an `affectedTo` attribute to the request entity and checking in the same NoSQL operation that it is empty before updating its value.

Let's try and find some more challenges for our API: we could perhaps have some traceability requirements. Well, if we have used the same persistence as for the data referential service, this feature would be embedded in the middle office service, since we keep a value date for every change. The date of the request is then deduced from the date of creation of the request in the database, and the date of the decision is the value date when the request received its first (and only) `patch` operation on the `decision` attribute.

Though we may find some other challenges after we show, in the next section, how the middle office service will be integrated into the information system, it looks like the currently planned implementation has an answer to all the possible additional requirements, which is a good sign of a well-thought-out, business-aligned architecture.

Plugging the middle office into the process of book contracting

Since the middle office service we have designed is activated only through API calls and only sends API requests to external actors, its inclusion in our API-first information system should not be very difficult. Calls behind buttons of the different GUIs would be the simplest way to go, but it is also possible to include validation steps inside workflows if needed. For example, here is how this would be realized with n8n, with a call to the middle office service in the middle of a workflow, followed by a special action that would put the workflow instance execution on hold until the webhook is called back by the middle office service:

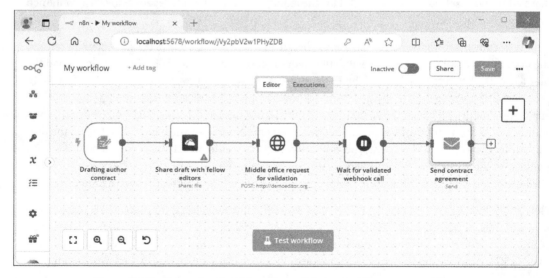

Figure 22.33 – Example of middle office validation inside an n8n workflow

A little bit of difficulty is involved in the dynamic aspect of the `Wait` callback, but this can be easily circumvented by adjusting the workflow like this, which completely separates the call of the middle office and the waiting for a decision to come out of it:

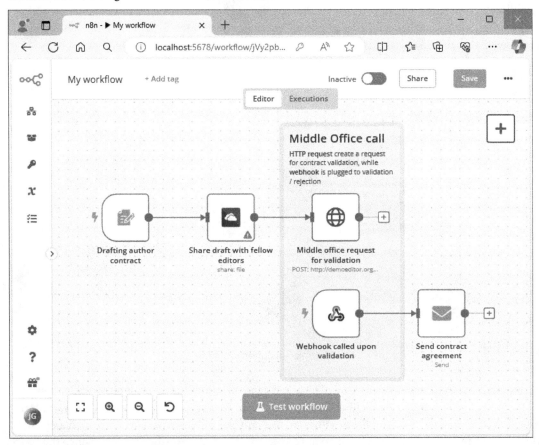

Figure 22.34 – Another way to represent the workflow

This second way of modeling the interaction may even be a bit closer to what we actually implemented in the API because there is no active correlation between the sending of the request and the receiving of the decision; only the identifier allows us to reconnect both, and it can even be skipped since the decision is inside the request, which contains the arguments such as the contract reference.

Since we were trying to find a solution to the architecture challenge previously, we could try one more time here, now that the integration of the middle office in the business processes has been explained. How would we handle the need for validation from two users in complex cases? We could, for example, consider that for an important commercial contract with a threshold of $100,000 turnover, we would need the risk assessment of the legal director or maybe the additional validation of the CEO. Or maybe, in some particular cases of a book published in several countries, several advisors would have to deal with the law in each country before both delivering their decision. Only two positive opinions would allow the process to go on. Sound complicated? In fact, it is not, since we have everything needed to handle parallel and successive tasks in n8n. Such a complex validation process could be realized with the following workflow, where you can see that the pattern with a message sent to validation and a trigger waiting for the corresponding response has simply been duplicated on two branches of the workflow that then join together:

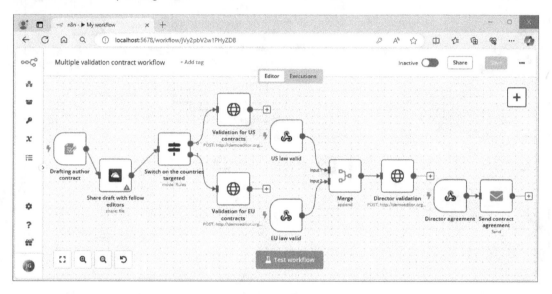

Figure 22.35 – Multiple validation workflow in n8n

What additional features could we think of? I would generally not think in these terms and rather let the functional people ask for something they actually need. But there are some cases where technical implementation people need to play the advising role and one of these cases is failure management. Most of the time, functional users will assume everything is going to work perfectly once the test period has validated the behavior of the system. How naive… Bugs always lurk, and even small network disconnections can derail a message from being delivered.

In *Chapter 15*, we showed a situation where this was efficiently solved by a retry policy coupled with an eventual consistency mechanism. In the present case, waiting for up to a full day for validation may become a business problem and the director of `DemoEditor` would probably ask us for a way to increase the robustness of delivery and ensure that, except if the information system is completely down because of electricity failure, the validation messages are always delivered as soon as possible, even if the network is temporarily down. Luckily, this is the reason why we added a RabbitMQ MOM to the list of external modules deployed in *Chapter 15*.

Instead of passing through a non-robust HTTP pipeline to deliver the decision, we will simply change the mechanism to go through a message queue. This first means that, instead of sending an HTTP callback, the template should be customized so as to send a message in a RabbitMQ queue:

```
{
    "id": "urn:org:demoeditor:mo:authoring",
    "title": [
        { "lang": "en", "value": "Validation of authoring contracts" }
        { «lang»: «fr», «value»: «Validation des contrats d'auteurs» }
    ],
    "fields": [ NOT REPEATED HERE FOR SIMPLIFICATION ],
    "possiblesDecisions": [
        {
            "id": "urn:org:demoeditor:mo:authoring:decisions:reject",
            "title": [ NOT REPEATED HERE FOR SIMPLIFICATION ],
            "actions": [
                { "href": "amqp://mom:15672/contract-rejection?contrac
tRef=${urn:org:demoeditor:mo:authoring:contract}" }
            ]
        },
        {
            "id":
"urn:org:demoeditor:mo:authoring:decisions:validate",
            "title": [ NOT REPEATED HERE FOR SIMPLIFICATION ],
            "actions": [
                { "href": "amqp://mom:15672/contract-validation?contra
ctRef=${urn:org:demoeditor:mo:authoring:contract}" }
            ]
        }
    ]
}
```

When meeting this kind of amqp:// protocol, the implementation of the middle office, instead of injecting an HTTP client and sending a request, would rather deposit a message in a queue, with some code like this:

```
using RabbitMQ.Client;

var factory = new ConnectionFactory() { HostName = this.MOMServer,
UserName = this.MOMUser, Password = this.MOMPassword };
using (var connection = factory.CreateConnection())
using (var channel = connection.CreateModel())
{
    channel.QueueDeclare(queue: "contract-rejection", durable: false,
exclusive: false, autoDelete: false, arguments: null);
```

```
    var body = Encoding.UTF8.GetBytes(JsonSerializer.
Serialize(requestWithDecision));
    channel.BasicPublish(exchange: "", routingKey: this.QueueLey,
basicProperties: null, body: body);
}
```

On the other side of the robust delivery pipe, an AMQP-capable receiver would retrieve the messages:

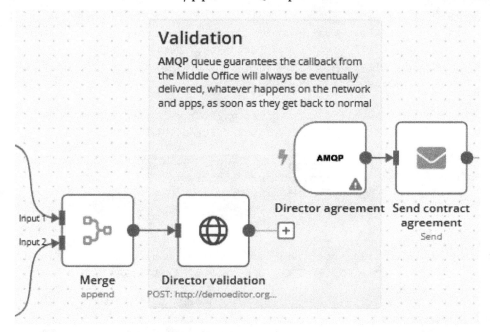

Figure 22.36 – Switching to queues to receive important validations

An **AMQP** queue guarantees the callback from the middle office will always be eventually delivered, whatever happens on the network and apps, as soon as they get back to normal.

The "store and forward" approach implemented by the MOM would guarantee that the message cannot be lost, and the fact that queues are actively observed would make for a delivery that is as quick as allowed by the state of apps and networks, without having to wait until the next day and the execution of an eventual consistency mechanism. And just in case you need to listen to a queue directly from a C# program, here is the kind of code you would have to use:

```
var factory = new ConnectionFactory() { HostName = this.MOMServer,
UserName = this.MOMUser, Password = this.MOMPassword };
using (var connection = factory.CreateConnection())
using (var channel = connection.CreateModel())
{
    channel.QueueDeclare(queue: Configuration["contract-rejection"],
durable: false, exclusive: false, autoDelete: false, arguments: null);
```

```
var consumer = new EventingBasicConsumer(channel);
consumer.Received += (model, ea) =>
{
    try
    {
        var body = ea.Body.ToArray();
        var message = Encoding.UTF8.GetString(body);

        Request? r = JsonSerializer.Deserialize<Request>(message);
        if (r is not null)
        {
            Console.WriteLine("Treating request {0}", r.Id);
            if (r.Decision is not null and r.Decision.Count() > 0)
            {
                if (r.Decision.Id ==
"urn:org:demoeditor:mo:authoring:decisions:reject")
                {
                    Console.WriteLine("{0}",
GetValueForCurrentCulture(r.Decision.Title));

                    // Continue rejection post-treatment
                }
            }
        }
```

If everything has gone as planned, you should have reached the end of this section without feeling the need to see the code. This is one of the most desirable aspects of a strong business alignment - being able to manipulate concepts and data schemas without having to worry too much about the actual code implementation. In fact, we could use any kind of language from here, and I trust the preceding explanation would allow almost every developer to create an API in their preferred platform and language.

We could go even further here by showing, for example, how a dynamic subscription to a webhook could be realized by a service at the same time it sends a request for validation, using the API approach that was detailed in *Chapter 16*. We could also show a few interesting Docker-capable reverse proxies such as Traefik or Caddy to organize the exposition of all the APIs on a single port of the demoeditor. org domain. We could plug in OpenTelemetry with https://github.com/open-telemetry/ opentelemetry-dotnet, show how to read logs in https://dozzle.dev/, or even how to use Vector (https://vector.dev/) to design observability pipelines on top of our existing system… but all these would merely be technical additions to a business-based strong basis, and the only important thing is the functional alignment. Once this foundational work is done, applying the right tool for the newly requested business requirements is a breeze, as this has hopefully been shown with the integration of n8n.

Summary

In this chapter, we have shown the very last part of creating the DemoEditor information system, at least in its first viable state. After dealing with data management, and then adding GUIs, this chapter was about orchestrating all the modules together and showing different methods to do so. Even if BPMN engines immediately come to mind, they are by far the most complex answer to the need to include business processes in the information system. GUI process integration, as well as simple webhooks, already go a long way, and, for the most complex cases, lightweight orchestration and choreography servers such as n8n cover much of the remaining path. If you had dedicated services embedding some particular tasks of a business process, as we did with the middle office service, there does not remain much for heavyweight, industrial BPMN engines such as Bonitasoft, Activiti, or other similar pieces of software.

This chapter has also given us the opportunity to validate that previous architecture choices correctly work with the final architecture of the information system. Nothing major had to be corrected and all necessary additional features were handled without major changes: authorization in the middle office service could be done with existing groups, EDM access was enough for the needs of the services, and so on. The only possible improvement that could have been put in place is an externally exposed clone of the EDM server, allowing us to handpick the documents that could be exposed to the outside of the information system, thus adding a second layer of security.

All in all, we can consider ourselves done with the initial construction of DemoEditor's information system, and we have essentially reached the version 1.0 milestone. In the rest of the book, we are going to show how changes can be applied to the information system. After all, this is the main goal we had right from the beginning: designing something that would stand the test of time and adapt as easily as possible to any business changes or decisions, of course, without degrading the performance and further capacity to evolve.

Just like we have roleplayed a scenario with a hypothetical DemoEditor director who gave directions on how the information system is supposed to work, we will, in the next chapter, show all the limits of the information system as it has been set up, simulate feedbacks from users, and adjust the information system in order to show that the initial structure makes it easy to evolve. Business/IT alignment will remain of the highest importance since the right understanding of functional concepts and how the information system implements them will be necessary to precisely target where the change has to be made in the system.

Part 4:
Validating
the Architecture

This final part of the book is about verifying that everything that has been thought of, designed, and implemented in the first three parts has indeed reached its objective, which is to obtain an information system in which there is very little technical debt, where business-/IT-alignment is good, and where changes of any kinds are easy to realize and do not make for complex technical modifications that harm the initial design. We will thus modify the data structure, the business rules, the security management, and even the business processes in this part, and hopefully show that all this can be handled very smoothly by the information system that has been put in place in its sample form.

This part contains only one chapter:

- *Chapter 23, Applying Modifications to the System*

23
Applying Modifications to the System

This is it! After a few iterations piloted by `DemoEditor`'s director, our information system has reached its version 1.0. The most important features have been tested and validated, so the company can start getting some functional value out of it. However, as change is the only constant, version 1.0 is not due to last long, and a few months later, there will be some new features to add to the system. This is when we are going to challenge our architecture to check that it is indeed capable of smooth evolution, without having to enter into complex technical projects to simply add a field or change a process. In short, this is what is going to judge the work we have done in every chapter that preceded this one.

This final chapter will show how different evolutions can be applied to the existing `DemoEditor` information system. It will start with changes that are simple, which does not mean they are small but, rather, that it is easy to know where precisely the information system has to be adjusted (and, generally, this also means that this is going to be in one place only). We will then work our way up until we reach business changes that will affect whole processes and, thus, need some adjustments, which we will carefully execute in different parts of the information system.

We will start by showcasing evolutions on the data part of the information system. The first section will explain how the modification of data content during the life cycle of the application should be handled, in particular the principle of forward compatibility and alternative ways to operate if such an approach is not possible. Examples for our sample business application will detail how the contract, the objects, and the implementation can – and should – be versioned independently from each other.

The following section will focus on adjusting IAM, which is a usual modification during the operation of an application. IAM is about authentication, so we will adjust this feature in the external client so that authors can authenticate with other methods than the ones that initially made it possible. IAM is also about authorization management, and the second example will focus on the way this can be adjusted in relation to the externalization that was introduced in the design of the information system of `DemoEditor`.

As the GUI is a way to implement some business processes, we will demonstrate an adjustment of the GUI. While the adjustment of the GUI can be easily explained, we will demonstrate its potentially significant impact on the GUI application itself. However, due to the careful planning of the entire system, this adjustment will not have any side effects on the data referential services.

Finally, we will deal with changes at the highest level, which are to the business processes themselves. Once the application is in production, we can use the BPM approach to modify how it works from a functional point of view, and we will show how this can be done using the orchestration tools that have been put in place, sometimes by simply adjusting the Webhooks' configuration. The dynamic nature of the system and the architectural decisions will face challenges when implementing more intricate adjustments within the processes.

In this chapter, we'll cover these topics:

- Evolving the data schemas used in the information system
- Dealing with forward and backward compatibility
- Adjusting the GUI to follow the business evolution of the system
- Changing business rules (in particular, those associated with IAM).
- Modifying the business processes and adjusting the technical implementations accordingly

However, before all these sections, we will start with an important warning on how to handle them. This chapter is indeed where a very good understanding of what has been done previously is important, and a map of the information system can be useful to remember the design choices and how they were implemented. But one thing will always beat documentation alone, and that is, again, business/ IT alignment – if that has been done correctly, anyone using the information system or studying the map of it will immediately spot where changes should take place.

The simple fact that names of concepts are shared between all actors of the system helps a lot. If commercials (business stakeholders) talk about "products" and editors talk about "projects," how could you know the changes that need to be done in such or such API in the directory of services without a good understanding of the workings of the system? However, when everything is called "books," even an intern or a newcomer will quickly conclude that there are good chances that modifications should be done in the "books" API. As for the rest of the information system, it follows the same lead – functional understanding has to be first to ensure that modifications and adjustments across the system are aligned with the overall operational objectives and requirements. This rule has been stated over and over, but it is even more important in this last chapter where we deal with the evolution of the information system. Realizing the benefits of this investment is truly valuable and justifies dedicating a section to cautioning about the importance of aligning the information system with business goals.

Technical requirements

You can find the code used in this chapter on GitHub: `https://github.com/PacktPublishing/Enterprise-Architecture-with-.NET`

A word of warning, and an anecdote

As explained in the introduction of this chapter, the following sections are going to show how we can apply modifications in the previously created information system and, more importantly, how the business alignment principles followed at design time help make these modifications easy. An extremely important warning at this point is that **the same functional approach that was followed to create the system must absolutely be used when evolving it**; otherwise, we will risk negating every care that was taken in the first place in a few bad moves. Again, I cannot stress enough this remark, since not following it may ruin all the efforts put into the design of the system beforehand and, thus, render the development process pointless.

Let me give you a real-life example that illustrates how important this piece of advice is (I actually believe this is the biggest risk warning in this book and the one that I hope you will remember as the most important). During an architecture project at a regional council in France, the company I was the CTO for implemented a kind of middle office service, a bit like the one that was shown in the previous chapter. This was part of a bigger project on the management of actors' data (*actors* are individuals or organizations that interact with the customer), where we had to clean the many duplications and uncontrolled point-to-point interactions that were making the company's system quite a mess when we started working on it. The following diagram below illustrates the initial state of the data streams between logical servers before the business/IT alignment project:

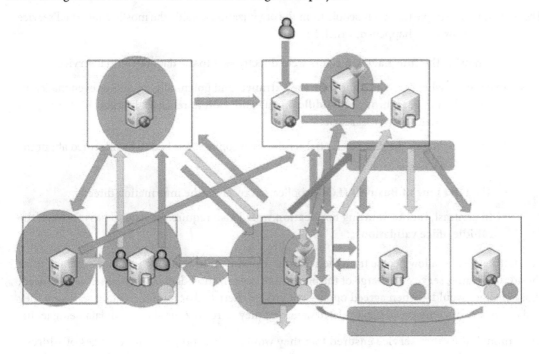

Figure 23.1 – The initial streams of data for actors

You might remember this diagram from *Chapter 3*, where it was used to illustrate misalignment between processes and technical implementations. It was the actual superposition of all the streams of data related to actors in the systems we were in charge of – a real spaghetti dish where adding new features was considered almost impossible. After many months of work, we finally turned this into something much more acceptable, where the main streams of data were completely aligned with the business processes:

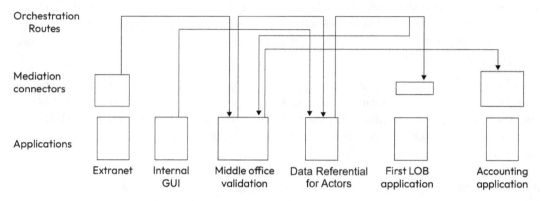

Figure 23.2 – A realigned sub-portion of the IT system for actors

The diagram is so simple that even people from the organization outside the mostly concerned service were able to follow what happened, which is:

- When using the **Internal GUI**, actors were directly sent to the data referential service

- In contrast, when the data came from the **Extranet** (and potentially careless or even malicious users), it had to pass through a **Middle office validation** to receive a human check and gain access to the data referential service

- Upon recording in the data referential service, two applications had to be informed about the actor's change:

  - The **First Line of Business (LOB)** application received the information directly

  - In contrast, the **Accounting application** would also require the data to pass through the **Middle office validation**

This last step was to allow people from the accounts department to validate the changes coming from the administrative service in charge of the people and organization data referential services. Indeed, those changes would – when agreed upon – enter their cherished accounting database, where they had taken care of verifying all bank-related data, and they were anxious about bad data being let in.

Even though the other service ensured that they would only send data with changes of address, or sometimes changes to the contacts of a company, the accounting service asked us to establish a validation page that they would check every morning, and they would validate that the changes were acceptable and reject some that were considered potentially harmful. The accounting team suggested a graphical interface that allowed them to review and approve or reject changes stored in a message queue. This customization was implemented within the middle office service for a new validation process, which resulted in a satisfactory outcome for all involved parties.

As time passed by, trust started to build between the two services, as the accounting people realized that the data was always clean, that there never was any attempt at modifying bank-related data, and that they progressively skipped through the changes without really analyzing them in depth as they would have when they started. The chief accountant then came to me and asked if it was possible to "auto-validate" the changes. His idea was to add a "check all" button at the top of the validation interface to streamline the process. This enhancement allowed users to quickly select all changes by using a single checkbox associated with each line. He even started to reflect upon how the pagination should be handled – if possible, it would be great to select all data, not only that which was visible, but he understood this would be a bit anti-ergonomic.

I let my counterpart continue for a while, and he even told me he had managed to procure a small budget to finance this evolution. This is where I stopped him and explained this new behavior was much simpler to implement than the approach that he had in mind, and I suggested that he rethink his need from a process-based perspective. It looked to me that what he wanted, essential, was just to skip the accounting validation phase:

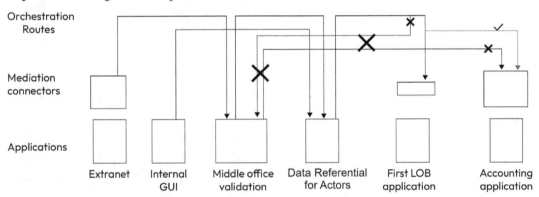

Figure 23.3 – A change in the actors' data orchestration

When presented with this schema, the chief accountant nodded and told me this is exactly what he meant. I repressed the urge to tell him that it would have been easier to simply tell me what he wanted to do functionally instead of trying to find a technical solution (which was my job, as the designer and implementer of the middle office and data referential service), and I went on to explain that it was possible to do so in just a few minutes.

He sounded surprised, as he was used to such changes in the system being complex and needing at least days, if not weeks, to be realized. Still, he offered me to try and do so on the pre-production system, as a test. I simply changed the callback address of the Webhook when a change was made to the data referential service to directly route the data to the accounting connector, rather than the middle office connector. Since both data streams followed the standardized format we had designed, the process change was seamless and immediate. We were able to make the switch without interrupting ongoing operations, thanks to using Apache Camel, based on Apache Karaf, which enabled us to make real-time changes to **Open Services Gateway Initiative (OSGI)** containers. OSGI is a Java-based framework that models the life cycle of an application and, in particular, defines how to update depending on modules while stopping a server application. Practically, the change involved changing the destination in an XML-based representation of Camel routes. The initial route definition was something like this (a simplified version):

```
<routes>
    <route>
        <from url="jms://ModifiedOrNewPerson"/>
        <to url="/api/middleoffice?validator=accounting">
    </route>
    <route>
        <from url="jms://AccountingValidated"/>
        <to url="/api/persons">
    </route>
    <route>
        <from url="jms://AccountingRejected"/>
        <log category="ChangesRejectedByAccounting">
    </route>
</routes>
```

It was simply changed to the following:

```
<routes>
    <route>
        <from url="jms://ModifiedOrNewPerson"/>
        <to url="/api/persons">
    </route>
</routes>
```

The change worked great the first time, particularly because I only had to change the orchestration routes. The mediation routes, which transformed the pivot format into a proprietary API or even database calls, were indeed already in place with mediation connectors. And it would, of course, have been a longer project if we had to modify some of them. In fact, in this particular case, the change was so simple and obvious once it was correctly analyzed that my customer eventually decided to push it into production right away, convinced of the total absence of any negative impact or side effects.

This experience was one of the most satisfying I had in my career and really showed a customer the value of all the standardization and business alignment we had done before. I often cite it at conferences or courses, but I also stress the point that it could have gone very badly if we had not discussed it.

Someone who didn't understand how the system was designed and didn't consider the big picture could have modified the middle office GUI to add the "check all" button that the customer initially asked for, and the results would have been really bad:

- First, the accounting team would have continued using the validation GUI every morning to perform a task that everyone would have considered more and more pointless and a waste of time

- The "check all" feature would have been hard to develop because of pagination – either it would have checked everything on the current page, requiring users to then process manually through all the pages, or it would have covered the whole data to validate without the user realizing how many lines they would need to validate, which would have been a risk

- Finally, and crucially, the "actors management" team (responsible for overseeing user roles and permissions) would still be under the impression that the accounting team did not trust them, even after all the data cleaning work they had done, thus creating silos and bad cooperation in the system.

All this means is that it is absolutely important that the people in charge of maintaining and evolving a system have the right level of knowledge. Otherwise, they may end up wasting all their efforts on clean design and prevent this initial investment from paying off. Ideally, of course, it is best to have the people who designed the system in charge, but this is often out of the question for financial reasons. Transferring competency is, thus, the real issue, and this has to be done in such a way that receivers of this knowledge understand precisely the why and not only the how. If the people maintaining the information system only know how it works, they will twist and modify it for future features in such a way that coupling will appear, and all the huge effort made to build clean architecture will not last for more than a few years at best. If they are conscious of the streams of data, the processes, and the exact role of each service, then not only will they keep the system clean, but they may even see how they can simplify it and make it even better, on the way to modifying its functions.

Modification at the data level

Adding some attributes to business entity models is one of the most obvious changes in an information system. After all, it is almost impossible to consider every type of data a business would need. It is often recommended to create concise yet accurate models, as this approach facilitates easy evolution in subsequent stages. The principle of business alignment underscores the importance of adaptability and seamless progression in system development. Let's check that the promise is kept by adding a few attributes to our models.

Adding a simple attribute

Let's take as an example a list of addresses that are placed in the `Contacts` petal of the author data referential service "flower." It happens that, for now, only the order in this list allows us to differentiate the addresses of the same author. This is enough to designate them uniquely, with URLs such as `http://demoeditor.org/authors/jpgou/addresses/0`, where the index of the subresource allows for a technical, but correct, manipulation. However, this makes it hard to understand the functional distinctions between the addresses. One of these distinctions is probably for the mail to be sent to the person, and there might be another one for the billing address. In the case of companies, there could also be addresses for delivery, a social headquarters' official address, and many other types. There certainly are business rules that state which of all these addresses should be used (at least by default) for specific use cases. For example, if we need to send a royalties check to an author, we might start by looking for a `billing` address and, if there isn't one, default to a `snailmail` contact address, and finally, a `living` address where the author is physically established.

The initial version of the schema for `addresses` was the following (without the `country` link, in order to simplify the display), in the OpenAPI contract:

```
ContactsPetal:
  type: object
  properties:
    addresses:
      type: array
      items:
        type: object
        properties:
          streetNumber:
            type: string
          streetName:
            type: string
          zipCode:
            type: string
          cityName:
            type: string
```

Adding a type of address is as easy as changing the contract to the following, compatible, version:

```
ContactsPetal:
  type: object
  properties:
    addresses:
      type: array
      items:
        type: object
        properties:
          addressType:
            type: string
          streetNumber:
            type: string
          streetName:
```

```
          type: string
      zipCode:
          type: string
      cityName:
          type: string
```

Since the order is not important in most serialization formats, adding the `addressType` attribute at the beginning of the schema does not create the same problem as it would have at the SQL database table definition level. In SQL databases, it is typically recommended to add new columns at the end of existing ones to prevent issues with clients that rely on the column order, especially in legacy systems still dependent on this order (and there are many legacy ones for SQL).

If you ever need to add some support for multiple addresses with different time periods, we could go on adding other attributes (such as `addressType`, `beginDate`, and `endDate`), as follows:

```
ContactsPetal:
  type: object
  properties:
    addresses:
      type: array
      items:
        type: object
        properties:
          addressType:
            type: string
          beginDate:
            type: string
            format: date
          endDate:
            type: string
            format: date
          streetNumber:
            type: string
          streetName:
            type: string
          zipCode:
            type: string
          cityName:
            type: string
```

This is generally when we realize that we are mixing two types of data that are not really the same thing – `streetNumber`, `zipCode`, and the like are really attributes of the address itself, but `addressType` and value dates for this address only have meaning when the address is associated with an author. In fact, they belong to the relationship between the address and the entity referring to it. It is too late to write something like this?

```
ContactsPetal:
  type: object
  properties:
    addresses:
      type: array
      items:
        type: object
        properties:
          addressType:
            type: string
          beginDate:
            type: string
            format: date
          endDate:
            type: string
            format: date
          address:
            $ref: '#/components/schemas/Address'
Address:
  type: object
  properties:
    streetNumber:
      type: string
    streetName:
      type: string
    zipCode:
      type: string
    cityName:
      type: string
```

It indeed is, as every client that has been designed to receive the following type of data will be broken:

```
{
    "contacts": {
        "addresses": [
            {
                "streetNumber": "12",
                "streetName": "Main Street",
                "zipCode": "PA 19004-3104",
                "cityName": "Philadelphia"
            }
```

```
            ]
        }
    }
}
```

Serialization frameworks, or any kind of code by the way, would not be able to adapt – without being rewritten – to the new version of data, which would be the following:

```
{
    "contacts": {
        "addresses": [
            {
                "addressType":
"urn:org:demoeditor:addressType:living",
                "beginDate": "1994-08-15T23:00:00",
                "endDate": "1995-06-15T09:00:00",
                "address": {
                    "streetNumber": "12",
                    "streetName": "Main Street",
                    "zipCode": "PA 19004-3104",
                    "cityName": "Philadelphia"
                }
            }
        ]
    }
}
```

This means that to stay compatible with the previous version of the Contacts petal, we have to serialize our data as follows:

```
{
    "contacts": {
        "addresses": [
            {
                "addressType":
"urn:org:demoeditor:addressType:living",
                "beginDate": "1994-08-15T23:00:00",
                "endDate": "1995-06-15T09:00:00",
                "streetNumber": "12",
                "streetName": "Main Street",
                "zipCode": "PA 19004-3104",
                "cityName": "Philadelphia"
            }
        ]
    }
}
```

It is less semantically correct, prevents reuse of the Address concept between entities that do not need the same attributes on the relationship, and will make it more difficult to prepare specific graphical components to handle addresses and lists of them. This simply shows how important it is to study carefully your business entities before signing a contract! Once you have made the mistake, if many people depend on your API, it is simply too late to correct it, and you will have to support it for a long time. This question of compatibility is, of course, a very important subject, and we will come back to it in a later section in order to detail what the remaining possibilities are. But for now, we will discuss another type of modification in data, which is when we adjust the possible values.

Extending the possible values of an existing attribute

Adding values makes sense in a data type that supports only a limited set of them. Of course, the data type itself never changes; otherwise, we would break compatibility. However, there are places for some improvement. For example, say your clients feel restrained about the use of ISO time values in dates and would like to be able to use more typical formats, such as the following:

```
{
    "contacts": {
        "addresses": [
            {
                "addressType":
"urn:org:demoeditor:addressType:living",
                "beginDate": "1994/08/15",
                "endDate": "1995/06/15",
                "streetNumber": "12",
                "streetName": "Main Street",
                "zipCode": "PA 19004-3104",
                "cityName": "Philadelphia"
            }
        ]
    }
}
```

In order not to break compatibility, the recommended way is to keep on supporting values such as 1994-08-15T23:00:00 and also support those such as 1994/08/15. In terms of schema definition, this accounts for a modification such as the following, using the oneOf keyword:

```
ContactsPetal:
    type: object
    properties:
        addresses:
            type: array
            items:
                type: object
```

```
properties:
  addressType:
    type: string
  beginDate:
    oneOf:
      - type: string
        format: date
      - type: string
        pattern: '^\d{4}\/\d{2}\/\d{2}$'
```

Generally speaking, it is a good idea to keep using simple `string` types because they are more open to changes. If you define something as `int`, for example, you need to be sure that it will always be this value. For example, an `age` value should be fine, except if someone wants at some point a decimal value, and you realize you should have named this age value `numberOfBirthdays`. However, that sounds a bit far-fetched, but I have a much better example – in France, we have numbers for "departments," which are a subdivision of the country below the "region" level. Historically, these numbers have been between 01 and 95. The leading zero was already a problem, but it remained possible to handle this with proper formatting. The first issue with the format occurred when the designation of overseas departments entered the list, with their 971 to 974 numbers. Luckily, they were still integers. But the real blow was when the department of Corsica was split into two numbers, and `20` became `2A` and `2B` departments, forcing every digital user to switch to strings.

Subsequently, restraining values through enumerations is something that you also need to pay particular attention to. For example, let's say it has been decided for books that it is only possible to bind and glue them, and that the cover can be soft or hard and nothing else, as the following schema extract shows:

```
components:
  schemas:
    Book:
      type: object
      properties:
        production:
          type: object
          properties:
            typeOfPaper:
              type: string
            assemblyMode:
              type: string
              description: mechanical assemblying mode used to bind
pages of the book in paper format
              enum:
                - bound
                - glued
            coverType:
```

```
              type: string
              enum:
                  - soft
                  - hard
```

It is important to realize that adding another value remains possible, but then the contract needs to be updated to a new version and redistributed. If developers have used this to reflect their serialization into enumerations themselves, then this means that a new version of the software will have to be deployed in the case of compiled, non-dynamic languages. So, again, be careful about using this possibility; it does not mean you should avoid it because it has its advantages. But you should only apply it when you are sure the list of values will only be modified in the long term or never.

A good way to balance between the two extremes of completely arbitrary strings and a predefined sets of values is to use links to subentities that can be user-defined, as we did for the countries for example:

```
public class Address
{
    //The rest of the class was omitted for simplicity

    [BsonElement("country")]
    public CountryLink? Country { get; set; }
}

public class CountryLink : Link
{
    [BsonElement("code")]
    public string ISOCode { get; set; }
}

public class Link
{
    [BsonElement("rel")]
    public string Rel { get; set; }

    [BsonElement("href")]
    public string Href { get; set; }

    [BsonElement("title")]
    public string Title { get; set; }
}
```

This way, the address can be serialized as this:

```json
{
    "streetNumber": "11",
    "streetName": "Vince Street",
    "cityName": "Paris",
    "zipCode": "75000",
    "country": {
        "isoCode": "FRA",
        "rel": "addressCountry",
        "href": "/api/authors/params/countryCodes/FRA",
        "title": "France"
    }
}
```

As far as the schema is concerned, `href` and `isoCode` simply are strings, without any restrictions. However, the implementation through a link to another entity allows the handling code to add some technical rules, stating that the link can only point to values that have been stored in the persistence. This way, the administrator of the application can add some countries when necessary. Of course, this should always be done in a compatible way (there is no way out of this), and country links that become irrelevant cannot be deleted. Instead, it must be deactivated in a manner that allows clients to recognize its inactive status. For example, the following added attributes could be used:

```json
{
    "country": {
        "isoCode": "YOU",
        "rel": "addressCountry",
        "href": "/api/authors/params/countryCodes/YOU",
        "title": "Yougoslavia",
        "active": false,
        "endDate": "2003-02-04T00:00:00",
        "reasonForEnd": "dissolution"
    }
}
```

Now we have seen what kind of schema changes can happen and how to deal with them, let's see how we can handle this at the persistence level.

Migrating a schema in a schemaless world

This section will certainly be the shortest of the book… and a true manifesto for NoSQL! Since we have not bothered to translate our structured data into tabular lines (and, in doing so, got rid of the complexity of an O/RM), there is no need to discuss commands such as ALTER TABLE ADD COLUMN... and the like.

Since we use NoSQL and MongoDB in the exercise, what do we have to do to accommodate new fields such as `addressType` and `beginDate`? Well, simply nothing – finding an author can be done with a code line such as the following:

```
var result = Database.GetCollection<Author>("authors-bestsofar").
Find(item => item.EntityId == entityId);
```

It stays exactly the same. As far as writing its best-so-far states is concerned (and the same is even more true for patches than for states), again, the code does not need any modification whatsoever and stays like this:

```
var collectionBestKnown = Database.GetCollection<Author>("authors-
bestsofar");
await collectionBestKnown.FindOneAndReplaceAsync<Author>(
    p => p.EntityId == entityId,
    author
);
```

This is why I enjoy NoSQL so much and recommend it strongly for any new LOB application. As time goes by, there is less and less justification for keeping persistence in SQL databases. It becomes evident that arguments in favor of doing so often stem from a fear of change or, more significantly, the burden of legacy systems.

Ascending compatibility and OpenAPI

I promised that we would return to compatibility, and this is the aim of this section. First off, we will briefly discuss forward and backward compatibility. Both are often said to be the same reality seen from one point of view (the old version) or another (the new version). As the article at `https://simplicable.com/en/backward-compatibility-vs-forward-compatibility` clearly explains, there is more to this, and – again – you should take the time to clearly understand the difference.

Backward compatibility is the concept of keeping everything that worked in the previous version working in the new one. In this case, we take the point of view of the new version, which means we (hopefully) know exactly what was in the old version. Then, ensuring compatibility is essentially a matter of following the rules that were covered previously, which can be summarized by "only add things" (values, attributes, etc.). If you want to ensure that the contract is not accidentally broken, a useful approach is to apply automated tests from the old version to the new one. Instead of modifying existing tests, simply add new tests for the new features. This method can significantly help in maintaining backward compatibility. Note that Dredd (`https://dredd.org/`) or other similar tools can help you in this, aided by its stronger API specialization compared to Postman or even Bruno (`https://www.usebruno.com/`).

Forward compatibility, conversely, is the concept of making the current version able to accommodate the changes of a new version. While it may initially seem like a direct reversal of the previous concept, considering the element of time reveals that in this case, the 'new version' refers to a **future** version. This means that we do not know what will be inside of it since it is not defined yet. All of a sudden, this makes it much harder because it doesn't need to adapt to changes but anticipate them. Luckily, some of those cases can be handled with simple methods. For example, the appearance of a new attribute in JSON or XML content is generally not destructive of the serialization code that was generated before it appeared, since most frameworks know how to avoid an error when additional, unknown data is passed to them, by simply ignoring it. A graphical interface may also be said to be forward-compatible if it generically displays unknown data. Of course, it will not be as ergonomic as the display of known data that has been carefully designed to occupy space and offer logical visual navigation, but at least it is compatible with some kind of display. And the simple fact of not crashing is already a kind of forward compatibility.

If we now return to our OpenAPI contracts, compatibility means versioning, and it is important to explain that there is not one versioning system but, in fact, three different versioning systems that work fluently together. If you want to correctly handle API versioning, you should master the following concepts:

- What we call the API is often the API contract, which defines how *any* implementation of such a contract should work. Versioning this contract means that any change to the operations, the content of the entities, or any other attribute of the contract (contact, license, etc.) should generate a change of version. If you use semantic versioning, a backward-compatible change would increase the minor number, while an incompatible change would increase the major version.

- As entities are assemblies of data attributes, they may change in their definition inside the contract. Of course, when they evolve, the contract that contains them is obliged to evolve in the version as well. But there is a need to separate both versions for the main reason that a contract may refer to several business entities, and they will then, of course, evolve at their own rhythms. The contract must then adjust its version as soon as one of the entities it contains evolves itself in the version. Also, the contract has to decide on whether to use a version of the entity definition or not. It may wait for a further version to accept many entity upgrades simultaneously if it feels it is better. In other words, it has its own life cycle, possibly independent from the entity's life cycle, even though it will generally be heavily guided by them.

- Implementations of the contract thus have a different versioning model, as you can very well evolve the software implementation (or any other aspect of a generic application that contains other things than simply the API) while still satisfying the same contract. This type of versioning is for the entire application that supports the API implementation.

An example will illustrate this better than words. Let's imagine that, in the April 2023 version of an application, we decided to expose an API for the first time, and that this application was not made of services but, rather, a monolithic approach that exposed both `authors` and `books` in the same 1.0 contract API. A few months later, the designer of `authors` realized that they needed an additional attribute and created a 1.1 version of the entity schema for `authors`. Since the API should reflect this immediately, the contract would be updated to 1.1 as well (but with the `books` schema still in 1.0). Everything is then backward compatible, and the implementation of this new service arrives in the June 2023 version of the implementing application.

After some time, some clients complain that, although version 1.1 of the `authors` schema is indeed compatible with version 1.0, they would like to be able to obtain version 1.0 nonetheless. The implementer of the application would, for example, publish a July 2023 version that would support version 1.1 *and* version 1.0 of the API contract. By default, clients calling `/api` would use the 1.1 updated version, but those wanting the older version would use `/api/1.0`.

The evolution can be represented as follows:

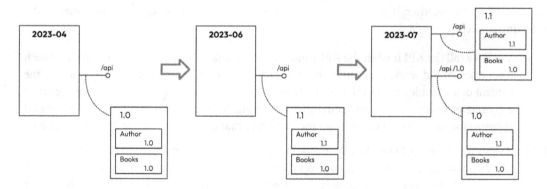

Figure 23.4 – The evolution of API versioning

Several versions follow, and, at some point, `authors` are between 1.0 and 1.3 and `books` are between 1.0 and 1.2. Since they always evolved separately, the API contract goes from 1.0 to 1.5, and it becomes difficult to know which version you should include in the `/api/[version]` URL when you want a given combination of authors and books contracts. In fact, this combination may even not exist at all. For example, there is no API version that serves `authors` in 1.0 and `books` in 1.2, and this becomes a problem for the clients that are used to having their required version.

Therefore, the API team gathers and decides that they will make it possible for a client to call an API with a header, `Accept,` that specifies the version requested. This way, when the API contract is 1.5 and supports, by default, authors in 1.3 and books in 1.2, it remains possible to call `/api/authors` by passing an `Accept=application/json+vnd/authors:1.1` header and receive some 1.1 version of an author.

This new approach can be schematized with this diagram:

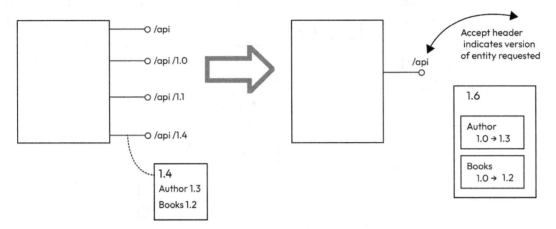

Figure 23.5 – A change in the support of entity versions in API

And while we are simplifying, how about doing what we have encouraged from the beginning of this book and providing a different application for each of the responsibilities? This is what we have done by creating two data referential services, one for the authors management and the second one for books management, since they are distinct business domains with separate life cycles for their main business entities:

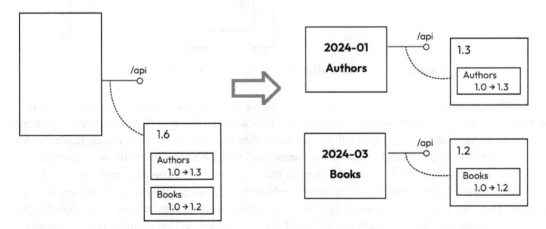

Figure 23.6 – Separating API contracts and implementation for independent versioning

Does this mean that you can reduce to two versioning systems only when aligning the contract with the main entity of the exposed domain? After all, since authors can go from **1.0** to **1.3** depending on the Accept header and the version of the contract is 1.3, they will evolve at the same time, right? Well, this would ignore all the other entities in the contract, which we have not represented for simplicity but are still present. If we were to represent some more content, the diagram would be more like this:

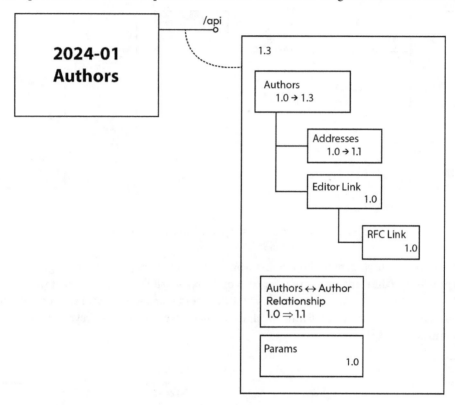

Figure 23.7 – The expanded content of the API contract for authors

And, although we could consider that each change of the sub-entities of authors, such as addresses or links to their editors, can become a new version of the author schema itself, thus moving together with the contract versioning, there still are some entities that are versioned independently, such as the author-to-author relationships or the parameters of the business domain (where we could indicate possible values for the countries used in the addresses, etc.). So, there really is a need for three versioning schemes – one for the software implementation, one for the contract, and one for the entities. The issue is that some implementations may support several contracts, and some contracts may assemble several versions of several entities. This is what makes API versioning a complex matter, and we have not even talked about major versioning and the declaration of future obsolescence in the contract, but this would lead us out of the scope of our main subject, so we will leave the subject of API compatibility here.

Modification in the GUI

Once the content of the data is adjusted, the GUI, of course, has to adjust to this change; otherwise, the additional attributes could only be handled through the API, and this is not a very ergonomic client. In most cases, updating the human interface means displaying additional data and, thus, adjusting screens. I will not go into ergonomic consideration, but not disturbing the user experience is definitely something to take into account, in particular when frequent inputs are needed. When someone has to input numerous business cases daily, even a minor change like the tabulation order in the form can lead to rejection. While this issue is beyond the scope of our current discussion, we will focus on examining how attribute additions and changes in ergonomics can be integrated into our sample GU.

Taking into account the addition of a new attribute

Apart from taking care of compatibility, there really is not much to say about how to deal with new attributes in the GUI. Let's take, as an example, the addition of the type of address in the schema, as explained in the first part of this chapter. We will start with the existing code to display the address (the SVG path content is omitted for clarity):

```
<p class="card-text">
@foreach (var address in @author.Contacts.Addresses)
{
    <svg xmlns="http://www.w3.org/2000/svg" width="16" height="16"
fill="currentColor" class="bi bi-building" viewBox="0 0 16 16">
    <path d="(...)"/>
    </svg>
    <span>@address.StreetNumber @address.StreetName @address.ZipCode @
address.CityName (@address.Country.Title)</span>
}
</p>
```

A naive approach would be to simply add the type of address to the code, like this:

```
<p class="card-text">
@foreach (var address in @author.Contacts.Addresses)
{
    <svg xmlns="http://www.w3.org/2000/svg" width="16" height="16"
fill="currentColor" class="bi bi-building" viewBox="0 0 16 16">
    <path d="(...)"/>
    </svg>
    <span>@address.AddressType address is:</span>
    <span>@address.StreetNumber @address.StreetName @address.ZipCode @
address.CityName (@address.Country.Title)</span>
}
</p>
```

However, it is recommended to make it possible to handle former versions of the address content, where the deserialization of the content may not provide any relevant information for the type of address. This could be done by indicating an "unknown" type of address in the code, but it is definitely better to simply revert to the older way of displaying an address (if you remember the SOLID rules, this is an application of the Liskov Substitution Principle):

```
<p class="card-text">
@foreach (var address in @author.Contacts.Addresses)
{
    <svg xmlns="http://www.w3.org/2000/svg" width="16" height="16"
fill="currentColor" class="bi bi-building" viewBox="0 0 16 16">
    <path d="(...)"/>
    </svg>
    @if (!string.IsNullOrEmpty(address.AddressType))
    {
        <span>@address.AddressType address is:</span>
    }
    <span>@address.StreetNumber @address.StreetName @address.ZipCode @
address.CityName (@address.Country.Title)</span>
}
</p>
```

Adding data is generally not a problem in a GUI, and keeping compatibility is neither. The only difficult thing, in this case, maybe to keep the interface as smooth as originally planned, particularly if the initial design is on another project and people without knowledge of UX now are in charge of evolving the GUI.

Modifying ergonomics of the GUI

Let's discuss a more complex way to take on evolution in the GUI and, in particular, an evolution of the behavior of an application and not just its display. In our example, we could imagine that the director of DemoEditor has seen some modern web applications where saving entered data is done along the flow of the UI, without clicking a Submit button, and they would like to see the same in our application. Luckily for us, we have not relied on the state of business entities, but the application is already capable of dealing with changes in the API and persistence. This request will only be a matter of evolving the GUI to provide changes to the API, and the "difficult" part will simply consist of defining JSON Patch content to send to the existing PATCH API each time there is a change. In fact, this request will create a better alignment of the GUI with the existing "record-all" data referential service (and I admit that this is the reason why I have only shown read operations in the GUI so far).

In the following sections, we will implement this way of working, using a brand-new page in Blazor in order to keep compatibility with the existing interface (after all, some users may be more comfortable with the existing interface and the good old Save button).

Be aware that the code will (as usual) be as unsophisticated as possible – names of attributes will be changed in the code from their serialized value, the URL will be hard-coded (even this is not a problem through the use of Docker Compose), no exception handling will be done, and so on. This is for two reasons – the first one is that I do not want to make the reading of the logic of code harder by interlacing it with technical code; the second one is more important and linked to the desire to keep our thinking where it should be. I truly believe a plague of software design is that we are fascinated with technology and spend way too much time programming things with elegant abstractions and highly sophisticated patterns that, most of the time, add no value to the code and even make it harder to read. Injection of classes that will never change, production of logs that offer no values in the monitoring or debugging of the application, and soft-coding values that will not change more frequently than a new version will be released – all of these (and other unnecessary sophistication) hinder code reading and evolution. Their use is often justified because of the growing perimeter of the code and the need to "keep things clean" or "think of the future," but this just does not make sense if you maintain small-perimeter, business-aligned services.

This simple approach would have the consequence that, if we were to calculate the patches to be sent to a server by comparing an initial business entity to the current one, we would have happily cloned the entity when it was read, and this would have been enough. However, since we are going to do something even more radical, which is sending a PATCH request for each atomic change of the business entity, we will not even need this! Yes, I know what you are thinking: "*But this will make for lots of server calls! Isn't that dangerous for performance?*" Do you remember the proverb, "*Premature optimization is the root of all evil*"? This applies perfectly here, and we will first get something that works fine, only afterward thinking of ways to optimize what can indeed cause problems.

Adding a change tracker in the form and implementing the auto-patch

The first thing we need to do to get this approach working is to find a way to track individual changes in the form representing an author. To do so, we will create a new Blazor page called Author-Diff. razor and use a different approach for EditForm, with InputText form units instead of simple HTML input tags:

```
<EditForm EditContext="authorContext" class="row g-3">
    <div class="col-md-6">
        <label for="firstName" class="form-label">First name</label>
        <InputText class="form-control" id="firstName" @bind-
Value="currentAuthor.FirstName" />
    </div>
    <div class="col-md-6">
        <label for="lastName" class="form-label">Last name</label>
        <InputText class="form-control" id="lastName" @bind-
Value="currentAuthor.LastName" />
    </div>
```

```
    <!-- rest of the form content omitted for readability -->

    <div class="col-4">
        <label for="inputState" class="form-label">Country</label>
        <InputSelect id="inputState" class="form-select" @bind-
Value="currentAuthor.MainAddress.Country.ISOCode">
            <option selected>Choose...</option>
            <option value="FRA">France</option>
            <option value="IND">India</option>
            <option value="USA">USA</option>
        </InputSelect>
    </div>
    <div class="col-4">
        <div class="form-check">
            <label class="form-check-label" for="gridCheck">
                Block author (prevents access to their books, in case
of proven misconduct)
            </label>
            <InputCheckbox class="form-check-input" id="gridCheck" @
bind-Value="currentAuthor.IsBlocked" />
        </div>
    </div>
    <h2>No need for a save button; every atomic change is sent to the
server</h2>
</EditForm>
```

Note that, although the `bind-Value` attributes still point to the `currentAuthor` business entity, there is no bound model to the form but an `EditContext` that connects to a field called `authorContext`. This field is defined in the `@code` region and filled like this:

```
private DTO.Author currentAuthor = new DTO.Author();

private EditContext authorContext;

protected override async Task OnInitializedAsync()
{
    if (id == null) return;
    currentAuthor = await Http.GetFromJsonAsync<DTO.Author>("http://
authors/Authors/" + id);
    authorContext = new EditContext(currentAuthor);
    authorContext.OnFieldChanged += Modification;
}
```

As you can see, `authorContext` is based on the `currentAuthor` model, which makes it possible for the form to continue to bind to the author's value (just like we did previously in the `Author.razor` page). But the addition of this class is that it records every change on the fly, and this is precisely the feature we will use when connecting its `OnFieldChanged` event to a method of our own, called `Modification`. It is inside this method body that we will take care of the changes that are done by the user in the form:

```
private async void Modification(object sender, FieldChangedEventArgs e)
{
    // Code to be detailed below
}
```

The first thing we will do in this function is retrieve the field that has been changed:

```
string path = e.FieldIdentifier.FieldName;
```

Then, since the names of the fields are not exactly the same as the ones expected in the JSON Patch, we will have to adjust the `path` value. In most cases (the `default` branch of our `switch` operation), we only have to adjust the case of the first letter. But for the phones, for example, since the GUI has decided to display separately the home and mobile phones and not all phones in a list, it is necessary to adjust the path. Also, since we display only the first address on the card, there are some adjustments to be made to point to the right path instead of just the name of the final field:

```
switch (e.FieldIdentifier.FieldName)
{
    case "HomePhone":
        path = "/Contacts/Phones/home/Number";
        break;
    case "MobilePhone":
        path = "/Contacts/Phones/cell/Number";
        break;
    case "StreetNumber":
    case "StreetName":
    case "ZipCode":
    case "CityName":
        path = "/Contacts/Addresses/0/" + path;
        break;
    case "ISOCode":
        path = "/Contacts/Addresses/0/Country/" + path;
        break;
    default:
        path = "/" + path.Substring(0, 1).ToLower() + path.Substring(1);
```

```
        break;
    }
```

Note that this approach needs a change of schema, like the one on the addresses that was presented at the beginning of the chapter as not backward-compatible. It would be great to be able to navigate through a JSON array using the selecting function, but JSONPath is not supported in JSON Patch, and the only way is to resort to keys or order in the array. This means that we have to evolve to types of phones as keys and use order for addresses (considering that the main one will always be the first in the array, by convention, and assuming there cannot be any secondary address as long as there is no primary one, which sounds acceptable from a functional point of view).

Returning to the in-flight modification of paths, this is just some technicality that could be handled by an automatic mapper (if sophistication is not at the cost of readability, of course – additions to the business model are not that frequent that they demand such a tool); the more interesting part is what we do with this `path`, and this is all about creating a JSONPatch entry:

```
    string jsonPatch = "[{\"op\": \"replace\", \"path\": \"" + path +
"\", \"value\": \"" + e.FieldIdentifier.Model.GetType().GetProperty(e.
FieldIdentifier.FieldName)?.GetValue(e.FieldIdentifier.Model) +
"\"}]";
    var content = new StringContent(jsonPatch, System.Text.Encoding.
UTF8, "application/json");
```

If someone changes the zip code of the main address, for example, the preceding code will generate JSON Patch content like the following:

```
[
    {
        "op": "replace",
        "path": "/Contacts/Addresses/0/zipCode",
        "value": "19024-6789"
    }
]
```

Finally, this content would simply be addressed to the PATCH API on the URL corresponding to the correct author resource, like this:

```
    await Http.PatchAsync("http://authors/Authors/" + currentAuthor.
EntityId, content);
```

Again, there is no exception handling there, but it would be quite easy to display a popup with the error content or a toaster, change the color of the interface, and so on.

Adjusting the behavior of the page

The top of the page contains the expected instructions:

```
@page "/authordiff/{id}"
@using Microsoft.AspNetCore.Authorization
@inject HttpClient Http


<PageTitle>Author (autopatch)</PageTitle>
```

However, there are some things to add for a complete working of our new page. In particular, until now, we would only take care of the nullity of [Parameter] public string? id { get; set; }. This time, we need also to take into account the fact that the model and context may be null during the loading of the page:

```
@if (currentAuthor == null || authorContext == null)
{
    <p><em>Loading...</em></p>
}
else if (id == null)
{
    <p>You should not reach this page without specifying an author
identifier</p>
}
else
{
    <div>
        <EditForm EditContext="authorContext" class="row g-3">

        <!-- Rest of the EditForm shown above -->
```

The page is now ready for testing. You can check that, each time you leave a form entry with a modification, a PATCH is sent to the server. The default style even highlights the content that has been modified while another color is used to show which element has the focus:

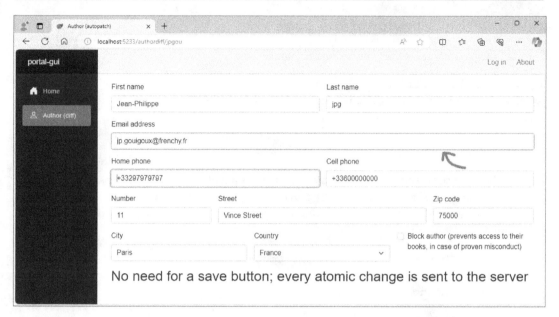

Figure 23.8 – Auto-patching following the change events in the GUI

This is, of course, a minimal implementation of the required change, but it works and has been made in such an easy way that no coupling or dependency to a complex framework has been added, which should make further evolution easy.

Considering further improvements

Let's return to performance first. As you will, "*premature optimization is the root of all evil*," but that does not mean we should not think of improvements afterward when problems actually arise. Let's imagine that we have detected a spike in API use, which we need to take care of. We know that, sometimes, users will fill in many fields simultaneously (typically, when creating a new entity). Since this puts stress on the server and semantically can be considered as one and only one big change of many attributes at once, it may indeed be better ergonomics to provide a way for the interface to "hold on," collecting all changes in a single JSON Patch unit and waiting for a user signal to send the whole package to the server. This way, there will be fewer server roundtrips that may create some user waits. Also, the history in the data referential service will be cleaner, as it will not record numerous individual field changes that essentially occur when many fields are filled out during the creation of a new entity.

This user signal can be a manual interaction – you could consider a **Hold my changes/ Send my changes** toggle button, which would be in the second mode by default; in this case, the interface would work just like the one shown in *Figure 23.8*. But it would be toggled in the first mode, and then the changes would be piled up in memory, and, only when the mode is changed again would the changes be sent back at once. The advantage of this is that generally speaking, the "buffered" mode is more adapted when creating the object. After that, changes are often small, and the first mode is then more adapted.

The semantics of the modification should also be taken into account. For example, there is generally no need to record atomic changes when an address is modified, and it is clearly changed to store at the same value date the change of a street name and number, and even a zip code and city name. Indeed, in this particular case, storing small changes would generate inconsistent data of "mixed," non-existent addresses. A simple but efficient mechanism, in this case, is to use graphical components called "cards" that act as a single entity, which is created apart and then only added to the main interface, and thus to the model, through data binding. As usual, all that matters is to keep aligned with the business intent.

The signal to record the changes could also be time-based – in this case, the interface would wait for a lack of interaction of a few seconds to consider whether the creation activity is done and send the cumulated changes to the server (you will not even have to store them by yourself; `EditContext` deals with this for you, and you can retrieve all the changes it stores whenever you are ready to handle them). Remember, however, to be extremely careful with ergonomic delays – users may feel pressured to enter additional information quicker than they would usually, and hurrying them will cause them to enter a lower quality of data, which is a much bigger problem than having a few changes very close in time in your database history.

Another further improvement could be to align the evolution that we have done previously with full state management of the application. As you certainly know, web applications are stateless by default, and Blazor and other **single-page applications (SPAs)** work extensively to simulate state maintenance within the browser. In fact, there is some state in the context of the browser, but by default, it does not survive a change of page in Blazor, for example. If you want to keep a context alive even when leaving a page and returning to it, it is necessary to add a state manager such as Fluxor, based on the Redux pattern (which we discussed in previous chapters).

This is not something that would be interesting to develop here because it would involve explaining how states, actions, and reducers interact with each other. However, one thing worth mentioning is that since the Redux pattern involves state management, all changes in the state are in fact the same as the actions that resulted in a `PATCH` call in the graphical interface we showed previously. This means that the two architectures (auto-patch change persistence and Redux pattern state management) should be closely aligned if you ever need to support state in an SPA and, simultaneously, have an auto-patch mechanism. In Redux terms, what would happen is the following:

1. An "action" would be sent to the state manager when a change is detected in a field of the form.
2. A "reducer" would transform the initial state of the page to the modified state, with the updated value of the field in the model bound to the form.
3. An "effect" would be invoked along this reduction that would be plugged into the `PATCH` call.
4. If necessary, a subsequent "action" would be run after receiving the response of the HTTP call (or, rather, an action of two possible kinds, depending on whether the `PATCH` request was successful or not).

5. This action in itself would create a change of state (a "reduction"), which would modify the model.

6. The binding would update the view to the new state of the model.

In our case, the PATCH call does not send a modified version of the entity (but it should, particularly because there are some business rules on the server that may affect other attributes of the business entity). But this addition of state management could be useful in another way – providing full state management makes it possible to return to a previous version of the state, which could be a good way to reassure users who might be afraid of making mistakes when the **Save** button disappears. Going back into the state would make it possible for them to send a revert order to the server without having to find out in the history what the previous value of the field was (and risking an error when canceling their incorrect manipulation). Although this would be quite an important development, which would need to be carefully examined in terms of total costs (not only development but also maintenance, documentation, and explaining to juniors), it shows how – when business is the driver of the technical aspects – everything falls nicely into place and works together when the application is improved, instead of it becoming more complex and continuously harder to evolve.

Adjusting IAM and business rules

After adjusting data and the GUI, business rules are another aspect that can change in an information system, and where a good alignment makes for easy adjustments. Centralizing business rules is of utmost importance when they change often and when a discrepancy may be harmful to your business (imagine a case where VAT is not calculated in the same way, depending on the path of the order in your IT department). A dedicated BRMS is a great way to handle this, but we have seen this is quite cumbersome. A much lighter yet efficient way to centralize business rules is to simply break down your information system into multiple services that have only a few responsibilities.

This is typically what we have done for IAM business rules, which are some of the most important rules, as they involve security and authorization. Also, they are the rules that change most quickly – orgcharts can change a lot in companies, due to newcomers, people retiring, interns working for a short period, employees changing jobs, authorizations being adjusted because they evolved in their jobs, and so on.

IAM is also a domain where it is easy to see that technical compatibility is not the most important; instead, it's behavior compatibility. Adjusting a role on a user is indeed easy, and from a technical point of view, it involves simply replacing an array of strings with another one. But imagine the francesca user is, all of a sudden, not in the director role anymore and that salaries cannot be paid, big contracts that necessitate the director's signature cannot be handled, and you start to realize that compatibility is also very much a functional issue.

Since IAM is such an important provider of business rules, we will discuss evolving them in this context alone, starting with common changes.

Common IAM changes

Those changes are, for example, a newcomer in a company, someone leaving, a change of role, and so on. Since we have put in place IAM, it is going to be quite easy, as you will see in the next sections.

Let's imagine someone is added to a company. If we use the internal directory of Keycloak, this is as easy as adding a user to the list (note that you can even plan some actions for them to realize upon account activation, such as forcing them to change their password at first log in or update their profile):

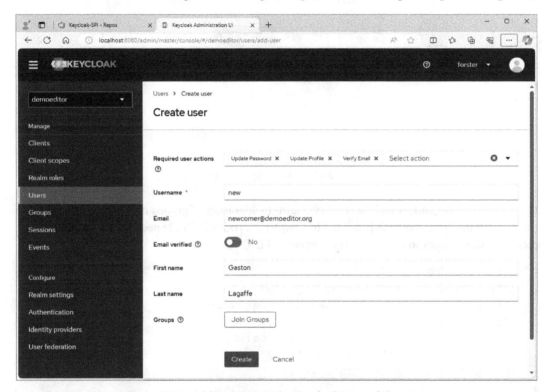

Figure 23.9 – Adding a user in the Keycloak internal directory

What happens if you have used an external identity provider such as the one we have shown with Azure AD/Entra? It's even simpler – you do not have to do anything in your information system, since the directory is managed outside of it, and you simply use the accounts and tokens it provides your system with.

How about adding some roles to the user? This is done in the **Role mapping** tab, where an **Assign role** button will let the administrator assign realm-based or client-based roles to the newcomer:

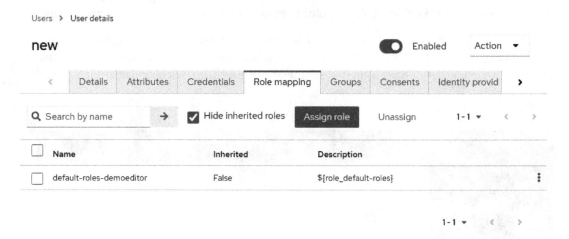

Figure 23.10 – Roles for the user

Let's say that your organizational chart evolves and the hierarchy of groups needs to be adjusted. If you have used the OPA approach proposed in *Chapter 13*, you will simply need to edit the following content and send it back to the appropriate service API:

```
"directory": {
    "frfra": {"groups": ["board"]},
    "frvel": {"groups": ["commerce", "marketing"]},
    "cemol": {"groups": ["commerce"]},
    "mnfra": {"groups": ["editors", "quality"]},
    "jpgou": {"groups": ["authors"]},
    "nicou": {"groups": ["authors"]}
}
```

The same thing goes for the roles associated by default to groups – if you ever consider that commerce should have the role of editor and all the associated authorizations (even when the role evolves), then you might affect the following change on the commerce item of the group mappings:

```
"group_mappings": {
    "board": { "roles": ["book-direction"] },
    "commerce": { "roles": ["book-sales", "book-edition"] },
    "editors": { "roles": ["book-edition"] },
    "authors": { "roles": ["book-writer"] }
}
```

In short, all the changes in the authorization rules are extremely easy to realize once everything has been correctly decoupled. However, there is one thing you must pay **absolute attention** to, and that is that everyone in charge understands precisely how the system works and where changes should be made. In the previous example, imagine that you trust someone with little knowledge of the system architecture; they might very well end up modifying the book-sales role by adding the authorizations that already exist in book-edition. At first, it looks the same, but it is in fact very different because the new authorizations added to the book-edition role will not appear automatically for the people in the commerce group. Even worse, imagine someone who has not been trained or even provided with documentation and hardcodes additional authorization in the data referential service, depending on the group created by the JWT, causing havoc in the system and destroying the nice architecture that was previously put in place.

This is an important warning because investment in low coupling can easily be reduced to nil if someone manipulates the information system without care or, the same, without knowing the business fundamentals that support all the technical bits and pieces. Documentation is of paramount importance. It is crucial for everyone in the company to understand that spending an hour to determine the perfect solution before changing a single line of code is far better than randomly adjusting parameters to achieve the desired behavior, which risks incurring technical or functional debt.

Industrializing Keycloak

The changes made to the IAM shown in the previous section were very easy because the business alignment was good. However, there are some changes that are a bit more complex – for example, those needed to add some robustness to ae system. Granted, these should not be considered functional changes that align with technical modifications, only purely technical additions. Still, they are part of the evolution of the system and, if everything has been designed correctly, should have zero impact on operations.

Since we have been in a sort of prototype phase of the information system so far, we were not very careful with IAM configuration persistence, for example. Of course, Docker Compose would create a temporary volume to keep this configuration of the demoeditor realm, but if the containers were dropped, it would be lost, and we would need to recreate the realm, the internal user, the client, and so on. This is, of course, not very clean, and one of the changes to be done, now we consider our information system is in version 1.0, is to assign persistence to the IAM. Keycloak happens to have everything ready to store its data in a PostgreSQL database, which is what we are going to do. Since we want to keep a low coupling, we will even create a dedicated database for this function. Also, while we work on Keycloak industrialization, I will quickly show you how to associate certificates for the HTTPS exposition (until now, we have only used HTTP exposition, which is not production-ready). This last evolution will enable us to comment out the following line in the code of the data referential service:

```
builder.Services.AddAuthentication(JwtBearerDefaults.
AuthenticationScheme).AddJwtBearer(options => {
    options.Authority = "http://iam/realms/demoeditor/";
```

```
        options.Audience = "account";
        //options.RequireHttpsMetadata = false;
}).AddCookie();
```

The industrialization is performed in the docker-compose.yaml file, first by modifying the iam service, like this:

```
iam:
  image: quay.io/keycloak/keycloak:23.0.7
  ports:
    - 8080:8080
  environment:
    - KEYCLOAK_ADMIN=forster
    - KEYCLOAK_ADMIN_PASSWORD=4AFXbm5vX7YFjN0rMYKK
    - KC_HTTPS_CERTIFICATE_FILE=/run/secrets/crt
    - KC_HTTPS_CERTIFICATE_KEY_FILE=/run/secrets/key
    - KC_DB=postgres
    - KC_DB_URL=jdbc:postgresql://dbiam/keycloak
    - KC_DB_USERNAME=postgres
    - KC_DB_PASSWORD=KY9CO8sHshzM5YxfLtzL
  command:
    - --proxy=edge
    - --hostname=iam.demoeditor.org:8443
  secrets:
    - crt
    - key
  networks:
    - front
    - back
```

The start-dev command can be removed if we are to operate in production, and hostname would correspond to the deployment host, associated of course with the certificates provided. The parameters starting with KC_DB will enable the database persistence for Keycloak and an additional service should be created in the Docker Compose list for this:

```
dbiam:
  image: postgres:14.5
  user: postgres
  environment:
    - POSTGRES_USER=postgres
    - POSTGRES_DB=keycloak
    - POSTGRES_PASSWORD=KY9CO8sHshzM5YxfLtzL
  volumes:
    - datadb:/var/lib/postgresql/data
  healthcheck:
```

```
    test: ["CMD", "pg_isready"]
    interval: 5s
    timeout: 1s
    retries: 6
    start_period: 5s
  networks:
  - back
```

Of course, for this to work, `user` and `password` should be similar to the ones provided in the `iam` service, and a common network (in our case, `back`) should exist. A volume should also be created under the name of `datadb` for the PostgreSQL data itself. In addition, we add a health check mechanism based on the `pg_isready` tool, which is specifically created for this purpose.

The database could be regularly backed up, just like we do for the Mongo databases that are used by the data referential services. In order to be industrialized, we would, for example, perform this kind of operation in `CronJob` so that it is performed automatically. The Kubernetes definition of such a job, which uses the `mongodump` tool and stores the dump on a mounted Azure Storage, would be something like this:

```
apiVersion: batch/v1beta1
kind: CronJob
metadata:
  name: monthly-backup-nosql-db
spec:
  schedule: "0 0 1 * *"
  jobTemplate:
    spec:
      template:
        spec:
          containers:
          - image: mongo:4.4
            name: backup-nosql-db
            command: ["mongodump","--host","db:27017","--
db","local","--out","/mnt/azure-storage"]
            volumeMounts:
              - name: azure-storage
                mountPath: /mnt/azure-storage
          restartPolicy: Never
          volumes:
            - name: azure-storage
              azureFile:
                secretName: azure-secret
                shareName: backupshare
                readOnly: false
```

Note that the image version used should be aligned with the one used for the db service, in order to make sure that the tools are perfectly compatible.

Modifying the identity sources

In the section about IAM (in *Chapter 15*), we showed how to plug the DemoEditor user directory into the Azure AD/Entra endpoint, but we kept the external users (mostly the ones created for the authors) in an internal database. Since we have a data referential service for the authors, with all information about their email address, their activity, and the fact that they should be blocked or not, how about making a nice evolution and asking Keycloak to use this data referential service as the backend for external user authentication?

Although it may seem quite a complex task, it is actually not that much of a big deal if you are ready to code a bit of Java. Indeed, Keycloak is developed with this platform and exposes an interface called SPI that enables programmers to create delegated IAM plugins. Adding an identity provider can be done by creating a class that implements some interfaces provided by the Keycloak SDK. For example, we could define a Java class like this:

```java
package org.demoeditor.keycloakprovider.user;

import org.keycloak.credential.CredentialInputValidator;
import org.keycloak.storage.UserStorageProvider;
import org.keycloak.storage.user.UserLookupProvider;
import org.keycloak.storage.user.UserQueryProvider;

// Other import operations omitted for simplicity

public class LegacyUserStorageProvider
        implements UserStorageProvider, UserLookupProvider,
CredentialInputValidator, UserQueryProvider
{
    private static final Logger log = LoggerFactory.
getLogger(LegacyUserStorageProvider.class);

    private KeycloakSession session;

    private ComponentModel model;

    public LegacyUserStorageProvider(KeycloakSession session,
ComponentModel model) {
        this.session = session;
        this.model = model;
    }
```

```
    // Rest of the class will be explained below
}
```

The rest of the function would consist of implementing the methods required by the chosen interfaces. For example, a method would be necessary to find users by email address (it would be the easiest way to associate a user with an author, and we could add some safety, demanding that the email_validated claim must be true):

```
@Override
public UserModel getUserByEmail(String email, RealmModel realm) {
    log.info("getUserByEmail({})", email);

    // Some code that would call the authors data referential service
to get the user
}
```

Another method would verify that the password provided by the connecting user to Keycloak is the one expected and validated by the authors data referential service (we could also let Keycloak deal with this, but in this hypothesis, we consider that the authors data referential service contains some legacy password hashes and a trusted Keycloak SPI plugin, with the salt and algorithm to calculate the hash condensate of passwords):

```
@Override
public boolean isValid(RealmModel realm, UserModel user,
CredentialInput credentialInput) {
        if (!this.supportsCredentialType(credentialInput.getType())) {
            return false;
        }
        StorageId sid = new StorageId(user.getId());
        String username = sid.getExternalId();

    // Calling the authors data referential service with the password
hash (using the credentialInput.getChallengeResponse() function) and
returning true if there is a match, false if not
    }
```

Of course, it is not necessary to implement all the methods, and some functions may be left without results. For example, we could confirm that the authors data referential service does not contain any groups with the following code (also note the interface supports the pagination of results, in order to maintain performance with high volumes):

```
@Override
public List<UserModel> getGroupMembers(RealmModel realm,
GroupModel group, int firstResult, int maxResults) {
        return Collections.emptyList();
}
```

It is even possible to create a class so that parameters for the provider are easily entered by the Keycloak administrator, by implementing another interface linked to the first one, which would define the three parameters needed to connect to the authors data referential service – namely, the URL of the service, a username, and a password:

```java
public class LegacyUserStorageProviderFactory implements
UserStorageProviderFactory<LegacyUserStorageProvider> {

    // Fields omitted for clarity

    @Override
    public void init(Config.Scope config) {
    ProviderConfigurationBuilder builder =
ProviderConfigurationBuilder.create();
        builder.property("config.key.authors.url", "Authors referential
URL",
            "AuthorsURL", ProviderConfigProperty.STRING_TYPE,
            "https://demoeditor.org/authors/", null);
        builder.property("config.key.authors.username", "Authors API
user",
            "UserName", ProviderConfigProperty.STRING_TYPE,
            "keycloak", null);
        builder.property("config.key.authors.password", "Authors API
password",
            "UserPassword", ProviderConfigProperty.STRING_TYPE,
            properties.get("config.key.authors.password"), null);
    configMetadata = builder.build();
    }
```

Once everything is ready in the code, a pom.xml file such as the following one (containing only the content relevant to our context) should be created in order to ease the compilation of the Java archive:

```xml
<?xml version="1.0" encoding="UTF-8"?>
<project xmlns="http://maven.apache.org/POM/4.0.0"
    xmlns:xsi="http://www.w3.org/2001/XMLSchema-instance"
    xsi:schemaLocation="http://maven.apache.org/POM/4.0.0 https://
maven.apache.org/xsd/maven-4.0.0.xsd">
    <modelVersion>4.0.0</modelVersion>

    <groupId>org.demoeditor</groupId>
    <artifactId>spi-demoeditor-authors</artifactId>
    <version>1.0.0</version>
    <name>Keycloak identity provider based on the Authors data
referential</name>
```

```
    <description>This Keycloak SPI implementation enables the IAM to
search for external users in the authors data referential service and
allow them to pass authentication if they have the right password and
a verified email address. the association between the user and the
author is done by the email address.</description>
    <packaging>jar</packaging>

    <dependencies>
        <dependency>
            <groupId>org.keycloak</groupId>
            <artifactId>keycloak-core</artifactId>
            <version>${keycloak.version}</version>
        </dependency>
        <dependency>
            <groupId>org.keycloak</groupId>
            <artifactId>keycloak-server-spi</artifactId>
            <version>${keycloak.version}</version>
        </dependency>
    </dependencies>

    <build>
        <finalName>spi-demoeditor-authors</finalName>
    <!-- Omitted for clarity -->
    </build>
</project>
```

As far as I am concerned, I prefer not to install too many diverse development tools on my computer, so Docker is called to the rescue again, and such a Dockerfile enables us to compile the JAR file without needing to install anything:

```
FROM maven:3.8.6
RUN mkdir -p /usr/src/app
WORKDIR /usr/src/app
ADD pom.xml /usr/src/app
RUN ["mvn", "--quiet", "verify"]
RUN ["mvn", "--quiet", "install"]
ADD . /usr/src/app
RUN ["mvn", "-e", "clean", "package"]
CMD ["/bin/bash"]
```

Finally, a script like the following shows the actions to perform in order to compile the JAR, retrieve it from the container that runs `maven`, copy the artifact inside the running Keycloak container, and then reload its configuration for the provider to appear in the dedicated list in the Keycloak administration console:

```
docker build -t spi-demoeditor-authors .
docker rm -fv temporary_container
docker run -d --name temporary_container spi-demoeditor-authors
docker cp temporary_container:/usr/src/app/target/spi-demoeditor-
authors.jar .
docker cp .\spi-demoeditor-authors.jar iam:/opt/keycloak/providers
docker exec iam /opt/keycloak/bin/kc.sh build
docker restart iam
```

Once the provider is selected, the Keycloak login page will validate external users with this plugin. Of course, that means we have created a bit of coupling between the external accesses and the data referential service, so this should only be done if this is acceptable in your context. Generally speaking, this is only advised when you have a massive number of accounts in a legacy system and you want to slowly evolve from it. In an information system such as `DemoEditor` that we start from scratch, we would, of course, not include this kind of "trick."

The challenges of time and compatibility in IAM adjustment

Before we conclude this part of the chapter on business rules and, in particular, IAM, let's just briefly remark on compatibility with respect to the lifetime of tokens. When you evolve a means of authentication, it is important to continue supporting the old one for at least the period of delay of activity associated with the tokens that were distributed; otherwise, some automated services might be broken.

This is a particular issue that happens when dealing with **dead letter queue** messages (which is a queue that stores messages that have been rejected in a transactional system because of a lack of validation) and the administrator tries to send them again to the system afterward; the associated token may be past its timeout, and even the administrator is not supposed to be able to renew it (otherwise, that would mean they could impersonate a user, which is not allowed, and they could, for example, undertake a bank transition in their name). The best approach in this case is for the administrator to provide a high-privilege token that can register the request in their own identity but for the benefit of the external user. This way, traceability is respected, although at the cost of a higher complexity. This is why banks always keep a strong separation between functional and technical administrators.

Now the last challenges of IAM have been explored, we will deal with the last part of the information system, where we want to check that our design efforts have allowed a smooth evolution – namely, the business processes.

Modifying the business processes

In fact, process modification may be the easiest one of all because it is at the highest level in the four-layer diagram; since we have done everything between the second and third layers to keep a strong alignment owing to the functional definition of the API contracts, changing how the business processes call functions in the second layer is bound to have a very limited effect.

Types of modifications

Since there are several ways to implement processes, there are as many ways to modify them (plus some more if you switch from one type to another); let's explore these types and check whether there are any particular challenges.

We have shown that BPMN engines or lower-tech orchestration engines such as n8n can be used to implement independent business process execution. Since they have graphical editors that make them easy to create, they are equally easier to modify, but one thing needs to be taken care of – the running instances of the workflow. Indeed, since these were created before a change in the workflow definition, it is expected that they align with the flow that was defined at their creation. Although it can seem weird in long-run workflows to observe in the list of instances some old-version items, it is the only way to ensure a nice upward compatibility.

In the particular case where a change would make the old workflow instances obsolete, it is important to have a plan to close them carefully. For example, if a reduced payment process was in place, there certainly is an end date for it. When the new process is put in place, it will take care of all the new instances, but the existing ones should remain untouched until they reach the end date. Then, a dedicated "fallback" operation should be realized in order to close the still-running workflows. Such operations could be the reimbursement of a pending order, the cancellation of a customer advantage, and so on. All necessary precautions should be taken care of, and we are talking functionally. In technical terms, most process engines support keeping an old template up even after it has been modified. In fact, most of them store a copy of the workflow template definition inside the instance to ensure it can always be completed, even if the template has been modified.

Another way to implement business processes is to use the GUIs. In this case, we will return to the first section of this chapter. In fact, this is why the example showed that the new form for author modification was created side by side with the current one. This way, there is no way that someone working on a given author sees their work canceled even if the application is deployed in the meantime; if the URL does not change and the application is indeed stateless, there will not be any change to the user. Once they go back to the menu, they will perhaps be directed to the new interface by default, but the current page will not move at the time they operate it. Again, this is all about compatibility in technical terms, but it is also in the behavior of the application.

The other way to implement the business process that we showed previously was by using event-driven architecture and Webhooks. This method is both easier and harder to upgrade in terms of its behavior. Indeed, the lowest coupling owing to the asynchronicity of the API callbacks makes it easier to change the implementation of the receiving end on the fly. However, at the same time, since the registering of the Webhook has been done by an old version, it is absolutely essential that the new one can handle the old callback contents, even if it registers to the Webhook differently and receives different data in the callbacks.

In this case, the best way to implement change is to follow these steps when deploying a new version of the service:

1. Start the process in a way that blocks the callbacks (either storing them for further use, if it can, or counting on the retry policies of the emitter to receive them later).

2. De-register old Webhooks.

3. Register again on the events, with the new parameters.

4. Open the gates again to receive the callbacks.

5. Hope that the emitter is well-designed enough to send the retries of the blocked events with the new grammar requested in the Webhook registration (which is almost never the case).

That last remark is the reason why hot deployment of an event-driven architecture is very hard to achieve and a solution with a bit of accepted downtime is almost always a better choice:

1. De-register Webhooks while the old version is still in place.

2. Upgrade the service.

3. Re-register the Webhooks with the new version.

Examples of business process modifications

Showing in detail the implementations of new versions of business processes would not add any value to the chapter, so we will only list a few examples of such changes and explain how they could be handled.

Changing the middle office validation for a true electronic signature would be as easy as changing the registration to the "new document" event and the registration to the "signed document," simultaneously of course, in order to avoid the misrouting of documents. Since an EDM is used, the signing or validation of the document would be completely decoupled from its storage, which would make this particularly easy. Beware of some EDM systems storing access tokens in the URL – there is a strong coupling between storage and access management, and the revoking of a token may leave the database storing the URLs of the document in a state where it cannot obtain documents anymore, the only solution being a mass migration of URLs.

In the process where we looped over potential authors for a book and waited for one to accept, an interesting change of business process would be to send proposals to many authors simultaneously, store the positive responses, and then organize an editor's vote to choose the right person. This is the kind of process change that can be designed purely on the BPMN diagram, which means it could have a guaranteed zero impact on the technical side of the information system, since the functional level acts in this case as an indirection layer, and the only changes in the technical layer would be the addition of business function implementations (which is necessarily upward compatible, as there was no preexisting version).

Another example of a business process change would be to request the author to fill in additional information about them *at the same time* they sign their contract (and not before). If the information system has been created with a hardcoded process, it would definitely break quite a lot of the code and operations put in movement when this happens. Since the two tasks have been separated in our design and they call different APIs, it is only a matter of changing the workflow of the initial version:

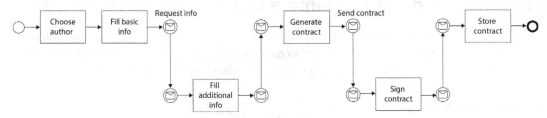

Figure 23.11 – The initial request to sign a new author

The newer version would be like this, and since the task already exists, we know that there will not be any impact on the IT department, but that this change of business process can be handled almost entirely by functional persons (and even completely by them if they have been trained to use NoCode tools):

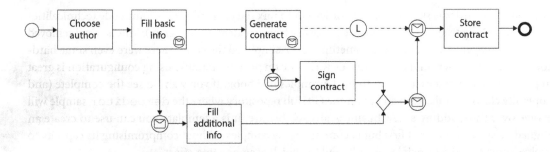

Figure 23.12 – The new process to request info and contract signing

Note that I did not illustrate the BPMN lanes for the editor and the author in these diagrams, as assigning tasks to one or the other is obvious.

Finally, changes in processes may be even bigger, particularly in the large projects of enterprise digitalization or when applying country-wide regulations such as GDPR and the like. In these cases, the use of four-layer diagrams (remember *Chapter 3*?) is important to clearly separate technical modifications (by adding some features) and process modifications (by modifying not only BPMN but also accompanying changes in the employee's way of working). Strong advice that is shared by many change consultants, as well as experts in digital transformation, is that these two kinds of changes must never be realized simultaneously; changing the software tools at the same time as changing the way to use them is, of course, seen as a nice shortcut, but it is one of the main reasons why so many software projects fail. If you want to successfully advance in your information system modernization, the only way is to make it "walk" (the functional leg moves, then the technical one, and then you repeat) and not make it "jump" (the functional and technical legs leave the ground at the same time). The only way you can "jump" ahead and land nicely is if your functional leg and technical leg already work perfectly together, which means you have performed business/IT alignment. Yes, we are back to it, since it is the foundational principle of all successful evolution of a software system in the long term.

A few last words on the sample software applications

We are almost at the end of this last chapter, and thus the entire book, which means the main subject would be normally covered by now. But I want to mention a few things about code and the way the sample application needs to be continued, since, of course, change is the only constant.

Code adjustments

First, there's the code itself. As mentioned previously, I have tried hard to simplify the examples in order to provide you with an easy read. This also means that the code may seem over-simplified at times, particularly for expert coders. This is done on purpose not only for the simplicity of reading but also to reiterate that the simplest, most readable code is often the best and that 80% of the technical sophistication we apply is hard to justify with actual numbers.

The code provided was skeleton only, in order not to mix the principles with some code technicalities (again, separating true, business-related complexity from accidental complexity). This is why I removed all error management and logs and sometimes over-simplified the code. There were even some hard-coded values I left here and there, but this was done on purpose because using configuration is great in production but makes it harder to read examples in a book. If you want to see the complete (and hopefully cleaner) code, keep an eye on the GitHub repository where the `demoeditor` sample will evolve over time, adding a new business-aligned feature. It is a template you can use to create an aligned IT system, simple at first but becoming highly complex without compromising its capacity to evolve, providing the principles are adhered to, which is an ongoing effort.

This remark may sound trivial, but there really is more than just looking at simple code. In fact, I think coding this way (i.e., focusing on the business alignment, even if it ignores all the best practices) can be interesting to ultimately get better architecture. Except when we are experts in a language or framework and technical sophistication comes up naturally, focusing on the technicalities of the code almost always takes up a bit of brain power that is not used for business alignment. It will, of course, depend very much on how people analyze and code, but personally, I think it is way easier to put all the code bricks in place while keeping them "low-tech," and then only add some technical sophistication afterward. In fact, in many cases, I do not even make this technical cleanliness effort afterward. Indeed, when all services have been correctly separated, they tend to be small and easy to change. In addition, repeating values are not necessarily a problem because the context is so small that they cannot be misunderstood.

For example, in the `portal` service, there are several calls to `http://authors/` and `http://books/`. Sure, I could create a config value for this and make it easier to change all values simultaneously. However, those values are located quite easily in the code, and replacing them with a `replace all` function in the text editor is trivial and bears no risk of confusion, Also, since those hostnames are, in fact, DNS entries for the Docker Compose application, there are not many cases that we could envisage where it would be necessary to change them.

The code provided in the GitHub link accompanying this book uses a tag to a state where the book code finishes and the commits have been made along with the book exercises in order for you to better understand the chapters. However, since we expect to have fully functional code, with features that would be expected even if not shown in the book (for example, we did not show any GUI to modify books), the GitHub code continues after the tag indicating the end of the book-related commits, and some more commits are included to make it a bit more industrial. My goal is that, by the time a `v1.0` tag is appended, the whole information system can be run by simply cloning the repository and entering the `docker compose up -d` command (a change in the `hosts` file will certainly be required, however, as well as some certificate creation if I expose only HTTPS).

Adjusting exposition

Talking about exposition, the way services are exposed, with different ports on the same host, would obviously be different in a real production environment. I did not want to complexify the sample, again, with some purely technical adjustments, but some tools such as Traefik or Caddy could be used to expose dedicated routes on the same host – for example, `http://demoeditor.org/iam` or `http://demoeditor.org/books`.

Traefik could be added as a service in the `docker-compose.yaml` file, like this, if you use a Docker Swarm orchestrator:

```
traefik:
  image: traefik:2.8
  ports:
    - "80:80"
    - "8080:8080"
  volumes:
    - /var/run/docker.sock:/var/run/docker.sock
  command: --api.insecure=true --providers.docker.swarmmode --log.
level=debug
  deploy:
    placement:
```

```
            constraints:
              - node.role == manager
```

The constraint used in the settings ensures that the Traefik reverse proxy is deployed on the orchestrator manager nodes, and the volume link is necessary for Traefik to spy on every change in the Docker containers to adjust its routes and load balancing. Labels are then added to the other services so that Traefik handles their exposition:

```
  books:
    image: demoeditor/books-api:0.1

(...) rest of the file omitted for simplicity

    deploy:
      labels:
        - "traefik.http.routers.server.rule=Host(`demoeditor.org`) &&
PathPrefix(`/books`)"
        - "traefik.http.services.server.loadbalancer.server.port=8080"
```

Caddy is a bit simpler to use but slightly less powerful than Traefik. Adding it as a service to the docker-compose.yaml file means inserting the following lines:

```
  caddy:
    image: caddy:2-alpine
    ports:
      - "80:80"
      - "443:443"
    volumes:
      - "./Caddyfile:/etc/caddy/Caddyfile"
      - caddydata:/data
      - caddyconfig:/config
    deploy:
      placement:
        constraints:
          - node.role == manager
```

And, as can be seen in the preceding settings, the content of the reverse proxy rules lies in a Caddyfile configuration file, which resembles this (I'm only showing the rule to re-expose the books data referential service API on the http://demoeditor.org/books URL):

```
demoeditor.org

reverse_proxy /books/* books:8080
```

The exposition of services will depend so much on your network, whether you have a dedicated domain or work on a local machine, if you want to use HTTPS or HTTP is enough (in case of pure exercise), whether you have an orchestrator or not, if interactions with an ingress should be taken into account, and so many other factors that I will end the subject here. I will conclude by saying that tools such as Traefik and Caddy might help you, in addition to non-Docker reverse proxies such as HAProxy or nginx.

Summary

That's it! We have finally reached the end of the book. This final chapter demonstrated the worthiness of everything we have done previously – thinking in terms of separation of responsibilities, decoupling the business and the technical implementation, analyzing functions in depth before drawing some technical contracts, ensuring API compatibility, choosing standard-aware modules and software applications, plugging everything together in a decoupled manner, and so on. All this was a lot of work, but the promise was that the evolution of the information system would later be guaranteed, and I hope you are convinced that it is indeed and that the additional exploration was really worth it.

We applied some changes to the data, in the GUI, and even deep inside the UX by adjusting how the whole application works; we also changed some business rules and authorization management, hardened the system, and modified the processes in many different ways, with none of these changes requiring a rethink of the technical implementation or even a change in architecture. This is proof that this investment in business/IT alignment and decoupling through norms and standards was justified. Again, this makes sense only in medium-to-large information systems, and the effort would be too high in small ones. But what we aimed at here was actual enterprise architecture.

As we reach the end of the book, let me summarize the most important message of the book once again – **always align the technology to the business**. One way to help you do so is to think of what you are trying to achieve in your information system as if no computers existed – imagine that you need to keep data on paper cards, that people in an old-fashioned office exchange paper folders for each business case, that contacts are kept on business cards in a drawer, that finishing a task and passing to the next actor is physically done by calling them on the phone and sending a secretary to their office with a batch of paper, and so on. Get everything clear in this no-computer domain – the business rules, the data, who is in charge of what, and how people work together. Now and now only, replace the paper cards with a database, the phone calls with electronic messages, the manually handled business processes by a BPMN or a LowCode orchestration mechanism, the conventions and business rules by some code or even a BRMS, and the chalkboard and reports by some nice GUI.

This is the way you ensure that technology remains at the service of the functionality and does not hinder its further development. In the past decade of my practice, I have seen both sides of the story – companies that cannot evolve because their business rules are so duplicated that they need months to adjust to a change of regulation, companies where data is so dirty that they have not extracted any value from business intelligence tools and do not even dream of making AI projects based on it, versus companies where automating processes through software was so badly done that people left the company. But I have also seen companies where good data governance and some nice referential services gave them a huge boost on their market, gaining two-digit additional market share; and companies where businesspeople started using NoCode tools to automate their tasks because the APIs provided by the IT department were really easy to use, bringing value to everyone in a matter of days. I even saw a particular project where the cost of adding a new feature was first estimated to take hundreds of days to develop, only to finally be realized in a much smarter and agile way in a few days because everyone around the table temporarily forgot about technical constraints and focused on the actual business value they needed to achieve.

I am certainly not the only one to have ended up using these best practices after a few decades of creating information systems, or parts of them, and advising companies in their business/IT alignment, but I humbly think this is one of the few books that has tried to gather all this experience in one single unit. There are many books on technical programming, but there are not many about translating the business point of view to a technical architecture, at least not from my point of view of the publishing market. I sincerely hope that this reading was worth your time and that you are now convinced there is a better way to practice software architecture.

Index

Y

Z

`packtpub.com`

Subscribe to our online digital library for full access to over 7,000 books and videos, as well as industry leading tools to help you plan your personal development and advance your career. For more information, please visit our website.

Why subscribe?

- Spend less time learning and more time coding with practical eBooks and Videos from over 4,000 industry professionals
- Improve your learning with Skill Plans built especially for you
- Get a free eBook or video every month
- Fully searchable for easy access to vital information
- Copy and paste, print, and bookmark content

Did you know that Packt offers eBook versions of every book published, with PDF and ePub files available? You can upgrade to the eBook version at `packtpub.com` and as a print book customer, you are entitled to a discount on the eBook copy. Get in touch with us at `customercare@packtpub.com` for more details.

At `www.packtpub.com`, you can also read a collection of free technical articles, sign up for a range of free newsletters, and receive exclusive discounts and offers on Packt books and eBooks.

Other Books You May Enjoy

If you enjoyed this book, you may be interested in these other books by Packt:

Enterprise Application Development with C# 10 and .NET 6

Ravindra Akella, Arun Kumar Tamirisa, Suneel Kumar Kunani, Bhupesh Guptha Muthiyalu

ISBN: 978-1-80323-297-3

- Design enterprise apps by making the most of the latest features of .NET 6

- Discover different layers of an app, such as the data layer, API layer, and web layer

- Explore end-to-end architecture by implementing an enterprise web app using .NET and C# 10 and deploying it on Azure

- Focus on the core concepts of web application development and implement them in .NET 6

- Integrate the new .NET 6 health and performance check APIs into your app

- Explore MAUI and build an application targeting multiple platforms - Android, iOS, and Windows

Real-World Implementation of C# Design Patterns

Bruce M. Van Horn II

ISBN: 978-1-80324-273-6

- Get to grips with patterns, and discover how to conceive and document them
- Explore common patterns that may come up in your everyday work
- Recognize common anti-patterns early in the process
- Use creational patterns to create flexible and robust object structures
- Enhance class designs with structural patterns
- Simplify object interaction and behavior with behavioral patterns

Packt is searching for authors like you

If you're interested in becoming an author for Packt, please visit `authors.packtpub.com` and apply today. We have worked with thousands of developers and tech professionals, just like you, to help them share their insight with the global tech community. You can make a general application, apply for a specific hot topic that we are recruiting an author for, or submit your own idea.

Share your thoughts

Now you've finished *Enterprise Architecture with .NET*, we'd love to hear your thoughts! Scan the QR code below to go straight to the Amazon review page for this book and share your feedback or leave a review on the site that you purchased it from.

https://packt.link/r/1-835-08566-0

Your review is important to us and the tech community and will help us make sure we're delivering excellent quality content.

Download a free PDF copy of this book

Thanks for purchasing this book!

Do you like to read on the go but are unable to carry your print books everywhere?

Is your eBook purchase not compatible with the device of your choice?

Don't worry, now with every Packt book you get a DRM-free PDF version of that book at no cost.

Read anywhere, any place, on any device. Search, copy, and paste code from your favorite technical books directly into your application.

The perks don't stop there, you can get exclusive access to discounts, newsletters, and great free content in your inbox daily

Follow these simple steps to get the benefits:

1. Scan the QR code or visit the link below

https://packt.link/free-ebook/9781835085660

2. Submit your proof of purchase
3. That's it! We'll send your free PDF and other benefits to your email directly